Field Theoretical Methods in Particle Physics

NATO ADVANCED STUDY INSTITUTES SERIES

A series of edited volumes comprising multifaceted studies of contemporary scientific issues by some of the best scientific minds in the world, assembled in cooperation with NATO Scientific Affairs Division.

Series B: Physics

RECENT VOLUMES IN THIS SERIES

Volume 46 – Nondestructive Evaluation of Semiconductor Materials and Devices
edited by Jay N. Zemel

Volume 47 – Site Characterization and Aggregation of Implanted Atoms in Materials
edited by A. Perez and R. Coussement

Volume 48 – Electron and Magnetization Densities in Molecules and Crystals
edited by P. Becker

Volume 49 – New Phenomena in Lepton-Hadron Physics
edited by Dietrich E. C. Fries and Julius Wess

Volume 50 – Ordering in Strongly Fluctuating Condensed Matter Systems
edited by Tormod Riste

Volume 51 – Phase Transitions in Surface Films
edited by J. G. Dash and J. Ruvalds

Volume 52 – Physics of Nonlinear Transport in Semiconductors
edited by David K. Ferry, J. R. Barker, and C. Jacoboni

Volume 53 – Atomic and Molecular Processes in Controlled Thermonuclear Fusion
edited by M. R. C. McDowell and A. M. Ferendeci

Volume 54 – Quantum Flavordynamics, Quantum Chromodynamics, and Unified Theories
edited by K. T. Mahanthappa and James Randa

Volume 55 – Field Theoretical Methods in Particle Physics
edited by Werner Rühl

Volume 56 – Vibrational Spectroscopy of Molecular Liquids and Solids
edited by S. Bratos and R. M. Pick

This series is published by an international board of publishers in conjunction with NATO Scientific Affairs Division

A Life Sciences B Physics	Plenum Publishing Corporation London and New York
C Mathematical and Physical Sciences	D. Reidel Publishing Company Dordrecht, Boston and London
D Behavioral and Social Sciences E Applied Sciences	Sijthoff & Noordhoff International Publishers Alphen aan den Rijn, The Netherlands, and Germantown U.S.A.

Field Theoretical Methods in Particle Physics

Edited by
Werner Rühl
University of Kaiserslautern
Kaiserslautern, Federal Republic of Germany

PLENUM PRESS • NEW YORK AND LONDON
Published in cooperation with NATO Scientific Affairs Division

Library of Congress Cataloging in Publication Data

Nato Advanced Study Institute on Field Theoretical Methods in Particle Physics, Universität Kaiserslautern, 1979.
Field theoretical methods in particle physics.

(Nato advanced study institutes series: Series B, Physics; v. 55)
Includes index.
1. Field theory (Physics)–Congresses. 2. Particles (Nuclear physics)–Congresses. I. Rühl, Werner. II. North Atlantic Treaty Organization. III. Title. IV. Series.
QC793.3.F5N37 1979 539.7'21 80-11773
ISBN-13: 978-1-4684-3724-9 e-ISBN-13: 978-1-4684-3722-5
DOI: 10.1007/978-1-4684-3722-5

Lectures presented at the NATO Advanced Study Institute on Field Theoretical Methods in Particle Physics, held at the University of Kaiserslautern, Kaiserslautern, German Federal Republic, August 13–24, 1979.

Softcover reprint of the hardcover 1st edition 1980
A Division of Plenum Publishing Corporation
227 West 17th Street, New York, N.Y. 10011

PREFACE

The Advanced Study Institute on Field Theoretical Methods in Particle Physics was held at the Universität Kaiserslautern in Kaiserslautern, Germany, from August 13 to August 24, 1979.

Twenty invited lectures and seminar-speakers and 100 other participants attended this Institute. The contributions of most of the lecturers and seminar-speakers are contained in this volume.

The revival of field theory in elementary particle physics that started about ten years ago has influenced all branches of elementary particle physics from fundamental research to pure phenomenology. The selection of field theoretical methods in particle physics appropriate for the Institute is therefore the first task for the organizers. We decided to have constructive problems of gauge field theories and solvable models as two major areas to be covered during the Institute. If one considers the concepts and terminology currently used by pure field theorists, one notices that many of them were introduced and discussed first by phenomenologists in comparing quite elementary models directly with experimental data. For this reason, it seemed worthwhile to reserve considerable time to phenomenological field theory.

The Institute was sponsored by the North Atlantic Treaty Organization whose funds made the Institute possible. It was cosponsored by the Bundes-Ministerium für Forschung und Technologie in Bonn and the Landes-Ministerium für Kultus in Mainz.

The City of Kaiserslautern made the Theodor Zink Museum available for a reception. Thanks are due in particular to its director, Dr. Dunkel.

The Institute would not have been possible without help and assistance from my colleagues at our department in particular Professor G. E. Hite, who assumed most of the organizational work and Professor A. Vancura and Dr. Lydia Heck who were responsible for the financial and travel problems during the Institute.

Thanks are due to Mrs.Chr.v.Aswegen and Mrs. B. Geib for typing and Dr. Burzlaff for his assistance in the proofreading of the manuscripts.

Last but not least I thank all the lecturers and seminar-speakers for their collaboration.

W. Rühl

CONTENTS

ON THE CONSTRUCTION OF QUANTIZED GAUGE FIELDS

Jürg Fröhlich

Institut des Hautes Etudes Scientifiques

91 440 Bures-sur-Yvette / France

ABSTRACT

We give a very elementary introduction to the geometry of classical gauge fields. The "observables" of classical gauge theory are isolated, and discrete approximations are discussed. We then present a general formulation of quantized Yang-Mills theory and state a reconstruction theorem. Subsequently we exemplify the general scheme in terms of lattice theories. Some basic properties - - confinement, phase transitions, etc. - of lattice theories are discussed, and connections to dual resonance models are sketched. We finally outline the main steps in the construction of the two-dimensional, abelian Higgs model in the continuum - and thermodynamic limit.

These lecture notes summarize a small portion of some recent work on the description and construction of quantized gauge fields [1 - 7] . For its major part that work has been done in collaboration with D. Brydges and E. Seiler. There are two excellent reviews [8,9] by E. Seiler which the reader who does not want to read the original publications is advised to consult. Some conceptual and foundational aspects of quantized Yang-Mills theory are discussed in [10,11]. Some of the results mentioned in these lecture notes are still unpublished and represent, in part, collaboration with K. Osterwalder and E. Seiler.

CONTENTS

Sections I.1. and I.2. have an elementary, introductory character. (The advanced reader should skip them). They are, however, quite useful as a piece of motivation of the basic concepts discussed in Sections I.3. and II.1. The remaining sections are sketchy, and the reader should consult [1-9] .

I. INTRODUCTION

In this section we try to introduce the main mathematical and physical notions concerning gauge fields.

I.1. Classical gauge fields

Classical abelian and non-abelian gauge fields have been used implicitly in physics for a long time, namely in the classical mechanics of rigid bodies; ("3 index symbols"). I illustrate this point by means of an example which I learnt from E. Seiler and which serves to explain the concept of a principal bundle.

Consider a spherical ball of radius ρ rolling on a two dimensional Riemannian surface, M, which we may choose for simplicity to be the Euclidean plane. ("Rolling" means that the point of contact with the plane on the ball is at rest at each instant). The orientation of the ball is described by a three-frame attached to the ball, the position of its center of mass by two coordinates (x^1,x^2).

We propose to describe the motion of that three-frame as the ball is rolling along an arbitrary curve $\gamma \subset M$.

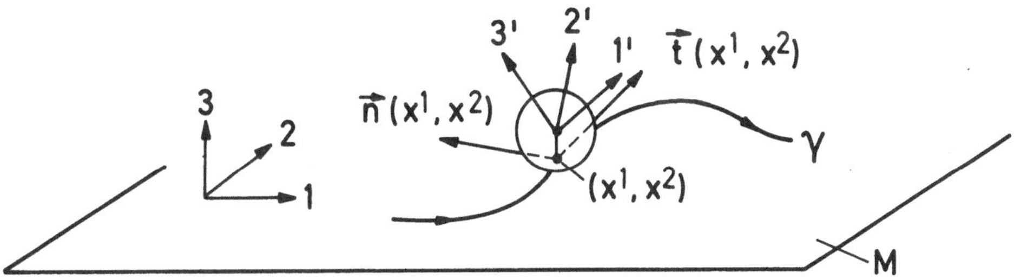

Fig. 1

The components of the vectors 1',2' and 3' in the basis 1,2,3 are given by the column vectors of an orthogonal matrix, B(x,y). At the point $p = (x^1,x^2) \in M$ the ball is rolling in the direction $\vec{t}\ (x^1,x^2)$ tangential to the curve γ. It thus rotates around the axis $\vec{n}\ (x^1,x^2)$, the unit vector orthogonal to $\vec{t}\ (x^1,x^2)$. If the total displacement of the center of mass is $d\ell$ the rotation angle is $\rho^{-1}\ d\ell$.

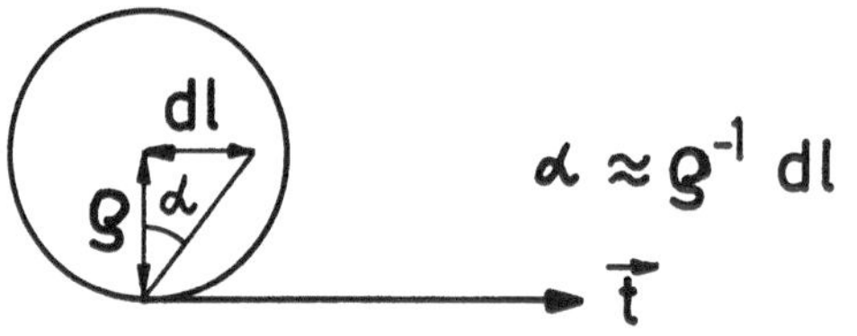

Fig. 2

Let $\vec{L} = (L_1, L_2, L_3)$ be the usual generators of rotations around the 1- , 2-, resp. 3- axis. Then the infinitesimal rotation of the ball is given by

$$B(x^1+dx^1, x^2+dx^2) = (\mathbb{1}+dR(x^1,x^2))\, B(x^1,x^2), \qquad (I.1)$$

where

$$dx^1 = t^1(x^1,x^2)d\ell \;,\; dx^2 = t^2(x^1,x^2)d\ell,$$

and

$$\begin{aligned}\mathbb{1} + dR(x^1,x^2) &= \mathbb{1} + \vec{n}(x^1,x^2) \cdot \vec{L}\,\rho^{-1}d\ell \\ &= \mathbb{1} + (t^1(x^1,x^2)L_2 - t^2(x^1,x^2)L_1)\,\rho^{-1}d\ell \\ &\equiv \mathbb{1} + \sum_{j=1}^{2} A_j\,(x^1,x^2)dx^j. \end{aligned} \qquad (I.2)$$

Thus

$$A_j(x^1,x^2) = \rho^{-1} \sum_{i=1}^{2} \varepsilon_{ji}\, L_i,\; j = 1,2. \qquad (I.3)$$

The 1-form $A=(A_1,A_2)$ with values in so(3), the Lie algebra of SO(3), given in (I.3), is called a connection (on a "principal SO(3) bundle with base space M").

The components $a^\alpha = (a_1^\alpha, a_2^\alpha, a_3^\alpha)$ defined by

$$\left.\begin{aligned} a_1^1 = o,\; a_2^1 = \rho^{-1},\; a_3^1 = o \\ a_1^2 = -\rho^{-1},\; a_2^2 = o, a_3^2 = o \end{aligned}\right\} \qquad (I.4)$$

are called vector potential.

Next, imagine the ball is rolling around a small rectangle with sides parallel to the 1- and the 2- axis of length ε, δ , respectively.

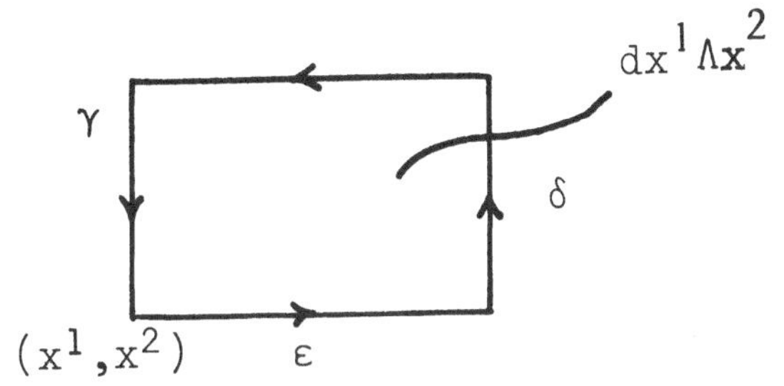

Fig. 3

We propose to determine the total rotation, ΔR, of the ball after one round trip along the curve γ depicted in Fig. 3 to second order in ε and δ . A simple calculation gives

$$\begin{aligned} \Delta R &= \mathbb{1} - [A_1, A_2]\, \varepsilon\cdot\delta \\ &= \mathbb{1} + \rho^{-2}[L_2, L_1]\, \varepsilon\cdot\delta \\ &= \mathbb{1} - \rho^{-2} L_3\, \varepsilon\cdot\delta \\ &\equiv \mathbb{1} - \frac{1}{2}\sum_{i,j=1}^{2} F_{ij}\, dx^i \wedge dx^j, \end{aligned} \tag{I.5}$$

i.e. $F_{ij} = [A_i, A_j] = \rho^{-2} L_3$

If the radius, ρ, of the ball depended on (x^1,x^2), i.e. $\rho = \rho(x^1,x^2) \neq$ const., we would find

$$F_{ij}(x^1,x^2) = \frac{\partial A_j}{\partial x^i}(x^1,x^2) - \frac{\partial A_i}{\partial x^j}(x^1,x^2) + [A_i, A_j]\,(x^1,x^2), \tag{I.6}$$

with $A_j(x^1,x^2) = \rho(x^1,x^2)^{-1} \sum_{i=1}^{2} \varepsilon_{ji} L_i$ (I.7)

The 2- form F is called <u>curvature</u>, its components F^{α}_{ij} in the basis (L_1, L_2, L_3) of so(3) are called <u>field strength</u>.

Suppose now that, at each point $p = (x^1,x^2)$ of the plane M, we introduce a new coordinate system $1''_p, 2''_p, 3''_p$ related to the system

1,2,3 by an orthogonal transformation $O_p = O(x^1,x^2)$. The orientation of the frame 1',2',3' relative to $1"_p, 2"_p, 3"_p$ is thus given by an orthogonal matrix

$$B^O(x^1,x^2) = O(x^1,x^2)^{-1} B(x^1,x^2).$$

By (I.1), (I.2) we have

$$B^O(x^1+dx^1,x^2+dx^2) = (1 + dR^O(x^1,x^2))\, B^O(x^1,x^2), \tag{I.8}$$

where

$$1 + dR^O(x^1,x^2) = O(x^1+dx^1,x^2+dx^2)^{-1}(1 + dR(x^1,x^2))O(x^1,x^2),$$

$$\equiv 1 + \sum_{j=1}^{2} A_j^O (x^1,x^2)dx^j. \tag{I.9}$$

Hence

$$A_j^O (x^1,x^2) = O(x^1,x^2)^{-1} A_j(x^1,x^2)\, O(x^1,x^2) - O(x^1,x^2)^{-1} \frac{\partial O(x^1,x^2)}{\partial x^j} \tag{I.10}$$

The mapping $O: M \to SO(3), (x^1,x^2) \mapsto O(x^1,x^2)$ is called a gauge transformation.

It follows easily from the definition of curvature that

$$F_{ij}^O(x^1,x^2) = O(x^1,x^2)^{-1} F_{ij}(x^1,x^2)O(x^1,x^2) \tag{I.11}$$

From the example discussed here the reader can, in principle, abstract most basic notions concerning principal bundles. But see [12].

Next we single out a vector $\vec{k} \in S^2$ (the unit sphere) attached to the ball, i.e. over each point $p \in M$ we have a two-sphere of possible positions of $\vec{k}$. The motion of the vector $\vec{k}$ as the ball is rolled from (x^1,x^2) to (x^1+dx^1,x^2+dx^2) is clearly described by

$$\vec{k}(x^1+dx^1,x^2+dx^2) = (1+A_j(x^1,x^2)dx^j)\vec{k}(x^1,x^2) \tag{I.12}$$

(We have started here to apply the summation convention).

Thus the connection A determines what one calls parallel transport of $\vec{k}$.

Under gauge transformations, O, $\vec{k}$ obviously transforms according to the equation

$$\vec{k}^{\,O}(x^1,x^2) = O^{-1}(x^1,x^2)\,\vec{k}\,(x^1,x^2). \tag{I.13}$$

This transformation law leaves (I.12) form-invariant if $A \to A^O$ is given by (1.10).

This example can be generalized as follows: Suppose the mass density of the ball is not rotation invariant. Then the ball will have a moment of inertia, Θ , which is a symmetric tensor of rank 2 not proportional to a multiple of the identity. With respect to rotations of the ball, Θ transforms according to a direct sum of the trivial (tr Θ) and a spin 2 $(\Theta - \frac{1}{3}\,\mathrm{Tr}\,\Theta \cdot \mathbb{1})$ representation. More generally, the ball may have some intrinsic properties described by a quantity Φ that transforms according to some representation U of SO(3) when the ball is rotated. It will be no surprise to learn that the parallel transport of Φ from (x^1,x^2) to (x^1+dx^1,x^2+dx^2) is given by

$$\Phi\,(x^1+dx^1,x^2+dx^2) = (\mathbb{1}+U(A_j(x^1,x^2))dx^j)\,\Phi(x^1,x^2), \tag{I.14}$$

and the gauge transformations by

$$\Phi^{\,O}(x^1,x^2) = U(O(x^1,x^2)^{-1})\Phi(x^1,x^2). \tag{I.15}$$

What we have discussed here can be extended to the case where M is a general two-dimensional manifold (surface). In this way one can picture many basic notions concerning fibre- and principal bundles with connection.

We end this section by briefly describing how the notions developed in the context of the rolling-ball example apply to classical field theory.

Let M be some manifold, physically the space-(imaginary) time manifold. We consider a classical, physical system described by some field Φ on M. The field Φ is supposed to have some internal "degrees of freedom" described as follows: For each point $x \in M$, $\Phi(x)$ is an element of some topological space V_x, homeomorphic to some fixed space V. Typically, V is a vector - or a homogeneous space. We also suppose that we are given a topological group G of homeomorphisms of V, physically speaking a group of internal symmetries. We are describing here what the mathematicians call a fibre bundle (with base space M, fibre V and group G), and Φ is called a cross-section of this bundle. For the moment (and in all examples discussed in subsequent sections) we may imagine that $V_x=V$, for all $x \in M$, and that the bundle is homeomorphic to $M \times V$. (This is however not so e.g. in the theory of the Wu-Yang magnetic monopole or the Yang-Mills instantons on the four sphere).

If one tries to make a dynamical theory of the field Φ one must be able to couple $\Phi(x)$ to $\Phi(y)$, for $x \neq y$, in other words, one must be able to compare $\Phi(x)$ and $\Phi(y)$, for $x \neq y$. However, a priori, the points in fibres over distinct points of the base space M cannot be compared, unless there is a notion of parallel-transporting $\Phi(x)$ from x to y along a curve γ_{yx} joining x to y. (In the example of the ball, parallel-transporting consisted of rolling). If M is a manifold, i.e. continuous, parallel transport can only be defined if G is a Lie group. In that case, suppose we are given a 1-form A on M with values in the Lie algebra $\mathfrak{g}$ of G.

Given $\Phi(x)$, let $\Phi_\gamma(y,x)$ denote the parallel transport (or - displacement) of $\Phi(x)$ from x to y along γ.

If x and y = x+dx are infinitely proximate the parallel displacement, $\Phi(x+dx,x)$, of $\Phi(x)$ from x to x+dx is defined by the formula

$$\Phi(x+dx,x) = (\mathbb{1}_V + A_j(x)dx^j)\ \Phi(x),\ j = 1,\ldots,\nu. \tag{I.16}$$

The 1-form A is the connection or gauge field. (In the example of the ball "parallel displacement" is the same as rolling). Equation (I.16) permits to calculate $\Phi_\gamma(y,x)$, without y being infinitely proximate to x, see Section II.2, and to define the covariant gradient: Let t be some vector in the tangent space at x. We set

$$\begin{aligned} t\cdot(\nabla_A\Phi)(x) &= \lim_{h\to 0} h^{-1}(\Phi(x+ht,x) - \Phi(x)). \\ &= (\nabla\Phi)(x) + A(x)\Phi(x). \end{aligned} \tag{I.17}$$

(If M is not flat the expression in the middle requires some obvious changes).

It is easy to see that, (with $\partial_i \equiv \partial/\partial x^i$),

$$F = [\nabla_A, \nabla_A],\ F_{ij} = \partial_i A_j - \partial_j A_i + [A_i, A_j] \tag{I.18}$$

corresponds to what was called curvature in the rolling-ball example. It is called curvature (2-form) or field strength.

Gauge transformations are homeomorphisms

$$h(x) : V_x \longrightarrow V_x,$$

Φ and A transform according to

$$\Phi(x) \longmapsto \Phi^h(x) = h(x)^{-1}\Phi(x),$$
$$A(h) \longmapsto A^h(x) = h(x)^{-1}A(x)h(x) - h(x)^{-1}(\nabla h)(x) \qquad (I.19)$$

The dynamics of Φ can be specified with the help of a field equation, e.g.

$$\nabla_A(\nabla_A \Phi)(x) = m^2 \Phi(x), \qquad (I.20)$$

(the covariant Klein-Gordon equation).

One may wish to introduce dynamics for the connection A, itself. A prominent example of field equations for A is the Yang-Mills equations

$$\nabla_A \cdot F = 0. \qquad (I.21)$$

(If $F = [\nabla_A, \nabla_A]$ the equations

$$\nabla_A \cdot (*F) = 0 \qquad (I.22)$$

are automatic. They are called Bianchi identities).

For discussions of classical field equations, see e.g. [13] and refs. given there. They will not be studied in the present notes.

The basic ansatz of present day elementary particle physics (without gravitational interactions) is to describe matter in terms of quantized versions of fields that are cross-sections of fibre bundles with connection and the fundamental interactions of matter in terms of quantized versions of those connections. The present choices for G are such that it equals or contains as a subgroup

$$SU(3)_{colour} \times SU(2)_{weak} \times U(1)_{electromag.} \qquad (I.23)$$

Although it is appealing that present day physics of matter and its fundamental interactions has become intrinsically geometrical it remains unsatisfactory that two kinds of geometries are involved, Riemannian (or affine) geometry in gravity, the geometry of fibre bundles in strong and electroweak interactions. Moreover, there is no convincing theoretical argument as to what the right fibre bundle (the right gauge group G) of elementary particle physics is.

We shall henceforth ignore these problems and proceed to sketch some rigorous results concerning quantum field theories that are based on fibre bundle geometry.

I.2. Some facts about the geometry of fibre bundles

The intuitive concept of fibre- and principal bundles has been developed in Section I.1. What the mathematicians understand by these words can be looked up e.g. in [12]. For our purposes the following may suffice:

Let M be the physical space-(imaginary)time manifold. Let V be a topological space with a topological group G of homeomorphisms of V into itself. Throughout these notes G will be a compact Lie group. Points in M are denoted x,y,..., Φ denotes a point in V, and h, g, ... elements of G. A <u>fibre bundle</u> $\mathfrak{F}$ over M with fibre V and group G consists roughly of a bundle space F with projection π such that, for all $p \in F$, $\pi(p) \in M$, and for all $x \in M$ $V_x := \pi^{-1}(x)$, the fibre over x, is homeormorphic to V.

For each $x \in M$, there is an open neighborhood $\Omega \subset M$ of x and a homeomorphism $\xi_\Omega : \Omega \times V \longrightarrow \pi^{-1}(\Omega)$ such that $\pi\xi_\Omega(x,\Phi)=x$, and $\xi_{\Omega,x}(\Phi) := \xi_\Omega(x,\Phi)$ is a homeomorphism from V to V_x. If ξ_y, ξ'_y, $y \in \Omega$, are two homeomorphisms from V to V_y then $h(y)=\xi_y^{-1}\xi'_y$ is supposed to be a continuous function of $y \in \Omega$ with values in G. The functions h are called <u>gauge transformations</u>. Finally, for

$$y \in \Omega \cap \Omega', \; g_{\Omega\Omega'}(y) := \xi^{-1}_{\Omega,y}\, \xi_{\Omega',y}$$

is supposed to be a continuous, G-valued function of y. It is called <u>transition function</u>.

If V happens to be the group G, we speak of a <u>principal bundle</u> (with base space M). The group G is called <u>gauge group</u>.

It follows from these definitions that bundles can be characterized by means of their transition functions:

Let $\{\Omega_i\}_{i \in I}$ be a cover of M by open neighborhoods with the property that for all $i \in I$ there exists a homeomorphism

$$\xi_{\Omega_i} : \Omega_i \times V \longrightarrow \pi^{-1}(\Omega_i)$$

with all the properties specified in the above definition. For

$$\Omega_i \cap \Omega_j \neq \emptyset, \text{ let } g_{ij} := g_{\Omega_i\Omega_j}$$

denote the transition function. Two sets of transition functions

$$\{g_{ij}\}, \{g'_{ij}\}$$

determine equivalent bundles iff

$$g'_{ij} = h_i^{-1} g_{ij} h_j, \tag{I.24}$$

with $h_j(x)$ a G-valued function on Ω_j.

This permits to associate with each fibre bundle $\mathfrak{F}$ = (F, M, V, G, π) a principal bundle $\mathfrak{P}$ = (P, M, V = G, G, π): $\mathfrak{P}$ is the principal bundle with the same transition functions as $\mathfrak{F}$. See [12] for details.

Examples: (1) Möbius strip (base space S^1= circle, fibre$[-1,1]$, group $\mathbb{Z}_2$); (2) F = M $\times$ V, G =$\{\mathbb{1}\}$, $\pi(\{x,\phi\}) = x$; this is called the product bundle; (3) The 3-sphere S^3 is a principle bundle with base space S^2, fibre $S^1 \cong U(1)$ and group U(1). Incidentally, this is the bundle space of the instanton of the two-dimensional $\mathbb{C}P^1\sigma$-model and of the Wu-Yang monopole.
(4) Interesting examples arise in the theory of functions of complex variables.

Next, we consider fibre bundles with connections, i.e. we reconsider the notion of parallel transport (or - displacement).

Let $\mathfrak{F}$ = (F, M, V, G, π) and $\{\Omega_i\}_{i\in I}$ be as above.

We suppose G is a (compact) Lie group with Lie algebra $\mathfrak{g}$. We assume that all transition functions are continuously differentiable on their domain of definition.

A connection, A, on $\mathfrak{F}$ is a family of 1-forms

$$\{A^{(i)}\}_{i\in I}$$

with values in $\mathfrak{g}$ such that $A^{(i)}$ is defined on Ω_i, $i\in I$, and for $x\in\Omega_i\cap\Omega_j\neq\emptyset$,

$$\begin{aligned} A^{(i)}(x) &= g_{ij}(x)A^{(j)}(x)g_{ij}^{-1}(x)-g_{ij}(x)(dg_{ij}^{-1})(x) \\ &= g_{ji}^{-1}(x)A^{(j)}(x)g_{ji}(x)-g_{ji}^{-1}(x)(dg_{ji})(x) \end{aligned} \tag{I.25}$$

Moreover, if h is a gauge transformation defined on Ω_i, $A^{(i)}$ transforms according to

$$A^{(i)} \longmapsto A^{h(i)} = h^{-1}A^{(i)}h - h^{-1}dh. \tag{I.26}$$

We have started, here, to use the notation

$$A = \sum_{j=1}^{\nu} A_j\, dx^j \tag{I.27}$$

We shall see that a connection is precisely what we need to define parallel transport on F.

Next, let $\Omega \subseteq \Omega_i$, for some $i \in I$, be some open subset of M.

Restricted to Ω, F is homeomorphic to

$$F'_\Omega = \Omega \times V.$$

First we define parallel transport on F'_Ω.

Let $x \in \Omega$, $\Phi(x) \in V$. We want to define the parallel transport,

$g_{\gamma_{yx}} \Phi(x)$, of $\Phi(x)$ from x to $y \in \Omega$

along a curve $\gamma_{yx} \subset \Omega$, with $g_{\gamma_{yx}} \in G$ a homeomorphism from V onto V, given the connection $A = A^{(i)}$. Suppose $y = x+dy$ is infinitely proximate to x. Then

$$g_{\gamma_{yx}} \Phi(x) \equiv g_{x+dx,x} \Phi(x) := (\mathbb{1}_V + A(x))\, \Phi(x), \qquad \text{(I.28)}$$

with

$$A(x) = \sum_{j=1}^{\nu} A_j(x)\, dx^j$$

This equation can be integrated along any oriented, continuous, piecewise smooth curve $\gamma_{yx} \subset \Omega$ connecting $x \in \Omega$ with $y \in \Omega$. To see this we may temporarily assume that Ω is flat, i.e. Ω is a subset of $\mathbb{R}^\nu$.

Let $\{x_i^k\}_{i=1}^{N_k} \subset \gamma_{yx}$ be a family of ordered sequences of points on γ_{yx} with the property that

$x_1^k = x$, $x_{N_k}^k = y$, for all $k = 1,2,3,\ldots$, and

$\operatorname{dist}(x_i^k, x_{i+1}^k) \longrightarrow 0$, as $k \longrightarrow \infty$, for all $i=1,\ldots,N_k-1$.

Then

$$g_{\gamma_{yx}} = \lim_{k\to\infty} \prod_{k=N_k}^{1} {}^{\circlearrowleft} \{\mathbb{1}_V + A_j(x_i^k)(x_{i+1}^k - x_i^k)^j\} \qquad \text{(I.29)}$$

The physicists like the following compact formula

$$g_{\gamma_{yx}} = P\,\{\exp \int_{\gamma_{yx}} A_j(z)\, dz^j\} \qquad \text{(I.30)}$$

as an abreviation for the r.s. of (I.29); (P:= "path ordering"). It follows from (I.26) and (I.29) that, under a gauge transformation h, g transforms according to

$$g_{\gamma_{yx}} \longmapsto g_{\gamma_{yx}}^{h} = h^{-1}(y)\, g_{\gamma_{yx}}\, h(x). \qquad \text{(I.31)}$$

This is the basic property that permits us to define parallel transport on F in terms of g. It is given by a homeomorphism

$$\Gamma_{\gamma_{yx}} : V_x \longrightarrow V_y, \quad \text{defined by}$$

$$\Gamma_{\gamma_{yx}} = \xi_{\Omega,y}\, g_{\gamma_{yx}}\, \xi_{\Omega,x}^{-1} \qquad \text{(I.32)}$$

Note that if ξ_Ω and ξ'_Ω are two homeomorphisms related by a gauge transformation, i.e. $h(y) \equiv \xi_{\Omega,y}^{-1}\, \xi'_{\Omega,y} \in G$ then

$$\Gamma_{\gamma_{y,x}} = \xi_{\Omega,y}\, g_{\gamma_{y,x}}\, \xi_{\Omega,x}^{-1} = \xi'_{\Omega,y}\, h^{-1}(y) g_{\gamma_{y,x}}\, h(x)\, \xi'_{\Omega,x}$$

$$= \xi'_{\Omega,y}\, g_{\gamma_{y,x}}^{h}\, \xi'_{\Omega,x}, \qquad \text{(I.33)}$$

i.e. $\Gamma_{\gamma_{y,x}}$ is independent of the choice of coordinates (the gauge). Equations (I.25) and (I.33) permit us to define $\Gamma_{\gamma_{yx}}$ for curves γ_{yx} that are not contained in a single coordinate neighborhood Ω_i: One cuts up γ_{yx} into curves

$$\gamma_{x_{\alpha+1}\, x_\alpha}$$

contained in $\Omega_{i(\alpha)}$, with

$$\Omega_{i(\alpha)} \cap \Omega_{i(\alpha+1)} \neq \emptyset \text{ and sets}$$

$$\Gamma_{\gamma_{yx}} = \Gamma_{\gamma_{x_N x_{N-1}}}\, \Gamma_{\gamma_{x_{N-1} x_{N-2}}} \cdots \Gamma_{\gamma_{x_2 x_1}}, \qquad \text{(I.34)}$$

with $x_N = y$, $x_1 = x$. By (I.25), (I.31) and (I.33) this is a consistent definition.

One may now ask the question under what conditions does

$$\Gamma_{\gamma_{yx}} \quad (\text{resp.}\, g_{\gamma_{yx}})$$

depend on the path γ_{yx} only through the endpoints x and y.
We first discuss this question locally, for γ_{yx} in a simply connected, open set $\Omega \subseteq \Omega_i$, for some $i \in I$.
In this case the answer is very simple: If and only if $g_{\gamma_{yx}}$ is of the form

$$g_{\gamma_{yx}} = h(y)\, h(x)^{-1} \qquad (I.35)$$

Clearly (I.35) is sufficient. To see that it is necessary one chooses a point $x_0 \in \Omega$ and sets $h(x_0) = \mathbb{1}_V$. One then chooses a family, L_{x_0}, of piecewise smooth, oriented curves, $\hat{\gamma}$, starting at x_0 with the property that each $x \in \Omega$ is contained in precisely one line $\hat{\gamma} \in L_{x_0}$. Let $\hat{\gamma}_{xx_0}$ be the portion of $\hat{\gamma}$ with endpoints x_0 and x. We set

$$h(x) = g_{\hat{\gamma}_{xx_0}} \quad .$$

Let x and y be arbitrary points in Ω and γ_{yx} connecting them. Since g_γ only depends on the endpoints of γ

$$g_{\hat{\gamma}_{yx_0}} = g_{\gamma_{yx}}\, g_{\hat{\gamma}_{xx_0}} \;, \qquad \text{i.e.}$$

$$g_{\gamma_{yx}} = g_{\hat{\gamma}_{yx_0}}\, g_{\hat{\gamma}_{xx_0}}^{-1} = h(y)\, h(x)^{-1}$$

which proves (I.35).

The <u>curvature</u>, F, of a connection A is defined by

$$F = dA + A \wedge A, \qquad (I.36)$$

i.e. $F_{ij} = \partial_i A_j - \partial_j A_i + [A_i, A_j]$; see also (I.18).

We now claim that <u>for</u> (I.35) <u>to hold it is necessary and sufficient that F vanishes on Ω</u>.
(In the example of the rolling ball this can only happen in the limit $\rho = \rho(x^1,x^2) \longrightarrow \infty$, for all $(x^1,x^2) \in \Omega$).
A proof of this last assertion can be obtained from the following consideration that is of independent interest: Pick a curve $\gamma_{yx} \subset \Omega$. Parametrize γ_{yx} by a function

$$x(s) = (x^1(s), \ldots, x^\nu(s)), \quad 0 \leq s \leq 1,$$

with $x(0) = x$, $x(1) = y$.

Now consider a function

$$x : [-1,1] \times [-1,1] \longrightarrow \Omega, \quad (s,t) \longmapsto x(s,t)$$

which is smooth in t and such that $x(o,t) = x, x(1,t) = y$, for all t, and $x(s,o) = x(s)$. Let $\gamma_{yx}(t)$ be the curve parametrized by $x(s,t)$. (It is a deformation of γ_{yx} leaving the endpoints fixed). We propose to calculate

$\frac{d}{dt} g_{\gamma_{yx}(t)}$. Let $\gamma_{x(s',t)x(s,t)}$ be the portion of $\gamma_{yx}(t)$ starting at $x(s,t)$ and ending at $x(s',t)$. Using (I.29) it is easy to see that

$$\frac{d}{dt} g_{\gamma_{yx}(t)} = \int_0^1 ds\, g_{\gamma_{yx(s,t)}} \left[\frac{\partial A_i}{\partial x^j}(x(s,t)) \frac{\partial x^i(s,t)}{\partial s} \frac{\partial x^j(s,t)}{\partial t}\right] g_{\gamma_{x(s,t)x}} \tag{I.37}$$

It is a simple exercise in integration by parts to show that

$$\int_0^1 ds \quad g_{\gamma_{yx(s,t)}} \left\{ \frac{\partial \Psi}{\partial x^i} (x(s,t)) - [\Psi, A_i] (x(s,t))\right\} \times \frac{\partial x^i(s,t)}{\partial s} g_{\gamma_{x(s,t)x}} = 0 \tag{I.38}$$

for any differentiable function Ψ on Ω. If we set

$$\Psi(x(s,t)) = A_j(x(s,t)) \frac{\partial x^j(s,t)}{\partial t}$$

in (I.38) and use (I.36),(I.37), we find

$$\frac{d}{dt} g_{\gamma_{yx}(t)} = \int_0^1 ds \quad g_{\gamma_{yx(s,t)}} \left[F_{ji}(x(s,t)) \frac{\partial x^i(s,t)}{\partial s} \cdot \frac{\partial x^j(s,t)}{\partial t}\right] \times g_{\gamma_{x(s,t)x}} \tag{I.39}$$

Incidentally, by differentiating both sides of (I.39) with respect to t, applying (I.38) with

$$\Psi (x(s,t)) = F_{jk}(x(s,t)) \frac{\partial x^j(s,t)}{\partial t} \cdot \frac{\partial x^k(s,t)}{\partial t},$$

and using the Bianchi identity

$$dF + [A,F] = 0 \tag{I.40}$$

one may calculate

$$\frac{d^2}{dt^2} g_{\gamma_{yx}}(t).$$

(For a slightly cumbersome way of calculating this see e.g.[14]. The physicists have been interested in equations for

$$\sum_{\ell} \frac{\partial^2}{\partial t_\ell^2} g_{\gamma_{yx}(t_1,\ldots,t_\ell,\ldots)},$$

because they suggest formal connections between Yang-Mills theory and dual resonance models [15]. To the author these connections appear, however, somewhat superficial).

As a simple corollary of equation (I.39) we have:

$g_{\gamma_{yx}}$ <u>depends only on</u> x <u>and</u> y <u>if and only if</u> $F = 0$. (I.41)

Suppose now we are given two connections A,A' on Ω such that $F \equiv F(A) = F(A') \equiv F'$.

Question: Are A and A' gauge-equivalent, in the sense of eqn.(I.26)? Unless G is abelian, the answer is: In general they are <u>not</u> gauge-equivalent. (The reader can find a simple example of this by studying the rolling-ball example!) This is an aspect of the <u>intrinsic non-linearity</u> of non-abelian gauge fields.

The "globalization" of the above considerations is only straightforward if the base space M is simply connected. (Recall the Bohm-Aharonov effect).

Next, we define the <u>holonomy groups</u> associated with a connection A: Let x be a point in M, and let L_x be the space of all bounded, continuous, piecewise smooth, oriented curves, $\mathcal{L}_x$, starting and ending at x which are called for short <u>loops</u>. (L_x may be equipped with the discrete topology). Given $\mathcal{L}_x \in L_x$, let $\mathcal{L}_x^{-1}$ be the same curve as $\mathcal{L}_x$ but with orientation reversed. On L_x we define multiplication to be the composition of oriented loops $(\mathcal{L}_x, \mathcal{L}_x' \longrightarrow \mathcal{L}_x \circ \mathcal{L}_x')$, so that $\mathcal{L}_x \circ \mathcal{L}_x^{-1} = \mathcal{L}_x^{-1} \circ \mathcal{L}_x = 1_x$, where 1_x is the identity element in L_x. In this way L_x becomes a topological group.

Given a connection A and an element $\mathcal{L}_x$ in L_x, let

$$g_{\mathcal{L}_x} = \xi_{\Omega,x}^{-1} \, \Gamma_{\mathcal{L}_x} \, \xi_{\Omega,x} \tag{I.42}$$

This is a homeomorphism from V onto V which belongs to the group G. Under gauge transformations, h, $g_{\mathscr{L}_x}$ transforms according to

$$g_{\mathscr{L}_x} \longmapsto g^h_{\mathscr{L}_x} = h(x)^{-1} g_{\mathscr{L}_x} h(x); \text{ see (I.31) - (I.34).}$$

We define the holonomy groups

$$H_x(A) = \{g_{\mathscr{L}_x} : \mathscr{L}_x \in L_x\}. \tag{I.43}$$

This is a closed subgroup contained in or equal to G, and $g : \mathscr{L}_x \in L_x \longmapsto g_{\mathscr{L}_x}$ defines a representation of L_x on V, with values in G.

If M is connected then $H_x(A)$ and $H_y(A)$ are obviously conjugate subgroups of G, for arbitrary x and y in M, (because L_x and L_y are isomorphic and conjugate). Thus $H_x(A)$ is independent of x, up to conjugation. If M is simply connected $H_x(A)$ is a connected subgroup of G. If $H_x(A)$ is conjugate to some fixed subgroup H properly contained in G, then the group of the bundle $\mathscr{F}$ can be chosen to be H. If a bundle $\mathscr{F}$ admits a connection A such that $H_x(A) = e$, the identity in G, then $\mathscr{F}$ is equivalent to the product bundle $M \times V$. All this is rather easily verified. In the remainder M is always assumed to be connected. Let χ be a character of G. Then Y, defined by

$$Y(\mathscr{L}_x) = \chi(g_{\mathscr{L}_x}), \tag{I.44}$$

is a character of L_x.

Given $\mathscr{L}_x$, let $A(\mathscr{L}_x)$ be the infimum of the areas of all smooth, surfaces bounded by $\mathscr{L}_x$. (It is assumed that M has a Riemannian metric).

Theorem: (I.45). (Reconstruction of the holonomy group). Suppose Y is a character of L_x with the following properties:

(1) Y is of positive type on L_x (of type I_n).
(2) $Y(1_x) = n$, for some natural number $n < \infty$.
(3) $|Y(\mathscr{L}_x) - n| \leq O(A(\mathscr{L}_x))$, as $A(\mathscr{L}_x) \longrightarrow 0$.

Then there exists a subgroup $H \subseteq U(n)$ and a representation $g : \mathscr{L}_x \in L_x \longmapsto g_{\mathscr{L}_x} \in H$ of L_x such that $Y(\mathscr{L}_x) = tr\ (g_{\mathscr{L}_x})$.

The representation g is unique, up to unitary equivalence. Moreover $g_{\mathscr{L}_x}$ is the parallel transport around $\mathscr{L}_x$ determined by an $\mathfrak{h}$ - valued connection A that is unique up to gauge transformations ($\mathfrak{h}$ is the Lie algebra of H), and $H_x(A) \simeq H$. □

This result is presumably wellknown. In any event, its proof is a fairly straightforward elaboration on the Gel'fand-Naimark-Segal construction which we skip. The great advantage of the objects

$$Y(\mathcal{L}_x) = \chi\,(g_{\mathcal{L}_x})$$

is that they are gauge-invariant and that if $\mathcal{L}_x = \gamma_{yx}^{-1}\mathcal{L}_y\,\gamma_{xy}$ for some $\mathcal{L}_y \in L_y$ and a curve γ_{xy} connecting x to y,

$$Y(\mathcal{L}_x) = Y\,(\mathcal{L}_y). \tag{I.46}$$

Let χ be a faithful, unitary character of G. By Theorem (I.45), knowledge of

$$\{Y\,(\mathcal{L}_x) : \mathcal{L}_x \in L_x\}$$

is equivalent to knowledge of the connection A and all parallel transporters

$$g_{\gamma_{yx}}$$

up to gauge equivalence on the connected component of M containing x.

<u>Definition</u>: A connection A is called irreducible if $H_x(A)=G$, (the group of the bundle).

Consider now a principal bundle $\mathcal{P} = (P, M, V{=}G, G, \pi)$. Let $\mathfrak{A}^I$ be the space of all irreducible connections on $\mathcal{P}$, and $\mathfrak{G}$ the group of all gauge transformations, modulo those which take values in the center of G. Clearly, $\mathfrak{G}$ acts as a transformation group on $\mathfrak{A}^I$. Let $\mathcal{O} = \mathfrak{A}^I/\mathfrak{G}$ denote the orbit space, i.e. with $A \in \mathfrak{A}^I$ we associate the class $[A]$ of all connections which are gauge equivalent to A(under gauge transformations in $\mathfrak{G}$).

It turns out that $\mathfrak{A}^I$ is a <u>principal bundle</u> with base space $\mathcal{O}$, fibre $\mathfrak{G}$ and projection π defined by $\pi\,(A) = [A] \in \mathcal{O}$, $A \in \mathfrak{A}^I$. See [16].Singer has shown that when the base space M of $\mathcal{P}$ is a sphere, S^3 or S^4, and G is non-abelian then $\mathfrak{A}^I$ is a <u>non-trivial bundle</u>, i.e. <u>not</u> equivalent to $\mathcal{O} \times \mathfrak{G}$. See [16]. (It is well-known [12] that $\mathfrak{A}^I$ does therefore not have any cross-sections, i.e. gauge fixing is impossible, globally. This is what is called by physicists <u>Gribov ambiguity</u>). It is therefore felt that the physics of non-abelian gauge fields should be described directly in terms of the objects in $\mathcal{O}$. If all connections in ^{I} are continuous, the geometry of $\mathcal{O}$ is fairly well known [16]. (E.g. $\mathcal{O}$ is a Hilbert manifold). Unfortunately, this does not seem to be very helpful for the understanding of quantized gauge fields, yet. In fact, even in classical Yang-Mills theory it is a nuisance that $\mathcal{O}$ is <u>not</u> a linear space (another aspect of the intrinsic non-linearity of Yang-Mills theory).

It is not hard to introduce different natural topologies on $\mathcal{O}$ and to show that, given such a topology, there is a natural definition of a space $C(\mathcal{O})$ of continuous functions on $\mathcal{O}$. (We permit ourselves to be a little unprecise here).

Let $A \in \mathfrak{A}^I$ be some connection on $\mathcal{P}$. Let

$$g_{\gamma_{yx}}(A)$$

denote the parallel transporters along γ_{yx} determined by A; see (I.29). Then

$Y_{\mathcal{L}_x}(A) = \chi\,(g_{\mathcal{L}_x}(A)),\quad \mathcal{L}_x \in L_x$ is gauge invariant, i.e. only depends on $\pi(A) = [A] \in \mathcal{O}$ and is "continuous in [A]".

Suppose now, χ is a faithful, unitary character of G .

<u>Theorem</u>. (I.47)
The algebra of functions generated by

$$\{Y_{\mathcal{L}_x}(A) : \mathcal{L}_x \in L_x\}$$

is dense in $C(\mathcal{O})$.

Proof.
By Theorem (I.45) the class of functions

$\{Y_{\mathcal{L}_x}(A) : \mathcal{L}_x \in L_x\}$ separates points in $\mathcal{O}$. Furthermore

$\overline{Y_{\mathcal{L}_x}(A)} = Y_{\mathcal{L}_x^{-1}}(A)$ belongs to that class, too, and finally

$Y_{1_x}(A) = \text{const.} > 0$. Thus, by the Stone-Weierstrass theorem

$\{Y_{\mathcal{L}_x}(A) : \mathcal{L}_x \in L_x\}$ is dense in $C(\mathcal{O})$. Q.E.D.

Remarks.

1.) This theorem serves as a motivation for viewing the functions $Y(\mathcal{L}) = \chi(g_{\mathcal{L}})$ as the basic observables of a Yang-Mills theory. For, classically, any gauge-invariant, continuous function of the gauge field A, i.e. every classical observable is a limit of sums of products of $Y(\mathcal{L})$'s.
2.) Let $\Lambda = \{\mathcal{L}_1, \mathcal{L}_2, \ldots\}$ be a denumerable set of closed, piecewise smooth, oriented loops, e.g. the closed loops

of a lattice superimposed on M. Then

$$\{Y(\mathcal{L}_1),\ Y(\mathcal{L}_1^{-1}),\ Y(\mathcal{L}_2),\ Y(\mathcal{L}_2^{-1}),\ \ldots.\}$$

generate a separable algebra, $C_\Lambda(\mathcal{O})$, of continuous functions on $\mathcal{O}$. If all connections

$A \in \mathfrak{A}^I$ are continuous then

$C_\Lambda(\mathcal{O}) \nearrow C(\mathcal{O})$, as $\Lambda \nearrow L_x$, for some $x \in M$, (in the topology of the supremum norm).

The algebra of functions $C_\Lambda(\mathcal{O})$, for Λ all closed loops in a lattice superimposed on M, is the starting point of the lattice approximation to Yang-Mills theory.

Problem.

Find other "nice" separable approximations to $C(\mathcal{O})$ which could be used as in starting point for a gauge-invariant regularization of Yang-Mills theory.

In the remainder of these notes we briefly summarize some new or recent results on quantized Yang-Mills theories. That part will be more sketchy. The reader is advised to consult the literature that is quoted.

I.3 A tentative general formulation of quantized Yang-Mills theory

Let $M = \mathbb{R}^\nu$ be Euclidean space-time. Let G denote the gauge group. Every principal bundle with base space $M=\mathbb{R}^\nu$ is equivalent to a product bundle, i.e. we may consider $M \times G$ to be the bundle space. Motivated by Theorem (I.47), Section I.2, we regard the functions $Y(\mathcal{L}) = \chi(g_\mathcal{L})$, where χ is an arbitrary unitary character of G as the basic "fields" of a Euclidean gauge theory on M with gauge group G. The purpose of this section is to propose a scheme for quantization of such a theory by which the $Y(\mathcal{L})$'s are converted into random fields on the space L of all oriented, smooth loops on M.

If one studies the example of free electromagnetism in $\nu=4$ dimensions as a theory of loop observables, the so called Wilson loops it becomes clear that one should require all loops in L to be at least twice continuously differentiable, oriented closed loops which are free of self-intersections, ("selfavoiding loops"). From now on L will be understood to be the space of all loops which have this property. (Classically, for a space $\mathfrak{A}^I$ of continuous, irreducible gauge fields on M, the algebra generated by $\{Y(\mathcal{L}) = \chi(g_\mathcal{L}) : \mathcal{L} \in L\}$ is still dense in $C(\mathcal{O})$, if χ is faithful).

We propose to discuss quantized gauge theories in terms of Euclidean Green's - or Schwinger functionals

$$S_n(Y_1(\mathcal{L}_1), \ldots, Y_n(\mathcal{L}_n)) \quad (Y_j(\mathcal{L}_j) = \chi_j(g_{\mathcal{L}_j})),$$

corresponding to "quantized versions" $y_j(\mathcal{L}_j)$ of the functions $Y_j(\mathcal{L}_j)$.

First, we describe this program heuristically.

Let $$U = \int d^\nu x \ \mathrm{tr}\ (F_{\mu\nu}(x)\ F^{\mu\nu}(x)) \tag{I.48}$$

denote the classical Euclidean Yang-Mills action. (It is assumed here that G is a subgroup of some unitary matrix group. Then the r.s. of (I.48) is well defined). Let d [A] denote a formal "Lebesgue measure" on the orbit space $\mathcal{O}$ of (very rough) gauge fields modulo gauge transformations. Consider the formal probability measure on $\mathcal{O}$ given by

$$d\mu\ ([A]) = Z^{-1}\ e^{-\beta U([A])} d[A], \tag{I.49}$$

with $Z = \int_{\mathcal{O}} e^{-\beta U([A])} d[A]$,

which, mathematically, is perfectly meaningless.

Heuristically, the Schwinger functionals are given by the <u>Euclidean Gell'Mann-Low formula</u>

$$S_n(Y_1(\mathcal{L}_1),\ldots,Y_n(\mathcal{L}_n)) = \int_{\mathcal{O}} d\mu\ ([A]) \prod_{j=1}^{n} N(Y_j(\mathcal{L}_j)) . \tag{I.50}$$

In the case of free electromagnetism, (I.49) and (I.50) can be given a rigorous, mathematical meaning if one defines $N(Y(\mathcal{L}))$ by

$$N(Y(\mathcal{L})) = \exp\ [i\oint_{\mathcal{O}} A_\mu(x)\ dx^\mu] e^{K|\mathcal{L}|} \tag{I.51}$$

where $K = \beta^{-1}$" $\lim_{|x|\to 0} 1/4\pi|x|$

is a divergent (normal ordering) constant, and $|\mathcal{L}|$ is the length of $\mathcal{L}$. (We have chosen A_μ to be real-valued here. (I.50) and (I.51) have to be understood as limits of regularized objects. See [5] for some general considerations concerning (I.51)). In this example one can see explicitly that $S_n(Y_1(\mathcal{L}_1), \ldots, Y_n(\mathcal{L}_n))$ diverges when

$$d(\mathcal{L}_i, \mathcal{L}_j) \equiv \min_{\substack{x \in \mathcal{L}_i \\ y \in \mathcal{L}_j}} |x-y| \text{ tends to } 0, \text{ unless } \nu=2.$$

The divergence is $\propto e^{O(d(\mathcal{L}_i,\mathcal{L}_j)^{-1})}$ for $\nu = 4$.

$$\text{Let } L^n_{\neq} = \{\mathcal{L}_1,\dots,\mathcal{L}_n \text{ in } L: d(\mathcal{L}_i,\mathcal{L}_j) > 0, \text{for } i \neq j\},$$
$$L^n_{>} = \{\mathcal{L}_1,\dots,\mathcal{L}_n \text{ in } L^n_{\neq} : \mathcal{L}_j \; \{x=(\vec{x},t): t>0\}, j=1,\dots,n\}. \tag{I.52}$$

Suppose that somebody sufficiently clever constructed for us Schwinger functionals $S_n(Y_1(\mathcal{L}_1),\dots,Y_n(\mathcal{L}_n))$ on $L_{\neq}^n$ satisfying the following properties

(YM1) $S_o=1$; for all $n=1,2,3,\dots$.

$S_n(Y_1(\mathcal{L}_1),\dots,Y_n(\mathcal{L}_n))$ is well defined and continuous in $\mathcal{L}_1,\dots,\mathcal{L}_n$ under small C^2 deformations of $\mathcal{L}_1,\dots,\mathcal{L}_n$ on $L_{\neq}^n$. Moreover

$$|S_n(Y_1(\mathcal{L}_1),\dots,Y_n(\mathcal{L}_n))| \leq K_n \exp\left[\text{const.}\min_{i\neq j} d(\mathcal{L}_i,\mathcal{L}_j)^{-\alpha}\right] \tag{I.53}$$

for some finite constants K_n and α.

(YM2) (Invariance). The S_n 's are Euclidean invariant, i.e.

$$S_n(Y_1(\mathcal{L}_1),\dots,Y_n(\mathcal{L}_n)) = S_n(Y_1(\mathcal{L}_{1,\beta}),\dots,Y_n(\mathcal{L}_{n,\beta})) \tag{I.54}$$

for arbitrary proper Euclidean motions $\mathcal{L} \longrightarrow \mathcal{L}_\beta$.

(YM3) (Osterwalder-Schrader positivity).

Let r denote reflection at $\{t=0\}$. Let $\mathcal{L}_r = \{x : r\,x \in \mathcal{L}\}$, and $\mathcal{L}^{-1}$ the loop $\mathcal{L}$ with orientation reversed. Then the N×N matrix C with matrix elements

$$C_{ij} = S_{n(i)+n(j)}(Y^i_{n(i)}((\mathcal{L}^i_{n(i)})_r^{-1}),\dots,Y^i_1((\mathcal{L}^i_1)_r^{-1}),$$
$$Y^j_1(\mathcal{L}^j_1),\dots,Y^j_{n(j)}(\mathcal{L}^j_{n(j)})), \; i,j = 1,\dots,N,$$

is positive semi-definite, for arbitrary loops

$$\{\mathcal{L}^k_1,\dots,\mathcal{L}^k_{n(k)}\} \subset L^{n(k)}_{>} \quad \text{(i.e. localized at positive times),}$$

for all $k = 1,\dots,N$, all N.

(YM4) (Symmetry)

$$S_n(Y_1(\mathcal{L}_1),\dots,Y_n(\mathcal{L}_n)) = S_n(Y_{\pi(1)}(\mathcal{L}_{\pi(1)}),\dots,Y_{\pi(n)}(\mathcal{L}_{\pi(n)})),$$

for arbitrary (cyclic) permutations π.

[(YM5) (Clustering)]
Specific, additional properties of Yang-Mills theories are (supposedly among others)

(YM6) $$\overline{S_n(Y_1(\mathcal{L}_1),\ldots,Y_n(\mathcal{L}_n))} = S_n(Y_1(\mathcal{L}_1^{-1}),\ldots,Y_n(\mathcal{L}_n^{-1}))$$

(YM3') (Extended O-S positivity) which we do not define here.

These two additional properties are not needed in the following. In Yang-Mills theory there appear to be further ("dual" and mixed) Schwinger functionals with interesting monodromy properties. [17] Their study is highly desirable, but we ignore them here. Sequences of functionals on the loop spaces $L_{\neq}^n$ yield a very rich mathematical structure. The main theorem about them is

Theorem. (I.55) If a sequence of functionals

$$S_n(Y_1(\mathcal{L}_1),\ldots,Y_n(\mathcal{L}_n)) \text{ on } L_{\neq}^n, \quad n = 0,1,2,\ldots,$$

satisfies properties (YM1)-(YM3) then one can reconstruct from them a separable, physical Hilbert space $\mathcal{H}$, a vacuum $\Omega \in \mathcal{H}$ and a unitary representation U of the Poincaré group $\mathcal{P}_+^\uparrow$ on $\mathcal{H}$ with

$$U((\Lambda,a))\,\Omega = \Omega, \quad \text{for all } (\Lambda,a) \in \mathcal{P}_+^\uparrow .$$

If in addition (YM5) holds the vacuum is unique, i.e. Ω is the only vector in $\mathcal{H}$ satisfying (I.55).

If, in addition to (YM1)-(YM3), (YM4) holds there exist "local" fields

$$y_i(\mathcal{L}), \mathcal{L} \in L, \mathcal{L} \subset \{x=(\vec{x},t) \in \mathbb{M}^\nu : t = \text{const.}\}$$

with the property that $[y_j(\mathcal{L}), y_{j'}(\mathcal{L}')] = 0$ if $\mathcal{L}$ and $\mathcal{L}'$ are space-like separated. □

A proper formulation of this theorem requires somewhat more precision than is possible here. The theorem is non-obvious in various respects. It represents collaboration with H. Epstein, K. Osterwalder and E. Seiler and will be discussed and proven elsewhere. May it suffice for us to emphasize that quantization of Yang-Mills theory is achieved when one is able to construct Schwinger functionals,

$$S_n(Y_1(\mathcal{L}_1),\ldots,Y_n(\mathcal{L}_n)) \text{ on } L_{\neq}^n, \quad n = 0,1,2,3,\ldots,$$

satisfying (YM1)-(YM4). In one case this has been done: For the free electromagnetic field ($G = U(1)$, no interactions) in arbitrarily many dimensions. Some other gauge theories have been quantized (i.e. Schwinger functionals satisfying (YM1)-(YM4) have been constructed): Spinor QED [18,19] and the abelian Higgs model [1,3,4] in two space-time dimensions. Some other models (QED, abelian Higgs and non-abelian pure Yang-Mills in three space-time dimensions) look promising, and progress has been reported by Balaban and Magnen and Sénéor in [20].

We end this section by drawing attention to some direct physical information coded into the Schwinger functionals of a Yang-Mills theory: It is assumed that (YM1)-(YM4) and (YM3') hold. Let $\mathcal{L}_{L\times T}$ be a (smoothed out) rectangular loop with sides of length L and T. Define

$$V_j(L) = \lim_{T \to \infty} -\frac{1}{T} \log S_1(Y_j(\mathcal{L}_{L\times T})) . \qquad (I.56)$$

The function $V_j(L)$ is supposed [21] to be a measure of the potential between two static (infinitely heavy) quarks, transforming under G according to a representation with character χ_j, resp. its conjugate, coupled to the gauge field and separated by a distance L.

If $V_j(L)$ diverges to $+\infty$, as $L \longrightarrow \infty$, this is interpreted as <u>confinement of those static quarks</u>. This can be justified, in part, for lattice theories which are discussed in the next section.

Confinement is only possible if the character χ_j is non-trivial on the center of the gauge group G, [5,22,23]. In many theories one may expect $V_j(L)$ does diverge to $+\infty$ when $L \nearrow \infty$. (For sufficiently light quarks this does <u>not</u> mean that confinement breaks down in the sense that the physical Hilbert space will contain states that are not invariant under time-independent, global gauge transformations; see Swieca's lectures and [24]). Seiler has shown, by a very simple argument [25], that

$$V_j(L) < \text{const. } L, \text{ as } L \longrightarrow \infty. \qquad (I.57)$$

It has been suggested in [5] how $S_1(Y_j(\mathcal{L}_{L\times T}))$ can be used to investigate the boundstate spectrum of very heavy quark-anti-quark pairs. Moreover,

$$S_2(Y_j(\mathcal{L}),\ Y_j(\mathcal{L}'))$$

contains information about the low-lying mass spectrum of pure Yang-Mills theory, [24].

II. LATTICE GAUGE THEORIES

One of the main virtues of lattice gauge theories is that they represent a gauge-invariant regularization of continuum gauge theories for which Schwinger functionals exist satisfying properties (YM1), (YM3), (YM3'), (YM4), (YM6), see Section I.3, and ($YM2_\ell$) Invariance under all those Euclidean motions which leave a lattice ℓ, typically

$$\varepsilon \mathbb{Z}^\nu \equiv \{x : \varepsilon^{-1} x \in \mathbb{Z}^\nu\}, \ \varepsilon > 0, \ \text{invariant.}$$

This is still enough for the reconstruction of a quantum mechanical system, as described in Theorem (I.55), with the exception of full Poincaré covariance of the resulting theory. As a consequence, only a weak form of locality is verified for lattice theories. See [1,21-23] .

In order to understand the basic structure and intrinsic properties of lattice gauge theories one is advised to go back to Remark 2), following Theorem (I.47), Section I.2: Let

$$\Lambda \equiv L_\ell = \{\mathscr{L}_1, \mathscr{L}_2, \ldots\}$$

be all finite, oriented closed loops composed of links of a lattice ℓ, e.g. $\varepsilon \mathbb{Z}^\nu$. Let

$$C_{L_\ell}(\mathcal{O})$$

denote the algebra of functions on the orbit space, $\mathcal{O}$, of continuous, classical connections (gauge potentials) on Euclidean space-time, $\mathbb{R}^\nu$, modulo gauge transformations, generated by

$$\{Y_j(\mathscr{L}) = \chi_{Y_j}(g_{\mathscr{L}}) : \mathscr{L} \in L_\ell\},$$

with χ_Y arbitrary irreducible, unitary characters of the gauge group G.

Clearly, $C_{L_\ell}(\mathcal{O})$ is a separable approximation to the space $C(\mathcal{O})$ of all "observables" of a classical gauge theory.

The idea is now to convert the elements of

$$C_{L_\ell}(\mathcal{O})$$

into random variables distributed according to a probability measure $d\mu$ (a positive, normalized, continuous linear functional on $C_{L_\ell}(\mathcal{O})$) with the property that the Schwinger functionals

$$S_n(Y_1(\mathcal{L}_1),\ldots, Y_n(\mathcal{L}_n)) = \int d\mu \prod_{j=1}^{n} Y_j(\mathcal{L}_j) \tag{II.1}$$

satisfy properties (YM1), ($YM2_\ell$),(YM3)-(YM6), (with the possible exception of (YM5) = clustering = uniqueness of the vacuum).

Thus, we are really trying to construct random fields Y_j on the loop space L_ℓ of a lattice ℓ having the mathematical structure determined by ($YM1_\ell$)-($YM6_\ell$), (with (YMn_ℓ)=(YMn), except for n=2). Given the values of all $Y_j(\mathcal{L})$, $\mathcal{L} \in L_\ell$, a simple variant of Theorem (I.45) shows that they determine a "lattice gauge field"

$$g = \{g_{xy} \in G : xy \in B_\ell\}$$

which is unique up to gauge transformations. Here B_ℓ is the set of all line segments, b(xy), whose endpoints, x and y, are "nearest neighbors" in ℓ. The b's are called bonds or links. A variant of Theorem (I.47) shows that the closure of

$$C_{L_\ell}(\mathcal{O})$$

is the space of all gauge-invariant functions of

$$g = \{g_{xy} : xy \in B_\ell\}.$$

(We know from Section II.2, (I.31), that a gauge transformation $g \longrightarrow g^h$ is given by a function, h, on ℓ with values in the gauge group G, and

$$g^h_{yx} = h(y)^{-1} g_{yx}\, h(x) \;).$$

Let

$$\prod_{xy \subset \mathcal{L}}^{\circlearrowleft}$$

denote an ordered product along an oriented loop (or curve) $\mathcal{L}$. From the above discussion we infer that

$$Y_j(\mathcal{L}) = \chi_{Y_j}(g_{\mathcal{L}}), \text{ where } g_{\mathcal{L}} = \prod_{xy \subset \mathcal{L}}^{\circlearrowleft} g_{xy}, \tag{II.2}$$

and g is the lattice gauge field determined (up to gauge transformations) by the values of the $Y_j(\mathcal{L})$'s, $\mathcal{L} \in L_\ell$. From now on the random variables $Y_j(\mathcal{L})$ are called Wilson loops.

Other examples of random fields on a loop-space, L_ℓ, are supplied by the lattice approximation to (Euclidean) dual resonance - or string models [26]. Among the main goals in the study of lattice gauge theories are

(A) Let $\mathcal{L}$= L×T be a rectangular loop with sides parallel to two axes of ℓ of length L, resp. T.

Let

$$V_j(L) = \lim_{T \to \infty} - \frac{1}{T} \log S_1(Y_j(L \times T)); \qquad \text{(II.3)}$$

see (I.56). Investigate the properties of $V_j(L)$, as $L \to \infty$, in particular for those characters χ_{Y_j} of G which are <u>non-trivial on the center</u> $\mathfrak{Z}_G$ of G.

This is the famous problem of <u>static quark confinement</u> (resp.-liberation). See [21,22,23,25,1,27,5,6,7] .

(B) Investigate the "excitation spectrum" (energy-momentum spectrum of low-lying "particles") in

$$S_1(Y_j(L \times T)) \text{ and } S_2(Y_j(\mathcal{L}),\ Y_j(\mathcal{L}'));$$

see e.g. [5,24,26]. This will supply information on the particle- and bound state content of lattice Yang-Mills theory.

(C) Improve the analysis in (A) and (B) in such a way that the results are uniform in the lattice spacing ε,($\ell = \varepsilon \mathbb{Z}^\nu$).

(D) For $\ell = \varepsilon \mathbb{Z}^\nu$ (and $\nu = 2, 3, (4?)$), exhibit lattice gauge theories (other than free electromagnetism) with the property that the limits as $\varepsilon \searrow 0$, of the Schwinger functionals $S_n(Y_1(\mathcal{L}_1), \ldots, Y_n(\mathcal{L}_n))$, $n = 1, 2, \ldots$, exist and satisfy (YM1)-(YM6) if the measures

$$\{d\mu = d\mu_\varepsilon\}_{\varepsilon > 0}$$

are correctly renormalized and the Wilson loops, $Y_j(\mathcal{L})$, are correctly normal ordered. See [1,3,4] and [20] for results or progress in this direction.

<u>Remark concerning matter fields.</u>

For pedagogical reasons we shall only consider bosonic matter in these notes; but see [1,22,18,19] . Given a gauge group G, a lattice matter field Φ is a random field on the lattice ℓ with values in a Hilbert space V (usually finite dimensional) that carries a unitary representation U^Φ, of G (as an endomorphism group). Thus

$$\Phi : x \in \ell \longmapsto \Phi(x) \in V.$$

Gauge transformations of Φ are of course defined by

$$\Phi \longrightarrow \Phi^h, \ \Phi^h(x) = U^\Phi(h(x))^* \ \Phi(x), \qquad \text{(II.4)}$$

where h takes values in G. The random variables

$$Y^{\Phi}(\gamma_{xy}) \equiv (\Phi(x), U^{\Phi}(\gamma_{xy})\ \Phi(y), \text{ with } g_{\gamma_{xy}} = \prod_{uv\in\gamma_{xy}} \circlearrowleft g_{uv}, \tag{II.5}$$

$\gamma_{xy} \subset B_{\ell}$ a connected, oriented curve starting at y and ending at x, are gauge-invariant. (These notions correspond to what is developed at the end of Section I.1 and in Section I.2).

II.1 Some of the basics about lattice gauge theories

General results may be found in [1,2,21,22,23,25]. The general ansatz for the measures $d\mu = d\mu_{\varepsilon}$ is the lattice version of the Euclidean Gell'Mann-Low formula (I.49) (including a matter field Φ):

$$d\mu_{\varepsilon}(\Phi,g) = Z_{\varepsilon}^{-1} e^{-\beta U_{\varepsilon}(\Phi,g)} Dg\ D\Phi_{\varepsilon}, \tag{II.6}$$

where

$$Z_{\varepsilon} = \int e^{-\beta U_{\varepsilon}(\Phi,g)} Dg\ D\Phi_{\varepsilon}\ ,\ Dg = \prod_{xy\in B_{\varepsilon}} dg_{xy},$$

with $B_{\varepsilon} = B_{\varepsilon \mathbb{Z}^{\nu}}$ and dg Haar measure on G,

$$D\Phi_{\varepsilon} = \prod_{x\in\varepsilon\mathbb{Z}^{\nu}} d\rho_{\varepsilon}(\Phi_x),$$

with $d\rho_{\varepsilon}$ a G-invariant probability measure on V, and

$$U_{\varepsilon}(\Phi,g) = U_{\varepsilon}^{YM}(g) + U_{\varepsilon}^{M}(\Phi,g) \tag{II.7}$$

a lattice action.

Wilson [21] was the first to propose lattice gauge theories and explicit expressions for (II.6) and (II.7). In the introduction to Section II we have proposed to view the lattice gauge field g_{xy} as arising from a "nice" continuum gauge field (connection)

$$A, \text{ via } g_{xy} \equiv g_{b(x,y)} = P\{\exp \int_{b(x,y)} A_j(z)dz^j\}, \tag{II.8}$$

see (I.30), Section I.2.

This is particularly useful if the continuum gauge field A is known to exist as a random field with the desired properties, as is the case for free electromagnetism ($G = U(1)$). In this case we may e.g. choose A to be a Gaussian random field with mean 0, $\langle A_i(x)\rangle_t = 0$, and covariance

$$\langle A_i(x)\ A_j(y)\rangle_t = D_{ij}^{t}(x-y), \tag{II.9}$$

where $D^t_{ij}(x)$, $t \geq 0$, is the Fourier transform of

$$(\delta_{ij} - {}^{p_i p_j}/(p^2 + \mu^2))\,(p^2+\mu^2)^{-1} e^{-t \sum_{j=1}^{\nu-1} p_j^2}. \qquad \text{(II.10)}$$

Here $t \geq 0$ labels an ultraviolet cutoff, and $\mu \geq 0$ is a bare mass introducing an infrared cutoff. As long as $t > 0$, g_{xy} given by (II.8) is well defined for A as in (II.9),(II.10), in arbitrary dimenion ν. For $\mu > 0$, the resulting lattice U(1) theory is not gauge-invariant, but when $\mu = 0$, gauge invariance is restored even for $t > 0$; see [1,3] . These observations are useful in the construction of the two-dimensional abelian Higgs model in the continuum limit [1,3,4] which we sketch in Section III. For G non-abelian and $\nu > 2$, no such construction of a lattice approximation is known. Instead one recurs to (II.6) and (II.7) with

$$U^{YM}_{\varepsilon} \quad \text{and} \quad U^{M}_{\varepsilon}$$

conventionally given by

$$U^{YM}_{\varepsilon} = - \sum_{p} \varepsilon^{\nu-4} \chi\,(g_{\partial p}),\; g_{\partial p} = \prod_{xy \subset \partial p}^{\circlearrowright} g_{xy} \qquad \text{(II.11)}$$

where χ is a faithful, unitary character of G, (e.g. the character of the fundamental representation if G is a unitary matrix group), and p denotes the unit squares (plaquettes, with boundary p = four links) in $\varepsilon\mathbb{Z}^{\nu}$;

$$U^{M}_{\varepsilon} = (1/2\beta) \sum_{yx \in B_{\varepsilon}} \varepsilon^{\nu-2} \left\| \Phi\,(x) - U^{\Phi}\,(g_{xy})\,\Phi\,(y) \right\|^2, \qquad \text{(II.12)}$$

and

$$d\rho_{\varepsilon}\,(\Phi) \overset{e.g.}{=} \exp\left[-\varepsilon^{\nu}\left(\left(\frac{m^2}{2}\right)\|\Phi\|^2 - \lambda : \|\Phi\|^4 :_{\varepsilon}\right)\right] d\Phi, \qquad \text{(II.13)}$$

where $:-:_{\varepsilon}$ denotes a Wick order, and $d\Phi$ is the Lebesgue measure on V.

In order to start with a well defined expression, one first restricts the summation on the r.s. of (II.11) to plaquettes p contained in some bounded set $\Lambda \subset \varepsilon\mathbb{Z}^{\nu}$ and the one on the r.s. of (II.12) to links $xy \subset \Lambda$. By (II.6) this yields a cutoff measure $d\mu_{\varepsilon\,\Lambda}(\Phi,g)$. If Λ belongs to a sequence of hypercubes and periodic boundary conditions are imposed at $\partial\Lambda$ then a weak limit, $d\mu_{\varepsilon}\ (\Phi,g)$, (the thermodynamic limit), of the measures $d\mu_{\varepsilon\,\Lambda}(\Phi,g)$ as $\Lambda \uparrow \varepsilon\mathbb{Z}^{\nu}$, can be constructed by a standard compactness argument. The lattice Schwinger functions

$$S^{(\varepsilon)}_{n,m}\,(Y_1(\mathcal{L}_1),\,\dots,Y^{\Phi}\,(\gamma^1_{x_1,y_1}),\dots)$$
$$= \int d\mu_{\varepsilon,\Lambda}\,(\Phi,g)\prod_{j=1}^{n} Y_i(\mathcal{L}_j)\prod_{k=1}^{m} Y^{\Phi}(\gamma^k_{x_k y_k}) \qquad \text{(II.14)}$$

obey properties (YM1), (YM2$_\ell$), (YM3), (YM4) and (YM6) (modified in the obvious manner to account for the

$Y^{\Phi}\,(\gamma^k_{x_k,y_k})$-variables).

Clustering (YM5) may fail in general, but is known to hold e.g. for small β[22] . Thus, lattice gauge fields exist, for arbitrary G and arbitrary space-time dimension ν.

Among numerous, very general results we mention the following two which turn out to be important.

(1) Universality of diamagnetism [1,2] :

Define

$$Z_{\varepsilon,\Lambda}\,(g) = \int e^{-\beta U^{M}_{\varepsilon,\Lambda}(\Phi,g)}\, D\Phi_{\varepsilon} \qquad \text{(II.15)}$$

Let Λ be a rectangle and impose periodic b.c. at $\partial\Lambda$. Then

$$|\,Z_{\varepsilon,\Lambda}\,(g)\,| \leq Z_{\varepsilon,\Lambda}(1), \qquad \text{(II.16)}$$

($g = 1$ means g_{xy} = identity in G, for all xy).

Inequality (II.16) holds no matter what gauge group G is chosen and even if Fermionic matter (leptons or quarks) is coupled to the gauge field. It expresses the fact that matter behaves diamagnetically under coupling to gauge fields. Inequality (II.16) does generally not survive ultraviolet renormalizations necessary for taking $\varepsilon \searrow 0$, unless the vacuum polarization is finite (i.e. $\nu \leq 3$). (There are related inequalities for pure Yang-Mills theories mentioned in [5] which appear to be renormalization-independent).

Next, suppose that G is abelian. Without loss of generality we may assume that $G = \mathbb{Z}_n$, $n=2,3,4\dots$, or $G = U(1)$. Then we may introduce polar coordinates

$$g_{xy} = e^{ia_{xy}}\,,\quad a_{xy} \in \mathbb{R}$$

$$\Phi(x) = r_x\, e^{i\Theta_x}, \qquad 0 \leq \Theta_x < 2\pi .$$

Let

$$\left.\begin{array}{l} dm_\varepsilon(a) = \lim_{\Lambda\uparrow\varepsilon\mathbb{Z}^\nu} Z_{\varepsilon,\Lambda}^{-1}\, e^{\beta\Sigma_{p\subset\Lambda}\cos(\delta\Sigma_{xy\subset\partial p} a_{xy})} \Pi\, da_{xy}, \\ \text{with } da \text{ the Lebesgue measure on } \left[o,\frac{2\pi}{\delta}\right], \text{ or let} \\ dm_\varepsilon(a) = dm_{\varepsilon,t,\mu}(a) \text{ be the restriction of the} \\ \text{Gaussian measure introduced in (II.9), (II.10)} \\ \text{to the variables } a_{xy} = \int_{b(x,y)} A_j(z)\, dz^j. \end{array}\right\} \quad \text{(II.17)}$$

(2) Correlation Inequalities [1,2,27,6] :

Let $G = \mathbb{Z}_n$ or $U(1)$, dm_ε as in (II.17), and $< - >$ the expectation given by the probability measure

$$d\mu'_{\varepsilon,\Lambda}(\Phi,a) = (Z'_{\varepsilon,\Lambda})^{-1}\, e^{-\beta U^M_{\varepsilon,\Lambda}(\Phi,a)}\, dm_\varepsilon(a)\, D\Phi_\varepsilon\,, \quad \text{(II.18)}$$

with $\Lambda \subseteq \varepsilon\mathbb{Z}^\nu$. Let F and G be in the multiplicative cone generated by $r(f)$, $f(x) \geqq o$, $\cos(a(g) + \theta(h))$.

Then

$$< F\, G > - < F >< G > \geqq o \quad \text{(II.19)}$$

For applications, see e.g. [2,4] .

Next, we consider a general lattice theory described by a measure as in (II.6) with action as in (II.7),(II.11) and (II.12). Suppose that the representation U^Φ of G on V is trivial on the center $\mathfrak{Z}_G$ of the gauge group. Let $< - >_G$ denote the expectation determined by the measure $d\mu_\varepsilon$ given in (II.6). Let $< - >_{\mathfrak{Z}_G}$ denote the expectation in the pure $\mathfrak{Z}_G$ lattice gauge theory with measure

$$d\mu_\varepsilon(\tau) = \lim_{\Lambda\uparrow\varepsilon\mathbb{Z}^\nu} (Z''_{\varepsilon,\Lambda})^{-1}\, e^{\beta\Sigma_p \varepsilon^{\nu-4}\chi(\tau_{\partial p})}\, D\tau, \quad \text{(II.20)}$$

where $\tau_{xy} \in \mathfrak{Z}_G$, for all xy, and $d\tau$ is Haar measure on $\mathfrak{Z}_G$. Then

$$< \Pi\, Y_j(\mathcal{L}_j) >_G \;\leq\; < \Pi\, Y_j(\mathcal{L}_j) >_{\mathfrak{Z}_G} . \quad \text{(II.21)}$$

Proof and applications are given in [6]. A special case of (II.21) was first proven in [27] .

The arguments used in the proofs of such inequalities are patterned on Ginibre's methods [28] .

II.2 On the phase diagram of some lattice gauge theories

Rigorous results on "high temperature"expansions (β small) in lattice gauge theories are established in [22]. It is proven there that if β is small enough and χ_{y_j} is non-trivial on the center $\mathfrak{z}_G$ of the gauge group G the "quark-antiquark potential" V_j defined in (II.3) satisfies

$$V_j(L) \geq \text{const.}\ L \tag{II.22}$$

Moreover, the Higgs mechanism is for lattice theories analyzed in that reference, too. In [5,23] there are general arguments suggesting that $V_j(L) \leq$ const., uniformly in L if Y_j is trivial on the center. A. Guth has announced that the four-dimensional pure U(1) lattice theory (in the so called Villain form) has a phase transition as β is varied: For β small (II.22) is valid, for β large $V_j(L) \leq$ const.. The proof is based on a combination of correlation inequalities (of the type proven in [1,2]) and a high temperature expansion. Similar results were previously proven for the $\mathbb{Z}_n$ theories in two and three dimensions and are discussed in Guerra's contribution where the reader can also find references to the original articles of Guerra et al.

In [6] the author has applied inequality (II.21) to prove that in all two-dimensional Yang-Mills theories $V_j(L) \geq$ const. L for all characters Y_j which are non-trivial on the kernel of the representation U^{Φ} used in the matter action (II.12).

This extends results of [2,30] . It is also shown in [6] that for three-dimensional U(n) theories with U^{Φ} trivial on $U(1) \subset \mathfrak{z}_{U(n)}$

$$V_j(L) > \text{const}\log(L+1), \tag{II.23}$$

if χ_{y_j} is non-trivial on U(1).

In [2,5] connections between lattice gauge theories on $\varepsilon\mathbb{Z}^{\nu}$ and non-linear σ-models on $\varepsilon\mathbb{Z}^{\nu-1}$ have been found. The following models are investigated there:

(i) Classical, two-component, neutral Coulomb gases and abelian σ-models (Ising-, $\mathbb{Z}_n$- and classical XY models).

(ii) Abelian lattice Higgs theories, in particular Landau-Ginzburg type theories.

(iii) Non-linear lattice-σ-models (e.g. the classical O(4) lattice model).

(iv) Pure, non-abelian lattice gauge theories.

The results are of the following kind:

(a) Rigorous connections between (ii) in ν dimensions and (i) in ν-1 dimensions, and between (iv) in ν dimensions and (iii) in ν-1 dimensions. E.g., $S_1(Y_j(\mathcal{L}))$ of a ν-dimensional gauge theory can generally be bounded above by (an integral of) a product of two-point functions of a (ν-1)-dimensional σ-model. As examples we mention:

If the two-dimensional Coulomb gas has a transition from a high temperature plasma phase with Debye screening [31] to a low temperature, dipolar phase with power low decay, as expected, then the three-dimensional Landau-Ginzburg (abelian Higgs) lattice theory has a transition from a superconducting phase without confinement of fractional charges, massive photons and vortices, at small electric charge, to a QED phase with massless photons and confined fractional charges, at large electric charge. This is shown in [2]. It is also shown there that Guth's result for U(1) implies the existence of a superconductor $\rightarrow$ QED transition in a four-dimensional Landau-Ginzburg lattice theory, with liberated magnetic monopoles in the QED phase.

For further results on phase transitions in lattice gauge theories see [5,27,32] and Guerra's contribution to these proceedings. Some other, general consequences of correlation inequalities in lattice gauge theories (confinement, Higgs mechanism,...) are given in [2,4].

II.3 Connections to dual resonance models

Recently many connections between (lattice) Yang-Mills theories and string (dual resonance) models have been proposed [33,14,15,26,34]. It has been suggested that lattice Yang-Mills theory is a theory of random surfaces [33,5,15,34] related to the lattice theory of dual strings (e.g. [11,34]). Such a connection would be useful as a starting point for an investigation of the particle spectrum of pure Yang-Mills theory.

In [5] an expansion of the n-loop Schwinger functionals $S_n^{(\varepsilon)}(Y_1(\mathcal{L}_1),\ldots,Y_n(\mathcal{L}_n))$ of pure lattice Yang-Mills theories in terms of random surfaces bounded by the loops $\mathcal{L}_1,\ldots,\mathcal{L}_n$ has been derived when the gauge group G is U(n) or O(n), n = 1,2,3,...,or SU(2).

For G = SU(2) this has provided rather powerful lower bounds on the potential $V_j(L)$ (χ_{Y_j} = spin 1/2 character) and has revealed an interesting connection with the theory of interacting random paths and non-relativistic strings. A method for obtaining upper bounds on $V_j(L)$ has also been suggested there.

III. REMARKS ON THE CONTINUUM LIMIT OF THE ABELIAN HIGGS MODEL IN TWO SPACE-TIME DIMENSIONS

The only continuum gauge theories satisfying properties (YM1) - (YM6) (except possibly (YM5)) of Section I.3 are

- free electromagnetism in arbitrary dimension
- massive spinor QED [18,19] } in two space-time dimensions.
- the abelian Higgs model [1,3,4] } in two space-time dimensions.

The situation concerning two- and three-dimensional, super-renormalizable (abelian and non-abelian) gauge theories looks fairly promising; see the contributions of Balaban and Magnen-Sénéor to [20] .

This situation is thus not overly encouraging. We present a few remarks on Higgs models. For some general information about constructive quantum field theory see [CQFT].

III.1 External (c-number) Yang-Mills fields

In [3] weak convergence of the measures

$$(Z_{\varepsilon,\Lambda}(g))^{-1}\, e^{-\beta U^{M}_{\varepsilon,\Lambda}(\Phi,g)}\, D\Phi_\varepsilon \;, \text{ as } \varepsilon \searrow 0, \qquad \text{(II.24)}$$

has been shown for $\nu = 2$ and

$$g_{xy} = P\,\{\exp \int_{b(x,y)} A_j\,(z)\, dz^j\,\},$$

with A(z) Hölder continuous in z, and

G = U(1), SU(2), etc.

The proof of convergence (for various boundary conditions) is rather complicated. In principle, it can be extended to $\nu = 3$, but this has only been done if the self-interaction of Φ vanishes, i.e. $\lambda = o$ in (II.13).

The following elements are crucial in the proof: Let $\Delta_A^{(\varepsilon)}$ be the finite difference covariant Laplacean on $\ell_2(\Lambda) \otimes V$ with periodic or O-Dirichlet boundary conditions at $\partial\Lambda$.

(a) $$\|(-\Delta_A^{(\varepsilon)} + m^2)^{-1}(x,y)\|_V \leqq (-\Delta^{(\varepsilon)} + m^2)^{-1}(x-y),$$

where $\Delta^{(\varepsilon)}$ is the usual finite difference Laplacean on $\ell_2(\Lambda)$ with the same b.c.. A proof of this "diamagnetic inequality" can be found in [1] and refs. given there.

(b) Convergence of

$$(-\Delta_A^{(\varepsilon)} + m^2)^{-1}$$

in various trace ideals, and L_p convergence of

$$(-\Delta_A^{(\varepsilon)} + m^2)^{-1}(x,y) \text{ for } p < \frac{\nu}{\nu-2} .$$

The proof involves showing real analyticity in A and using a Neumann series expansion in A for "small" A; see [3] .

(c) $$\det\left((-\Delta_A^{(\varepsilon)} + m^2)^{-1} (\Delta^{(\varepsilon)} + m^2) \right) \leqq 1;$$

this is a special case of the diamagnetic inequality (II.16) due originally to R. Schrader and R. Seiler.

(d)

$$\det\left((-\Delta_A + m^2)^{-1} (\Delta + m^2) \right)$$

$$\equiv \lim_{\varepsilon \searrow 0} \det\left((-\Delta_A^{(\varepsilon)} + m^2)^{-1} (\Delta^{(\varepsilon)} + m^2) \right)$$

exists for Hölder-continuous A, $\nu=2$; see [3].

These elements somewhat cleverly combined with the diamagnetic inequality (II.16), the original Nelson-Glimm method (proving stability of $P(\Phi)_2$ theories, see [CQFT] and refs.given there) and numerous,lengthy estimates yield a proof of (II.24).
For

$$\nu = 2,\ G = U(1),\ dm_\varepsilon(A) = d\,m_{\varepsilon,t,\mu}(A)$$

the Gaussian measure defined in (II.9),(II.10),one derives from (II.24) that the weak limit of the measures

$$d\mu_{\varepsilon,\Lambda,t,\mu}(\Phi,a)$$

exists, as $\varepsilon \searrow 0$, for $t > 0, \mu \geqq 0$. This follows from the diamagnetic inequality (II.16) and (II.24) by Lebesgue dominated convergence.

III.2 Removal of cutoffs

In order to show that the weak limit of the measures

$$d\mu_{\Lambda,t,\mu} = W - \lim_{\varepsilon \searrow 0} d\mu_{\varepsilon,\Lambda,t,\mu}, \text{ as } t \searrow 0,$$

exists one must do an ultraviolet expansion [4], involving a truncated (high-momentum) perturbation expansion which exhibits cancellations of divergent Feynman diagrams with counterterms.

The ultraviolet expansion is applied to unnormalized expectations

$$Z_{\Lambda,t,\mu} < F >_{\Lambda,t,\mu},$$

where $< - >_{\Lambda,t,\mu}$ is the expectation obtained from $d\mu_{\Lambda,t,\mu}$, and $Z_{\Lambda,t,\mu}$ is the natural continuum partition function.

In the following Λ and μ are suppressed temporarily. The initial form of the expansion is roughly

$$Z_{t_N} < F >_{t_N} = \sum_{n=1}^{N} (Z_{t_n} < F >_{t_n} - Z_{t_{n-1}} < F >_{t_{n-1}}), \quad \text{(II.25)}$$

with $t_0 > 0$ some suitable constant.

The differences,

$$Z_{t_n} < F >_{t_n} - Z_{t_{n-1}} < F >_{t_{n-1}},$$

are then interpolated in a somewhat sophisticated way that depends on n and involves "changes of A-covariance" and "integrations by part on function space" with subsequent cancellations of divergent diagrams; see [3,4] .

One obtains an upper bound on

$$\left| Z_{t_n} < F >_{t_n} - Z_{t_{n-1}} < F >_{t_{n-1}} \right|$$

of the form:

$$c^n \prod_{i=1}^{n} t_i^{\delta} \; e^{c(\log t_n)^2} (n!)^r (\log t_n)^n$$

This proves convergence of (II.25), for $t_n \overset{e.g.}{\propto} \exp(-n^{\gamma}), 0<\gamma<1$.

A technically subtle part in the proof of the upper bounds is the estimation of large Feynman diagrams. (There one makes use, among many other things, of (a)). Convergence of the ultraviolet expansion suffices to show that

$$< F >_{\Lambda,\mu} = \lim_{t \searrow o} < F >_{\Lambda,t,\mu}$$

exists, for Λ bounded and $\mu^2 > o$.

Subsequently one uses fairly standard methods to establish upper bounds on

$$< F >_{\Lambda,\mu}$$

that are uniform in Λ and μ. Thanks to the correlation inequalities (I.19) one has monotonicity in Λ and μ, for a total set of random variables, F. Thus, the limits $\Lambda \uparrow \mathbb{R}^2$ and $\mu \downarrow o$ exist. The existence of the 0-bare-mass limit, $\mu \downarrow o$, is yet another manifestation of the well established experience that constructive field theory methods never create artificial infrared problems (which might be regarded as one of its modest triumphs).

To date it is only known that the Schwinger functionals

$$S_{n,m} (Y_1(\mathcal{L}_1), \ldots, Y^{\Phi}(\gamma^1_{x_1 y_1}), \ldots)$$

$$= < \prod_{j=1}^{n} Y_i(\mathcal{L}_j) \prod_{k=1}^{m} Y^{\Phi}(\gamma^k_{x_k y_k}) >$$

of the limiting expectation $< \text{---} >$ satisfy properties (YM1) - (YM4), (YM6) (without normal ordering of Y_j's, Y^{Φ}'s) so that they determine a relativistic quantum field theory (Theorem(I.55), Section I.3), but detailed, physical information is lacking, (e.g. Higgs mechanism ?).

IV. A LOOK INTO THE FUTURE OF THE SUBJECT

In the Euclidean approach to quantized Yang-Mills theory one proposes to convert (the traces of) holonomy operators on a principal bundle into random fields on a loop space over physical space-time. Thus, one attempts, in fact, to construct stochastic processes and random fields on spaces of geometrical objects, the closed loops in physical space-time. This is an instance of combining geometry and probability theory, i.e. a problem in random geometry. Random geometry still appears to be an underdeveloped branch of mathematics. (For other examples in random geometry see e.g. [10,11] and refs. given there).

A number of conceptual problems arises: E.g. is there a reasonable notion of "distribution-valued connections", or, in other words, is there a geometric interpretation of "normal-ordered" holonomy operators, (see Section I.3), etc..

Gauge fields (i.e. the gauge orbits, [A], of connections, A, or the traces of holonomy operators) are intrinsically non-linear fields (at least in the non-abelian case). Constructive quantum field theory methods have so far not had much success as a means of studying non-linear fields. One of the main reasons might be that non-linear fields cannot be localized on classical phase space, a technical device that has so far appeared to be crucial for non-perturbative renormalization, [35]. In the analysis of [1,3,4] outlined in Section III and in [8,9] the non-linearily of gauge fields has been circumvented in a somewhat unnatural way. Presumably, this is only possible if the gauge group is abelian, and the gauge field couples to a conserved current. Even then the price to be paid is a fairly clumsy and tedious analysis.

We have tried to explain the underlying geometric reasons why the lattice approximation is a natural gauge-invariant regularization of continuum Yang-Mills theory (End of Section I.2, introduction to Section II). What remains to be seen is how one can do hard analysis (non-perturbative renormalization) starting from lattice theories. The popular magic word is: Renormalization group ("block spin") transformations. This has first been advertized by Kadanoff and Wilson. A rigorous program of this sort has been described by Balaban in [20]. The program can only be regarded as <u>really successful</u> if one eventually achieves a non-perturbation renormalization of a four-dimensional, non-super-renormalizable, asymptotically free gauge theory.

Another approach, due to Jimbo, Miwa and Sato [17] is based on analyzing the monodromy structure of the Schwinger functionals of the loop variables, $Y_j(\mathcal{L})$, and the dual ("disorder") variables. The general monodromy properties of the Schwinger functionals follow from "topological commutation relations". One then studies monodromy preserving deformations and uses the Schwinger-Dyson equations for the Schwinger functionals.

In some examples (e.g. the two-dim. Ising model), with free Schwinger-Dyson equations, Jimbo, Miwa and Sato have carried out their program, with impressive success. One might hope that there exist "non-local" conserved currents in Yang-Mills theory yielding relations between Schwinger functionals which reinforce the J-M-S program in a suitable way.

But this is mere speculation.

In conclusion, the author wishes to thank D. Brydges and E. Seiler for the joy of collaboration and H. Epstein and E. Seiler for many most valuable discussions. He thanks the organizers of the Kaiserslautern school for inviting him to give lectures.

REFERENCES

1. D. Brydges, J. Fröhlich, E. Seiler, On the Construction of Quantized Gauge Fields (CQGF), I, to appear in Ann.Phys. (N.Y.) 1979.
2. D. Brydges, J. Fröhlich, E. Seiler, Nucl.Phys. B152, 521 (1979).
3. D. Brydges, J. Fröhlich, E. Seiler, CQGF, II, to appear in Comm.Math.Phys.
4. D. Brydges, J. Fröhlich, E. Seiler, CQGF, III, Preprint (Univ.of Virginia) to appear.
5. B. Durhuus and J. Fröhlich, A Connection Between ν-Dim. Yang-Mills Theory and (ν-1) Dim., Non-Linear σ-Models, to appear in Comm.Math.Phys.
6. J. Fröhlich, Physics Letters 83B, 195 (1979).
7. J. Fröhlich, A New Look at Generalized, Non-Linear σ-Models and Yang-Mills Theory, Proceedings of the Bielefeld Symposium, Dec. 1978, L. Streit, ed., to appear.
8. E. Seiler, CQGF, contribution to the proceedings of the "Colloquium on Random Fields", Esztergom, Hungary, June 1979, to be published.
9. E. Seiler, Quantized Gauge Fields: Results and Problems, contribution to ref. 20 .
10. J. Fröhlich, Random Geometry and Yang-Mills Theory, contribution to the proceedings in ref. 8 .
11. J. Fröhlich, unpublished lectures at "Strasbourg Rencontres" May 1979, and at the Cargèse summer school, Sept. 1979; publication in preparation.
12. N. Steenrod, "The Topology of Fibre Bundles", Princeton University Press, Princeton, N.J. 1951.
13. M. F. Atiyah, Fermi Lectures, Scuola Normale Superiore, Pisa, 1978; M. Lüscher, Nucl. Physics B140, 429(1978); T. Jousson, J. Hubbard, O. McBryan, F. Zirilly, Comm.Math. Phys. 68, 259 (1979); I. Segal, J. Funct.Anal. 33, 175 (1979).
14. E. Corrigan, B. Hasslacher, Phys.Letters 81B, 181 (1979).
15. Y. Nambu, Phys.Letters 80B, 372 (1979); J. L. Gervais and A. Neveu, Phys.Letters 80B, 255 (1979); A. M. Polyakov, Phys.Letters 82B, 247 (1979).
16. I. M. Singer, Comm.Math.Phys. 60, 7 (1978); I. M. Singer, Lectures, Cargèse, 1979; P. K. Mitter, Lectures, Cargèse , 1979; M. Lévy and P. K. Mitter, eds.,to be published.
17. J. Fröhlich, Comm.Math.Phys. 47, 269 (1976);Comm.Math.Phys. 66, 223 (1979);Jimbo, Miwa, Sato, Preprints,Research Inst.f.Math. Sciences,Kyoto,1978/79; G.'t Hooft, Nucl.Phys. B138, 1 (1978).

18. J. Fröhlich and E. Seiler, Helv.Phys.Acta 49, 889 (1976).
J. Fröhlich, in "Renormalization Theory", G. Velo and A. S. Wightman, eds., Reidel, Dordrecht-Boston, 1976.
19. J. Challifour, D. Weingarten, Preprint, University of Indiana, 1979.
20. Proceedings of the I.A.M.P. congress, Lausanne, August, 1979, K. Osterwalder, ed., Springer Lecture Notes in Physics, to appear.
21. K. Wilson, Phys.Rev. D10, 2445 (1974).
22. K. Osterwalder, E. Seiler, Ann.Phys. (N.Y.) 110, 440 (1978).
23. J. Glimm, A. Jaffe, Nucl. Phys. B149, 49 (1979).
24. J. Fröhlich, unpublished.
25. E. Seiler, Phys.Rev. D18, 482 (1978).
26. "Dual Theory", M. Jacob, ed., North Holland, Amsterdam, 1974.
R. Giles, C. B. Thorn, Phys.Rev. D16, 366 (1977).
F. Gliozzi, T. Regge, M. A. Virasoro, Phys.Letters 81B, 178 (1979).
27. G. Mack, V. B. Petkova, Preprints, DESY, 1979.
28. J. Ginibre, Comm.Math.Phys. 16, 310 (1970).
29. A. Guth, Preprint, Cornell University, to appear.
30. G. Mack, Comm.Math.Phys. 65, 91 (1979).
31. D. Brydges, Comm.Math.Phys. 58, 313 (1978).
32. J. Drouffe, G. Parisi, N. Sourlas, Preprint, C.E.N. Saclay, 1979.
33. D. Förster, Preprint, Cornell University, 1979, and Preprint to appear.
34. I. Bars, F. Green, Preprint, I.A.S.-Princeton, 1979.
35. J. Glimm, A. Jaffe, Fortschr.Physik 21, 327 (1973).
CQFT "Constructive Quantum Field Theory", G. Velo and A. S. Wightman, eds., Lecture Notes in Physics 25, Springer Verlag, Berlin - Heidelberg - New York, 1973.

GAUGE FIELDS ON A LATTICE

SELECTED TOPICS II

Francesco Guerra

Institute of Physics

University of Salerno, Italy

I. INTRODUCTION

Models of quantum gauge fields on a lattice, introduced some years ago by Wilson [15] and others, continue to provide interesting structures to study both from a physical and mathematical point of view. While their main motivation comes from elementary particle theory, in relation with the problem of infrared behaviour and quark confinement, these models, formulated in the Euclidean discrete version, provide also interesting examples of statistical mechanics systems, whose behaviour is completely different from the more familiar one of systems related to the conventional ferromagnetic structure of the Ising model.

In particular gauge fields on a lattice have been actively studied in the framework of the program of constructive quantum field theory. Basic references about work done until 1978 can be found in [5]. For more recent work we refer to the nice review given by J. Fröhlich in this same volume [8].

Here we consider some selected topics related to recursive equations, leading to cluster expansions or contour expansions, and to recent developments in the analysis of the phase space structure of gauge field models on a lattice.

Lack of time during the lectures prevented us from presenting some of the results obtained about the problem of lattice spacing removal. Therefore we refer to Fröhlich's lectures and to [2,6] for the consideration of these topics.

We work in the framework of the Euclidean formulation of quantum field theory, whose basis was given by Symanzik [14] and Nelson [13], and consider ultraviolet cut off quantum gauge field theories according to the fully gauge invariant prescription of Wilson [15] and others, with gauge field variables taking values on the gauge group. For the general structure of these theories we refer to [5].

This report is organized as follows. In Section 2 we introduce the basic ideas of recursive equations and consider in particular "contour" expansions of the type studied by Minlos and Sinai [12] . As a definite example we consider the Z_2 model of gauge fields and give the main ideas of the contour expansion introduced by Marra and Miracle [11]in their study of this model in the unconfined region of phase space.

In Section 3 we consider the complementary case of recursive equations giving rise to expansion of cluster type. As a definite example we consider the cluster expansion of pure SU(2) Yang-Mills field given bj Immirzi, Guerra and Marra [10].

The examples given in Section 2 and 4 give a nice picture of the elementary excitations contained in gauge field theory both in the unconfined and confined phase.

In Section 4 we begin our review about known features of phase space for gauge models with a short introduction on methods based on duality, giving the non-elementary example of a non-abelian gauge group. Finally in Section 5 we discuss the phase space structure of the Z_2 models, in general d-dimensional space-time, with a particular emphasis on the nature of the transition between the confined and the unconfined regimes, by relying also on conjectures motivated by recent very surprising numerical investigations of Creutz, Jacobs and Rebbi [3] , giving evidence of a first order transition for Z_2 in four dimensions.

Our treatment will be completely rigorous from a mathematical point of view, but without any pretention of particular mathematical sophistication and deepness. In fact our main objective is to familiarize people working in high energy physics with some of the techniques and methods of modern mathematical physics (and possibly convince some of them about the usefulness of the latter).

In conclusion the author would like to thank the Organizing Committee of the 9th International Summer Institute on Theoretical Physics, and in particular Prof. V. E. Müller and Prof. W. Rühl for the kind hospitality extended to him in Kaiserslautern.

II. RECURSIVE EQUATIONS
CONTOUR EXPANSIONS

The main ideas of recursive equations can be explained as follows. Let us suppose that we can describe completely the physical content of a theory through a set u of functions in some abstract space (for example all correlation functions of quantum field theory or of the Ising model). Assume that u_0 is a known functional and that by effect of the interaction there is an "operator" S such that u must satisfy an equation of the type

$$u = u_0 + Su \qquad \text{(recursive equation)}.$$

It is tempting to solve the equation through the iterative expansion

$$u = u_0 + Su_0 + S^2u_0 + \dots \quad .$$

But in order to get sensible results and avoid possible disasters (try to expand $u = u_0 + Su$, with u an unknown real number, $u_0 = 1$ and $Su = 2u$ by definition, and get the Euler paradox $-1 = 1 + 2 + 4 + 8 + \dots$!). We must find the way to get some control on the meaning of the expansion and its convergence.

The most favourable situation is when we can assume that u belongs to some well defined Banach space B, with norm $\|u\|$. This assumption can have either a direct physical meaning (for example all correlation functions of the Ising model must be bounded by one) or be postulated (with its physical soundness verified *a posteriori*!).Then we can try to define rigorously the operator S and estimate its norm $\|S\|$. If it happens that for some number $K<1$ we have $\|S\| \leq K$ then elementary considerations give immediately the convergence of the iteration and the uniqueness of the solution (in the given Banach space B!).

The success of this strategy relies on the strict interlinking of physical and mathematical considerations as the right choice of the functional u, the Banach space B, and the estimate on the norm $\|S\|$ of S.

For historical reasons the expansion arising from the iteration of the basic equations are called either cluster expansions or contour expansions.

In this section we give some examples of contour expansions. Let us begin with some preliminaries about Minlos-Sinai type expansions [12].

Let us consider some model of statistical mechanics living on a given geometrical arena (for example a cubic unit lattice $\mathbb{Z}^d$ in d dimensions).

Assume that the dynamical properties of the model allow us to define the concept of elementary objects γ , called "molecules" with the following properties. First of all a concept of compatibility between two different molecules γ,γ' must be defined. We write $\gamma\sim\gamma'$ (or $\gamma\not\sim\gamma'$) if γ and γ' are (or are not) compatible. We consider (possibly empty) collections of compatible molecules

$$\Gamma \equiv \{\gamma_1,\ldots \gamma_s\}$$

called "clusters", and families of clusters, G. We associate to each molecule a real number $\mu(\gamma)$, the "activity" of the molecule. Usually (but not in all models!) $\mu(\gamma) \geq 0$, in this case the construction in the following has a direct probabilistic meaning. The activity of a cluster Γ is defined by

$$\mu(\Gamma) = \prod_i \mu(\gamma_i),$$

where γ_i are all components of Γ.

The main assumption, which gives the real motivation to all concepts introduced before, is that for the given model of statistical mechanics the partition function in a finite volume Λ can be expressed as

$$Z_\Lambda = \sum_{\Gamma\in G_\Lambda} \mu(\Gamma)$$

where G_Λ is the family of all clusters made of compatible molecules contained in the volume Λ .

It is convenient to define for any $M \in G$ (the index Λ is suppressed in the following) the correlation functions

$$\rho(M) = Z^{-1} \sum_{\Gamma'\in G,\Gamma'\supset M} \mu(\Gamma')$$

If $\mu \geq 0$ then $\rho(M)$ can be interpreted as the total probability of formation of clusters Γ' containing M. If μ can take negative values such probabilistic interpretation will be no longer possible but the equation can also be controlled in this case.

The conditions $\Gamma'\in G$ and $\Gamma'\supset M$ imply the existence of Γ such that $\Gamma \in G$, $\Gamma' = \Gamma\cup M$ and $\Gamma \sim M$, where the compatibility of two clusters Γ and M means that $\gamma_i \sim m_j$ for any $\gamma_i \in \Gamma$ and $m_j \in M$. Therefore we can also write

$$\rho(M) = Z^{-1} \sum_{\Gamma\in G,\ \Gamma\sim M} \mu(\Gamma)\ \mu(M)$$

Our goal is to find recursive equations for ρ's. First of all notice that $\rho(\emptyset) = 1$, for the empty set $\emptyset$, and define

$$\rho_0(M) = \begin{cases} 1 & \text{if } M = \emptyset \\ 0 & \text{otherwise} \end{cases}$$

Let $M = \emptyset$ and m_1 be the first molecule in M, in some given order, for example lexicographic, so that

$$M = \{m_1, M'\}$$

Consider the set $\{\Gamma|\Gamma \sim M'\}$ of all clusters in the family G compatible with M', it can be written as the disjoint union of $\{\Gamma|\Gamma \sim M'\}$ and $\{\Gamma|\Gamma \sim M', \Gamma \sim m_1\}$. Notice that any Γ in this last set can be written in the form $\Gamma = \Gamma' \cup N$ where $\Gamma' \sim M', \Gamma' \sim N$ and $\Gamma' \sim m_1$, while $N \neq \emptyset$, $N \sim M'$ and $M \nsim m_1$, the meaning of the symbol $\nsim$ being that for any $n_j \in N$ we have $n_j \sim m_1$. On the basis of these considerations we can split the sum over $\Gamma \sim M$ appearing in the definition of $\rho(M)$ as a difference of a sum over $\Gamma \sim M'$ and a sum over N and Γ', where N and Γ' have the properties mentioned before. Therefore we have

$$\rho(M) = \mu(m_1)\rho(M') - Z^{-1} \sum_N \sum_{\Gamma'} \mu(\Gamma')\,\mu(N)\,\mu(m_1)\,\mu(M')$$

Notice that if $\mu \geq 0$ then we have immediately $\rho(M) \leq \mu(m_1)\rho(M')$ and by recursion $\rho(M) \leq \mu(M)$, which is a useful bound in some cases.

Let us write the sum over N in the form

$$\sum_N \ldots = \sum_{n=1}^{\infty} \frac{1}{n!} \sum_{\gamma_1 \ldots \gamma_n}$$

where we must have

$$\gamma_i \simeq \gamma_j,\ i \neq j,\ \gamma_i \sim M',\ \gamma_i \sim \Gamma',\ \gamma_i \nsim m_1$$

and the factor $n!$ is inserted in order to avoid overcounting of clusters.

By the fundamental theorem of arithmetics we have $(1-1)^n = 0$, for $n > 0$ (this is the only really deep mathematical fact exploited in these lectures!), and by the binomial theorem

$$0 = \sum_{k=0}^{n} (-1)^k \binom{n}{k} \quad \text{or} \quad -1 = \sum_{k=1}^{n} (-1)^k \binom{n}{k}$$

Therefore the second term in the previous expression of ρ can be written as

$$z^{-1} \sum_{n=1}^{\infty} \sum_{k=1}^{n} (-1)^k \binom{n}{k} \frac{1}{n!} \sum_{\gamma_1 \ldots \gamma_n} \sum_{\Gamma'} \mu(\Gamma')\mu(N)\mu(M')\mu(m_1)$$

At this point it is convenient to express the sum over n as a sum over h, where $k + h = n$, and put $\bar{\gamma}_1 = \gamma_{k+1}, \ldots, \bar{\gamma}_h = \gamma_n$. Then the sum over $\bar{\gamma}$'s can be done explicitely in the form of some ρ and we end up with the equations

$$\rho(M) = \mu(m_1) \left[\rho(M') + \sum_{k=1}^{\infty} \frac{(-1)^k}{k!} \sum_{\{\gamma\}} \rho(M', \gamma_1, \ldots, \gamma_k) \right]$$

where the sum over γ's must take into account only clusters

$$\Gamma = \{\gamma_1 \ldots \gamma_k\}$$

for which $\gamma_i \nparallel m_1$.

Clearly these equations do not bring any reference to the finite volume and have meaning for any finite cluster M also in the thermodynamic limit. They will be the starting point of our treatment.

In terms of the renormalized quantities

$$\hat{\rho}(M) = \rho(M)/\mu(M)$$

the equations have the form

$$\hat{\rho}(M) = \sum_{k=o}^{\infty} \frac{(-1)^k}{k!} \sum_{\{\gamma\}} \mu(\Gamma) \; \hat{\rho}(M' \cup \Gamma), \; M \neq \emptyset$$

Notice also that

$$\hat{\rho}(\emptyset) = 1$$

Let us now introduce the Banach space B made of all functionals $u \equiv \{u(M)\}$ with norm

$$\|u\| = \sup_M \frac{|u(M)|}{A(M)}$$

where $A(\emptyset) = 1$ and $A(M) = \prod_i A(m_i)$ with $A(m) > o$ to be specified later. Introduce the operator $\underline{K}$ on B given by

$$(\underline{K}u)(\emptyset) = 0$$

$$(\underline{K}u)(M) = \sum_{k=o}^{\infty} \frac{(-1)^k}{k!} \sum_{\{\gamma\}} \mu(\Gamma) \; u(M' \cup \Gamma), M \neq \emptyset$$

and the functional $\hat{\rho}_0 \equiv \{\hat{\rho}_0(M)\}$ with

$$\hat{\rho}_0(\emptyset) = 1 \text{ and } \hat{\rho}_0(M) = 0 \text{ for } M \neq \emptyset$$

Then for the functional $\hat{\rho} \equiv \{\hat{\rho}(M)\}$ the previous equations can be written in the synthetic form

$$\hat{\rho} = \hat{\rho}_0 + \underline{K}\hat{\rho}$$

which is of the general type of recursive equations.

An upper bound on the norm of $\underline{K}$ can be easily found by estimating $|(\underline{K}u)(M)|/A(M)$. This can be done by enlarging the sum over $\{\gamma\}$ in the definition of $\underline{K}$ to all $\gamma_i \not\vdash m_1$, not necessarily incompatible. Then each term in k factorizes and the sum can be done in the form of an exponential, so that

$$||\underline{K}|| \leq \frac{1}{A(m_1)} \quad \exp \sum_{\gamma \not\vdash m_1} |\mu(\gamma)|A(\gamma)$$

Therefore if $A(m)$ can be chosen for each molecule m such that

$$\exp \sum_{\gamma \not\vdash m} |\mu(\gamma)|A(\gamma) \leq K\, A(m)$$

for some $K < 1$, then we have $\|\underline{K}\| \leq K < 1$ and the convergence (in the given Banach space B!) of the iterative expansion is assured.

Applications of this scheme to the Ising model in the two phase region are very well known [12] . Here we would like to give the example of the Z_2 gauge model by following the treatment of Marra and Miracle [11].

First of all let us introduce the Z_2 model. See also [1,16,5] .

On the unit cubic lattice $\mathbb{Z}^d$ in d dimensions, let us consider a variable $\sigma(\ell)$, for each link ℓ, taking values on Z_2, i.e. $\sigma(\ell) = \pm 1$, with equal a priori probability. Assume that the $\sigma(\ell)$ are distributed according to the Gibbsian factor

$$e^{\beta_1 \sum_\ell \sigma(\ell)} \quad e^{\beta_2 \sum_P \sigma_P}$$

where the sums over ℓ and P extend respectively over all links ℓ and plaquettes P contained in some finite volume Λ ,β_1 and β_2 are non-negative constants and

$$\sigma_P = \prod_{\ell \in P} \sigma(\ell),$$

where the product is over all links ℓ belonging to a given plaquette P.

Consider the statistical mechanics system enclosed in a large finite volume Λ with all $\sigma(\ell) = +1$ on the boundary (we call these (N) boundary conditions, where (N) is for Neumann, in analogy with the case of the elliptic boundary value problem for second order partial differential equations). Then the partition function is given by

$$Z_\Lambda = \int \prod_\ell d\sigma(\ell) \exp\left[\beta_1 \sum_\ell \sigma(\ell) + \beta_2 \sum_P \sigma_P\right]$$

where the integral extends over all configurations $\{\sigma(\ell)\}$ in Λ. Notice that for an Ising variable σ we write

$$\int f(\sigma)\, d\sigma \equiv \frac{1}{2}\left[f(1) + f(-1)\right]$$

For each configuration of the σ's in Λ let us consider the surface Σ made of plaquettes P for which $\sigma_P = -1$ in the given configuration.

For an easier visualization from now on we consider the case of a three-dimensional system, d = 3, but all considerations extend easily to the general case.

On the dual lattice (see Section 4) consider the set Σ' of links dual to Σ. It is clear that Σ' is a set of closed contours. This comes from the fact that only an even number of faces of each cube c on $\mathbb{Z}^d$ can have $\sigma_P = -1$, because

$$\prod_{P \subset C} \sigma_P = 1$$

and from the chosen (N) boundary conditions. Let

$$\Sigma' \equiv \{s_1, \ldots, s_j \ \ldots\}$$

be the unique decomposition of Σ' in disjoint connected contours on the dual lattice. By neglecting unessential constant terms, the partition function can be written as

$$Z_\Lambda = \sum_{\Sigma'} \int_{\Sigma'} \prod_\ell d\sigma(\ell) e^{-2\beta_2 |\Sigma'|} \quad e^{\beta_1 \sum_\ell \sigma(\ell)}$$

where the integral extends now to all configurations of σ's compatible with each Σ' and $|\Sigma'|$ is the length of Σ'. Now we exploit the expansion

$$e^{\beta\sigma} = \text{ch}\,\beta\,(1 + \sigma\,\text{tgh}\,\beta)$$

in order to write

$$e^{\beta_1 \sum_{\ell} \sigma(\ell)} = (\text{ch } \beta_1)^L \sum_M (\text{tgh } \beta_1)^{|M|} \sigma_M$$

where L is the number of all links in Λ , M is a subset of links in Λ , made of $|M|$ elements, and

$$\sigma_M = \prod_{\ell \in M} \sigma(\ell) \ .$$

If we introduce

$$\varepsilon(\Sigma', M) = \int_{\Sigma'} \sigma_M \prod_{\ell} d\bar{\sigma}(\ell)$$

where Σ' is made of connected closed contours on the dual lattice, and M' is a subset of links on the lattice, then the partition function,but for unessential terms, can be expressed as

$$Z_\Lambda = \sum_{\Sigma'} \sum_M \varepsilon(\Sigma',M)\, e^{-2\beta_2|\Sigma'|} (\text{tgh}\beta_1)^{|M|}$$

It is elementary to verify that $\varepsilon(\Sigma',M)$ is zero unless M is made of closed contours on the lattice. Moreover given the decomposition of M in disjoint connected closed contours, $M \equiv \{ m_1,\dots, m_i,\dots\}$ we have

$$\varepsilon(\Sigma',M) = \prod_{i,j} \varepsilon(s_j,m_i)$$

where $\varepsilon(s_j,m_i) = -1$ if the contours s_j (on the dual lattice) and m_i (on the lattice) are simply interlinked and $\varepsilon(s_j,m_i)=+1$ otherwise.

Now we are ready for the definition of molecules in this case. Consider any couple $\{\Sigma', M\}$, call connected two elements

$$s_j \in \Sigma', \ m_i \in M,$$

if $\varepsilon(s_j,m_i) = -1$ then each $\{\Sigma',M\}$ is decomposed into disconnected couples $\{\Sigma'_k,M_k\}$ each made of connected contours. These couples are molecules. Therefore a molecule $\gamma \equiv \{s_1,\dots; m_1,\dots\}$ is made of two collections of disjoint closed contours on the dual lattice and on the lattice, such that each couple (ss), (sm) or (mm) can be joined by a chain of elements in γ for which $\varepsilon = -1$. Two molecules

$$\gamma = \{ s_1,\dots ; m_1,\dots\} \quad \text{and} \quad \gamma' = \{s'_1,\dots ; m'_1,\dots\}$$

are compatible,

$$\gamma \sim \gamma', \text{ if } \ s_i \neq s'_j \ , \ m_i \neq m'_j, \ \varepsilon(s_i,m'_j) = \varepsilon(s'_i,m_j) = 1,$$

for any i, j.

We can define the activity of a molecule by

$$\mu(\gamma) = (\mathrm{tgh}\ \beta_1)^{\sum_i |\partial m_i|} e^{-2\beta_2 \sum_j |\partial s_j|} \prod_{i,j} \varepsilon(s_j, m_i)$$

where $|\partial m_i|$ and $|\partial s_j|$ are the lengths of the contours in γ.

Clearly the partition function can be written as

$$Z_\Lambda = \sum_\Gamma \mu(\Gamma)$$

where Γ are clusters of compatible molecules, and the whole general analysis previously given can be immediately applied.

In the paper [11] it is shown how to get convergence of the iteration expansion in a region of phase space where β_2 is sufficiently large and β_1 sufficiently small. This result will be one of the inputs of an analysis of the phase space diagram of the Z_2 model given in Section 5.

Notice that the Z_2 can be considered as a gas of compatible molecules as described before. Unfortunately the fact that their activity may be negative makes a strict probabilistic interpretation impossible: the contributions to the partition function of different clusters can partially cancel each other.

3. CLUSTER EXPANSIONS

We will give an example of cluster expansion for gauge fields on a lattice taking SU(2) as gauge group (Yang-Mills field) by following the method outlined in [10]. Analogous expansions for an Abelian (Maxwell) gauge field were presented in [4] . See also [5] for additional information and further references.

In order to describe the configurations of the model we associate to each oriented link ℓ, on the unit lattice $\mathbb{Z}^d$ a variable $g(\ell)$ taking values on the gauge group SU(2). To each plaquette P made of the oriented links $\ell_1, \ell_2, \ell_3, \ell_4$ we assign a potential energy $U_P = -\beta\,\chi(g_P)$, where β is some coupling constant, $g_P = g(\ell_1)g(\ell_2)g(\ell_3)g(\ell_4)$ and $\chi(g) = \mathrm{Tr}(g)$.

For the bounded region $\Lambda \subset \mathbb{Z}^d$ we consider the Gibbs measure

$$d\mu_\Lambda = Z_\Lambda^{-1} \exp[-U_\Lambda]\, dg_\Lambda$$

where dg_Λ is the product measure of the invariant Haar measures for the variables g's associated to links in Λ, $U_\Lambda = \sum_{P \subset \Lambda_m} U_P$ and Z_Λ is the partition function

$$Z_\Lambda = \int \exp\,[-U_\Lambda]\, dg_\Lambda .$$

We are interested in the infinite volume limit, $\Lambda \to \mathbb{Z}^d$ of the averages

$$< A >_\Lambda = \int A \, d\mu_\Lambda$$

where A is any sufficiently regular function of a finite number of variables g.

Obviously the theory is locally gauge invariant, i.e. invariant under transformations of the type

$$g(\ell) \longrightarrow g_\varphi(\ell) = \varphi(n)\, g(\ell)\, \varphi(n')^{-1}$$

where $n \to \varphi(n)$ is some field on $\mathbb{Z}^d$ taking values on SU(2) and n,n' are the sites of the oriented link ℓ. Any variable A will be accordingly transformed $A \to A_\varphi$. Clearly the averages $< \quad >_\Lambda$ given before are gauge invariant, i.e. $\langle A\rangle_\Lambda = \langle A_\varphi\rangle_\Lambda$ and the same will happen for the thermodynamic limit, $\Lambda \to \mathbb{Z}^d$ if it exists. But a simple argument, see for example [5], shows that, for any choice of boundary conditions and for any value of the coupling constant, gauge invariance is preserved in the infinite volume limit. Therefore no spontaneous breaking of the local symmetry is possible. Obviously this does not mean that the system cannot have multiple phases, it only means that any phase is locally gauge invariant.

It is clear that any function of the SU(2) variables can be approximated through linear combinations of the matrix elements $R^{(j)}_{mm'}(g)$ of the irreducible representations of the gauge group SU(2), where $j = 0, \frac{1}{2}, 1, \ldots$ labels the representation and $m, m' = -j, -j+1, \ldots, j$. On the other hand we are interested only in gauge invariant quantities, therefore we consider the following concepts.

Let $J: \mathbb{L}^d \to \mathbb{Z}^+/2$, $\ell \to j(\ell)$ be some function which associates to each link ℓ of $\mathbb{L}^d$ (the set of all links of $\mathbb{Z}^d$) some irreducible representation $j(\ell)$ of SU(2). We assume that J is of compact support, in the sense that $j(\ell) = 0$ for ℓ outside some bounded region depending on J, and <u>admissible</u> according to the definition that the zero representation $(j = 0)$ must be contained in the tensor product $j_1 \otimes \ldots \otimes j_{2d}$ of the representations $j_1, \ldots j_{2d}$ associated to all links starting from each given site of $\mathbb{Z}^d$. For a given J we define coupling schemes A by associating to each site n of $\mathbb{Z}^d$ a normalized vector $a(n)$ taking values in the zero isotopic spin space of the tensor product of the representations associated before to links starting from n, as explained before.

Given (J,A) let us associate to each link ℓ the functions of

$g(\ell)$ given by

$$(-1)^{j-m} R^{(j)}_{mm'}(g(\ell))$$

where $j = j(\ell)$ and $m,m' = -j, \ldots, j$, and consider the functions

$$R(J,A) = \Sigma(\prod_{\ell}(-1)^{j-m} R^{j}_{mm'}(g(\ell)))\prod_{n} a_{m_1 \ldots m_{2d}}(n)$$

where the products extend over all links and sites in the support of J and the sum saturates in the appropriate way all matrix indices of R's and vector indices of a's, in such a way that R(J,A) is a gauge invariant field variable.

We are interested in the averages

$$u(J,A) = \langle R(J,A)\rangle$$

where $\langle\ \rangle$ is the infinite volume limit of the averages $\langle\ \rangle_\Lambda$ introduced before. The knowledge of $u(J,A)$ for all J and A allows us to calculate the average of any gauge invariant variable.

By a simple generalization of the methods employed in [4] it is not difficult [10] to derive recursive equations for the set $\{u(J,A)\}$ in the form

$$u(J,A) = \frac{\beta}{2j+1} \sum_{P} \sum_{i} c(P,i)u(J(P,i),A(P,i))$$

where J is a non-trivial field such that there is a link ℓ (for example the first in lexicographic order) for which $j = j(\ell)>0$, Σ_P extends over all $2(d-1)$ plaquettes having ℓ as a side, Σ_i extends over at most sixteen terms coming from the application of the Clebsch-Gordan (CG) decomposition to the representations associated to each side of the plaquette P. Moreover $J(P,i)$ is a new field such that for each side ℓ' of the plaquette P one has $J(P,i)(\ell') = j(\ell')\pm\frac{1}{2}$ according to the value of i, $A(P,i)$ are new coupling schemes obtained from A through multiplication with CG coefficients and normalization and finally $c(P,i)$ are some constants which can be explicitely calculated through more or less elegant CG gymnastics [10].

Clearly if $J \equiv 0$ then $u(J,A) = 1$. Therefore it is convenient to introduce $u_0(J,A) = 1$ if $J \equiv 0$ and $u_0(J,A) = 0$ otherwise and an appropriate operator S such that the recursive equation can be written in the form

$$u = u_0 + Su.$$

This equation can be interpreted in the Banach space B made of functionals $v \equiv \{v(J,A)\}$ with norm

$$\| v \| = \sup_{J,A} \frac{|v(J,A)|}{\prod_{\ell} q(j(\ell))}$$

where $q(0) = 1$ and $q(j)$ is appropriately defined for $j>0$.

Then a tedious calculation [10] gives $\|S\| <1$ provided $\beta<2/5(d-1)$. Therefore we have uniqueness of the state for β small enough. It is not known whether the limitation on β is essential or introduced by the method of proof. In other words it is not known whether there are singularities for some value of β. In the case of the Ising model, as it is very well known, or of the Z_2 model as it is explained for example in [5] and in Section 2, expansion of the cluster type converge at small β and of the contour type at large β. Since we have already obtained in the SU(2) case expansions of the cluster type which converge for β small, it would be interesting to explore the large β region in the SU(2) case through expansions of contour type. In principle these can be obtained from duality arguments as it will be explained in the next section, but their complete control has not been obtained yet, so that no general conclusion can be derived on rigorous mathematical basis for the existence of thermodynamic singularities for SU(2) Yang-Mills fields on a lattice.

4. DUALITY

Concepts based on duality have been very useful in the study of statistical mechanics models and more recently they have also been applied to gauge fields on a lattice, see [16, 1].

Here we give a rough sketch of the main ideas involved taking into account the non-elementary case of a non-Abelian gauge group (we choose SU(2) for the sake of simplicity). Then we consider the Z_2 case for Ising type models or gauge type models.

First of all we need some geometrical considerations. Let $\mathbb{Z}^d$ be the unit square lattice in d dimensions with lattice sites n and basic directions $i_1,..,i_d$. We call V_0 the set of all sites, i.e. $V_0 \equiv \mathbb{Z}^d$, V_1 the set of all links, V_2 the set of all plaquettes and in general V_s the set of all elementary simplexes of dimensions s, with $s \leq d$. A general simplex in V_s is characterized by its minimal site n (with respect to the natural partial ordering on $\mathbb{Z}^d$) and s directions $j_1,..,j_s$, chosen among $i_1,..,i_d$. We write

$$\{n;j_1,..j_s\} \in V_s .$$

All sites of the simplex are given by $n + \varepsilon_1 j_1 +..+\varepsilon_s j_s$, where each $\varepsilon_i = 0,1$ independently.

The dual lattice $\mathbb{Z}'^d$ is defined by

$$\mathbb{Z}'^d = \mathbb{Z}^d + \frac{1}{2}(i_1 + .. + i_d),$$

while the dual of a given simplex is given by

$$\{n; j_1, .., j_s\}' = \{n + \frac{1}{2} j_1 + .. + \frac{1}{2} j_s - \frac{1}{2} j_{s+1} - .. - \frac{1}{2} j_d;\ j_{s+1}, .., j_d\}$$

Notice that if $\{\} \in V_s$ then $\{\}' \in V_{d-s}$.

The following table gives the duality relations for d = 0,1,2, 3,4.

d = 0	site ↔ site (trivial)
d = 1	site ↔ link
d = 2	site ↔ plaquette, link ↔ link
d = 3	site ↔ cube, link ↔ plaquette
d = 4	site ↔ hypercube, link ↔ cube, plaquette ↔ plaquette

We will discuss duality transformations for the SU(2) model. First of all let us notice that the basic Fourier decomposition for the Z_2 group $\exp \beta\sigma = \mathrm{ch}\ \beta + \sigma \mathrm{sh}\ \beta$ has its counterpart in the SU(2) case written in the form

$$\exp \beta\chi(g) = \sum_j c(\beta; 2j)\chi^{(j)}(g),$$

where

$$\chi(g) = \chi^{(\frac{1}{2})}(g) = \mathrm{Tr}\ (g) \text{ and } \chi^{(j)}(g) = \mathrm{Tr}\ (R^{(j)}(g))\quad j = 0, \frac{1}{2}, 1...$$

The functions $c(\beta;2j)$ are given by the elementary integrals

$$c(\beta;2j) = \int e^{\beta\chi(g)} \chi^{(j)}(g)\, d\mu(g) =$$

$$= \frac{1}{\pi} \int_{-\pi}^{+\pi} e^{2\beta \cos \frac{1}{2}t} \frac{\sin(j+\frac{1}{2})t}{\sin \frac{1}{2} t} \sin^2 \frac{1}{2}t\, dt$$

The partition function of the SU(2) gauge model is given by

$$Z_\Lambda = \int e^{\beta \sum_{P\subset\Lambda} \chi(g_P)} \prod_\ell d\mu(g(\ell))$$

as we have explained in the previous section. Let ℓ be one selected link inside Λ. Consider the 2(d-1) plaquettes P_i having ℓ as a side and write $g_{P_i} = g\, g_{C_i}$, where $g \equiv g(\ell)$.

Let us integrate with respect to $d\mu^i(g)$ the contribution to the Gibbsian factor coming from the plaquettes P_i. By expanding each exponential with the formula given before we get

$$\int e^{\beta \sum_{P_i} \chi(g_{P_i})} d\mu(g(\ell)) =$$

$$= \int \sum_{j_1,\dots,j_{2(d-1)}} \prod_i c(\beta;2j_i)\, \chi^{(j_i)}(g_{P_i})\, d\mu(g)$$

$$= \sum_{\{j\}} c \sum_{\{m\}} \left[\int \prod_i R^{(j_i)}_{m_i m_i'}(g)\, d\mu(g) \right] \prod_i R^{(j_i)}_{m_i' m_i}(g_{c_i})$$

The term in square brackets is nothing but the projection operator

$$E^{(j_1 \dots j_{2(d-1)})}_{m_1 m_{1'},\, m_2 m_2', \dots}$$

on the zero subspace of the tensor product of the representations $j_1,\dots,j_{2(d-1)}$. The same procedure can be applied to all other links and we see that general configurations of $\{j\}$ give zero contribution unless the field $J \equiv \{j\}$ is admissible.

Therefore the partition function is expressed in the form

$$Z = \sum_{\{j\}} \prod_P c(\beta;2j_P) \sum_{\{m\}} \prod_\ell E(\ell)$$

It is convenient to go to the dual lattice, taking for example d=3. Then the general structure of the dual model of SU(2) can be described as follows.

The basic variables are elements of the dual of the gauge group (i.e. the set of all its irreducible representations) each associated to each link of the dual lattice. Call $J \equiv \{j(\ell)\}$ a generic configuration, it must be admissible in the sense that the zero representation must be contained in the tensor product of representations associated to the links of each plaquette. For each plaquette P consider the projection E(P) on the zero subspace, E(P) will have matrix indices associated to each dyhedron having P as a side (by a dyhedron we mean a couple of orthogonal plaquettes having one link in common). Then we form the product $\prod_P E(P)$ by saturating all matrix indices for each dyhedron and express the partition function as

$$Z = \sum_{\{j\}} \prod_{\ell} c(\beta;2j_\ell) \prod_P E(P)$$

where the sum runs over all admissible $\{j\}$ configurations.

Systems of this type can be analyzed in the general framework given in Section 2. It is not difficult to identify molecules and their activities, but it seems very difficult to prove convergence of the resultant contour expansions. We give a rough sketch of the procedure.

First of all let us consider the system enclosed in a large finite volume Λ putting $j(\ell) = 0$ for all links ℓ on the border. Notice, that if for some configuration of the j's an inner link ℓ has $j(\ell) = 0$ then all plaquettes belonging to dyhedra having the link ℓ as edge are disconnected. Let $\{j\}$ be a particular admissible configuration, two plaquettes belonging to a dyhedron are considered connected if the edge ℓ has $j(\ell) > 0$. In this way we can split the set of all plaquettes into disjoint connected components. Each of these components can be assumed as "molecules". The activity of a molecule can be defined through summation over all configurations $\{j\}$, which do not change the splitting of plaquettes into disjoint connected components. In such a way the partition function assumes the standard form

$$Z = \sum_{\Gamma} \mu(\Gamma)$$

where Γ are all clusters of molecules. Notice, that a molecule formed by only one plaquette has activity $\mu(\gamma) = 1$, so that it does not appear in the expansion of the partition function.

We were not able to control the convergence of the resulting contour expansion, but surely this problem deserves careful consideration because it is strictly connected with the problem of the phase space structure for the SU(2) gauge field theory.

Let us now end this section with a short review of duality arguments applied to models with elementary variables taking values on the group Z_2, see also [16].

First of all let us define the following duality map of the interval $(0, \infty)$ on itself

$$\beta \rightarrow \beta'(\beta) = -\frac{1}{2} \log \operatorname{tgh} \beta$$

and notice that we have equivalently

$$\text{sh } 2\beta \text{ sh } 2\beta' = 1$$

$$\text{ch } 2\beta \text{ tgh } 2\beta' = 1$$

We consider also the map of $[0,\infty)$ on $[0,1)$ given by

$$\beta \longrightarrow F(\beta) = 1 - \log\,(1+e^{-2\beta})/\log 2$$

so that the duality relation can be written as

$$F(\beta) + F(\beta') = 1.$$

In general we will graduate axis of the phase diagrams so that $F(\beta)$ will appear instead of β . This procedure is useful because the point $\beta=\infty$ can be represented on the diagram and the duality transformations becomes simply a reflection with respect to the point 0.5 of the interval $(0,1)$.

We consider classes of models by $(s,s+1)_d$ where $0\leq s\leq d-1$ and characterized by two non-negative coupling constants β_s and β_{s+1}. They are defined as follows.

To each simplex x of dimension s we associate an independent variable $\sigma(x)$ taking values on Z_2, i.e. $\sigma(x) = \pm 1$, with equal a priori distribution. On the other hand to each simplex y of dimension s + 1 we associate the variable σ_y defined by

$$\sigma_y = \prod_{x\in y} \sigma(x)$$

Then the interaction is assumed to be

$$e^{\beta_s \sum_x \sigma(x) + \beta_{s+1} \sum_y \sigma_y}$$

Clearly the models $(0,1)_d$ are of the Ising type, while $(1,2)_d$ are of the gauge type.

The duality transformation (see [16]) proves that the thermodynamic properties of the models $(s,s+1)_d$ and $(d-s-1,d-s)_d$ are the same, provided that the coupling constants

$$\{\beta_s,\beta_{s+1}\} \quad \text{and} \quad \{\beta'_{d-s-1},\ \beta'_{d-s}\}$$

are related by

$$\beta'_{d-s-1} = \beta'\ (\beta_{s+1}),\ \beta'_{d-s} = \beta'(\beta_s)$$

where β' is the duality map introduced before.

It is also convenient to introduce the notations $(\underline{s},s+1)_d$ and $(s,\overline{s+1})_d$ to mean $\beta_s = 0$ and $\beta_{s+1} = \infty$ respectively. Elementary considerations of cohomology on the lattice show that the thermodynamics of the model $(s,\overline{s+1})_d$ is identical to $(\underline{s-1},s)_d$. For example, if in the gauge model $\beta_2 = \infty$ then only configurations for which all plaquette variables $\sigma_P = 1$ contribute to thermodynamic quantities. But if $\sigma_P = 1$ for all plaquettes then the link variables $\sigma(\ell)$ become stochastically equivalent to products of site variables, $\sigma(\ell) \simeq \sigma(n)\sigma(n')$.

Therefore the duality relation for models with zero external field, $\beta_s = 0$, is the following

$$(\underline{s},s+1)_d \quad \leftrightarrow \quad (\underline{d-s-2}\ ,\ d-s-1)_d$$

$$\beta'_{d-s-1} = \beta'(\beta_{s+1})$$

Some models are selfdual. Lattices with an even number of dimensions d can host selfdual models in zero external field. For odd d selfdual models have generic external field. For example the Ising model and the gauge model are selfdual in zero external field for d=2 and d=4 respectively, while selfduality holds for d=1 and d=3 respectively, if the external field is non zero.

5. THE PHASE SPACE STRUCTURE

The study of the phase space structure for gauge field theory models is extremely interesting both from a physical and mathematical point of view. In fact all problems of confinement and non-confinement, infrared behaviour, asymptotic freedom, etc. are re-related to some aspects of the phase diagram.

Here we take into account only the case of the Z_2 gauge field model. It is believed that the Z_2 model can be considered as a prototype for most of the main aspects of general gauge field models.

Let us briefly recall the definition of the model and its main properties. For a review of the situation until 1978, see [5]. Some new aspects will be discussed in the following.

Let us consider the d-dimensional unit square lattice $\mathbb{Z}^d$ and let us associate to each link ℓ a random variable $\sigma(\ell)$ taking values on $Z_2 \cong \{-1,1\}$ with equal *a priori* probabilities. Define also as usual

$$\sigma_P = \prod_{\ell \in P} \sigma(\ell)$$

for each plaquette and consider the Gibbsian distribution

$$\exp\left[\beta_1 \sum_{\ell} \sigma(\ell) + \beta_2 \sum_{P} \sigma_P\right]$$

in the infinite volume limit, with $\beta_1, \beta_2 \geqq 0$.

The construction of the infinite volume limit of the correlation functions

$$< \sigma_A >, \ \sigma_A = \prod_{\ell \in A} \sigma(\ell),$$

A finite set of links can be done either by monotonicity, exploiting correlation inequalities of the Griffiths-Kelly-Sherman type, or through converging expansions. In fact the finite volume expectations

$$<\sigma_A>_\Lambda^{(N)}$$

with all spin $\sigma(\ell) = 1$ on the boundary of Λ, are monotonically decreasing to

$$<\sigma_A>^{(N)}$$

as Λ is enlarged and let cover all $\mathbb{Z}^d$, while the analogous expectations

$$<\sigma_A>_\Lambda^{(D)}$$

with free boundary conditions are monotonically increasing to

$$<\sigma_A>^{(D)}$$

as $\Lambda \to \mathbb{Z}^d$.

On the other hand cluster expansions converge [5] in the small β_2 large β_1 region and contour expansions [11] converge in the large β_2 small β_1 region, obviously they give uniqueness of the infinite volume state and absence of thermodynamic singularities in these regions, but in general the states $< >^{(N)}$ and $< >^{(D)}$ could be different.

The following diagram gives an idea of the regions covered by convergent expansions. As mentioned in section 4 we plot $F(\beta)$ on each axis.

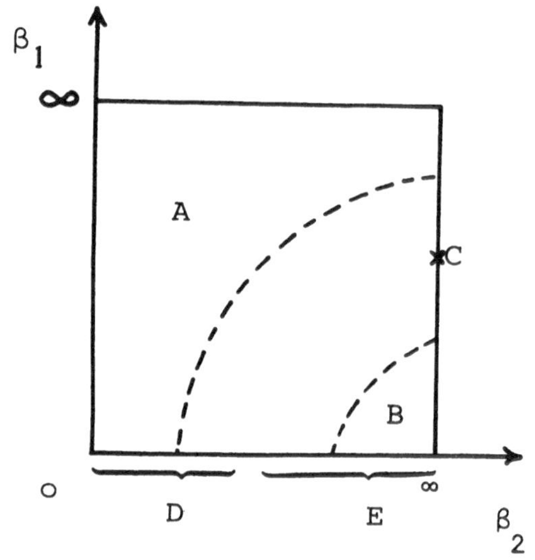

Figure 1 The phase space diagram of the Z_2 gauge model

In the region A cluster expansions are convergent [5]. Notice that there is convergence at small β_2 for any β_1 and at any value of β_2 if β_1 is large enough. At the boundaries of the region A the model becomes trivial, in fact if $\beta_2 = 0$ we have independent factorization for each link, i.e. each variable $\sigma(\ell)$ is distributed independently from the others and we have $\langle\sigma(\ell)\rangle = \text{tgh } \beta_1$, while for $\beta_1 = \infty$ all variables become identically equal to one.

The contour expansions, see [11] and Section 2, are convergent in the region B. The investigation of the behaviour of the model in the strip between regions A and B is one of the main problems of the theory.

For $\beta_2 = \infty$ we have the Ising limit (see Section 4), in fact all σ_P must be equal to one and therefore each $\sigma(\ell)$ becomes a "gradient" of the form $\sigma(\ell) \simeq \sigma(n)\ \sigma(n')$ where the new independent variables $\sigma(n)$ interact through the two body interaction $\beta_1\sigma(n)\sigma(n')$. Therefore on the boundary $\beta_2 = \infty$ we find a critical point C at some value β_{1C} of β_1.

For $\beta_1 = 0$ the two regions of uniqueness at small and large β_2, D and E respectively, are characterized by the area or perimeter behaviour of the Wilson loop. See [5] and [9] for additional information.

Of course, the phase space structure depends on the number of dimensions.

For d=2 the duality relation between the gauge model and Ising model in external field of the previous section gives immediately the following picture of the superposed phase diagrams of the two models.

Here (β_1, β_2) and (h, β) are respectively the coupling constants of the models (we should put $h = \beta'_o, \beta = \beta'_1$ in order to be coherent with the general notation of section 4). From the known analyticity in h, consequence of the Lee-Yang - Theorem, we see that the Z_2 model shows no singularities all over the phase plane, there is only a critical point at $\beta_2 = \infty$ and $F(\beta_1) = \frac{1}{2}$. The figure shows also selfduality of the Ising model in two dimensions.

For d=3 the gauge model is selfdual. The Ising critical point C has a mirror image C'. It is a still unproved conjecture that the two critical points C and C' are joint by a critical line separating the phase diagram in two regions, B the baryon or confined region and Q the quark or unconfined region. The behaviour of the critical line could be similar to those given in Fig.3_1 and 3_2. Should a behaviour of the type given in 3_3 be verified then our ideas on confinement or non-confinement would be submitted to a drastic revision.

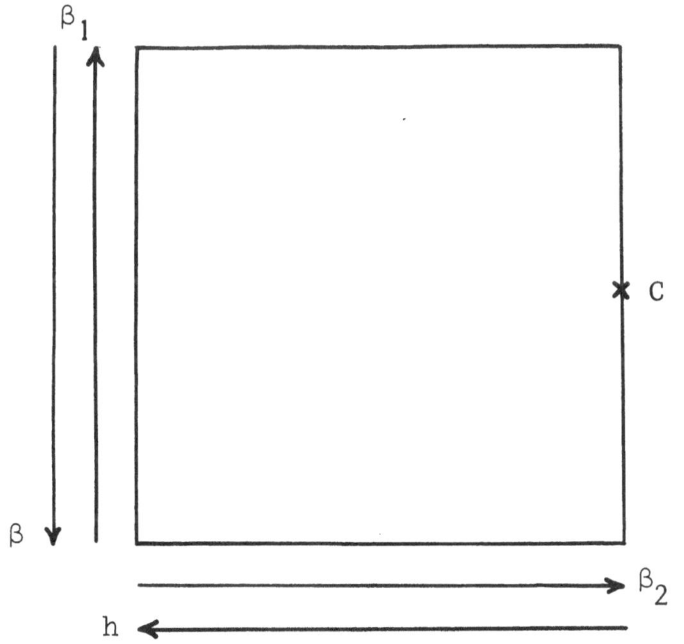

Figure 2 The phase space diagram for the gauge and Ising models, d=2.

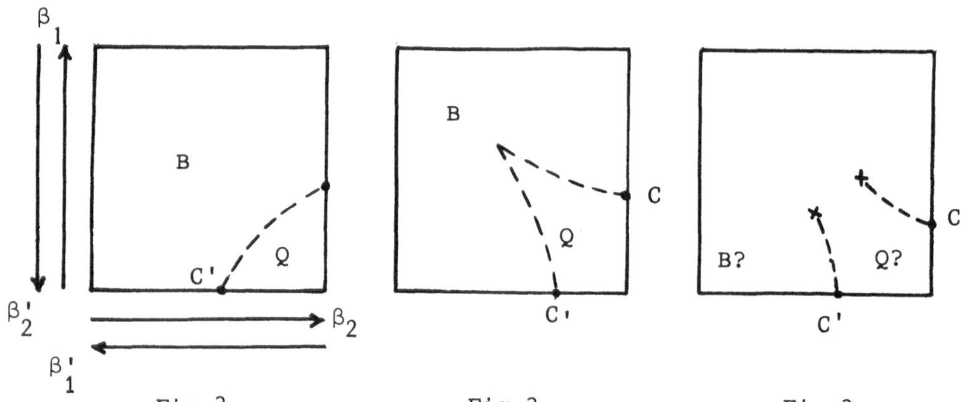

Fig.3_1 Fig.3_2 Fig.3_3

Figures 3_1 - 3_3 Phase space diagrams for gauge models, d=3

For d=4 we still have a critical point C, which is the same as that for I_4 (Ising model in four dimensions). On the other hand we expect also some singularity point C' at $\beta_1 = 0, F(\beta_2) = \frac{1}{2}$ separating the two regions of area behaviour of the Wilson loop (static confinement) and perimeter behaviour. Obviously, since for $\beta_1 = 0$ the model is selfdual, the two regions can be mapped one into the other. By analogy with the U(1) (Maxwell) field on the lattice we could say that, at small β_2, static "electric charges" are confined, while they are not confined at large β_2. Here, by duality, the confinement should be a property of "magnetic charges". What is the nature of the singular point C', is it critical or not? Is there a singularity line connecting the two singular points C and C', what is its shape? The classification of these open problems, even in the case of such a simple model, would greatly improve our understanding of the problem of quark confinement in elementary particle theory.

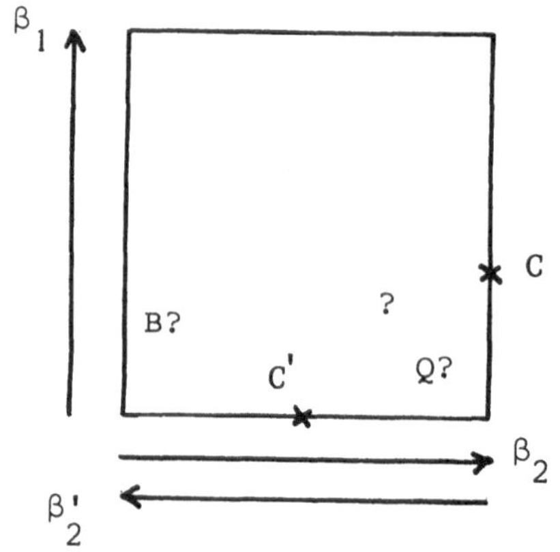

Figure 4 Phase space diagram for the gauge model in d=4

In a series of recent extremely interesting papers Creutz, Jacobs and Rebbi [3] , by exploiting Monte Carlo methods, have found clear evidence that in C' a first order transition takes places. In particular there is a jump in the average value $<\sigma_P>_\beta$ as a function of

$$\beta \equiv \beta_2 \text{ at } F(\beta) = \frac{1}{2}.$$

In view of the interest and possible further implications of these results we give a close look at the general structure of the $(Z_2)_4$ model.

First of all let us notice that, when applying duality arguments for correlation functions, we must correctly take into account boundary conditions and their transformation under duality. General boundary conditions do not transform into general boundary conditions (states with negative probability can appear!), but if we restrict our attention to N b.c. (all on the border), D b.c. (or free boundary conditions), or P b.c. (periodic boundary conditions) then we see immediately through simple explicit computation that N and D are related by duality and P are selfdual.

The $(Z_2)_4$ model is completely specified by the correlation functions $<\sigma_{P_1} \ldots \sigma_{P_n}>$, where $P_1, \ldots, P_n$ are generic different plaquettes. The duality transformation gives in a finite volume Λ

$$<\sigma_{P_1} \sigma_{P_2} \ldots>^{(X)}_{\beta,\Lambda} = < e^{-2\beta'(\sigma_{P_1'} + \sigma_{P_2'} + \ldots)} >^{(X')}_{\beta',\Lambda}$$

where P_1' $P_2'\ldots$ are dual plaquettes to $P_1 P_2 \ldots, \beta' = \beta'(\beta)$ and the boundary conditions can be X=N,P,D and X'=D,P,N respectively, i.e. N' = D,D' = N,P' = P.

As $\Lambda \longrightarrow \mathbb{Z}^4$ the correlations with N or D b.c. converge monotonically, while for P b.c. convergence cannot be proven in general but only through subsequences. From correlation inequalities we have

$$<\sigma_{P_1} \ldots >^{(D)} \leq <\sigma_{P_1} \ldots >^{(P)} \leq <\sigma_{P_1} \ldots >^{(N)}$$

in the finite or infinite volume.

It is nice to see that for the one plaquette expectation for P b.c. in a "cube" we have, starting from

$$<\sigma_P>^{(P)}_{\beta,\Lambda} = <e^{-2\beta'\sigma_{P'}}>^{(P)}_{\beta',\Lambda}$$

and exploiting

$$e^{-2\beta' \sigma_{P'}} = \text{ch } 2\beta' - \text{sh } 2\beta' \sigma_{P'}$$

the following relation

$$\text{tgh } 2\beta \langle\sigma_P\rangle^{(P)}_{\beta,\Lambda} + \text{tgh}2\beta' \langle\sigma_P\rangle^{(P)}_{\beta',\Lambda} = 1$$

where the symmetry of the "cube" has been exploited in order to substitute $\sigma_{P'}$ with σ_P.

In particular at the singular point

$$F(\beta_C) = \frac{1}{2}, \text{ sh } 2\beta_C = 1, \quad \text{ch}2\beta_C = \sqrt{2}, \text{ tgh } 2\beta_C = \frac{1}{\sqrt{2}}$$

and therefore

$$\langle \sigma_P\rangle^{(P)}_{\beta_C,\Lambda} = \frac{1}{\sqrt{2}}$$

which is independent of Λ. Therefore the infinite volume limit can be trivially done! On the other hand duality for D,N b.c. gives us at the singular point

$$\langle\sigma\rangle^{(D)}_{\beta_C} + \langle\sigma_P\rangle^{(N)}_{\beta_C} = \sqrt{2}$$

while Griffiths inequalities state

$$\langle \sigma_P\rangle^{(D)}_{\beta_C} \leq \langle \sigma_P\rangle^{(P)}_{\beta_C} \equiv \frac{1}{\sqrt{2}} \leq \langle \sigma_P\rangle^{(N)}_{\beta_C}$$

We see immediately that a jump in $\langle\sigma_P\rangle$, as found by Creutz, Jacobs and Rebbi amounts to

$$\langle\sigma_P\rangle^{(D)}_{\beta_C} \neq \langle\sigma_P\rangle^{(N)}_{\beta_C}$$

and the first order transition appears associated to a spontaneous breaking of the symmetry given by duality.

We can see also immediately that if we introduce auxiliary variables

$$\tau_P = \text{tgh } 2\beta\, \sigma_P - \frac{1}{2}$$

the duality relation becomes

$$\langle\tau_{P_1} \tau_{P_2} \ldots \tau_{P_n}\rangle^{(X)}_{\beta} = (-1)^n \langle\tau_{P'_1} \tau_{P'_2} \ldots \tau_{P'_n}\rangle^{(X')}_{\beta'}$$

i.e. it appears as reflection covariance of the variables τ . Then the first order transition should not be surprising: it is very much related to the analogous first order transition for the Ising model when the external field is varied and crosses the value zero when the temperature is kept fixed below the critical value. In a sense the coupling parameter β plays the same role of the external magnetic field in the Ising model, while the reflection symmetry $\sigma(n) \to -\sigma(n)$ in Ising is replaced by the duality symmetry expressed in similar form $\tau_p \leftarrow \tau_{p'}$ through the variables τ.

Just as in the Ising case we can move along the first order transition line by varying the temperature until the two magnetized phases become identical at the critical point. It should be possible to add some (duality invariant?) interaction to the $(Z_2)_4$ model with a coupling constant allowing us to move along the first order line, while β allows to cross the line. But this is an open problem, as open (and extremely difficult) as is the problem of proving rigorously the results of [3] . In fact the available methods based either on contour expansions or infrared bound techniques [7] seem to require modifications in order to be effective.

REFERENCES

1. R. Balian, J. M. Drouffe and C. Itzykson, Phys.Rev. D11, 2098 (1975).
2. J. Bellissard and G. F. De Angelis, Gaussian Limit of Compact Spin Systems, Marseille-Salerno Preprint, 1979.
3. M. Creutz, L. Jacobs and C. Rebbi, Phys.Rev.Lett. 42, 1390 (1979), Monte Carlo Study of Abelian Lattice Gauge Theories, Brookhaven Preprint 1979, and following papers.
4. G. F. De Angelis, D. de Falco and F. Guerra, Lett.Nuovo Cimento 19, 55 (1977).
5. G. F. De Angelis, D. de Falco, F. Guerra and R. Marra, Acta Physica Austriaca, Suppl. XIX, 205 (1978).
6. G. F. De Angelis, S. De Martino, S. De Siena, Reconstruction of Euclidean Fields from Plane Rotator Models, Phys.Rev. 1979, to appear.
7. J. Fröhlich, B. Simon and T. Spencer, Comm.Math.Phys. 50, 79 (1976).
8. J. Fröhlich, Lectures given at this Institute.
9. G. Gallavotti, F. Guerra and S. Miracle-Solé. A Comment to the Talk by E. Seiler, in: Mathematical Problems in Theoretical Physics, G. Dell'Antonio, S. Doplicher and G. Jona-Lasinio, eds., Springer Verlag, Berlin, Heidelberg (1978).
10. G. Immirzi, F. Guerra and R. Marra, Lett. Nuovo Cimento 23, 237 (1978) and paper in preparation.
11. R. Marra and S. Miracle-Solé, Comm.Math.Phys. 67, 233 (1979).
12. R. A. Minlos and Ya. G. Sinai, Trans.Moscow Math.Soc. 17, 237 (1967); 19, 121 (1968).

GRAVITATIONAL INSTANTONS, G-INDEXTHEOREM AND NON-LOCAL BOUNDARY TERMS

H. Römer

Fakultät für Physik
Universität Freiburg
Freiburg/Br.

In the last few years, Euclidean functional integrals have turned out to be a very useful tool both technically and - more important - also conceptually in relativistic quantum field theory on flat space-time.

The starting point is the Wick rotated Feynman-Kac integral

$$Z = \int_C D\varphi \; e^{-I\,[\varphi]} \,, \tag{1}$$

where I $[\varphi]$ denotes the Euclidean action, which may contain a coupling to an external source, and the functional integration runs over all field functions which obey a certain constraint condition, here symbolized by C. The precise nature of this constraint is given by the physical situation to be investigated.

For instance, considering a thermal ensemble with temperature T one has to impose periodic boundary conditions in Euclidean time:

$$\varphi\,(\tau,\vec{x}) = \varphi(\tau+\beta,\vec{x}) \qquad \text{with} \qquad \beta = \frac{1}{T}\,. \tag{2}$$

Then Z becomes a (grand) canonical partition function, which upon differentiation with respect to external sources will give thermal Green's functions. Also other boundary or subsidiary conditions like fixed angular momentum, charge or volume may be incorporated. In these cases the use of Lagrange multipliers is sometimes advantageous.

For the Wick-rotated vacuum persistence amplitude one will normally impose no boundary condition at all, except finiteness of the Euclidean action.

The Euclidean functional approach motivates or allows a convenient formulation of several interesting questions: One can decompose the set of field configurations into topological equivalence classes, possibly subject to the subsidiary conditions C

a) for fixed Euclidean (or Minkowskian) time or
b) for Euclidean space time.

In case a) one is lead to the concept of topological conservation laws, whereas in case b) "topological winding numbers" will serve as labels for different equivalence classes.

Very often it will be fruitful to interprete the field configurations as sections of appropriate bundles and try to classify these bundles. Then the "topological winding numbers" will reappear as characteristic numbers, which distinguish non-equivalent bundles.

Another interesting question is the investigation of the stationary points of the Euclidean action, the so-called instanton configurations. A whole literature has been devoted to their study in the last years.

The standard example for the above mentioned scheme is of course Euclidean Yang-Mills theory. Here the "winding" number is

$$k = \frac{1}{8\pi^2} \int \mathrm{tr}\, F \wedge F \quad , \tag{3}$$

where F is the Yang-Mills field strength.

Assuming, as it seems quite likely now that all stationary points of the Euclidean action correspond to (anti) self-dual field strength $*F = {}^{+}_{(-)}F$ we see that finiteness of the action will already imply (anti)selfduality of F for every stationary point.

Using the fact that every self-dual configuration solution of the Yang-Mills equations on the punctured sphere $S^4 - \{pt\}$ with finite Euclidean action can be continued to a self-dual solution on S^4 [1] , we conclude that for every stationary point with finite action on Euclidean space-time the winding number k has to be integer valued.

Trying to apply a Euclidean functional formalism to quantum gravity one encounters several peculiar features [2] . We here restrict ourselves to pure gravity without additional matter fields. First of all it is straightforward to write down a Euclidean analogue of the Einstein action:

$$I[g] = -\frac{1}{16\pi}\int_M R_{sc}\sqrt{g} - \frac{1}{8\pi}\int_{\partial M}[K]\sqrt{k} \ . \qquad (4)$$

Here M is a Riemannian (rather than pseudoriemannian) manifold with Riemannian metric g. R_{sc} denotes the Ricci scalar. The boundary term [3] is necessary for noncompact M or $\partial M \neq \emptyset$ in order to cancel the second derivatives of g, which enter into R_{sc}, and is given by

$$[K] = \text{tr } L - \text{tr } L_0 , \qquad (5)$$

where L is the second fundamental form of ∂M imbedded into M and L_0 the second fundamental form of ∂M with the induced metric h imbedded into $\partial M \times R$. The stationary points of I are given by $R_{ik}=o$, where R_{ik} is the Ricci tensor.

The following problems, unknown in ordinary field theory on flat space-time arise:

- There is no natural analogue of a Wick rotation for the transition between pseudoriemannian and Riemannian manifolds. The procedure which offers itself [2] in this situation is to perform all the calculations in the Riemannian regime and try an analytic continuation only for the final results (n-point functions etc.). For a discussion of this point see e.g. ref. [2].
- The specification of physical boundary conditions becomes even more crucial for quantum gravity. We shall exhibit below some possible choices.
- The functional integration runs over whole sets of equivalence classes of Riemannian manifolds. One may ask oneself whether one should include manifolds which admit no spin structure and whether one should impose topological subsidiary conditions. Fluctuations into non equivalent topologies are possible otherwise.
- The action is not bounded from below. Hawking and coworkers have given a prescription [2] how to use the positive action theorem [4 , 6] or the generalized positive action conjecture [2] to overcome this difficulty at least for certain special boundary conditions.

Let us now come to the discussion of some possible natural boundary conditions in Euclidean quantum gravity.

A very natural requirement would be to require M to be compact without boundary. This boundary condition is for instance employed by S. Hawking in his discussion of "space-time foam" [5], where a determination of the average Euler number per volume of the gravitational vacuum is carried out. We shall here be interested in the gravitational analogue of Yang-Mills instantons, whose effect tends to zero at infinity. So, we are forced to consider non-compact manifolds. To keep the analogy with Yang-Mills instantons as close as possible, the boundary condition of asymptotic euclidicity (AE-condition) suggests itself:

M is called asymptotically Euclidean [6] (AE) if outside some compact subset K it is diffeomorphic to $R\times S^3$, where R represents the proper distance from K and the metric on S^3 approaches the standard metric on S^3 for $R\to\infty$ at least as fast as r^{-2}. Now the positive energy theorem [4,6] states that an AE manifold with vanishing Ricci scalar (hence a candidate for a stationary AE point of the action) has non-negative action I, and that I=o is equivalent to M being flat Euclidean.

This means that under AE conditions the action has no non-trivial minima.

A richer structure is obtained by slightly weakening the AE condition to the ALE boundary condition [7,2] (for asymptotically locally Euclidean). This time one only demands

M-K is diffeomorphic to $R\times S^3/\Gamma$, where Γ is a discrete subgroup of SO(4), which acts freely as isometries.

The metric on S^3/Γ tends to the standard metric at least as fast as r^{-2}.

This ALE boundary condition, although apparently only slightly weaker than the AE condition admits many interesting instanton solutions, whose consistency and effect on vacuum tunneling and related phenomena can be investigated.

We shall restrict ourselves to the following special class of operations [8] of SO(4) on S^3 (more general cases are under investigation :)

Observe that $SO(4) = SU(2)\times SU(2)/_{Z_2}$

and via the map

$$S^3 = \{ (x_0,x_1,x_2,x_3) \mid x_0^2 + x_1^2 + x_2^2 + x_3^2 = 1 \} \to SU(2) \qquad (6)$$

$$(x_0,x_1,x_2,x_3) \to x_0 \mathbb{1} + i(x_1\sigma_1 + x_2\sigma_2 + x_3\sigma_3)$$

the group SU(2) and the unit sphere S^3 can be identified with the unit quaternions.

Then consider only discrete groups $\Gamma \subset SU(2)$ and let every $q \in SU(2)$ act on S^3 by left multiplication:

$$\tilde{x} = x_o + i\,\vec{x}\,\vec{\sigma} \rightarrow q\,\tilde{x}\,. \tag{7}$$

(One could, of course, as well consider right multiplication, for a detailed discussion see ref.[8] , in the most general case one had to consider simultaneous left and right multiplications).

Now the discrete subgroups of SU(3) which act freely on S^3 are all known, they are just (the doubles of) the symmetry groups of the Platonic polyhedra :

1) Cyclic group Z_s of order s
2) Binary dihedral group $\tilde{D}_s$ of order 4s
3) Binary tetrahedral group $\tilde{T}$ of order 24
4) Binary octahedral group $\tilde{O}$ of order 48 (8)
5) Binary icosahedral group $\tilde{I}$ of order 120

For instance for $\Gamma = Z_s$ the spaces S^3/Γ are the special lens spaces L(s,1), where the lowest member L(2,1) is just the real projective space $\mathbb{R}P(3)$, the 3-sphere with antipodal points identified. More general lens spaces can only be obtained by simultaneous left- and right multiplication.

For $\Gamma = Z_s$ and all s = 1,2,... there are explicitely known gravitational instantons:

s = 1 is the flat space
s = 2 is the Eguchi-Hanson ALE space [9] and
s > 2 corresponds to the multi Eguchi-Hanson spaces [7] .

All these spaces have a purely (anti) self-dual Riemannian curvatur tensor : $R = \underset{(-)}{+} R^*$.

According to Hawking's generalized positive action conjecture [2] this is not a coincidence:

Every ALE space with vanishing Ricci scalar has non-negative action; and zero action is equivalent with a (anti) selfdual curvature tensor.

So, for ALE boundary conditions self duality of the curvature tensor would be automatic for every minimum of the action, quite in analogy to Yang-Mills theory.

For the other discrete subgroups of SU(2) N. Hitchin [10] has sketched a construction of the corresponding gravitational instantons, which makes their existence very likely. Below we shall see that indeed also for these boundary conditions very strong consistency requirements are fulfilled, which provides additional evidence for their existence and, furthermore, gives information on the behaviour of matter fields in these gravitational background fields and on their role in vacuum tunneling. Before entering into these discussions we briefly have to mention the gravitational analogues for thermal boundary conditions. The closest analogue is provided by the AF (for asymptotically flat) boundary condition [8] :

M is called asymptotically flat, if
1) M - K is diffeomorphic to $R \times S^1 \times S^2$
2) for $r \to \infty$ the metric tends to standard metric on $S^1 \times S^2$

The only stationary points of the action under this boundary condition are the flat space $S^1 \times R^3$ and the Euclidean Schwarzschild space. This is by the Euclidean version of Hawking's "no hair" theorem for black holes. The curvature tensor of the Euclidean Schwarzschild space is not self dual and, hence, there are no non-trivial AF solutions of the Einstein equations with self-dual curvature.

Again there is a local version of the boundary conditions, which admits more structure:

M is called asymptotically locally flat [8] (ALF), if
- M - K is diffeomorphic to $\mathbb{R} \times S^3/\Gamma$, where Γ is a discrete subgroup of SO(4) which acts by isometries and freely on S^3
- The metric tends to

$$dr^2 + r^2 (\sigma_1{}^2 + \sigma_2{}^2) + \sigma_3{}^2 \tag{9}$$

at least as fast as r^{-1}, where $\sigma_1, \sigma_2, \sigma_3$ are the standard invariant one-forms S^3.

This ALF condition is really a local version of AF because, via the Hopf fibration S^3 is locally $S^2 \times S^1$.

For $\Gamma = Z_s \subset SU(2)$ acting by left multiplication there are ALF instanton solutions; they are just Hawking's Euclidean multi Taub NUT spaces [11] . Their boundaries at infinity are lens spaces L(s,1) with a distorted metric according to eq.(9). For the other discrete subgroups of SU(2) the existence of ALF instantons is an open question. The following investigation will give necessary conditions for their existence. Also the problem of other group actions and possible non self-dual ALF instantons is under investigation.

The index theorem of Atiyah and Singer [12]is the central tool for our investigations about gravitational instanton manifolds. Its main content for us is that the index of an elliptic operator D between complex vector bundles E and F over a manifold M is for M compact and $\partial M \neq o$ connected to the topology of the bundles E,F and the tangent bundle TM of M and expressible by their characteristic classes. Hence, natural elliptic operators on M provide topological information about the manifold M. The following presentation is based on the results of ref. [8] and, as far as the generalized Dirac operator is concerned, of ref. [13].

For our purposes we need the generalized version of the index theorem for compact manifolds with boundary by Atiyah, Patodi and Singer [14] . The non-compact gravitational instanton manifolds will be realized by shifting the boundary to infinity. We only need to consider dim M = 4.

Let M be an oriented four-dimensional compact Riemannian manifold with boundary ∂M. Let E and F be complex vector bundles over M and let D be an elliptic differential (or pseudo differential) operator from the sections $\Gamma(E)$ of E to the sections $\Gamma(F)$ of F.

$$\text{Define kernel } D = \{f \varepsilon \Gamma(E) \mid Df = o\}$$

$$\text{cokernel } D = \Gamma(E)/D\Gamma(E) = \text{kernel } D^* \tag{10a}$$

with respect to any adjoint of D.
Then by definition index D is the finite integer

$$\text{index } D = \dim \text{ kernel } D - \dim \text{ cokernel } D. \tag{10b}$$

This apparently is a purely analytical quantity, related to the solution of certain partial differential equations. The Atiyah-Singer theorem now states that it is expressible by invariants of the underlying bundles.

In the cases we are going to consider it reads

$$\text{index } D = A^D + B^D + \xi^D[\partial M] \quad , \tag{11}$$

where the meaning of the three terms on the right-hand-side is the following:

$$A^D = \{ \frac{\text{ch } E - \text{ch } F}{e(TM)} \text{ td } TM \otimes \mathbb{C} \} [M] \; . \tag{12}$$

Here ch, td, e are the Chern character, the Todd class and the Euler class [12,15] respectively, certain well defined characteristic classes, the formal division can really be performed, and the result is a differential form, which in turn is a polynomial in the Riemannian curvature and the curvature quantities of connections in the bundles E and F. The part of degree four of this differential form has to be integrated over the compact manifold M to give A^D. The physical importance of this term is illustrated by the fact that the integrand of A^D gives the anomaly of a field theoretical current canonically associated to D [16] .

The other two contributions are absent for $\partial M = \emptyset$. B^D is a local boundary term [14] , which arises by integrating an integrand, which, in addition to the curvature quantities also contains the second fundamental form,over the boundary ∂M.

The local contributions A^D and B^D behave additively under covering of the base space M. This is different for the third contribution $\xi^D[\partial M]$.

This is a spectral invariant [14] of the tangential part of the operator D. If D is of first order and decomposed into a normal part and a tangential part $\tilde{D}$:

$$D = \sigma\left(\frac{\partial}{\partial n} + \tilde{D}\right) \quad \text{with } \tilde{D} \text{ hermitean,}$$

then, in terms of the eigenvalues $\{\lambda\}$ of $\tilde{D}$ one defines

$$h_D = \dim \ \text{kernel } \tilde{D}$$

$$\eta_D(s) = \sum_{\lambda \neq o} \text{sign } \lambda |\lambda|^{-s} \tag{13}$$

and one has [14]

$$\xi_D[\partial M] = -\frac{1}{2}(\eta_D(o) + h_D). \tag{14}$$

Here $\eta_D(o)$ is obtained from $\eta_D(s)$ by analytic continuation. The notation $\xi_D[\partial M]$ indicates that the non local boundary term only depends on the Riemannian manifold M with its induced metric and on D but not on the interior. In principle $\xi_D[\partial M]$ can be evaluated by diagonalizing $\tilde{D}$. This, however, is in general not feasible, and the G-index theorem [12,17] to be described below, will give a more convenient method for calculating ξ_D.

We shall consider five types of canonical elliptic differential operators on gravitational instanton manifolds:

1) The Euler operator D_E: $\Omega^e(M) \rightarrow \Omega^o(M)$, an elliptic operator from differential forms of even degree to differential forms of odd degree. Its index equals the Euler characteristic $\chi(M)$, and the density of A_{D_E} is related to the

anomaly of the Euler current introduced in ref. [16] .

2) The signature operator $D_S : \Omega^+(M) \to \Omega^-(M)$, an elliptic operator from differential forms which are even under the involution $\alpha = (\ \ )*$ ($*$ is the Hodge star-isomorphism $\Omega^p \to \Omega^{4-p}$) to forms odd under α . Its index equals the signature $\tau(M)$ and the integrand of A_{Ds} gives the anomaly of the " Hirzebruch current" [16].

3) The Dirac operator $D_{1/2}:\Delta^+(M) \to \Delta^-(M)$ from spinor fields of positive to spinor fields of negative chirality. The density of $A_{D_{1/2}}$ gives [16] the gravitational axial anomaly for spin 1/2.

The last two types of operators are related to higher spins. Their index densities yield the gravitational axial anomalies for higher spin:

4) The Rarita-Schwinger operator[18,19,20]

$$D_{3/2} : \Delta^+_{3/2}(M) \to \Delta^-_{3/2}(M),$$

actually an operator between virtual bundles.

5) The generalized Dirac operator [13,18]

$$D_{mn} : \Delta(m,n) \to \Delta(n,m)$$

between spinor fields ot type (m,n) and (n,m).

The index theorem of Atiyah Singer and Patodi reads for ALE and ALF instantons with $\Gamma \subset SU(2)$ acting to the left as described above for these operators[8]:

1) Euler Operator

$$\text{index } D_E = \chi(M) = \frac{1}{16\pi^2}\int_M \text{tr } R\wedge *R + \frac{1}{|\Gamma|} + 0 \tag{15}$$

2) Signature operator

$$\text{index } D_s = \tau(M) = \frac{1}{24\pi^2}\int_M \text{tr } R\wedge R + 0 + $$

$$+ (\tilde{\xi}_s - \frac{2}{3}\frac{\varepsilon}{|\Gamma|}) \tag{16}$$

3) Dirac operator

$$\text{index } D_{1/2} = I_{1/2} = -\frac{1}{192\pi^2}\int_M \text{tr } R \wedge R + 0 + $$

$$+ (\tilde{\xi}_{1/2} - \frac{1}{12}\frac{\varepsilon}{|\Gamma|}) \tag{17}$$

4) Rarita-Schwinger operator

$$\text{index } D_{3/2} = I_{3/2} = \frac{21}{192\pi^2} \int_M \text{tr } R \wedge R + o + (\tilde{\xi}_{3/2} - \frac{7}{4} \frac{\varepsilon}{|\Gamma|}) \tag{18}$$

5) Generalized Dirac operator

$$\text{index } D_{mn} = I_{mn} = \frac{c_{mn}}{8\pi^2} \text{ tr} \int_M R \wedge R + o + \xi_{mn}^{\text{ALE ALF}} . \tag{19}$$

with $c_{mn} = -\frac{1}{720} (m+1)(n+1)\{ n(n+\varepsilon)(3n^2+6n-14) - m(m+\varepsilon)(3m^2+6m-14)\}$.

Here $\varepsilon = o$ refers to ALE and $\varepsilon = 1$ to ALF.

Notice that for $R = \pm *R$ the volume terms are all proportional and can, for instance, be evaluated by calculating $\xi_{1/2}$ and using Lichnerowicz's theorem [21] which states in our case that $I_{1/2} = o$.

If the gravitational instanton manifolds really exist all these indices as computed from (15) - (19) really have to be integers, which is an extremely strong consistency requirement.

We now briefly sketch, how one computes the non local boundary contribution [14] by means of the G - index theorem [12, 17] .

For the G-index theorem one considers elliptic operators $D: E \to F$, which in addition commute with the action of a discrete group Γ . Then kernel D and cokernel D are finite dimensional representation spaces of Γ and for every $g \in \Gamma$ one defines

$$\text{index}_g D = \text{tr } g \,|_{\text{kernel } D} - \text{tr } g \,|_{\text{cokernel } D} . \tag{20}$$

Then for compact Riemannian manifolds with boundary one has

$$\text{index}_g D = A_g^D + B_g^D + \xi_g^D [\partial M] . \tag{21}$$

Here [12],[17]

$$A_g^D = \frac{i^*(\text{ch}_g E - \text{ch}_g F)\,\text{td}(TM^g \otimes C)}{\text{ch}_g (\Lambda_{-1} N^g \otimes C)\, e(TM^g)} [M^g] , \tag{22}$$

where ch_g is the so-called equivariant Chern character, X^g is the fixed-point set of g, $N^g \otimes C$ the complexified normal bundle of X^g in X, Λ_{-1} means the alternating sum over the exterior powers and i^* denotes restriction to X^g. Notice, that only an integral over the fixed points has to be evaluated, and only the action of g in the neighbourhood of its fixed points enters.

For the non-local boundary term one has the following relation for covering manifolds [14]

$$\xi_D\,[Y/\Gamma] = \frac{1}{|\Gamma|} \sum_{g\in\Gamma} \xi_g^D[\,Y\,]$$

$$= \frac{1}{|\Gamma|}\{\xi^D\,[Y] + \sum_{g\neq e}\xi_g^D\,[Y]\} \tag{23}$$

$$:= \frac{1}{|\Gamma|}\;\xi^D\,[Y] + \tilde{\xi}_D[Y/\Gamma]$$

In the ALE case, on which we shall concentrate for illustration, $Y = S^3$ with standard metric and $\xi_D(S^3) = 0$.
Hence
$$\xi^D\,[\,S^3/\Gamma] = \tilde{\xi}[\,S^3/\Gamma].$$

It remains to calculate the quantities $\xi_g^D\,[S^3]$.

To achieve this, one uses [14] a suitable continuation of S^3 and of the action of Γ on S^3 inwards to the disk D^4. For the standard metric on S^3 it suffice to endow D^4 with its flat Euclidean metric and to let Γ act as Euclidean rotations in four dimensions with the only fixed point $O \in D$. Here the action around O is given by two angles Θ_1 and Θ_2 of rotations in two-dimensional planes. For $\Gamma \subset SU(2)$ these angles are equal (up to a sign depending on whether the action is to the left or to the right). The G-index is zero for D^4, $B_g^D = 0$ because there is no fixed point on the boundary, and the G-index theorem yields[8],[13],[20]

$$\tilde{\xi}_{1/2} = \frac{1}{4|\Gamma|} \sum_{g\neq e} \frac{1}{\sin^2 \frac{\Theta_g}{2}}$$

$$\tilde{\xi}_s = 4\,\xi_{1/2} - 1 + \frac{1}{|\Gamma|} \tag{24}$$

$$\tilde{\xi}_{3/2} = 3\,\xi_{1/2} - 2 + \frac{2}{|\Gamma|}$$

The remaining task is to find the angles Θ_g from the geometrical properties of the groups Γ, and one arrives at the following results [8] for the ALE case.

In the table we have indicated the relationship of the intersection of the instanton manifolds to the Cartan matrices of simple Lie groups as used by N. Hitchin [10] .

So, we find all indices integer, as necessary for consistency, a non trivial fact in view of the large denominators which occur in $\tilde{\xi}_{1/2}$.

Γ	$\xi_{1/2}$	χ	τ	$I_{1/2}$	$I_{3/2}$
$Z_s \leftrightarrow A_{s-1}$	$\frac{s^2-1}{12\,s}$	s	s-1	o	2(s-1)
$\tilde{D}_s \leftrightarrow D_{s+2}$	$\frac{1}{48s}(4s^2+12s-1)$	s+3	s+2	o	2(s+2)
$\tilde{T} \leftrightarrow E_6$	$\frac{167}{288}$	7	6	o	12
$\tilde{O} \leftrightarrow E_7$	$\frac{383}{576}$	8	7	o	14
$\tilde{I} \leftrightarrow E_8$	$\frac{1079}{1440}$	9	8	o	16

For the generalized Dirac operator it may here suffice to give the boundary term ξ_{mn} [13] for the Eguchi-Hanson metric [9], corresponding to $Z_2 \subset SU(2)$:

$$\xi_{mn}[\partial M] = \frac{1}{32}(m+1)(n+1)\,[(-1)^m-(-1)^n], \tag{26}$$

which, indeed together with (19) gives I_{mn} integer for all m and n. For the ALF case we only give some results [8] here.

For the explicitely known multi Taub-NUT family, corresponding to Z_s one finds consistency with $\tau=s-1$; $I_{1/2}=0$; $I_{3/2}=2\tau$.

For ALF the only other consistent $\Gamma \subset SU(2)$ acting to the right is $\tilde{D}_s$, which yields $\chi=s+3, \tau=s+2$, $I_{1/2}=0, I_3=2(s+2)$.

If, instead the action is to the left, $\xi[\partial M]$ must change sign. This gives the theorem [8] :

There are no self-dual ALE or ALF metrics with $\Gamma \subset SU(2)$ acting to the left.

More general results for possible non self-dual metrics and more general actions on S^3 will be discussed elsewhere.

REFERENCES

1. See e.g. J. M. Singer, Cargese lectures 1979, for a review of Yang-Mills instantons.
2. For a review see S. W. Hawking: Euclidean Quantum Gravity, D.A.M.P.T. preprint 1979.
3. G. W. Gibbons, S. W. Hawking, Phys.Rev. D15,2752 (1977).
4. G. W. Gibbons, S. W. Hawking, M. J. Perry, Nucl.Phys. B (1979) to appear.
5. S. W. Hawking, D.A.M.T.P. preprint 1979.
6. G. W. Gibbons, C. N. Pope, "The Positive Action Conjecture and Asymptotically Euclidean Metrics in Quantum Gravity", D.A.M.T.P. preprint and Comm.Math.Phys. (1979) to appear.
7. G. W. Gibbons, S. W. Hawking, Phys.Lett. 75B, 430 (1978).
8. G. W. Gibbons, C. N. Pope, H. Römer: "Index Theorem Boundary Terms for Gravitational Instantons" D.A.M.T.P. preprint 1979. This paper also contains a thorough discussion of the various boundary conditions.
9. T. Eguchi, A. Y. Hanson, Phys.Lett. 74B, 249 (1978).
10. N. Hitchin: "Polygons and Gravitons" to appear in Math.Proc. Comb.Phil.Soc.
11. S. W. Hawking, Phys.Lett. 60A, 81 (1977).
12. M. F. Atiyah, J. M. Singer, Bull.Ann.Math.Soc. 69, 422 (1963); Ann.Math. 87, 484 and 546 (1968).
13. H. Römer, Phys.Lett. 83B, 172 (1979).
14. M. F. Atiyah, W. K. Patodi, J. M. Singer, Bull.London Math. Soc. 5, 229 (1973); Math.Proc.Cambridge Philos.Soc. 77, 43 (1975); 78, 405 (1975); 79, 71 (1976).
15. R. Palais, ed., Seminar on the Atiyah-Singer Index Theorem, Annals of Math.Studies No. 76 (Princeton University Press 1965).
16. N. K. Nielson, H. Römer, B. Schroer, Nucl.Phys. B136, 475 (1978) and references therein.
17. M. F. Atiyah, J. B. Segal, Ann.Math. 87, 531 (1968).
18. S. M. Christensen, M. Y. Duff, Phys.Lett. 76B, 571 (1978).
19. M. T. Grisaru, N. K. Nielson, H. Römer, P. van Nieuwenhuizen, Nucl.Phys. B140, 477 (1978).
20. A. J. Hanson, H. Römer, Phys.Lett. 80B, 58 (1978).
21. A. Lichnerowicz, Comptes Rendues 257A, 5 (1968).

ON NONLOCAL CHARGES IN QUANTUM FIELD THEORY

Jan T. Łopuszanski

Institute of Theoretical Physics
University of Wrocław
Wrocław, Poland

1. The topics of my lecture are the nonlocal charges. Before I shall enter into the explanation of this notion of somewhat misleading name (it will turn out that the local properties of nonlocal charges are quite important in my presentation) let me say a few words about conventional charges in the context of symmetries in quantum field theory.

What does one mean by symmetry of the quantum field theory? One has usually in mind a mapping which leaves the Wightman functions invariant. Then this mapping is unitarily (or antiunitarily) implementable in a Hilbert space.

To make things as simple as possible and still leave the door open for including the supersymmetric charges into my considerations, I shall restrict myself to symmetries which depend on continuous parameters, and even - in the later part of this talk - to infinitesimal transformations.

Let me first of all call to your attention that not every symmetry of the Wightman functional is of physical interest, at least as far as the S-matrix approach is concerned. To see that let me present the following example. Take unitary operators forming a group

$$U(\alpha_1 \ldots \alpha_k) = U(\underline{\alpha}) \quad , \quad U(\underline{\alpha})^+ = U(\underline{\alpha})^{-1} \tag{1a}$$

where $\alpha_1, \ldots \alpha_k$ are the group parameters, such that

$$U(\underline{\alpha})\Omega = \lambda(\underline{\alpha})\Omega \quad , \quad |\lambda(\underline{\alpha})| = 1 \tag{1b}$$

where Ω is the (unique) vacuum; take a set of relativistically covariant, local, massive, real, interacting scalar quantum fields

$$\varphi_i(x),\ i = 1,\dots n,\ x \equiv (x_0, x_1, x_2, x_3).^{*)}$$

Then the mapping

$$\varphi_i(x) \to U(\underline{\alpha})\, \varphi_i(x) U(\underline{\alpha})^{-1} = \varphi_i(x, \underline{\alpha})$$

leaves the Wightman functional invariant. As I insisted that the fields φ_i interact with each other, we should expect that the S-matrix

$$S \neq 1.$$

Then, in general,

$$U(\underline{\alpha}) S\, U(\underline{\alpha})^{-1} = S(\underline{\alpha}) \neq S$$

From pragmatic point of view one would like to deal with one S-matrix and not with infinitely many S-matrices depending on the parameters of the group. Therefore it is reasonable to require

$$S(\underline{\alpha}) = S$$

or

$$[S, U(\underline{\alpha})] = 0\,. \qquad (2)$$

In this way we exclude from our considerations all symmetries of the Wightman functions which do not conform with (2).

What are the conditions for the fields which would secure the validity of (2)?

The sufficient but not necessary condition for (1) is that:

i) $E_m\, U(\underline{\alpha})\, \varphi_i(f)\, \Omega \neq 0$

when

$$E_m\, \varphi_i(f)\Omega \neq 0$$

where E_m is the projection operator on the Hilbert space of one-particle states of mass m and f(x) is a test function,

ii) φ_i should be in some sense local

*) We use the metric (+ - - -)

with respect to $\varphi_j(\alpha)$, $i,j = 1, \ldots n$[1].

The most favourable situation is, when we have the sharp causality relation

$$[\varphi_i(x,\underline{\alpha}), \quad \varphi_j(y)] = 0 \text{ for } (x-y)^2<0; \tag{3}$$

we may, however, relax the locality condition considerably; e.g. for translations one has

$$\varphi_i(x,\underline{\alpha}) = \varphi_i(x+a)$$

and then (3) holds for $(x+a-y)^2 < 0$ (consider also the case of quasilocal fields etc.).

It is, however, worthwhile to notice that there can still exist symmetries which satisfy (2) but do not satisfy (3) even in its weakened form. This class of transformations was not yet thoroughly explored.

2. Let me turn now to infinitesimal transformations.

The symmetry group $U(\underline{\alpha})$ is induced by generators

$$Q_\ell, \quad \ell = 1, \ldots k$$

which are selfadjoint operators .*) I shall call these generators "charges". For these charges the relation (3) reduces to

$$[\,[Q_\ell, \varphi_i(x)\,], \varphi_j(y)] = 0 \text{ for } (x-y)^2 < 0 \tag{4}$$

$$\ell = 1, \ldots k, \quad i,j = 1, \ldots n.$$

Hence it follows that also

$$[\,Q_\ell\,, S\,] = 0. \tag{5}$$

It is a right place to include in our considerations fermionic fields as well as supersymmetric charges. For non-super-symmetric charges and fields of arbitrary spin relation (4) has to be changed to

$$[\,[Q_\ell, \varphi_i\,(x)\,], \varphi_j(y)\,]_{\underline{+}} = 0 \text{ for } (x-y)^2 < 0 \tag{4a}$$

---

*) Usually one assumes that they have a common dense domain of definition, provided this cannot be proven.

for supersymmetric charges this becomes

$$[\, [\, Q_\ell,\ \varphi_i(x)]_{\pm},\ \varphi_j(y)]_{\pm} = 0 \quad \text{for } (x-y)^2 < 0 \tag{4b}$$

in compliance with the Spin- and-Statistics Theorem.

It is a lore that each charge (also the supersymmetric one) which satisfies (4)(and (5)) can be presented in the form*)

$$Q = \int j_o(x_o,\vec{x})\, d^3x \tag{6}$$

where the fields $j_\mu(x)$, $\mu=0,1,2,3$ which I shall call "current", satisfy the continuity equation

$$\partial^\mu j_\mu = 0. \tag{7}$$

We do not need to specify the transformation properties of the current with respect to the Lorentz group, but we should require that the current is in some sense local with respect to the underlying fields. To my knowledge the conjecture, expressed by formula (6), was not yet proven.

For Q's given by (6) we may relax the requirement of self-adjointness condition demanding only that Q has to be a Hermitian operator.

What was proven is that for massive field theories and translationally covariant current the condition (7) is necessary and sufficient for the existence of the charge (6), [2] . This statement was then extended to non-translationally covariant currents which have definite transformation under the Lorentz group [3].**)

---

*) The notation in formula (6) should be understood in a symbolic way; as it stands it does not make sense mathematically.

**) A non-translationally and non-Lorentz covariant current which is not locally conserved may still yield a well defined charge. Let us consider the following example [3] : Take a translationally and Lorentz covariant, locally conserved current j_μ which gives rise to a charge $Q = \int j_o(x_o,\vec{x})d^3x$ and define

$$J_\mu(x) \equiv \sum_{j-1}^{3} x_j\, \partial^j j_\mu(x) = \partial^j(x_j j_\mu) - 3\, j_\mu$$

Then $$\partial J_\mu(x) = \partial^i j_i(x) \neq 0$$

but $$-\frac{1}{3}\int J_o(x_o,\vec{x})d^3x = Q.$$

For such Q's

$$Q \, \Omega = 0. \tag{8}$$

The proof of (8) is trivial as soon as

$$[\, Q, P_0 \,] = 0 \tag{9}$$

where P_μ, $\mu = 0,1,2,3$ are the generators of the translation group. However, (9) does not need to occur for each charge.

3. In connection with (9) the following observation seems to be in order: it is advisable to distinguish between constants of motion R characterized by the relation

$$[\, R, P_0 \,] = 0 \tag{10}$$

and conserved quantities C for which

$$[\, C, S \,] = 0. \tag{11}$$

Not every constant of motion is a conserved quantity and vice versa. To see that consider the generators of the Lorentz boost M_{0i}, $i = 1,2,3$, which satisfy (11) but do not satisfy (10). On the other hand the asymptotic particle number operators in an interacting, massive scalar field theory in four space-time dimensions

$$N_{ex} \equiv \int \frac{d^3p}{2\omega_p} a^+_{ex}(\vec{p}) \, a_{ex}(\vec{p}) ; \quad \omega_p = +\sqrt{\vec{p}^{\,2}+m^2}$$

where the suffix "ex" stands either for "in" or "out", commute not only with P_0 but with all generators of the Poincaré group, but do not commute with the S-matrix, since the absence of non-elastic scattering implies absence of any scattering [4] .*)

Notice that each constant of motion R found in the theory of the incoming and outgoing free field is automatically a constant of motion of the interacting field, interpolating between the asymptotic free fields.**)

---

*) Not true for two space-time dimensions.

**) This follows from the following equalities

$$P_0 = H_0(\varphi_{in}) \equiv \frac{1}{2} \int d^3p \; a^+_{in}(\vec{p}) \, a_{in}(\vec{p}) = H_0(\varphi_{out})$$

$= H_0(\varphi) + V(\varphi) \equiv H(\varphi)$ where $H(\varphi)$ denotes the Hamiltonian for the interacting field and $V(\varphi)$ the interaction part of it.

For a free field the number of constants of motion is not limited, as the system has infinitely many degrees of freedom. Therefore the number of constants of motion for the interacting field is also infinite.

Be, however, aware of the fact that not all of these constants of motion contribute essentially to the description of the dynamical structure of the system. Usually only few of them are of dynamical relevance. This becomes clear e.g. in case of classical statistical mechanics, where the majority of the constants of motion describes merely the initial values of the coordinates and only few univalent integrals of motion are of importance (e.g. energy). In field theory e.g. the aforementioned quantities N_{ex} do not supply any useful information as far as the interaction is concerned.

It is therefore important to find a way to characterize those constants of motion which can be of some value for examining the physical system. To propose such a criterion is the main task of my talk.

However, before I enter into more detailed presentation of this problem let me make a few more remarks about conserved quantities and constants of motion. Although it is true that conserved quantities (i.e. those which commute with the S-matrix) like regular charges giving rise to symmetries, are useful in characterizing the dynamics of the system, the use of some of the constants of motion which are not conserved can be even more effective in exploring the S-matrix and consequently in describing the interaction mechanism of the physical system. This can be clearly seen on the example of the quantum nonlinear σ-model in two space-time dimensions [5] . For this model new nonlocal constants of motion were found which have some interesting properties not shared by the regular charges. It turns out, for example, that they are not conserved (they do not commute with the S-matrix), thereby providing a direct and elegant proof that there is interaction in the nonlinear model. It was also found that the restriction arising from the existence of these quantities exclude particle production (which in two-space-time dimensions does not lead to a trivial S-matrix) and are consistent with the factorization of the multiparticle S-matrix into products of two-particle S-matrices. The latter fact would make it possible to calculate the S-matrix exactly.

The nonlocal constants of motion for the nonlinear σ-model are of the form

$$Q^{ab} = \sum_{c} \int\int_{x_o=y_o=t} \varepsilon(x_1-y_1) j_o^{ac}(x_o,x_1) j_o^{bc}(y_o,y_1)\, dx_1\, dy_1 - Z \int j_1^{ab}(t,x_1)dx_1 \tag{12}$$

where

$$j_\mu^{ab} = - j_\mu^{ba} \qquad a,b = 1,\ldots m, \ \mu = 0,1$$

are the locally conserved currents linked to the O(n)-symmetry of the model and Z is a renormalization constant. One infers from (12) that the new quantities are obtained by multiple spaceintegration of products of local fields. In some sense the new quantities are a generalization of the notion of a charge.

In view of this nontrivial example it is interesting to find out what can be said about such nonlocal constants of motion, which I shall call "nonlocal charges", in the general setting of local quantum field theory in more than one space dimension, say, in four space-time dimensions.

At first sight such a task might seem to be hopeless for, at least, two reasons:

first, the existence of nonlocal charges depends on nonlinear relations between local fields (in contrast to the linear relation (7) in case of a standard charge),

second, the proper definition of a formal expression as in (12) requires a careful analysis of the short distance behaviour of products of local fields (e.g. by application of the Wilson-Zimmermann expansion) [6] and, as such, is extremely model dependent.

Nevertheless, there is a way to circumvent these difficulties which I shall try to present to you now. This is - as I mentioned before - the main goal of this talk.

4. To begin with I list relevant assumptions

i) positive definite metric in the Hilbert space;
ii) unique vacuum state Ω ;
iii) the presence of the mass gap ($m \neq 0$);
iv) finite number of independent, relativistically covariant, interacting, (anti) local quantum fields;
v) irreducibility of the asymptotic free fields of mass $m \neq 0$;
vi) strict interaction among the fields.

Ad iii) I restrict myself, for simplicity, to one mass multiplet.
Ad iv) by independence of fields I mean that e.g. in case of scalar fields

$$(\Omega, \varphi_i(x)\, E_m\, \varphi_j(y)\, \Omega) = c\, \Delta^{(+)}(x-y)\delta_{ij},$$

$$c \neq 0,$$

provided

$$E_m \; \varphi_i(f)\,\Omega \neq 0.$$

Ad vi) by strict interaction among the fields I mean that the Wightman functional is not a product of two Wightman functionals or - in physical terms - there is a "good mixing" of all the components of the physical system.

Let me now return to the subject of standard charges. They were characterized by

$$[\,[Q,\varphi_i(x)]_{\pm},\varphi_j(y)\,]_{\pm} = 0 \quad \text{for} \quad (x-y)^2 < 0. \qquad (4)$$

Let us restrict our considerations to charges which are constants of motion, i.e. satisfy

$$[\,Q,P_o\,] = 0, \qquad (9)$$

Relation (4) is closely related to

$$Q = \int j_o(x_o,\vec{x})\,d^3x \qquad (6)$$

as taking the commutator of the charge with the field results in suppressing one three-dimensional integration over space.

This observation makes it plausible to try to find a similar analogy for a nonlocal charge which - as was said before - is obtained for the nonlocal σ-model by multiple space integration of products of local fields. The relation which can be regarded as a counterpart of an N-fold three-dimensional integration over space reads [7]

A) $$[\ldots[Q,\varphi_{i_1}(x_1)]\,,\varphi_{i_2}(x_2)]_{\pm}\ldots\;\varphi_{i_{N+1}}(x_{N+1})] = 0$$

for any collection of (anti) local fields $\varphi_i(x)$ and points $x_i, i=1,\ldots N+1$, taken from (N+1) arbitrary double-cones with disjoint bases at time t=o.*) We shall refer to N as to the genus of the nonlocal charge Q[8] and shall denote it hereafter by $Q^{(N)}$. (13)

In addition, we demand - similarly as in the case of ordinary charge - that

---

*) For $N \leq 2$ this is the same as the condition $(\vec{x}_i-\vec{x}_k)^2 < 0$ for $i \neq k$, $i,k = 1,\ldots N+1$.

B) $Q^{(N)}$ is a constant of motion i.e. it satisfies

$$[\, Q^{(N)}, P_o \,] = 0, \tag{9a}$$

and

C) is a Hermitian operator.

Relation (13) shows clearly that - in spite of the name - the nonlocal charge displays local properties.

In our - Buchholz's and my [7] - opinion, relation (13) can be used as one of the criteria to discriminate between relevant and irrelevant constants of motion.

Indeed, the nonlocal charges of Lüscher, presented in formula (12), satisfy relation (13) and can be classified as charges of genus 2.

The standard charges have genus 1.

The advantage of the approach, characterized by assumptions A) through C), as compared with the tedious procedure presented for the nonlinear σ-model by [5] lies in the following:

i) we do not need to know how the nonlocal charges are built out of field operators,
ii) we have a general, model-independent formulation which is suitable for an analysis in a more general, axiomatic framework.

It remains to check whether the assumptions A),B) and C) proposed by us are stringent enough to produce a handy tool in examining the S-matrix.

It turns out that this is, really, the case.

5. In case of a standard charge, which we shall from now on write $Q^{(1)}$, relation (4) tells us that the field

$$[\, Q^{(1)}, \varphi_j(x) \,] \equiv A_j(x)$$

is (anti) local with respect to all φ_i (which form an irreducible set, assumption v)) and therefore is (anti) local with respect to itself [9]. Consequently, taking into account known theorems [1] we have

$$A_{j,ex}(x) = \sum_{\ell} k^{ex}_{j\ell}(x,\partial)\, \varphi_{\ell,ex}(x).$$

As $Q^{(1)}$ is a global quantity, satisfying (9), we have also

$$[\, Q^{(1)},\ \varphi_{j,ex}(x)\ ] = A_{j,ex}(x).$$

Comparing both these formulae, we get

$$[\, Q^{(1)},\ \varphi_{j,ex}(x)\ ] = \sum_{\ell} k^{ex}_{j\ell}\,(x,\partial)\varphi_{\ell,ex}(x). \tag{14}$$

Applying (14) to the vacuum state and taking into account (8) as well as the stability of the one-particle state we get

$$Q^{(1)}\ \psi(x,j) = \sum_{\ell} k^{ex}_{j\ell}(x,\partial)\ \ \psi(x,\ell) \tag{15}$$

where

$$\psi(x,j) = \varphi_{j,ex}(x)\ \Omega\ . \tag{15a}$$

Relation (15) entails that

$$k^{in}_{j\ell} = k^{out}_{j\ell} \equiv k_{j\ell}. \tag{16}$$

From (14) and (16) follows immediately

i) $Q^{(1)}$ is a bilinear form in the asymptotic free fields "in" and "out",

ii) $Q^{(1)}$ satisfies

$$[\, Q^{(1)},\ S\ ] = 0 \tag{5a}$$

[10] ; $Q^{(1)}$ is always a conserved quantity.

As a consequence of assumption iv) follows that the number of independent charges, satisfying (4), (5) and (9), is limited and each charge is a linear combination of a certain finite set of basic charges [11] .

Let me now go over to nonlocal charges $Q^{(N)}$, $N > 1$, defined by assuming A), B) and C) [7] . Surprisingly enough, it turns out that these assumptions supply sufficient amount of information about the local properties of $Q^{(N)}$ to establish its form in terms of the irreducible set of finite number of asymptotic free fields $\varphi_{i,ex}(x)$; in more detail: the expansion of $Q^{(N)}$ in terms of asymptotic creation and annihilation operators terminates at a finite sum; the kernels in this expansion are completely determined by the matrix elements of $Q^{(N)}$ between states with an asymptotic particle number less than or equal to N.

In, at least, two respects the behaviour of the nonlocal charges may differ from that of the regular charges:

$Q^{(N)}$ may change the number of asymptotic particles but not more than by $N-2 \geq 0$; therefore $Q^{(2)}$ (like $Q^{(1)}$) preserves the number of particles,

$Q^{(N)}$ may not commute with the S-matrix, i.e. may not be a conserved quantity, the last property may be helpful in obtaining informations about the structure of the S-matrix.

In particular, for N = 2 the nonlocal charge defined on the dense domain of nonoverlapping asymptotic states reads

$$Q^{(2)} = \sum_{(\underline{m})} \int \frac{d^3p}{2\omega_p} \sum_i \sum_j A_{ij}^{(\underline{m})}(\vec{p})\, a^+_{i,ex}(\vec{p})\, \partial_p^{(\underline{m})}\, a_{j,ex}(\vec{p}) \; +$$

$$+ \sum_{(\underline{m})} \sum_{(\underline{n})} \int \frac{d^3p}{2\omega_p} \int \frac{d^3q}{2\omega_q} B_{ijk\ell}^{(\underline{m})(\underline{n})ex}(\vec{p},\vec{q}) \tag{17a}$$

$$\times\, a^+_{i,ex}(\vec{p})\, a^+_{j,ex}(\vec{q})\, \partial_p^{(\underline{m})}\, \partial_q^{(\underline{n})}\, a_{k,ex}(\vec{p})\, a_{\ell,ex}(\vec{q})$$

$+$ h.c. ,

where $\omega_p \equiv +\sqrt{p^2 + m^2}$

$(\underline{m}) \equiv (m_1, m_2, m_3)$, $(\underline{n}) \equiv (n_1, n_2, n_3)$

$\partial_p^{(\underline{m})} \equiv \partial_{p_1}^{m_1}\, \partial_{p_2}^{m_2}\, \partial_{p_3}^{m_3}$

the sum over $(\underline{m})$ and $(\underline{n})$ is finite, (17b)
the kernels A and B^{ex} are locally square integrable functions respectively in p and q provided $\vec{p} \neq \vec{q}$.

The procedure to obtain (17) is much more involved than that for a regular charge; fine mathematical tools as well as assumptions about the domain of $Q^{(2)}$ are needed. I shall not report on these things here. They will be published soon elsewhere.

Notice that, although the kernel A is the same for the incoming and outgoing fields, it may well happen that

$$B^{in} \neq B^{out} \tag{18}$$

and therefore $Q^{(2)}$ does not need to commute with the S-matrix.

Notice also that although every nonlocal charge $Q^{(2)}$ must have the form (17), it is not obvious that every operator of the shape (17) has to be a nonlocal charge of genus (2), satisfying the requirements A), B and C); a proof of such a conjecture seems to be difficult.

The question whether there are in four space-time dimensions nonlocal charges $Q^{(2)}$ satisfying (18) is difficult to answer (definitely there are such in two space-time dimensions [5] but this can be due to peculiar properties of this Minkowski space). It seems that in case of one scalar field or one gauge invariant complex field and with absence of derivatives under the integrals in (17) this special kind of the nonlocal charge, which we denote by $Q_o^{(2)}$, is a conserved quantity; however, for two real scalar fields which do not display the gauge invariance of the first kind, conformity of $Q_o^{(2)}$ with (18) is not excluded. This problem is still under investigation.

6. So far the kernels in the expressions (14), (15) and (16) for $Q^{(1)}$ as well as in (17) for $Q^{(2)}$ were not determined. It is interesting to find out which type of kernels are admissible in a field theory. In these considerations the assumption about strict interaction among the fields, listed as vi), is of paramount importance.

In case of the standard charge $Q^{(1)}$ the answer is well known (Coleman and Mandula, Lopuszanski, Gerber and Reeh [12]). It is spelled out in the so called "No-go" Coleman-Mandula Theorem. Its extended version, encompassing also the supercharges [13] reads: *)

i) the only charges $Q^{(1)}$ which conform with (4) and (9) are scalar charges linked to internal symmetries, finite number of spinorial charges linked to supersymmetric transformations, one vector charge, P_μ, $\mu = o,1,2,3$, and three generators of the spatial rotations $M_{ik} = - M_{ki}$, $i,k = 1,2,3$;

ii) there is no nontrivial mixing of internal and geometrical transformations as far as the tensorial charges are concerned; at this level there is no mixing of spins and masses in the multiplets;

iii) however, there is a relation between spinorial charges and the energy-momentum vector - on one hand - and

*) Keep in mind that we are dealing with massive fields.

so-called central charges, belonging to the generators of internal symmetries and commuting with all charges - on the other hand; at this level there is an interrelation between the internal and geometrical charges mediated by spinorial charges and the mixing of spins in the multiplets is allowed.

This severe selection of admissible charges results - as was stressed before - from assumption vi). To make this clear imagine that we are dealing with two sets of fields which do not interact with each other, then each set has its own conservation laws e.g. each set has its own energy-momentum vector, say $P_\mu^{(1)}$ and $P_\mu^{(2)}$. Obviously, the operators

$$a\,P_\mu^{(1)} + b\,P_\mu^{(2)}$$

for arbitrary real numbers a and b are also conserved quantities for the system as a whole; thus, already at this stage, we have infinitely many (trivial) conservation laws.

One concludes from the above remark that the extended Coleman-Mandula Theorem is no longer true for free fields.

When one has a strictly interacting system of fields, then there are many transition amplitudes

$$(\psi_{out}, \psi'_{in}) \neq 0 \tag{19}$$

present which would vanish in case of a noninteracting system (in particular, the amplitudes for nonelastic scattering). One may make use of (19) as well as of the hermiticity of $Q^{(1)}$ inspecting the expressions [14]

$$(\psi_{out}, Q^{(1)}\,\psi'_{in}) = (Q^{(1)}\psi_{out}, \psi'_{in}). \tag{20}$$

As $Q^{(1)}$ acts additively on ψ'_{in} as well as on ψ_{out}, one gets infinitely many relations which allow to evaluate the kernel of the charge and in this way to solve the problem.

Equation (19) as well as (20) adopted to the case of a nonlocal charge of genus N in the form

$$(\psi_{out}, Q^{(N)}\psi'_{in}) = (Q^{(N)}\psi_{out}, \psi'_{in}) \tag{20a}$$

is the main tool in evaluating the kernels of $Q^{(N)}$.

Let me restrict myself to the case of N=2.

The difficult part of the problem is to find all admissible $Q^{(2)}$, expressed by (17), in which derivatives of the annihilation operators

$$\partial_{\underline{p}}^{(m)} \alpha_{j,ex}(\vec{p}) \quad \text{and} \quad \partial_{\underline{q}}^{(n)} \alpha_{j,ex}(\vec{q})$$

will appear. Let me, for convenience, denote the number of derivatives

$$\sum_{i=1}^{3} m_i$$

in $Q^{(2)}$ by the suffix $D, Q_D^{(2)}$.

To make things as simple as possible let me restrict myself further to the case of one scalar field (and discard the supersymmetric charges, of course). Let us now compare the procedure applied for $Q_D^{(1)}$ with that for $Q_D^{(2)}$.

For the regular charge $Q_o^{(1)}$ the problem is rather easy to solve. We find

$$Q_o^{(1)} = \sum_{\mu=o}^{3} a^{\mu} P_{\mu}$$

where a_{μ} are arbitrary real constants. To evaluate $Q_1^{(1)}$ we take advantage of assumption B)

$$[Q_1^{(1)}, P_o] = o \tag{21a}$$

as well as of the fact that

$$[Q_1^{(1)}, P_i] = \sum_{\mu=o}^{3} a_{i\mu} P^{\mu} \in Q_o^{(1)} . \tag{21b}$$

From (21) one is able to establish in a unique way that

$$Q_1^{(1)} = \sum_{\mu=o}^{3} a^{\mu} P_{\mu} + \sum_{i,k=1}^{3} a^{ik} M_{ik} .$$

We conclude that $Q_1^{(1)}$ is completely determined by $Q_o^{(1)}$. It can also easily be shown that

$$Q_D^{(1)} = Q_1^{(1)} \quad \text{for} \quad D \geq 1.$$

The problem becomes, however, more involved when we go over to an estimate of $Q_D^{(2)}$. Although it is again relatively easy to find that

$$Q_o^{(2)} = \sum_{\mu=0}^{3} a^{\mu} P_{\mu} + \sum_{\mu,\nu=0}^{3} b^{\mu\nu} P_{\mu} P_{\nu}$$

where a^{μ} and $b^{\mu\nu} = b^{\nu\mu}$ are arbitrary real constants, for $Q_1^{(2)}$ we are confronted with the rather difficult problem that $Q_o^{(2)}$ does not uniquely determine $Q_1^{(2)}$, as there exist nonlocal charges of type $Q_1^{(2)}$, say $\hat{Q}_1^{(2)}$, which commute with the whole translational group

$$[\hat{Q}_1^{(2)}, P_{\mu}] = 0, \mu = 0,1,2,3.$$

A well known example is the conserved Pauli-Lubanski vector

$$W_{\varkappa} = \sum_{\lambda,\mu,\nu} \varepsilon_{\varkappa\lambda\mu\nu} P^{\lambda} M^{\mu\nu} \in Q_1^{(2)}$$

$$\varkappa,\lambda,\mu,\nu = 0,1,2,3.$$

So far we - Buchholz and I - were not able to settle the problem whether the class of nonlocal charges $Q_1^{(2)}$ consists of products of P_{μ}, M_{ik} and of W_{μ} only. The work is in progress.

The case of $Q_D^{(2)}$, D > 2, has not yet been investigated by us.

Should it turn out that $Q^{(2)}$ for one scalar field consist only of the enveloping algebra of P_{μ} , M_{ik} and W_{μ} , this result - in my opinion - should not be considered trivial as it would merely indicate that our approach yields an elegant way to recover all the relevant constants of motion in a theory of one scalar field in which one does not intuitively expect any spectacular results. This result does not preclude that in theories with greater variety of fields there are hidden nonlocal charges which do not belong to the aforementioned enveloping algebra enlarged possibly by internal and spinorial charges and even do not commute with the S-matrix.

ACKNOWLEDGMENT

I wish to express my gratitude to I. Bialynicki-Birula, D. Buchholz and J.R. Klauder for valuable discussions and remarks.

REFERENCES

1. H. J. Borchers, Nuovo Cimento 15, 784 (1960); H. Araki and R. Haag, Comm.Math.Phys.4, 77 (1967).
2. D. Kestler, D.W. Robinson and A. Swieca, Comm.Math.Phys.2, 108 (1966); B. Schroer and J. Stichel, Comm.Math.Phys.3, 258 (1966).
3. W. D. Gerber and H. Reeh, J.Math.Phys.19, 59 (1978) J. T. Lopuszanski, Comm.Math.Phys.38, 317 (1974).
4. S. Ø. Aks, J.Math.Phys.6, 516 (1965); A. Martin, unpublished result.
5. K. Pohlmeyer, Comm.Math.Phys.46, 207 (1976); M. Lüscher and K. Pohlmeyer, Nucl. Phys.B137, 46 (1978); M. Lüscher, Nucl. Phys.B135, 1 (1978); A.B. Zamolodchikov and Al.B. Zamolodchikov, Nucl.Phys. B133, 525 (1978); C. H. Woo, preprint ZIF Bielefeld 1979.
6. K. G. Wilson and W. Zimmermann, Comm.Math.Phys.24, 87 (1972) and references quoted there.
7. D. Buchholz and J.T. Lopuszanski, Letters Math.Phys.3, 175 (1979).
8. D. Buchholz, Proceedings of the International Conference on Operator Algebra, Ideas and Their Application in Theoretical Physics, Leipzig, 1977.
9. H. J. Borchers, Nuovo Cimento 15 784 (1960).
10. J. T. Lopuszanski, Comm.Math.Phys.12, 80 (1969), 14, 158 (1969), Springer Tracts in Modern Physics Springer Verlag, Berlin-Heidelberg, 1970, Vol.52; L. J. Landau, Comm.Math.Phys.17, 156 (1970); K. Kraus and L. J. Landau, Comm.Math.Phys. 24, 243 (1972).
11. J. T. Lopuszanski and J. Lukierski, Reports Math. Phys.1, 265 (1971).
12. S. Coleman and J. Mandula, Phys.Rev. 159, 1251 (1967); J. T. Lopuszanski, J.Math. Phys. 12, 2401 (1971); W. D. Gerber and H. Reeh, preprint Göttingen, 1978.
13. R. Haag, J.T. Lopuszanski and M. Sohnius, Nucl.Phys. B88, 257 (1975).
14. C. A. Orzelesi, J. Sucher, C.H. Woo, Phys.Rev.Letters 21, 1550 (1958); C.A. Orzelesi, Revs.Mod.Phys.42, 381 (1970).

SEMICLASSICAL METHODS IN FIELD THEORIES

Shau-Jin Chang

Physics Department, University of Illinois at Urbana-Champaign
Urbana, Il. 61801

Most of the materials presented in these lectures are taken from a series of publications listed in Refs. 1-4. The first three lectures are dealing with generalized WKB methods. In the last lecture, we discuss the physical origin of the instability associated with a constant B field in a classical Yang-Mills field theory.

1. GENERALIZED WKB METHOD

The WKB method is well-known in a one-dimensional quantum system. However, there are some subtleties when one tries to generalize the WKB method to systems with more than one degrees of freedom [1,5]. In the following we shall start with a system of two degrees of freedom. Consider a system described by

$$L = \frac{1}{2} m (\dot{x}^2+\dot{y}^2) - V(x,y), \tag{1.1}$$

$$H = \frac{1}{2m} (p_x^2+p_y^2) + V(x,y). \tag{1.2}$$

The time independent Schrödinger equation gives

$$[- \frac{\hbar^2}{2m} (\frac{\partial^2}{\partial x^2} + \frac{\partial^2}{\partial y^2}) + V(x,y)]\psi(x,y) = E\psi(x,y). \tag{1.3}$$

In the WKB method, we expand

$$\psi = Ae^{iS/\hbar}, \quad S = S_0+\hbar S_1+\hbar^2 S_2+ \tag{1.4}$$

For a scattering problem E is usually given. For a bound state problem however, we should expand E as powers of $\hbar$ as well,

$$E = E_0 + \hbar E_1 + \ldots . \tag{1.5}$$

Substituting (1.4), (1.5) into (1.3) and equating the coefficients of $\hbar$, we have a set of equations

$$\frac{1}{2m}(\nabla S_0)^2 + V(x,y) = E_0, \tag{1.6}$$

$$-i\nabla^2 S_0 + 2\vec{\nabla} S_0 \vec{\nabla} S_1 = 2mE_1, \ldots \tag{1.7}$$

We shall concentrate on the lowest order eq. (1.6). Just as in optics, for every solution to (1.6) we can construct a set of rays (Fig. 1).

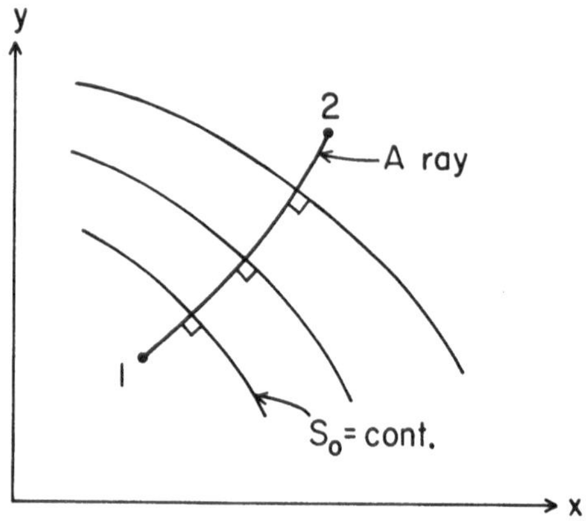

Figure 1 Construction of a ray from S_0 = constant contours.

Along a ray, the phase S_0 is stationary

$$\delta S_0 = \delta \int_1^2 ds \sqrt{2m(E-V)} = 0, \tag{1.8}$$

where $ds \equiv \sqrt{(dx)^2+(dy)^2}$ is the arc length along the path. It is easy to see that a ray obeys the equation for a classical trajectory,

$$m\ddot{x} = -\frac{\partial V}{\partial x}, \quad m\ddot{y} = -\frac{\partial V}{\partial y}. \tag{1.9}$$

For the lowest energy bound state (i.e. the ground state) we start at the minimum of $V(x,y)$. Suppose that $V(x,y)$ has a minimum at (x_0,y_0) with

$$V(x_0,y_0) = V_0, \text{ a minimum.} \tag{1.10}$$

It is trivial to see that the classical ground state energy is

$$E_0 = V_0, \tag{1.11}$$

and that for an arbitrary neighboring point (x,y)

$$V(x,y) > E_0. \tag{1.12}$$

We refer to regions with V>E as tunneling regions. In a tunneling region, S is imaginary, and we can write

$$\psi = Ae^{-R/\hbar}, \; R = R_0 + \hbar R_1 + \dots . \tag{1.13}$$

Equations (1.6) and (1.7) become

$$-\frac{1}{2m}(\nabla R_0)^2 + V(x,y) = E_0, \tag{1.14}$$

$$\nabla^2 R_0 - 2\vec{\nabla} R_0 \vec{\nabla} R_1 = 2mE_1, \dots . \tag{1.15}$$

In analogy to the ray representation, we can construct paths perpendicular to the family of curves R_0= constant. Along such a path,

$$R_0 = \int_1^2 ds \sqrt{2m(V-E_0)} \tag{1.16}$$

is stationary, and is often a minimum. These minimal-R_0 paths are referred to as "the most probable escape (tunneling) paths" (MPEP) [5].

We can show from the variational principle $\delta R_0 = 0$ that a most probable escape path $x = x(\tau)$, $y = y(\tau)$ obeys the Euclidean equation of motion [6],

$$m\frac{d^2x}{d\tau^2} = \frac{\partial V}{\partial x}, \; m\frac{d^2y}{d\tau^2} = \frac{\partial V}{\partial y} \tag{1.17}$$

where τ is an appropriate parametrization which may be naively identified as the "Euclidean time".

Now we are in a position to determine the WKB ground state wave function. Coleman [6] pointed out that Eq.(1.17) provides the true equation of motion for a related system with an "inverted potential" $V_I \equiv -V(x,y)$, and a total energy $E_I \equiv -E_0$, (See Fig.2)

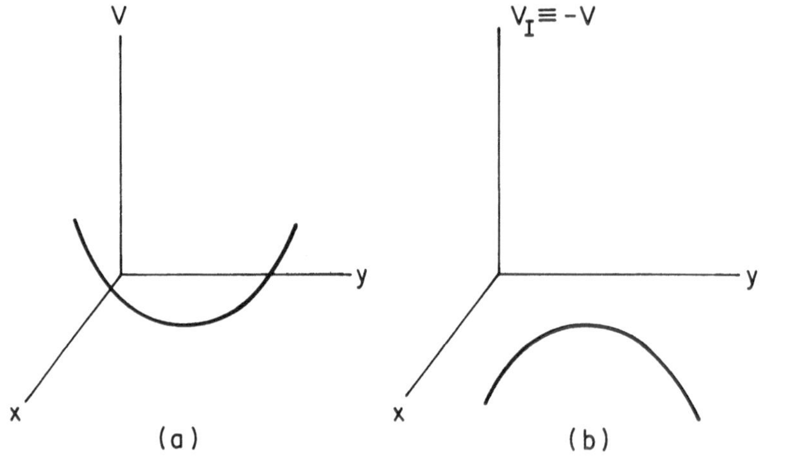

Figure 2 a) A typical potential $V(x,y)$
b) The inverted potential $V_I(x,y) \equiv - V(x,y)$

In this related problem, V_I has a local maximum at (x_o,y_o). The MPEPs in the original problem become the classical trajectories in this related problem. For $E_I = V_I(x_o,y_o) = - E_o$, the particle rests on the top of the potential V_I. To find the MPEP between (x_o,y_o) and a neighboring pont (x,y), we let the particle move from (x_o,y_o) to (x,y) under the influence of V_I (as illustrated in Fig. 3).

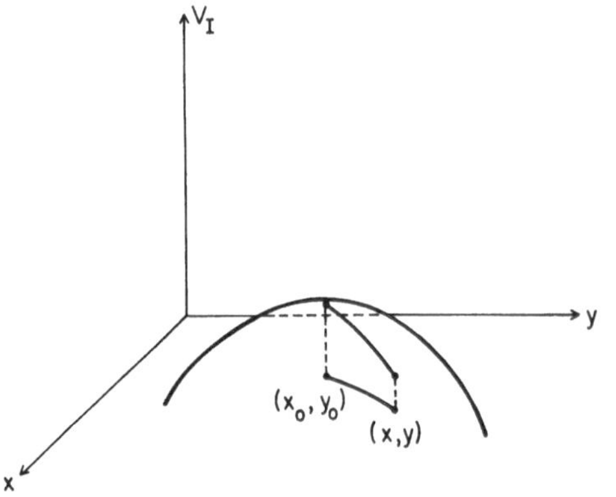

Figure 3 The MPEP between (x_o,y_o) and (x,y) emerges as a classical trajectory under the influence of V_I.

Since the particle is at rest at a local maximum, we can select an arbitrary initial direction of motion by giving the particle an infinitesimal push in the chosen direction. By a proper adjustment of the initial direction, we can make the particle pass through any given neighboring point. Hence, the MPEP always exists between (x_O, y_O) and an arbitrary neighboring point (x,y) at $E = E_O$. The WKB wave function at (x,y) is then given by

$$\psi(x,y) = A\exp\left[- \int_{(x_O,y_O)}^{(x,y)} ds \sqrt{2m(V-E_O)} \right]_{MPEP} . \tag{1.18}$$

A remark is now in order. I wish to point out that we can obtain the WKB exponent R_O for any path without solving (1.14). To obtain R_O for a given path, we can start with any parametrization $x = x(\lambda), y = y(\lambda)$, and substitute it into L to give an effective one-dimensional Lagrangian

$$L_{eff} = \frac{1}{2} m(\lambda)\dot{\lambda}^2 - V(\lambda) \tag{1.19}$$

with

$$m(\lambda) = m\left[\left(\frac{dx}{d\lambda}\right)^2 + \left(\frac{dy}{d\lambda}\right)^2 \right] \tag{1.20}$$

$$V(\lambda) \equiv V(x(\lambda), y(\lambda)). \tag{1.21}$$

We can obtain R_O from this L_{eff} trivially as

$$R_O = \int_{\lambda_1}^{\lambda_2} d\lambda \sqrt{2m(\lambda)(V(\lambda) - E_O)}$$

$$= \int_1^2 ds \sqrt{2m(V-E_O)} \tag{1.22}$$

as required. We can now determine the MPEP by imposing $\delta R_O = 0$. This is a rather trivial observation. But it is a useful observation when we try to generalize our results to systems with many degrees of freedom.

Let us generalize our results to a system with n degrees of freedom,

$$L = \frac{1}{2} m \dot{\vec{r}}^2 - V(\vec{r}) \tag{1.23}$$

with

$$\vec{r} = (x_1, x_2, \ldots x_n). \tag{1.24}$$

We assume that V has a minimum at $\vec{r}_0$. We can describe a path between $\vec{r}_0$ and $\vec{r}_1$ as a one-parameter set of equations

$$x_i = x_i(\lambda), \ \lambda_0 < \lambda < \lambda_1 \tag{1.25}$$

such that

$$\vec{r}(\lambda_0) = \vec{r}_0, \ \vec{r}(\lambda_1) = \vec{r}_1. \tag{1.26}$$

Along this path, we have an effective Lagrangian,

$$L_{eff} = \frac{1}{2} m(\lambda)\dot{\lambda}^2 - V(\lambda) \tag{1.27}$$

with

$$m(\lambda) = m \left(\frac{d\vec{r}}{d\lambda}\right)^2 \tag{1.28}$$

$$V(\lambda) = V(\vec{r}(\lambda)). \tag{1.29}$$

The WKB wave function along the path is

$$\psi = Ae^{-R/\hbar}, \ R = R_0 + \hbar R_1 + \ldots \tag{1.30}$$

with

$$R_0 = \int_{\lambda_0}^{\lambda_1} d\lambda \sqrt{2m(\lambda)(V(\lambda) - E_0)} \tag{1.31a}$$

$$= \int_{\vec{r}_0}^{\vec{r}_1} ds \sqrt{2m(V - E_0)}. \tag{1.31b}$$

The condition for the path to be a MPEP, $\delta R_0 = 0$, gives rise to the Euclidean equations of motion

$$\frac{d^2\vec{r}}{d\tau^2} = \frac{\partial V}{\partial \vec{r}} \tag{1.32}$$

where τ is an appropriate parametrization. The ground state WKB wave function can now be written compactly as

$$\psi(\vec{r}) = A \exp \left[- \int_{\vec{r}_0}^{\vec{r}} ds \sqrt{2m(V-E_0)}\right]_{MPEP}. \tag{1.33}$$

We now reduce the task of finding a WKB wave function to solving a set of Euclidean equations (1.32).

We can now generalize our result to quantum field theories [1]. Consider a scalar field theory

$$\mathcal{L} = \frac{1}{2}\dot{\phi}^2 - \frac{1}{2}(\vec{\nabla}\phi)^2 - U(\phi) \tag{1.34}$$

$$\mathcal{H} = \frac{1}{2}\pi^2 + \frac{1}{2}(\vec{\nabla}\phi)^2 + U(\phi) \tag{1.35}$$

where $\pi \equiv \dot{\phi}$ is the momentum conjugate to $\phi(x)$. The Schrödinger equation now becomes a functional differential equation

$$\int dx\left[-\frac{1}{2}\hbar^2\frac{\delta^2}{\delta\phi^2} + \frac{1}{2}(\vec{\nabla}\phi)^2 + U(\phi)\right]\psi(\phi) = E\psi(\phi). \tag{1.36}$$

A path in the function space joining two given configurations ϕ_1 and ϕ_2 is a one-parameter family of field configuration,

$$\phi(x,\lambda),\ \lambda_1 \leqq \lambda \leqq \lambda_2,$$

such that

$$\phi(\vec{x},\lambda_1) = \phi_1(\vec{x}),\quad \phi(\vec{x},\lambda_2) = \phi_2(\vec{x}). \tag{1.37}$$

Along this path, the effective Lagrangian takes the same form as (1.27) with

$$m(\lambda) = \int d^3x \left(\frac{\partial\phi}{\partial\lambda}\right)^2 \tag{1.38}$$

$$V(\lambda) = \int d^3x \left[\frac{1}{2}(\nabla\phi)^2 + U(\phi)\right]. \tag{1.39}$$

The WKB exponent obeys (1.31a) as well. The condition of MPEP, $\delta R_0 = 0$, implies that $\phi(\vec{x},\lambda)$ obeys the Euclidean field equation

$$\left(\frac{\partial^2}{\partial\tau^2} + \nabla^2\right)\phi(\vec{x},\tau) - U'(\phi) = 0 \tag{1.40}$$

with an appropriate parametrization $\tau = \tau(\lambda)$.

If ϕ_0 is a local minimum of $U(\phi)$ and $\phi(x)$ is a neighboring configuration then, the ground state WKB wave function around ϕ_0 is

$$\psi(\phi) = A\exp\left[-\int_{\phi_0}^{\phi} d\lambda\sqrt{2m(\lambda)\,(V(\lambda) - E_0)}\right]_{MPEP.} \tag{1.41}$$

In the third lecture I shall work out some $\psi(\phi)$'s explicitly.

2. TUNNELING IN SYSTEMS WITH MANY DEGREES OF FREEDOM

When a system of many degrees of freedom has two or more minima of equal energy at $\vec{r}_1, \vec{r}_2, \ldots,$, we can construct WKB wave functions from each of the minima, say, $\psi_1(x)$ from $\vec{r}_1$, $\psi_2(x)$ from $\vec{r}_2$, etc. The true WKB ground state wave function is a linear combination of ψ_i's,

$$\psi(x) = \Sigma\, C_i\, \psi_i(x). \tag{2.1}$$

Usually, the wave function ψ_1 constructed from $\vec{r}_1$ has non-vanishing values at the locations of other minima, $\vec{r}_2, \vec{r}_3$, etc. The overlappings of the ψ_i's give rise to the tunneling rates among these minima. The coefficients C_i's can be determined by diagonalizing the matrix generated from the ψ's.

The tunneling amplitude between WKB states located at $\vec{r}_1$ and $\vec{r}_2$ can be computed once the MPEP between $\vec{r}_1$ and $\vec{r}_2$ is known. It is given by

$$\text{Tunneling amplitude} = e^{-R_o/\hbar} \tag{2.2}$$

with

$$R_o = \left[\int_{\vec{r}_1}^{\vec{r}_2} ds \sqrt{2m(V-E_o)}\,\right]_{MPEP}. \tag{2.3}$$

In the following, I shall demonstrate the tunnelings in two examples.

We begin with a simple system [2]. Consider an ammonia molecule shown in Fig. 4

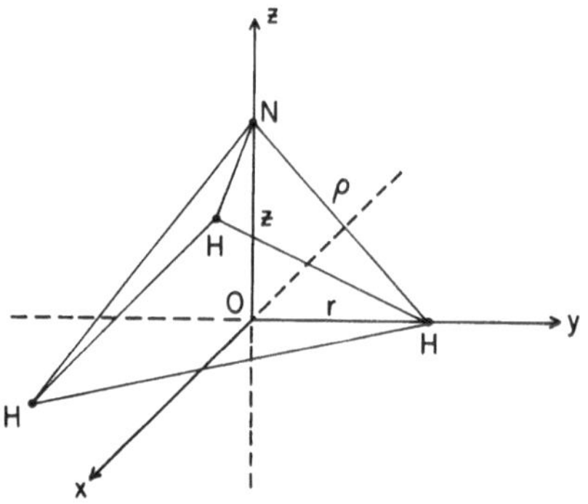

Figure 4 An ammonia molecule

The molecule consists of three hydrogen atoms H located at the vertices of an equilateral triangle and a nitrogen atom N located at a distance z from the hydrogen plane and at equal distances ρ from the hydrogens. For simplicity, we assume that the forces among hydrogens and between a hydrogen and the nitrogen are all harmonic forces. Let r $(\equiv\sqrt{\rho^2 - z^2})$ be the distance of hydrogen from their common center O. Let r = a and z = b be the equilibrium distances. The kinetic and potential energies of this symmetric configuration can be written as

$$T = \frac{1}{2} M \dot{r}^2 + \frac{1}{2} m \dot{z}^2 \tag{2.4}$$

$$V = \frac{1}{2} k_1 (r-a)^2 + \frac{1}{2} k_2 (\rho-c)^2 \tag{2.5}$$

where $c \equiv \sqrt{a^2 + b^2}$ is the equilibrium distance between a hydrogen and the nitrogen. M is the total mass of hydrogen atoms, m is the reduced mass of nitrogen relative to hydrogens, and k_1, k_2 are spring constants associated with the coordinates r and ρ.

In Fig. 5, we study the potential V(r,z) as a function of both r and z.

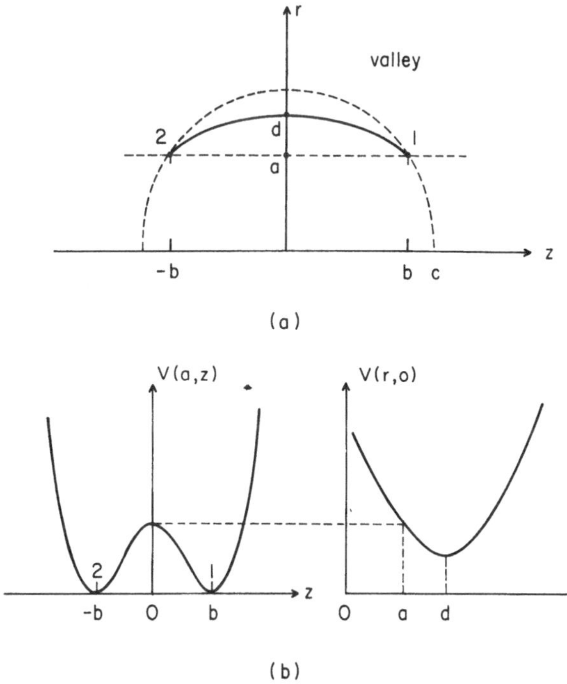

Figure 5 a) Potential V(r,z) defined over the rz plane. Points 1 and 2 are locations of the ground states (minima).
b) The potential evaluated along the straight path r = a and along the r-axis

The potential V has two minima. One is at $r = a$, $z = b$, and is denoted by 1 in Fig. 5. The other is at $r = a$, $z = -b$, and is denoted by 2 in Fig. 5. The first term of V in Eq.(2.5) is zero at the straight line $r = a$, and the second term of V is zero at the circle $r^2 + z^2 (\equiv \rho^2) = c^2$. The minimum of V as we vary z continuous from -b to b follows the "valley" as shown in Fig. 5a. Fig. 5b provides V for a straight path $r = a$, $(V(a,z))$, and along the r-axis $(V(r,o))$. Note that the saddle point is at

$$r = d \equiv \frac{k_1 a + k_2 c}{k_1 + k_2}, \quad z = 0.$$

Physically, the valley corresponds to the minimum energy configuration when we move atom N by force slowly from point 1 to point 2. The hydrogen atoms get pushed away as atom N passes through the origin. We would like to know what is the most probable tunneling path as atom N tunnels from point 1 to point 2.

In particular, do the hydrogen atoms move outward simultaneously during the tunneling?

We can determine the MPEP by solving the classical trajectory for the inverted potential $-V(r,z)$.

In Fig.6 we find that the original minima 1 and 2 now become local maxima. The valley becomes a ridge. It is easy to convince oneself that the classical trajectory joining points 1 and 2 lies between the straight path $r = a$ and the ridge. Its precise trajectory is determined by the balance of the "gravitational" force, $-\vec{\nabla} V$, and the centrifugal force. Thus, the hydrogen atoms will move outward as the nitrogen atom tunnels through, but they do not expand as much as that required by minimizing $V(r,z)$ alone.

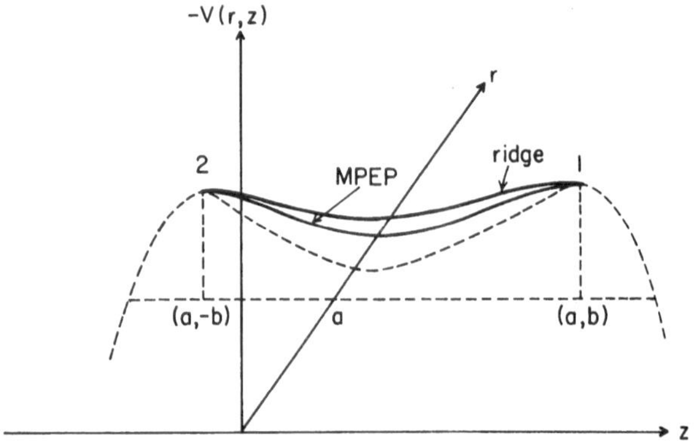

Figure 6 The "inverted potential" - V(r,z). The original minima now become two peaks at positions 1 and 2.

As a second example, we examine vacuum tunnelings in a Yang-Mills field theory [1,7]. A Yang-Mills field theory is described by

$$\mathcal{L} = -\frac{1}{2g^2}\,\mathrm{Tr}(F_{\mu\nu}^{\;2}) = \frac{1}{g^2}\,\mathrm{Tr}(\vec{E}^2 - \vec{B}^2) \tag{2.6}$$

$$\mathcal{H} = \frac{1}{g^2}\,\mathrm{Tr}(\vec{E}^2 + \vec{B}^2) \tag{2.7}$$

where

$$F_{\mu\nu} \equiv \partial_\mu A_\nu - \partial_\nu A_\mu + \frac{1}{i}[A_\mu, A_\nu] \tag{2.8}$$

$$E_i \equiv F_{oi},\quad B_i \equiv \frac{1}{2}\varepsilon_{ijk}\,F_{jk} \tag{2.9}$$

In Eqs.(2.6) - (2.9), $F_{\mu\nu}$ and A_μ are 2 x 2 matrices. They are related to the field tensor $F_{\mu\nu}^{(a)}$ and vector potential $A_\mu^{(a)}$ through

$$F_{\mu\nu} = \frac{1}{2}\,g\,F_{\mu\nu}^{(a)}\,\tau^{(a)} \tag{2.10}$$

$$A_\mu = \frac{1}{2}\,g\,A_\mu^{(a)}\,\tau^{(a)} \tag{2.11}$$

where $\tau^{(a)}$ are 2 x 2 Pauli matrices, and a = 1,2,3 stands for color index.

I shall remind you about the following important facts: (1) The classical Yang-Mills ground states are given by pure gauges

$$A = iU^{-1}\partial_\mu U \tag{2.12}$$

Eq. (2.12) implies that

$$F_{\mu\nu} = 0. \tag{2.13}$$

(2) There are topologically different ground states which can be characterized by different winding numbers. In the temporal gauge $A^o = 0$, we can express the winding number as

$$Q = \frac{1}{24\pi^2 i}\int d^3x\ \mathrm{Tr}\,(\vec{A}\mathrm{x}\vec{A}\cdot\vec{A}). \tag{2.14}$$

The change of winding numbers between two spacelike surfaces σ_1, σ_2 is,

$$\Delta Q = Q(\sigma_2) - Q(\sigma_1) = \frac{1}{4\pi^2} \text{ Tr} \int_{\sigma_1}^{\sigma_2} d^4x (\vec{E}\cdot\vec{B}). \tag{2.15}$$

To apply the WKB method, we start with the construction of a tunneling path in function space. We shall use the temporal gauge $A^0 = 0$. Let $\vec{A}^{(1)}(\vec{x})$ and $\vec{A}^{(2)}(\vec{x})$ be two ground state configurations. A path in the function space joining $\vec{A}^{(1)}$ and $\vec{A}^{(2)}$ is a family of field configurations $\vec{A}(\vec{x},\lambda), \lambda_1 \leq \lambda \leq \lambda_2$, such that

$$\vec{A}(\vec{x},\lambda_1) = \vec{A}^{(1)}(\vec{x}), \; \vec{A}(\vec{x},\lambda_2) = \vec{A}^{(2)}(\vec{x}). \tag{2.16}$$

Along such a path and treating λ as a dynamical variable, we have

$$E_i = \frac{\partial Ai}{\partial \lambda} \dot{\lambda} \tag{2.17}$$

$$B_i = \frac{1}{2} \varepsilon_{ijk} (\partial_j A_k - \partial_k A_j + \frac{1}{i} [A_j, A_k]). \tag{2.18}$$

The effective one-dimensional Lagrangian is

$$L_{eff} = \frac{1}{2} m(\lambda) \dot{\lambda}^2 - V(\lambda). \tag{2.19}$$

with

$$m(\lambda) \equiv \frac{2}{g^2} \int d^3x \text{ Tr } (\frac{\partial \vec{A}}{\partial \lambda})^2 \tag{2.20}$$

$$V(\lambda) \equiv \frac{1}{g^2} \int d^3x \text{ Tr } (\vec{B}^2). \tag{2.21}$$

The WKB wave function becomes $\psi = Ae^{-R_0/\hbar}$ with

$$R_0 = \int_{\lambda_1}^{\lambda_2} d\lambda \sqrt{2m(\lambda)(V(\lambda) - E_0)}. \tag{2.22}$$

The condition that the path be a MPEP is once again given by $\delta R_0 = 0$. With an appropriate parametrization $\tau = \tau(\lambda)$, it is straightforward to show that $A_\mu(\vec{x},\tau)$ obeys the Euclidean field equations [1]

$$F_{\mu\nu} = i\,[D_\mu, D_\nu] = \partial_\mu A_\nu - \partial_\nu A_\mu + \frac{1}{i}\,[A_\mu, A_\nu] \tag{2.23}$$

$$[D_\mu, F_{\mu\nu}] = 0 \tag{2.24}$$

with

$$D_\mu \equiv \partial_\mu + \frac{1}{i} A_\mu \tag{2.25}$$

$$\mu = (1,2,3,\tau). \tag{2.26}$$

In fact, one can show that if R_0 is a minimum, $F_{\mu\nu}$ must be either self-dual (instanton) or anti-self-dual (anti-instanton). The inverse is also true: From every instanton or anti-instanton solution, we can construct a MPEP in function space such that the vacuum tunneling is maximal.

There are several advantages of using WKB method to describe vacuum tunnelings. Among them are: (1) It provides us with most probable intermediate field configurations in Minkowski-space during the tunnelling; (2) It presents us with an explicit potential barrier $V(\lambda)$ through which the tunneling is taking place; and (3) It gives rise to a probability amplitude $\psi(\lambda)$ for intermediate field configurations associated with non-integer winding numbers.

To give an explicit example, we choose the $Q = 1$ instanton solution and obtain the MPEP as

$$A_0 = \frac{\vec{x}\cdot\vec{\tau}}{\vec{x}^2+\lambda^2+a^2}\,\dot{\lambda} \tag{2.27}$$

$$\vec{A} = \frac{\lambda\vec{\tau}+\vec{x}\times\vec{\tau}}{\vec{x}^2+\lambda^2+a^2}\,. \tag{2.28}$$

In (2.27) and (2.28), $\vec{\tau}$ stands for Pauli matrices, and $\underline{a}$ denotes the size of the instanton. From (2.27), (2.28), we obtain

$$\vec{E} = -\frac{2a^2}{(\vec{x}^2+\lambda^2+a^2)^2}\,\vec{\tau}\,\dot{\lambda} \tag{2.29}$$

$$\vec{B} = -\frac{2a^2}{(\vec{x}^2+\lambda^2+a^2)^2}\,\vec{\tau}\,. \tag{2.30}$$

One can check easily that the winding number indeed changes by one,

$$\Delta Q = \frac{1}{4\pi^2} \mathrm{Tr} \int d^4x \, \vec{E}\cdot\vec{B} = 1. \tag{2.31}$$

The effective mass and potential are

$$m(\lambda) = \frac{6\pi^2 a^4}{g^2 (\lambda^2+a^2)^{5/2}} \tag{2.32}$$

$$V(\lambda) = \frac{3\pi^2 a^4}{g^2 (\lambda^2+a^2)^{5/2}}. \tag{2.33}$$

Even though we can describe the tunneling in terms of the variable λ, it is more convenient to introduce a winding number density variable $q(\lambda)$ through

$$m(\lambda)\, d\lambda^2 = m\, dq^2 \quad , \ m = \text{constant} \tag{2.34}$$

together with the normalization conditions

$$q(-\infty) = 0 \ , \ q(\infty) = 1. \tag{2.35}$$

Eqs. (2.34) and (2.35) determine $q(\lambda)$ and m uniquely,

$$q(\lambda) = \frac{\Gamma(5/4)}{\Gamma(1/2)\Gamma(3/4)} \int_{-\infty}^{\lambda/a} \frac{dt}{(t^2+1)^{5/4}} \tag{2.36}$$

$$m = \frac{6\pi^2 a}{g^2} \left[\frac{\Gamma(1/2)\Gamma(3/4)}{\Gamma(5/4)} \right]^2$$

$$= 340.037 \ a/g^2. \tag{2.37}$$

In Fig. 7, we plot the barrier potential $V(q)$ as a function of q. The above results can be extended to describe multi-instanton solutions.

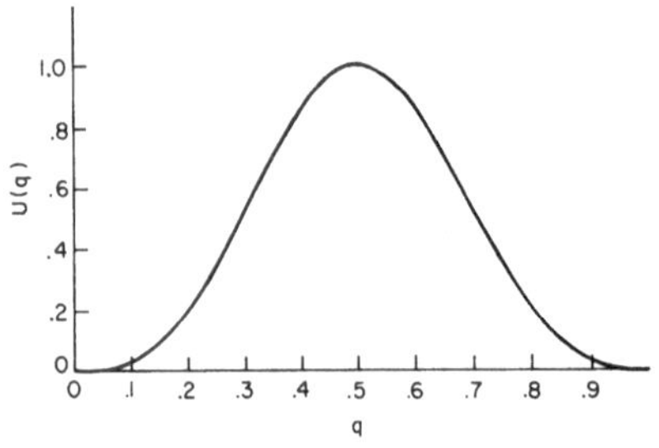

Figure 7 Normalized potential $U(q) \equiv V(q)/V(o)$ as a function of the winding number q.

3. GROUND STATE WAVE FUNCTIONALS

In lecture 1, we have developed a method to construct the WKB wave functional in the neighborhood of a local minimum. In this lecture, I shall work out some simple example to illustrate the method in detail.

Example 1. Free Field

Consider a free scalar field described by

$$\mathcal{L} = \frac{1}{2} (\partial_\mu \phi)^2 - \frac{1}{2} m^2 \phi^2 \tag{3.1}$$

$$\mathcal{H} = \frac{1}{2} \dot{\phi}^2 + \frac{1}{2} (\nabla\phi)^2 + \frac{1}{2} m^2 \phi^2. \tag{3.2}$$

The classical ground state is at $\phi = 0$ with the energy $E_0 = 0$. Let $\phi(\mathbf{x})$ be an arbitrary field configuration, and let $\phi(\vec{x},\tau)$ be the MPEP joining $\phi = 0$ and $\phi(\vec{x})$. This MPEP $\phi(\vec{x},\tau)$ obeys the Euclidean equation

$$(-\frac{\partial^2}{\partial\tau^2} - \nabla^2)\,\phi + m^2\,\phi = 0 \tag{3.3}$$

with the boundary conditions

$$\phi(\vec{x},-\infty) = 0 \tag{3.4}$$

$$\phi(\vec{x},0) = \phi(\vec{x}) \tag{3.5}$$

The choice of $\tau = -\infty$ as the initial boundary point (3.4) is a consequence of $E_0 = 0$, and the choice of $\tau = 0$ as the final boundary point (3.5) is pureley for convenience. Using the Fourier representation,

$$\phi(x,\tau) = \int \frac{d^3k}{(2\pi)^3}\, e^{i\vec{k}\vec{x}}\, \tilde{\phi}(\vec{k},\tau), \tag{3.6}$$

we find that $\tilde{\phi}(\vec{k},\tau)$ obeys

$$(-\frac{\partial^2}{\partial\tau^2} + k^2 + m^2)\,\tilde{\phi}(\vec{k},\tau) = 0. \tag{3.7}$$

Equation (3.7), together with the boundary condition (3.4), gives

$$\tilde{\phi}(\vec{k},\tau) = e^{\sqrt{k^2+m^2}\,\tau}\,\tilde{\phi}(\vec{k}). \tag{3.8}$$

Hence, we have

$$\phi(\vec{x},\tau) = \int \frac{d^3k}{(2\pi)^3} e^{\sqrt{k^2+m^2}\,\tau} e^{i\vec{k}\vec{x}} \tilde{\phi}(\vec{k}). \tag{3.9}$$

We can obtain $\tilde{\phi}(\vec{k})$ from $\phi(\vec{x})$ by the boundary condition at $\tau = 0$, giving

$$\phi(\vec{x}) = \int \frac{d^3k}{(2\pi)^3} e^{i\vec{k}\vec{x}} \tilde{\phi}(\vec{k}). \tag{3.10}$$

Equations (3.9) and (3.10) describe the MPEP explicitly. We can now compute $m(\tau)$ and $V(\tau)$ along the MPEP. They are

$$m(\tau) \equiv \int d^3x \left(\frac{\partial\phi}{\partial\tau}\right)^2$$

$$= \int \frac{d^3k}{(2\pi)^3} (k^2+m^2) e^{2\sqrt{k^2+m^2}\,\tau} (\tilde{\phi}(\vec{k}))^2 \tag{3.11}$$

$$V(\tau) = \int d^3x \left[\frac{1}{2}\left(\frac{\partial\phi}{\partial x}\right)^2 + \frac{1}{2} m^2\phi^2\right] = \frac{1}{2} m(\tau). \tag{3.12}$$

Hence, the WKB exponent is

$$R_0 = \int_{-\infty}^{0} d\tau \sqrt{2m(\tau)\ V(\tau)}$$

$$= \int_{-\infty}^{0} d\tau \int \frac{d^3k}{(2\pi)^3} (k^2+m^2)\ e^{2\sqrt{k^2+m^2}\,\tau} (\tilde{\phi}(\vec{k}))^2$$

$$= \int \frac{d^3k}{(2\pi)^3 2\omega} (k^2+m^2) (\tilde{\phi}(k))^2 \tag{3.13}$$

with

$$\omega \equiv \sqrt{k^2+m^2}. \tag{3.14}$$

Example 2. Kink - antikink state

Consider a two-dimensionel ϕ^4 theory described by

$$\mathcal{L} = \frac{1}{2}\dot{\phi}^2 - \frac{1}{2}\left(\frac{\partial\phi}{\partial x}\right)^2 - \frac{1}{4} g(\phi^2-c^2)^2 \tag{3.15}$$

$$\mathcal{H} = \frac{1}{2}\dot{\phi}^2 + \frac{1}{2}\left(\frac{\partial\phi}{\partial x}\right)^2 + \frac{1}{4}g(\phi^2 - c^2)^2. \tag{3.16}$$

The classical ground states are $\phi = c$ and $\phi = -c$.

We wish to study the WKB amplitude for producing a kink-antikink pair located at $\pm$ a, (See Fig.8), giving

$$\phi(x) = c\left[\tanh\frac{\mu}{2}(x+a) - \tanh\frac{\mu}{2}(x-a) - 1\right] \tag{3.17}$$

with

$$\mu \equiv \sqrt{2gc^2} \tag{3.18}$$

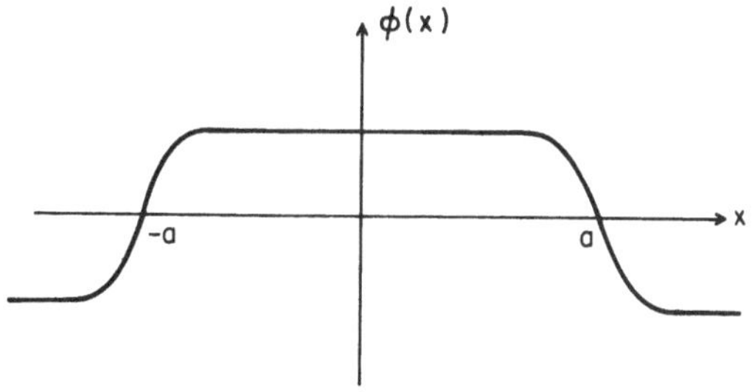

Figure 8 A kink-antikink configuration

For simplicity, we assume $\mu a \gg 1$ so that the kink and antikink are widely separated.

For general μ and a, we do not have an exact solution to the MPEP. However, for $\mu a \gg 1$, we can first construct a trial family of configurations (path) which is very instructive. The trial path that we wish to study is a circular symmetric configuration in $x\tau$ plane as shown in Fig. 9.

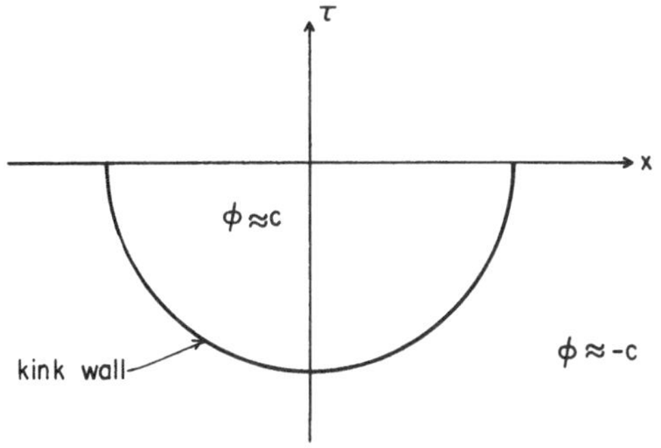

Figure 9 A "bounce" configuration. This configuration does not correspond to a MPEP.

It is made of a domain of $\phi = c$ region imbedded in the $\phi = -c$ region. The two regions are separated by a semi-circular kink wall. This configuration is an analog of the "bounce" solution studied by S. Coleman in the decay of false vacuum [6]. We can express this trial path, in the region $-\infty < \tau < 0$, as

$$\phi(x,\tau) = -c \tanh \frac{\mu}{2} (r-a) \tag{3.19}$$

with

$$r \equiv \sqrt{x^2+\tau^2} . \tag{3.20}$$

The WKB wave functional for the kink-antikink configuration along this path is $\psi = A\, e^{-R_0/\hbar}$ with

$$R_0 = \int_{-\infty}^{0} d\tau \sqrt{2m(\tau)V(\tau)}$$

$$\approx \int_{-\infty}^{0} d\tau \int dx \left[\frac{1}{2}\left(\frac{\partial\phi}{\partial\tau}\right)^2 + \frac{1}{2}\left(\frac{\partial\phi}{\partial x}\right)^2 + \frac{1}{4} g \left(\phi^2-c^2\right)^2\right]$$

$$= \int_{0}^{\infty} r dr \int_{\pi}^{2\pi} d\Theta \left[\frac{1}{2}\left(\frac{\partial\phi}{\partial r}\right)^2 + \frac{1}{4} g\left(\phi^2-c^2\right)^2\right]$$

$$\approx a\, \pi\, M \tag{3.21}$$

where

$$M \equiv \int dr \left[\frac{1}{2}\left(\frac{\partial\phi}{\partial r}\right)^2 + \frac{1}{4} g \left(\phi^2-c^2\right)^2\right] \tag{3.22}$$

is the kink mass. Note that R_0 is given approximately by

$$R_0 = M\times(\text{length of the kink wall}). \tag{3.23}$$

This is a general result, and is true as long as $\phi(x,\tau)$ obeys the Euclidean equation approximately. Now, we are in a position to guess the true MPEP. It is given in Fig. 10.

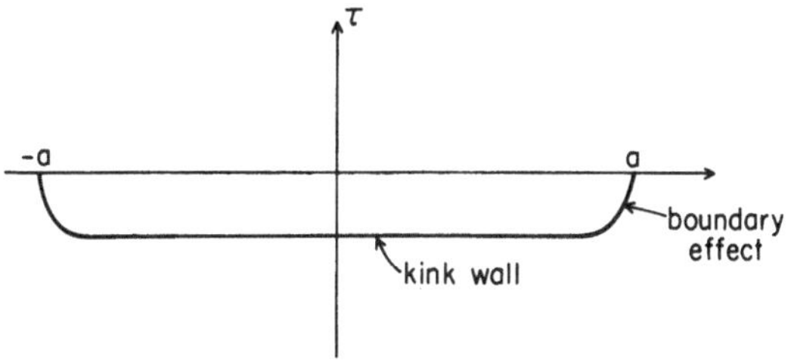

Figure 10 Configuration which corresponds to the MPEP leading to a kink-antikink pair.

Except for small boundary effects near $x = \pm a$, the kink wall follows a horizontal line between -a and a with a total length $L \approx 2a$. This configuration provides a much larger WKB amplitude $e^{-R_0/\hbar}$ with

$$R_0(\text{MPEP}) \approx 2\, a\, M. \tag{3.24}$$

In Fig. 11 we plot the intermediate tunneling configurations associated with this MPEP.

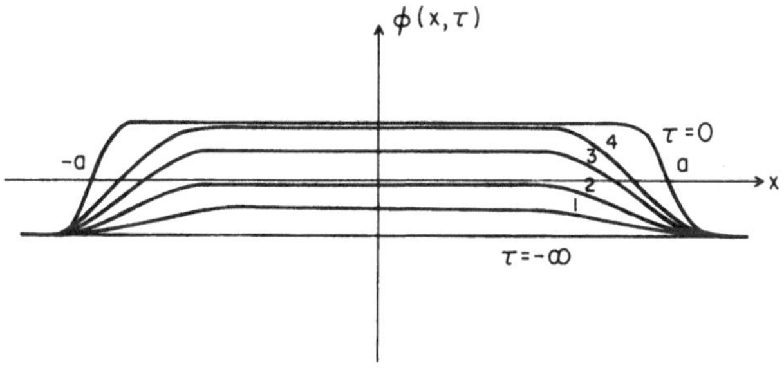

Figure 11 The most probable intermediate configurations which give rise to a virtual kink-antikink pair.

Example 3. Fluctuations in xy model

We shall estimate the WKB amplitude for a certain class of fluctuations in xy model which are special cases of spin waves [3]. The xy model can be expressed as

$$\mathcal{L} = \frac{1}{2g^2} \left[\left(\frac{\partial \vec{\sigma}}{\partial t}\right)^2 - \left(\frac{\partial \vec{\sigma}}{\partial x}\right)^2 \right] \tag{3.25}$$

with

$$\vec{\sigma}^2 \equiv \sigma_1^2 + \sigma_2^2 = 1. \tag{3.26}$$

The classical ground state is $\vec{\sigma}$ = const., independent of x and t. There is no analog of kink and antikink in xy model due to the continuous nature of the symmetry. However, we can consider field fluctuations which change spin directions. Introducing polar coordinates

$$\sigma_1 = \sin \Theta$$

$$\sigma_2 = \cos \Theta , \tag{3.27}$$

we have

$$\mathcal{L} = \frac{1}{2g^2} (\partial_\mu \Theta)^2 = \frac{1}{2g^2} \left[\left(\frac{\partial \Theta}{\partial t}\right)^2 - \left(\frac{\partial \Theta}{\partial x}\right)^2 \right]. \tag{3.28}$$

Now, we wish to find the WKB wave functional joining the classical ground state $\Theta = 0$ to the following configuration,

$$\Theta = 0, \quad \text{for} \quad |x| > b \tag{3.29}$$

$$\Theta = \pi, \quad \text{for} \quad |x| < a \tag{3.30}$$

where $a < b$. The most probable tunneling path which links $\Theta = 0$ at $\tau = -\infty$ to Θ given in (3.29) - (3.30) is

$$\Theta = 0 \qquad r > b$$

$$\Theta = \frac{\pi}{\ln\frac{b}{a}} \ln \frac{b}{r} \qquad a \leq r \leq b$$

$$\Theta = \pi \qquad r < a \tag{3.31}$$

with

$$r = \sqrt{x^2 + \tau^2}. \tag{3.32}$$

This MPEP is shown in Fig. 12. It is easy to verify that Θ in (3.31) obeys the Euclidean eq. $\partial^2 \Theta = 0$. From (3.31), we can compute the WKB wave functional as $\psi = Ae^{-R_0/\hbar}$ with

$$R_o = \frac{\pi^3}{2g^2} \frac{1}{\ln\frac{b}{a}} . \tag{3.33}$$

R_o is a function of b/a only. It does not damp out as a and b increase if b/a is fixed. Thus, both large and small size fluctuations are equally probable.

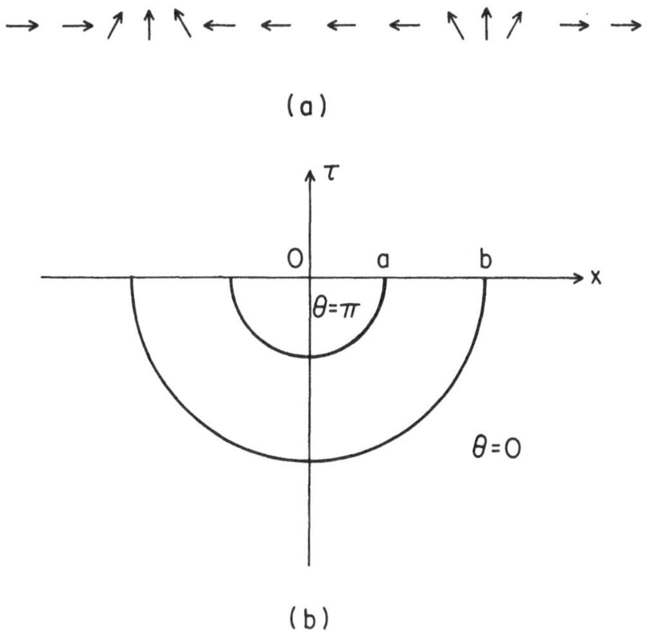

Figure 12 (a) The field configuration described in Eqs.(3.29) (3.30); (b) The MPEP associated with the configuration defined in (a). In this configuration, Θ is a function of $r \equiv \sqrt{x^2+\tau^2}$ only. The explicit r dependence of Θ(r) is given in (3.31).

We shall refer to such a configuration as a "patch", and parametrize it by the position x of the center of the patch and the radii a and b. We denote it as a state $\langle x,a,b|$. The ground state wave functional for such a configuration is

$$\langle x,a,b|\psi\rangle \approx e^{-\xi/2} \tag{3.34}$$

with

$$\xi \equiv \frac{\pi^3}{g^2} \frac{1}{\ln\frac{b}{a}} . \tag{3.35}$$

We can also introduce a similar configuration which changes Θ from 0 to $-\pi$, and shall refer to such a configuration as an "antipatch". The WKB wave functional for an anti-patch is equal to that of a patch. In the following I shall concentrate on the contribution due to independent patches. The generalization to include anti-patches is straightforward.

A two-patch configuration is defined by

$$\Theta(x) = \Theta_1(x) + \Theta_2(x) \tag{3.36}$$

where Θ_1 and Θ_2 are angles associated with patch 1 at x_1 with radii a,b, and patch 2 at x_2 with radii a_2,b_2. We denote this two-patch configuration by $<x_1,a_1,b_1;x_2,a_2,b_2|$ and obtain the ground state wave functional

$$<x_1,a_1,b_1;x_2,a_2,b_2|\psi> = e^{-\xi_1/2}\, e^{-\xi_2/2} \tag{3.37}$$

where ξ_1,ξ_2 are ξ's defined relative to (a_1,b_1) and (a_2,b_2).

We can generalize the above construction to n-patch (or anti-patch) configurations straightforwardly.

From these multipatch (or antipatch) configurations we can construct a trial ground state,

$$|\psi> = N \sum_{n=0}^{\infty} \sum_{\text{size}} \int \frac{dx_1}{a_1} \cdots \frac{dx_n}{a_n} e^{-\Sigma\, \xi_i/2}$$

$$\times |x_1,a_1,b_1;x_2,a_2,b_2;\ \ldots;\ x_n,a_n,b_n> . \tag{3.38}$$

We can use this trial ground state to evaluate the vacuum expectation value of the spin $<\sigma_i>$, and the correlation function $<\sigma_i(x)\sigma_j(y)>$. I shall not go into the details for evaluating the matrix elements. They are given in Ref.[3]. However, I would like to point out that in evaluating the matrix element of $<\sigma_2>$, different size patches in $|\psi>$ contribute multiplicatively. The expectation value of σ_2 due to a given size (a,b) is

$$<\sigma_2>_{(a,b)} = 1 - \varepsilon e^{-\xi} < 1 \tag{3.39}$$

where ε is a constant of order 1 and ξ is defined in (3.35). The expectation value of σ_2 due to ψ is

$$\begin{aligned}\langle\psi|\sigma_2|\psi\rangle &= \prod_{\text{all sizes}} (1 - \varepsilon e^{-\xi_i}, \\ &\approx \exp\left[- \sum_{\text{all sizes}} \varepsilon e^{-\xi_i}\right] \\ &= \exp(-\infty) = 0. \end{aligned} \tag{3.40}$$

By including fluctuations of all sizes, the expectation value of spin vanishes as required by general considerations. Thus, the WKB method enables us to construct a trial ground state which is consistent with this general requirement. The correlation function calculation also reproduces the general power law dependence as predicted by the spin wave consideration alone.

4. INSTABILITY OF CONSTANT YANG-MILLS FIELDS

In this last lecture, I shall work on the problem of stability of a constant Yang-Mills field [4,8]. It is known in quantum electrodynamics that a constant E field is unstable under field emissions. Electron-positron pairs appear spontaneously in the presence of an E field, and it leads to instabilities. Since Yang-Mills fields also carry charges, such instabilities occur too. In the following, I shall concentrate on a constant B-field which is stable in quantum electrodynamics, but is no longer stable in the Yang-Mills field. Classical Yang-Mills fields A_μ^a , $F_{\mu\nu}^a$ obey

$$F_{\mu\nu}^{\ a} = \partial_\mu A_\nu^a - \partial_\nu A_\mu^a + g\varepsilon^{abc} A_\mu^b A_\nu^c \tag{4.1}$$

$$(D_\mu F_{\mu\nu})^a \equiv \partial_\mu F_{\mu\nu}^{\ a} + g\varepsilon^{abc} A_\mu^b F_{\mu\nu}^c = 0 \tag{4.2}$$

where

$$D_\mu^{ac} \equiv \delta^{ac}\partial_\mu + \varepsilon^{abc} A_b^\mu \ . \tag{4.3}$$

In eqs. (4.1) (4.2), a,b,c, stand for color indices, and μ,ν stand for Lorentz indices. Eq. (4.1) is equivalent to the matrix equation (2.8). Under small fluctuations,

$$A_\mu \to A_\mu + \delta A_\mu \tag{4.4}$$

$$F_{\mu\nu} \to F_{\mu\nu} + \delta F_{\mu\nu} \tag{4.5}$$

we obtain from (4.1) (4.2) equations linear in δA_μ^a and $\delta F_{\mu\nu}^a$:

$$\delta F_{\mu\nu}^{\ a} = (D_\mu \delta A_\nu)^a - (D_\nu \delta A_\mu)^a \tag{4.6}$$

$$(D_\mu \delta F_{\mu\nu})^a + g\varepsilon^{abc} F_{\mu\nu}^{\ c} \delta A_\mu^b = 0. \tag{4.7}$$

We shall solve for the normal modes

$$\delta A(x,t) = \delta A(x)\, e^{-i\omega t}. \tag{4.8}$$

If ω is real for all normal modes, δA will remain small for all time. We shall refer to the original solution as stable. If one or more modes have complex ω, some δA will grow exponentially in time. Then, the original solution is unstable against small fluctuations [9].

We now study the stability of a constant B-field. For specific, we assume that the B-field lies in the third color as well as third spatial direction. A possible choice of vector potential for such a B-field is,

$$A_1^3 = -\frac{1}{2} By,\ A_2^3 = \frac{1}{2} Bx \text{ all other A's} = 0 \tag{4.9}$$

$$F_{12}^3 = B, \text{ all other } F_{\mu\nu} = 0. \tag{4.10}$$

In the following we shall use the temporal gauge

$$A_0^a = \delta A_0^a = 0. \tag{4.11}$$

We consider separately the variation δA in the third or in the (1.2) color direction.

i) Variation δA_i^3

Since only A_μ^3 and $F_{\mu\nu}^3$ are non-vanishing, the covariant derivatives on δA_μ^3 and $\delta F_{\mu\nu}^3$ are the same as ordinary derivatives. Hence, we obtain

$$(D_\mu \delta A_\nu)^3 = \partial_\mu \delta A_\nu^3 \tag{4.12}$$

$$\delta F_{\mu\nu}^{\ 3} = \partial_\mu \delta A_\nu^3 - \partial_\nu \delta A_\mu^3 \tag{4.13}$$

and

$$(D_\mu \delta F_{\mu\nu})^3 = \partial_\mu \delta F_{\mu\nu}^{\ 3} = 0. \tag{4.14}$$

Thus, δA_μ^3 and $\delta F_{\mu\nu}^3$ obey the same set of equations as in QED. There is no instability for these variations.

ii) Variation δA_i^a, a = 1,2

Substituting Eq.(4.6) into (4.7), we have

$$[D_\mu(D_\mu \delta A_\nu - D_\nu \delta A_\mu)]^a + g\varepsilon^{abc} F_{\mu\nu}{}^c \delta A_\mu^b = 0 . \tag{4.15}$$

Next, we move D_μ through D_ν such that it can operate on δA_μ. We can achieve this by noting that

$$\begin{aligned}(D_\mu D_\nu)^{ac} &= (D_\nu D_\mu)^{ac} + [D_\mu, D_\nu]^{ac} \\ &= (D_\nu D_\mu)^{ac} + g\ \varepsilon^{abc}\ F_{\mu\nu}{}^b .\end{aligned} \tag{4.16}$$

Hence, we have

$$[D_\mu^2 \delta A_\nu - D_\nu(D_\mu \delta A_\mu)]^a + 2g\varepsilon^{abc} F_{\nu\mu}{}^b \delta A_\mu^c = 0. \tag{4.17}$$

The factor 2 appearing in the last term of Eq.(4.17) is significant. Setting $\nu = 0$ in (4.17) and making use of

$$\delta A_o = F_{o\mu} = 0, \qquad D_o = \partial_o ,$$

we have

$$-\partial_o(D_\mu \delta A_\mu)^a = 0. \tag{4.18}$$

Eq.(4.18) is known as Gauss' law. With appropriate boundary conditions, Eq.(4.18) can be written in $\delta A_o = 0$ gauge as

$$D_\mu \delta A_\mu = D_i \delta A_i = 0. \tag{4.19}$$

With the help of (4.19), Eq.(4.17) reduces to

$$(D_\mu^2 \delta A_\nu)^a + 2g\varepsilon^{abc} F_{\nu\mu}{}^b \delta A_\mu^c = 0. \tag{4.20}$$

Introducing the spin matrix,

$$i(S_{\alpha\beta})_{\nu\mu} = -i(g_{\alpha\nu} g_{\beta\mu} - g_{\alpha\mu} g_{\beta\nu}) \tag{4.21}$$

we can rewrite (4.20) as

$$(D_\mu^2 \delta A_\nu)^a + 2ig\varepsilon^{abc} \frac{1}{2} F_{\alpha\beta}^{\ b} (S_{\alpha\beta})_{\nu\mu} \delta A_\mu^c = 0. \tag{4.22}$$

Remember that F_{12}^3 is the only non-vanishing component of $F_{\mu\nu}^{\ a}$. Hence, we have

$$F_{\alpha\beta}^{\ b} S_{\alpha\beta} = \delta^{b3} B S_3 \tag{4.23}$$

where $S_3 = S_{12}$ is the spin operator in the vector form. Defining non-hermitian combinations which are eigenstates of the charge operator,

$$\delta A_i \equiv (\delta A_i^{(1)} + i\delta A_i^{(2)})/\sqrt{2} \tag{4.24}$$

$$\delta A_i^+ \equiv (\delta A_i^{(1)} - i\delta A_i^{(2)})/\sqrt{2} \tag{4.25}$$

we have from (4.22)

$$(\partial_\mu + igA_\mu^{(3)})^2 \delta A_i - 2gB(S_3)_{ij} \delta A_j = 0, \tag{4.26}$$

$$(\delta_\mu - igA_\mu^{(3)})^2 \delta A_i^+ - 2gB(S_3)_{ij} \delta A_j^+ = 0. \tag{4.27}$$

Eqs.(4.26), (4.27) describe charged particles with magnetic moment 2g moving in a constant B field. Since both the energy ω and the third component of the momentum k_3 are constants of motion, we can rewrite (4.26) as

$$[-\omega^2 + k_3^2 - (\partial_\perp + igA_\perp^{(3)})^2]\delta A = 2gB(S_3 \delta A)\cdot \tag{4.28}$$

Eq.(4.28) has been solved in standard quantum mechanics text book [10]. It can be recasted as a harmonic oscillator equation,

$$(-\nabla_\perp^2 + \frac{1}{4} g^2 B^2 \rho^2)\delta A = (\omega^2 - k_3^2 + mgB \pm 2gB)\ \delta A \tag{4.29}$$

where

$$\rho^2 = x^2 + y^2, \tag{4.30}$$

and m is the magnetic quantum number. The eigenvalue to (4.29) is wellknown. In cylindrical coordinates, we have

$$(2n+1)gB + |m|gB = \omega^3 - k_3^2 + mgB \pm 2gB \tag{4.31}$$

and consequently,

$$\omega^2 = k_3^2 + (|m|-m)gB + (2n+1)gB \mp 2gB. \tag{4.32}$$

In (4.31) and (4.32), n is a non-negative integer. For $|m| = m$, $n = 0$, and taking the negative sign in (4.32), we obtain

$$\omega^2 = k_3^2 - gB. \tag{4.33}$$

Thus, ω^2 is negative if $k_3^2 < gB$. This implies an imaginary ω and consequently an exponential increasing solution,

$$\omega = i\sqrt{gB - k_3^2} \tag{4.34}$$

$$\delta A = \delta A(o)\exp\left(\sqrt{gB - k_3^2}\; t\right) \tag{4.35}$$

Hence, the constant B-field is unstable.

The simplest unstable mode located at the origin has the eigenvalue $m = n = k_3 = 0$, $S_3 = 1$. Its δA is given by

$$\delta A_1^1 = -\delta A_2^2 = \varepsilon e^{-gB\rho^2/4}\, e^{\sqrt{g\beta}\, t}$$

$$\text{all other } \delta A\text{'s} = 0. \tag{4.36}$$

The field tensor has the time dependence,

$$E_1^1 = -E_2^2 = \varepsilon\sqrt{gB}\; e^{-gB\rho^2/4}\, e^{\sqrt{gB}\, t} \tag{4.37}$$

$$B_3^3 = B - g\varepsilon^2\, e^{-gB\rho^2/2}\, e^{2\sqrt{gB}\; t} \tag{4.38}$$

all other components of $F_{\mu\nu}$ remain zero. Note that both E^2 and B^2 are gauge invariant. The instability described above is a genuine physical effect, not a gauge dependent artifact.

In the following, I shall mention several interesting points:

i) Magnetic moment contribution

We have shown that eigenvalue ω^2 for small fluctuations obeys

$$\omega^2 = \underbrace{k_3^2}_{(1)} + \underbrace{(2n-1)gB + (|m|-m)gB}_{(2)} - \underbrace{\mu B\, S_3}_{(3)} . \tag{4.39}$$

On the right hand side of (4.39), the first term denotes the contribution due to motion in z-direction, the second term represents the contribution due to the circular motion, and the last term describes the magnetic moment contribution. For an ordinary electrodynamics, the magnetic moment associated with a charged vector field is $\mu = g$, and hence, the total ω^2 is always positive

($\omega^2 \geq k_3$). For a Yang-Mills field, the magnetic moment associated with δA is $\mu = 2g$. Then, we find that $\omega^2 \geq k_3^2 - gB$, and it may become negative. This gives rise to instability.

ii) Number of unstable modes

We have demonstated the existence of an unstable mode located at the origin. Obviously, there are many such unstable modes at different space points. We wish to count the number of unstable modes inside a given volume. To achieve this, we consider unstable solutions for $m \neq 0$.

For each $m \neq 0$, there is one unstable mode. The spatial and angular dependence of δA for such an unstable mode is,

$$\delta A = \text{const. } e^{im\Theta} \rho^{|m|} e^{-\frac{1}{4} gB\rho^2}$$

$$= \text{const. } \exp\left[f(\rho)\right] \quad (4.40)$$

where

$$f(\rho) \equiv |m| \ln \rho - \frac{1}{4} gB\rho^2 . \quad (4.41)$$

We find that $|\delta A|$ vanishes both at $\rho = 0$ and at $\rho \to \infty$. Hence, δA peaks at some radius ρ_m. To determine ρ_m, we set

$$f'(\rho) = \frac{|m|}{\rho} - \frac{1}{2} gB\rho = 0 \quad (4.42)$$

and obtain

$$\rho_m = \sqrt{\frac{2|m|}{gB}} . \quad (4.43)$$

For large m, δA is narrowly peaked at ρ_m. Hence, we can include in ρ all unstable modes whose ρ_m is smaller then ρ. Hence, the number of unstable modes within $\rho = \rho_m$ ist $2|m|$. From ρ_m in (4.43) we can compute the area per unstable mode (assuming k_3=0) as

$$\frac{\pi\rho_m^2}{2|m|} = \frac{\pi}{gB} \approx L^2 \quad (4.44)$$

where L is approximately the minimal size for the instability to occur. From (4.44), we can characterize the stability of a system by a "Reynolds-like" number, gBL^2. If $gBL^2 \ll 1$, the system is stable. If $gBL^2 \gg 1$, the system becomes unstable.

iii) Other configurations

We can apply the above criterion to study the stability of other interesting systems such as plane waves and Coulomb field solutions. Consider the Coulomb field produced by a magnetic charge,

$$B = Q/r^2. \tag{4.45}$$

The Coulomb field has no intrinsic length. At any point r, the distance over which B is approximately constant is

$$L \approx r. \tag{4.46}$$

Using L = r as the size parameter, the "Reynolds number" becomes

$$gBL^2 = \frac{gQ}{r^2} r^2 = gQ \tag{4.47}$$

which is independent of r. Thus, a Coulomb field is stable if $gQ \ll 1$, and is unstable if $gQ \gg 1$ [11].

Similarly, for a plane wave, we can identify B as the amplitude of the wave and $L = \lambda$ as the wave length of the wave. The stability of a plane wave depends on the magnitude of $gB\lambda^2$.

iv) Shielding of the B-field

As we can see in Fig. 13, the B fields (δB) generated by the orbital motion of the unstable modes tend to reduce the original B field. These unstable modes generate a shielding mechanism of the original B field, and prevent the external B field from penetrating into the interior region.

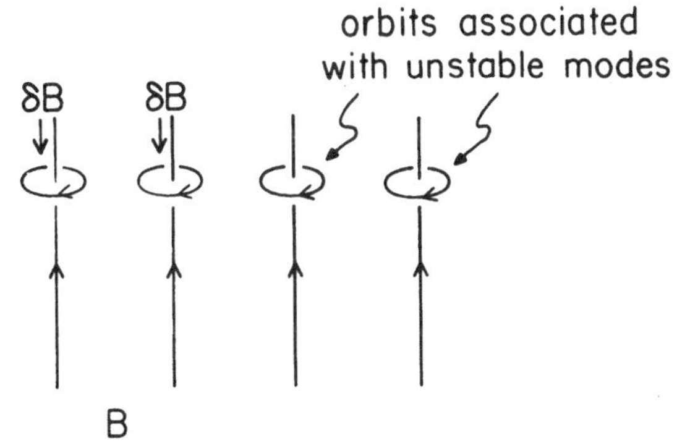

Figure 13 The magnetic field, δB, produced by the orbital motion of an unstable mode tends to cancel the original B field.

ACKNOWLEDGEMENT

I wish to thank Professor W. Rühl, Professor G.E. Hite, and the organizers of the Summer Institute for their hospitality during my visit to Kaiserslautern. This work was supported in part by the U.S. National Science Foundation under Grant Nr. 79-00272.

REFERENCES

1. K.M. Bitar and S.J. Chang, Phys.Rev. D17, 486 (1978), K.M. Bitar and S.J. Chang, Phys. Rev. D18, 435 (1978).
2. S.J. Chang, "An Example of Quantum Tunneling in Molecular Physics", Physica 96A, 183 (1979). The above volume of Physica was also published separately as a book "Themes in Contempory Physics", North-Holland Publ. Comp.,Amsterdam 1979.
3. K.M. Bitar, S.J. Chang, G. Grammer and J.D. Stack, "A mechanism for destruction of order in the 2-d nonlinear σ model" Preprint. Ill-(TH)-78-17.
4. S.J. Chang and N. Weiss, "Instability of constant Yang-Mills fields", Phys.Rev. D (in press).
5. T. Banks, C.M. Bender, and T.T. Wu, Phys.Rev. D8, 3346 (1973); T. Banks and C.M. Bender, Phys. Rev. D8, 3366 (1973).
6. S. Coleman, Phys. Rev. D15, 2929 (1977); 16, 1248 (E) (1977); C.G. Callan and S. Coleman, ibid 16, 1762 (1977).
7. J.L. Gervais and B. Sakita, Phys.Rev. D16, 3507 (1977).
8. N.K. Nielsen and P. Olesen, Nuclear Physics B144, 376 (1978).
9. There are several earlier and interesting works in this area. See, e.g. J.E. Mandula, Phys. Letters, 67B, 175 (1977); M. Magg, Phys. Letters 78B, 481 (1978); P. Sikivie and N. Weiss, Phys.Rev. Letters 40, 1411 (1978); Phys.Rev. D18, 3809 (1978). See also Ref.[8].
10. See e.g. L.D. Landau and E.M. Lifshitz, Quantum Mechanics - Non-Relativistic Theory, Pergamon Press, New York, 1965, Chapter XV.
11. This is analogous to what Mandula found in the case of an electric source. See Ref.[9].

RECENT RESULTS ON THE INSTANTON APPROXIMATION

B. Berg

II. Institut für Theoretische Physik

Universität Hamburg

ABSTRACT

Recent results on the two-dimensional $\mathbb{C}P^{N-1}$ models suggest that the QCD instanton gas is dense and infrared finite. These results and the computation of quantum fluctuations of quarks in a multi-instanton background are reviewed.

I. INTRODUCTION

In the Euclidean region the action of QCD is defined by

$$S = \int d^4x L$$

with

$$L = -\frac{1}{2g^2} \cdot \mathrm{tr}(F_{\mu\nu}F_{\mu\nu}) - i\bar{\psi}(D_\mu\gamma_\mu - M)\psi$$

The gauge fields A_μ are antihermitean 3 × 3 matrices which belong to the adjoint representation of the SU(3) color group and

$$F_{\mu\nu} = \partial_\mu A_\nu - \partial_\nu A_\mu + [A_\mu, A_\nu].$$

The quark fields ψ are in the fundamental representation of the color group. Color as well as an unknown number (≧4) of flavor indices are suppressed. $D_\mu = \partial_\mu + A_\mu$ is the covariant derivative, M the mass matrix. We shall deal with massless quarks and admit a SU(n) color group.

Since 1975 one knows about instantons [1,2], i.e. finite action

$$S = -\frac{1}{2g^2}\int d^4x\,tr(F_{\mu\nu}F_{\mu\nu})$$

solutions of the Yang-Mills (Y.M.) equations. The instantons carry a topological charge K. For a SU(2) gauge group the 1-instanton solution [2] (K=1) is given by

$$A_\mu(x) = \frac{1}{2}\bar{\sigma}_{\mu\nu}\partial_\nu \ln\rho(x)$$

with

$$\rho(x) = 1 + \frac{\lambda}{(x-y)^2}$$

Here

$$\bar{\sigma}_{\mu\nu} = \bar{\eta}^a_{\mu\nu} i\sigma^a \quad \text{where} \quad \bar{\eta}^a_{\mu\nu}$$

is the anti-selfdual 't Hooft [3] tensor and σ^a are the Pauli matrices. λ is the scale size and y the position of the instanton. For higher SU(n) gauge groups the 1-instanton solution can be simply obtained by imbedding the SU(2) instanton.

The quantum theory is obtained from the classical action by means of functional integration. In the saddle point approximation to the functional integral one has to sum over all finite action solutions in the Euclidean region and to calculate the quantum fluctuations around these. Therefore for this approximation the relevance of instantons is evident. Taking into account that their action is S[inst.]>0, we see that instanton effects are of order

$$\exp\{-\frac{1}{2g^2}\}$$

i.e. exponentially small in the coupling. This is the reason why instantons could be responsible for physical effects which cannot be seen in ordinary perturbation theory.

Important physical questions which have been discussed in connection with instantions are:

1. Is there a dynamical mass gap generation in the Y.M. theory?
2. Quark confinement.
3. The U(1) problem.

Quantum fluctuations around the 1-instanton solution have been calculated in Ref.[3] It turns out that the integration over the instanton scale size is infrared divergent like

$$\int_0^\infty \frac{d\lambda}{\lambda^5} (\lambda m)^{\frac{11 \cdot n}{3}} \tag{1}$$

A popular approximation for the computation of instanton effects is the dilute gas approximation. In this approximation one replaces the instanton gas by an ensemble of well separated 1-instanton configurations. This exponentiates essentially the infrared divergence of the 1-instanton contribution (1). Introducing by hand a phenomenological parameter λ_c the integration over the instanton scale size is cut off at a size much smaller then the assumed mean instanton separation. At the end of the calculation the limit $\lambda_c \to \infty$ does not exist, in other words the dilute instanton gas does not behave thermodynamically. Within this approximation instanton effects cannot account for a mass gap or quark confinement. Concerning the U(1) problem instantons have given important hints how this problem may be resolved. They give, however, not a complete understanding of the chiral symmetry breaking [4] and in contrast to recent results [5] based on the 1/N expansion they cannot account for the mass of the η'.

So far the dilute gas approximation. Recently, however, it has turned out that the Y.M. instanton gas is presumably dense and infrared finite. If this is correct instantons could eventually give an answer to the raised physical questions.

The suggestion that the Y.M. instanton gas is dense and infrared finite comes from recent investigations [6,7] of the $\mathbb{C}P^{N-1}$ models in two dimensions. This is reported in Section II. Some speculations concerning the Y.M. case (based on Ref.[8]) and the compution of quantum fluctuation of quarks in an arbitrary Y.M. instanton background [9] are reported in Section III.

II. THE $\mathbb{C}P^{N-1}$ INSTANTON GAS

The $\mathbb{C}P^{N-1}$ model in Euclidean space describes n complex fields (n=N)

$$z_\alpha(x) \qquad (\alpha = 1,\ldots,n;\ x = (x_1,x_2))$$

which are subject to the constraint $\bar{z}_\alpha z_\alpha = 1$.
Fields related by

$$z'_\alpha(x) = e^{i\Lambda(x)} z_\alpha(x)$$

are considered to be gauge equivalent. The Euclidean action is given by

$$S = \frac{n}{2f} \int d^2x \bar{D}_\mu z D_\mu z$$

where

$$D_\mu = \partial_\mu + A_\mu , \; A_\mu = -\bar{z}\partial_\mu z \; , \quad (\mu=1,2)$$

is the covariant derivative. The $\mathbb{C}P^{N-1}$ models have a non-trivial topology and the topological charge is

$$Q = \frac{-i}{2\pi} \int d^2x \varepsilon_{\mu\nu} \partial_\mu A_\nu \; ; \quad \varepsilon_{12} = 1.$$

The instanton solutions of the $\mathbb{C}P^{N-1}$ model are known [10] and the multi-instanton z^0 with topological charge $Q = K$ and action $S[z^0]$ can (up to solutions which are of measure zero) be given by

$$z^0_\alpha = \frac{p_\alpha}{|p|}, p_\alpha = c_\alpha \prod_{j=1}^{k} (s-a^j_\alpha), (\alpha=1,\ldots,n), c_n=1, s=x_1-ix_2. \tag{2}$$

The $n(K+1)-1$ complex parameters c_α, a^j_α will be labelled by $\lambda_i (i=1\ldots n(K+1)-1)$.

The functional integral for the expectation of a (gauge invariant) observable is

$$\langle \mathcal{O} \rangle = Z^{-1} \int D[z] \, \lambda^{-S}$$

where the partition function Z normalizes the expectations: $\langle 1 \rangle = 1$. The functional integral may be approximated +) by summing over all saddle points and taking into account the quantum fluctuation around each saddle point. The pure (anti-) instanton gas arises by taking as saddle points all the (anti-) instanton solutions.

Let $z(x)$ be a $\mathbb{C}P^{N-A}$ field with topological charge K. It can be written as

$$z_\alpha = e^{i\Lambda}\{(1-|\eta|^2)^{\frac{1}{2}} z^0_\alpha + \eta_\alpha\}$$

$$\bar{z}^0\eta = 0 \qquad \text{("background gauge")}$$

For fluctuations around z^0 we neglect terms of order $\mathcal{O}(\eta^3)$ and obtain the Gaussian approximation to the action

$$S = \frac{n\pi}{f} K + \frac{2n\pi}{f} (\eta, \Delta\eta) \; .$$

+) This is an approximation to function space and not a systematic expansion in some coupling constant.

Here we have introduced the scalar product

$$(\phi,\psi) = \frac{2n\pi}{f}\int d^2x\bar{\phi}\psi$$

and Δ is the fluctuation operator defined by this expansion of the action.

Now I like to proceed in a formal way. Suppose that

$$e^i_\alpha(x,\lambda) \qquad (i = 1,2,\ldots..)$$

denotes a complete set of orthonormal eigenvectors with eigenvalues $E_i \neq 0$ of the fluctuation operator Δ.+) Then by expanding

$$\eta_\alpha = \sum_{i=1}^{\infty} \xi_i e^i_\alpha(x,\lambda)$$

the Gaussian approximation to the action becomes

$$S = \frac{n\pi}{f} K + \frac{2n\pi}{f}\sum_{i=1}^{\infty} E_i|\xi_i|^2.$$

Approximating furthermore $\mathcal{O}$ by its value $\mathcal{O}(\lambda)$ for the instanton $z^0(x,\lambda)$, the instanton gas expectation of $\mathcal{O}$ becomes

$$<\mathcal{O}>_{inst} = Z^{-1}\sum_{k=0}^{\infty}(K!)^{-n}\int \prod_{j,i} d^2\lambda_j d^2\xi_i J(\lambda)\mathcal{O}(\lambda)$$

$$\cdot\exp -\{\frac{2\pi}{f}K + \frac{2n\pi}{f}\sum_{i=1}^{\infty}E_i|\xi_i|^2\} \qquad (3)$$

where $J(\lambda)$ is the Jacobian defined by

$$D[z^0] = \prod_j d^2\lambda_j J(\lambda) .$$

The factor $(K!)^{-n}$ in (3) is obtained by recognizing that the multi-instanton solution as given by (2) remains the same under interchange of the parameters a^i_α $(i=1,\ldots\ldots,K)$ within a fixed component α . Therefore the factor $(K!)^{-n}$ ensures that each multi-instanton solution is counted only once.

For the $\mathbb{C}P^1$ [6] and $\mathbb{C}P^{N-1}$ [7] models the instanton gas can be computed rigorously. The result is

+) This makes of course only sense if the spectrum of the fluctuation operator is discrete. This is the case if one works on a compact space. Because of conformal invariance a sphere with radius R is appropriate.

$$\langle \mathcal{O} \rangle_{inst} = Z^{-1} \sum_{k=0}^{\infty} (K!)^{-n} z^{n\,K} \,. \tag{4}$$
$$\int \frac{\prod_{\alpha=1}^{n-1} d^2 c_\alpha}{(\bar{c}_\gamma c_\gamma)^n} \prod_{\beta=1}^{n} \prod_{j=1}^{K} d^2 a_\beta^j \, \mathcal{O}(c_\alpha, a_\beta^j) \exp\{-U(c_\alpha, a_\beta^j)\}$$

Here

$$z = m \frac{2n}{f} e^{-n+2/2n}$$

is the fugacity and $m = \mu \exp\left(-\frac{\pi}{f_R(\mu)}\right)$

is the renormalization group invariant mass.
By choosing an appropriate mass-scale the fugacity can be scaled to $z = 1$ and therefore the physics of the instanton gas does not depend essentially on z. The many body potential

$$U = \frac{n}{2\pi} \int d^2x \ln|p|^2 \partial_{\bar{s}} \partial_s \ln|p|^2 + \frac{1}{2} nk(\ln(\bar{c}_\alpha c_\alpha) - 1)$$
$$- \sum_{\alpha=1}^{n} [k \ln|c_\alpha|^2 + \sum_{i<j} \ln|a_\alpha^i - a_\alpha^j|^2] \tag{5}$$

accounts for the interaction (p is defined by (2)).

For the $\mathbb{C}P^{N-1}$ model the potential U describes the interaction of n different types of particles, which I will call "instanton constituents". In the sector K there are $n \cdot K$ particles, K particles of each constituent type. Their position is described by the parameters a_α^j $(j=1,\ldots\ldots,k;\ \alpha=1,\ldots,n)$; α counts the different constituents. For $N = 2$ the gas simplifies to the Coulomb gas at temperature $T = 1$ (cf. Ref. [6,7]).

The instanton gas (4) is dense and infrared finite, whereas the dilute gas approximation for the $\mathbb{C}P^{N-1}$ models is infrared divergent. I like to clarify this point further.

What we are computing is the grand canonical ensemble. Let Z be its partition function, then the pressure is proportional to P,

$$P = \lim_{V \to \infty} \frac{1}{V} \ln Z \,.$$

Here V is a (box) volume. Infrared finite means that the partition function Z behaves like $\exp P.V$ for $V \to \infty$, but not worse then that. It is instructive to investigate this first for the free gas, i.e.

let us set $U \equiv 0$ for the moment. Introducing a temperature parameter via

$$U \to \frac{1}{T} U$$

this corresponds to the high temperature limit $T = \infty$.
Now we easily read off from (4) that Z behaves like

$$\Sigma \frac{V^{nK}}{(K!)^n}, \qquad V \to \infty$$

and the simple estimate

$$\exp\{V\} < \Sigma \frac{V^{nK}}{(k!)^n} < \exp\{nV\}$$

proves the infrared finiteness. This is the kinematical aspect of the problem. The dynamical aspect amounts to including the potential U. This is non-trivial. For the Coulomb gas in two dimensions Fröhlich [11] has proven that the thermodynamic limit exists for $T > 1$ and for $T = 1$ it becomes equivalent to the massive Dirac field. Therefore the $\mathbb{C}P^1$ instanton gas is well-defined. The similar behaviour of the potential U for the separation of all instanton constituents largely suggest that the thermodynamic limit of the instanton gas exists also for $N \geq 3$.

Let us compare this with the dilute gas approximation. Introducing the scale size

$$\lambda^2 = \frac{1}{\bar{c}_\gamma c_\gamma} \sum_{i<j} |c_i|^2 |c_j|^2 |a_i - a_j|^2$$

of a single instanton, we can calculate from (4) the 1-instanton contribution, which is proportional to

$$V \int_0^{\lambda_c} d\lambda \lambda^{n-3} \tag{6}$$

Here λ_c cuts off the scale size integration at a finite value. Assuming the instanton gas to be dilute we have to exponentiate the 1-instanton contribution

$$Z_{dilute} = \sum_K \frac{1}{K!} \left(V \int_0^{\lambda_c} d\lambda \lambda^{n-3}\right)^K$$

and after carrying out the infinite volume limit $V \to \infty$ we are left with a divergent scale size integration in the physical quantities, i.e. the limit does not exist.

The correct interpretation of (6) is that large instantons are more probable then small instantons and therefore the instanton gas is dense. In the exact instanton gas (4) dilute configurations are of such a negligible weight that they cannot destroy the infrared finiteness. Concluding this section I like to remark that one can compute the internal energy of the instanton partition function for all N in the $T = \infty$ limit. The result has a nice 1/N expansion

$$\lim_{V\to\infty} \frac{U}{V} = \text{const.}\ N^2 \ (1 - \frac{1}{N} + O\ (\frac{1}{N^2})) .$$

By this heuristic argument we expect instanton effects in the $\mathbb{C}P^{N-1}$ models to be neither exponentially small nor exponentially large in 1/N.

III. STATUS OF THE QCD INSTANTON GAS

Relying on the results for the $\mathbb{C}P^{N-1}$ models the infrared divergence of the 1-instanton contribution (1) suggests that the QCD instanton gas may be dense and infrared finite. I will further comment on this in Section III.2. In Section III.1 the Atiyah, Hitchin, Drinfeld, Manin [12] (AHDM) construction of the multi-instanton solution is given for SU(n). For establishing explicitly the QCD instanton gas one has to calculate the quantum fluctuations of quarks and gluons in a multi-instanton background. For the determinant of the corrresponding Dirac operator quite explicit formulas were derived recently [9] . This is reported in Section III.3.

III.1 THE YANG-MILLS INSTANTONS

We consider the multi-instanton solution of AHDM [12]for the case of SU(n) in the notation relying on Ref. [13].
The gauge potential of the multi-instanton solution is given by

$$A_\mu = V^+(x)\, \partial_\mu V(x) \qquad (7)$$

where V is an (n + 2K)xn complex matrix

K denotes the instanton number, viz.

$$K = -\frac{1}{16\pi^2}\int d^4x \mathrm{Tr}(F_{\mu\nu}F^*_{\mu\nu})$$

where

$$F^*_{\mu\nu} = \frac{1}{2}\varepsilon_{\mu\nu\rho\sigma}F_{\rho\sigma} \quad , \quad \varepsilon_{0123} = 1$$

is the dual field.

V is normalized to $V^+(x)V(x) = 1$ and behaves as $V(x) \to V(x)g(x)$, $g(x) \in SU(n)$ under gauge transformation. V has to be determined as solution of a set of 2K complex linear equations

$$V^+\Delta_{A'} = 0 \qquad (A' = 1,2) \tag{8}$$

with $\Delta_{A'}$ a SU(2) spinor (cf.e.g. Ref.[9]) of $(n+2K) \times K$ matrices depending linearly on x_μ :

$$\Delta_{A'} = a_{A'} + b^A x_{AA'} \quad ; \quad x_{AA'} = x_\mu (e_\mu)_{AA'} \tag{9}$$

$e_\mu = (1, -i\sigma_j)$ and σ_j the Pauli matrices.

$a_{A'}$ and b_A are constant $(n + 2K) \times K$ matrices

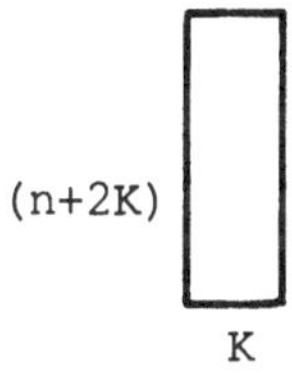

which parametrize the multi-instanton solution. The instanton parameter matrices are not unconstrained, however, but must be chosen such that

$$\Delta^+_{A'}\Delta_{B'} = -\varepsilon_{A'B'}f^{-1} \qquad \text{for all } x. \tag{10}$$

Here Δ^+_A is the adjoint in the sense of spinors defined by

$$\Delta^+_1 = -\bar{\Delta}^T_2 \quad , \quad \Delta^+_2 = \bar{\Delta}^T_1$$

f^{-1} is a complex $K \times K$ matrix and the gauge potential (7) constructed via (8) is non-singular if f^{-1} is invertible for all x. From (9) it follows that (10) holds if and only if

$$a^+_{A'}a_{B'} + a^+_{B'}a_{A'} = 0 \tag{11a}$$

$$a^+_{A'}b_A + b^+_A a_{A'} = 0 \tag{11b}$$

$$b^+_A b_B + b^+_B b_A = 0 \tag{11c}$$

The most general transformations of the $a_{A'}$ and b_A parameter matrices which leave the gauge potential A_μ and the constraints (11) invariant+) are

$$a_{A'} \longrightarrow U a_{A'} K \; ; \qquad b_A \longrightarrow U b_A K. \tag{12}$$

$$(U \in U(n+2K) \;,\; K \in GL(K,\mathbb{C}))$$

The invariance (12) can be used to transform b_A into its "normal form"

$$b^1 = \left.\begin{array}{|c|} \hline 0 \\ \hline \begin{array}{l} 10 \\ 00 \\ 01 \\ 00 \ldots \\ \\ \\ \hfill 1 \\ \hfill 0 \end{array} \\ \hline \end{array}\right. \begin{array}{l} \}\, n \\ \\ \\ \}\, 2K \\ \\ \\ \end{array} ; \qquad b^2 = \left.\begin{array}{|c|} \hline 0 \\ \hline \begin{array}{l} 00 \\ 10 \\ 00 \\ 01 \ldots \\ \\ \\ \hfill 0 \\ \hfill 1 \end{array} \\ \hline \end{array}\right. \begin{array}{l} \}\, n \\ \\ \\ \}\, 2K \\ \\ \\ \end{array} \tag{13}$$

Now all instanton parameters are "sitting" in $a_{A'}$. They may be counted with the result that

$$4nK - n^2 + 1 \quad \text{if} \quad K \geq \frac{1}{2} n$$

$$4K^2 + 1 \quad \text{if} \quad K < \frac{1}{2} n$$

in agreement with results obtained by applying [14] the Atiyah-Singer index theorem.
The self-duality

$$F^*_{\mu\nu} = F_{\mu\nu}$$

of the field strength can be proven by using some properties of the projection operator

$$P = VV^+ = 1 - \Delta^{A'} f \Delta^+_{A'} \qquad \text{(cf. Ref. [13]).}$$

+) Up to gauge transformations of the potential of course

III.2 IS THE QCD INSTANTON GAS INFRARED FINITE?

If the QCD instanton gas exists in the thermodynamic sense, then it is dense, because this means that the infrared divergent dilute configurations have a negligible statistical weight. In Section II we have seen that the kinematical aspect of the questions of infrared finiteness amounts in counting the number of instanton parameters which can be interchanged without changing the instanton. The ADHM construction is, however, not as explicit as the $\mathbb{C}P^{N-1}$ instanton solution (2). The question is, how to introduce convenient parameters. An interesting proposal has recently been made by Belavin et al. [8].

Let the gauge group be SU(2) and N be a (K+1) dimensional quaternionic vector, i.e. having components of the form $n_i = n_i^\mu(e_\mu)$. Then Manin (unpublished) has proven that the equation

$$N^+V(x) = 0$$

has K solutions if the number of solutions is finite at all. Therefore the authors of Ref. [8] suggest to fix two quaternionic vectors N_1, N_2 and to define instanton parameters $a_1 \ldots a_K$ as the solutions of $N_1^+ V = 0$ and $b_1 \ldots b_K$ as the solutions of $N_2^+ V = 0$. Clearly translations and dilatations of the instanton introduce translations and dilatations of these parameters. The hope is that a convenient choice of N_1 and N_2 can be done such that the parameters $a_1 \ldots a_K$ and $b_1 \ldots b_K$ describe positions of instanton constituents analog to the $\mathbb{C}P^1$ case. Of course there has to be some dependence between these parameters, because there are only 8K-3 independent ones. A heuristical intuition can be gained by studying the 't Hooft solutions. For example the choice $N_1 = V(\infty)$ gives as parameters $a_1 .. a_K$ the centers of the instanton.

Within each set of solutions the instanton parameters can be interchanged without changing the instanton and, if the functional determinant coming from this parametrization does not depend on K,

$$Z = \sum_K \frac{1}{(K!)^2} \int z^{2k} \exp - U(a_1 \ldots a_K, b_1 \ldots b_K) d^4a_1 \ldots d^4a_K d^4b_1 \ldots d^4b$$

The kinematical problem $U = 0$ is then analog to the $\mathbb{C}P^1$ instanton gas (cf. Section II).

To set up the complete instanton gas one major problem is to determine the potential U. This has not yet been carried out. However, the methods which enable [9] the computation of quantum fluctuations of quarks in a general instanton background (cf. next section) can be generalized for treating the quantum fluctuation of gluons.

III.3 QUANTUM FLUCTUATIONS OF QUARKS IN AN ARBITRARY YANG-MILLS INSTANTON BACKGROUND

We like to compute

$$\Gamma = \ln\det{}'D \underset{def}{=} \tfrac{1}{2} \ln\det{}'D^2$$

where D is the Dirac operator and the prime indicates that zero modes have to be omitted.
Explicitly:+)

$$D = 2i\gamma_\mu D_\mu = \begin{pmatrix} 0 & T^+ \\ T & 0 \end{pmatrix}$$

with

$$\gamma_\mu = \begin{pmatrix} 0 & e_\mu \\ e_\mu^+ & 0 \end{pmatrix} , \quad D_\mu = \partial_\mu + A_\mu$$

and

$$T^+ = 2ie_\mu D_\mu \quad , \quad T = 2ie_\mu^+ D_\mu$$

The UV divergence of Γ can be regularized by adding for Pauli-Villars regulator fields with large masses M_i and alternating "metric" e_i , such that

$$\sum_{i=1}^{4} e_i = -1 \quad ; \quad \sum_{i=1}^{4} e_i M_i^{2p} = 0 \qquad (p = 1,2,3).$$

The regularized determinant

$$\Gamma_{reg} = \frac{1}{2} \mathrm{Tr} \{ \ln(D^2+P_0) + \sum_{i=1}^{\nu} e_i \ln(D^2 + M_i^2) \}$$

is finite and perfectly well-defined, (P_0 is the projector onto the zero modes of D).

Γ_{reg} is calculated by computing the variation $\delta\Gamma_{reg}$ with respect to the instanton parameters and integrating subsequently. The first part of the calculation can be traced back to the known [13,15] Green's function of the Dirac operator. This part is rather lengthy but nevertheless quite straightforward. The result, which has been obtained independently by several authors [8,9,16], is

$$\delta\Gamma_{reg} = \frac{1}{6\pi^2} \int d^4x \mathrm{Tr}(\delta A_\mu j_\mu) \;{}^{++)} \qquad (14)$$

where

+) For the actual calculation we have worked on the sphere S^4, here I give all formulas in the flat Euclidean space.

++) $\mathrm{Tr} = \mathrm{Tr}_{color}$

$$\delta A_\mu = v^+\{d^{A'} f\partial_\mu \Delta_A^+ - \partial_\mu \Delta^A f d_{A'}^+\} v \ , \quad d^{A'} \underset{\text{def}}{=} \delta a_{A'}$$

$$j_\mu = v^+\{b^A f b_A^+ \Delta^{A'} f \partial_\mu \Delta_{A'}^+ - \partial_\mu \Delta^{A'} f \Delta_{A'}^+ b^A f b_A^+\} v$$

and the matrices b^A have been taken in their normal form (13). The main difficulty was to integrate $\delta\Gamma_{reg}$, a step which involved a lot of guesswork. I like therefore to give some details how this was actually done. Integrating $\delta\Gamma_{reg}$ amounts to write $\mathrm{Tr}(\delta A_\mu j_\mu)$ as a sum of total variations and total derivatives:

$$\mathrm{Tr}(\delta A_\mu j_\mu) = \sum_i v_i \delta \mathrm{Tr}(V_i) + \sum_j x_j \partial_\mu \mathrm{Tr}(X_\mu^j). \tag{15}$$

If we could classify all possible total variations $\delta\mathrm{Tr}(V_i)$ and all possible total derivatives $\mathrm{Tr}(X_\mu^j)$ and find a set of linear independet "basic traces" F_K such that we can expand the left-hand side and right-hand side of (15) in terms of basic traces, then we would end up with a set of linear equations for the coefficients v_i and x_j which has to be solved.

Classifying the total variation and total derivatives one is finally led (after trying some larger sets) to

$$\begin{aligned}
\mathrm{Tr}(V_1) &= \mathrm{Tr}\{f\partial_\mu f^{-1} f\partial_\mu f^{-1} f\partial_\nu f^{-1} f\partial_\nu f^{-1}\} \\
\mathrm{Tr}(V_2) &= \frac{1}{4}\,\mathrm{Tr}\{f b^{+A} b_A f b^{+B} b_B\} \\
\mathrm{Tr}(X_\mu^1) &= \mathrm{Tr}\{f\delta f^{-1} f\partial_\mu f^{-1} f\partial_\nu f^{-1} f\partial_\nu f^{-1}\} \\
\mathrm{Tr}(X_\mu^2) &= \mathrm{Tr}\{f\delta f^{-1} f\partial_\nu f^{-1} f\partial_\mu f^{-1} f\partial_\nu f^{-1}\} \\
\mathrm{Tr}(X_\mu^3) &= \mathrm{Tr}\{f\partial_\mu \delta f^{-1} f\partial_\nu f^{-1} f\partial_\nu f^{-1}\} \\
\mathrm{Tr}(X_\mu^4) &= \mathrm{Tr}\{f\partial_\nu \delta f^{-1} f\partial_\mu f^{-1} f\partial_\nu f^{-1}\} \\
\mathrm{Tr}(X_\mu^5) &= -\frac{1}{2}\,\mathrm{Tr}\{f\delta f^{-1} f b^{+A} b_A f\partial_\mu f^{-1}\} \\
\mathrm{Tr}(X_\mu^6) &= -\frac{1}{2}\,\mathrm{Tr}\{f b^{+A} b_A f\partial_\mu \delta f^{-1}\}
\end{aligned} \tag{16}$$

In the search for basic traces one first recognizes, that by the use of the matrix algebra of the AHDM construction all operators in Eq. (15) may be expanded into linear combinations of traces of products of the form

$$d_{A'} f b_A^+ \ , \quad b_A f d_A^+ \ , \quad \Delta_{A'} f d_{B'}^+ \ ;$$

$$\Delta_{A'} f \Delta_{B'}^+ \ , \quad \Delta_{A'} f b_A^+ \ , \quad b_A f \Delta_A^+ \ , \quad b_A f b_B^+ \ ,$$

where all spinor indices are contracted. All terms have to contain precisely one matrix $d_{A'}$ or $d_{A'}^+$ and four matrices b_A or b_A^+ .

From (10) we note

$$f^{-1} = -\frac{1}{2} \underline{\Delta^+ \, \Delta} \tag{17}$$

where we have introduced the graphical notation

$$\underline{\Delta^+ \, \Delta} \underset{\text{def}}{=} \Delta^{+A'} \Delta_{A'}$$

for spinor contractions. By (17) the maximum number of f's in each basic trace is limited to 5. Therefore the system of basic traces closes. Already using a reality property (see below) we end up with 66 operators F_K . E.G. the 5f-terms give $5 \cdot 3 \cdot 3 = 45$ different operators from the following possible contractions:

$$\overset{3}{\underset{5 \cdot 3}{\overrightarrow{\underline{dfb}^+}}} \; \Delta fb^+ \; \Delta fb^+ \; \Delta fb^+ \; \Delta f\Delta^+$$

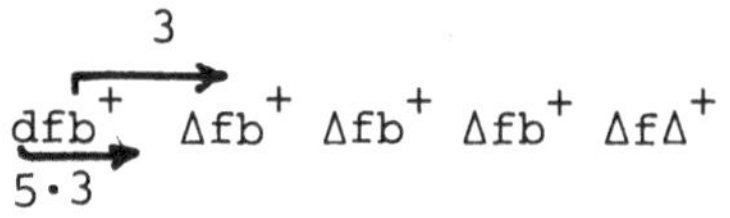

We have used an algebraic computer program to expand the l.h.s. of (15) and a set of total variations and total derivatives like (16) in terms of these basic traces, and found out that this overdetermined system of linear equations has no solution. This is not a big surprise, since the basic traces so far defined are not linearly independent.

In our paper [9] we have used 3 types of linear relations between the basic traces. First the identity

$$\varepsilon_{A'B'} \, \varepsilon_{C'D'} = \varepsilon_{A'C'} \, \varepsilon_{B'D'} - \varepsilon_{A'D'} \, \varepsilon_{B'C'}$$

which reads graphically

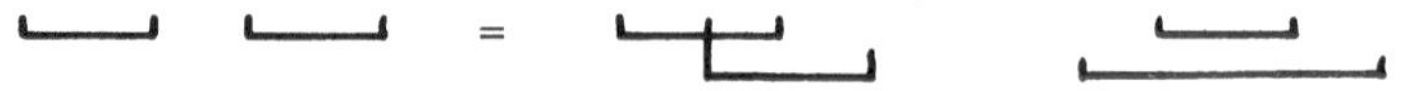

Second the relations

$$d_{A'}^+ \Delta_{B'} + \Delta_{A'}^+ d_{B'} + d_{B'}^+ \Delta_{A'} + \Delta_{B'}^+ d_{A'} = 0$$

$$d_{A'}^+ b_A + b_A^+ d_{A'} = 0$$

Finally reality of the traces which relied on the use of SP(r) as gauge group instead of SU(r) . We were at the end led to a "minimal" basis of 10 operators given in the following:

$$T_1 = \varepsilon_{\mu\nu\rho\sigma} \; \text{Tr}\{f \delta f^{-1} f \partial_\mu f^{-1} f \partial_\nu f^{-1} f \partial_\rho f^{-1} f \partial_\sigma f^{-1}\}$$

$$T_2 = \mathrm{Tr}\{f\delta f^{-1} f\partial_\mu f^{-1} f\partial_\mu f^{-1} f\partial_\nu f^{-1} f\partial_\nu f^{-1}\}$$

$$T_3 = \mathrm{Tr}\{f\delta f^{-1} f\partial_\mu f^{-1} f\partial_\nu f^{-1} f\delta_\mu f^{-1} f\partial_\nu f^{-1}\}$$

$$T_4 = \mathrm{Tr}\{f\delta f^{-1} f\partial_\mu f^{-1} f\partial_\nu f^{-1} f\delta_\nu f^{-1} f\partial_\mu f^{-1}\}$$

$$T_5 = \mathrm{Tr}\{f\partial_\mu \delta f^{-1} f\partial_\mu f^{-1} f\partial_\nu f^{-1} f\partial_\nu f^{-1}\}$$

$$T_6 = \mathrm{Tr}\{f\partial_\mu \delta f^{-1} f\partial_\nu f^{-1} f\partial_\mu f^{-1} f\partial_\nu f^{-1}\}$$

$$T_7 = \mathrm{Tr}\{f\delta f^{-1} f\overbrace{b^+ b} f\partial_\mu f^{-1} f\partial_\mu f^{-1}\}$$

$$T_8 = \mathrm{Tr}\{f\delta f^{-1} f\partial_\mu f^{-1} f\overbrace{b^+ b} f\partial_\mu f^{-1}\}$$

$$T_9 = \mathrm{Tr}\{f\partial_\mu \delta f^{-1} f\partial_\mu f^{-1} f\overbrace{b^+ b}\}$$

$$T_{10} = \mathrm{Tr}\{f\delta f^{-1} f\overbrace{b^+ b} f\overbrace{b^+ b}\}$$

In this basis we obtained the equation:

$$\mathrm{Tr}(\delta A_\mu j_\mu) - 2\partial_\mu \mathrm{Tr}\{df\partial_\mu \Delta^+ \Delta f b^+ \Delta f b^+ \Delta f \Delta^+$$

$$- \sum_{i=1}^{2} v_i \delta \mathrm{Tr}(V_i) - \sum_{j=1}^{6} x_j \partial_\mu \mathrm{Tr}(X_\mu^j) =$$

$$\begin{array}{lllllll}
 & & & & & & \frac{1}{4}\, T_1 \\
+\{-\frac{3}{4} + 2V_1 & +2X_1 & & & & \} & T_2 \\
+\{\ \frac{3}{4} & +X_1 + 2X_2 & & & & \} & T_3 \\
+\{\ \frac{1}{4} + 2V_1 & +X_1 + 2X_2 & & & & \} & T_4 \\
+\{\ 1 - 4V_1 & -X_1 & & +2X_3 + 2X_4 & & \} & T_5 \\
+\{-\frac{1}{2} & & -X_2 + & X_3 + X_4 & & \} & T_6 \\
+\{\ 2 & -10X_1 - 4X_2 & & & +2X_5 & \} & T_7 \\
+\{-2 & -2X_1 - 8X_2 & & & +X_5 & \} & T_8 \\
+\{\ 6 & & & -4X_3 - 10X_4 & -X_5 + 2X_6 & \} & T_9 \\
+\{\ \ -2V_2 & & & & +8X_5 & \} & T_{10}
\end{array}$$

One immediately recognizes that T_1 decouples completely from the V_i and X^j and therefore has to be itself a total divergence. In fact

$$T_1 = \delta \int_0^1 dt \varepsilon_{\mu\nu\rho\sigma} \, T\{K^{-1}\partial_t K K^{-1}\partial_\mu K \ldots K^{-1}\partial_\sigma K\} + \partial_\mu \Sigma_\mu$$

with

$$K = (1 - t)(1 + x^2)\, 1 + t f^{-1}.$$

The other nine equations have a solution given by

$$V_1 = \frac{1}{4}, \quad V_2 = -5$$

$$X_1 = \frac{1}{8}, \; X_2 = -\frac{7}{16}, \; X_3 = \frac{1}{16}, \; X_4 = 0, \; X_5 = -\frac{5}{4}, \; X_6 = -\frac{7}{2}.$$

This solves the problem of integrating $\delta\Gamma_{reg}$.

We have made use of the gauge group SP(r). By imbedding SU(r) into SP(r) it is not difficult to prove that the final answer for Γ_{reg}, as given in the following formula, is also true for SU(r), i.e. f defined by (10,17). The result is:

$$\Gamma_{reg} - 2r\Gamma^{o}_{reg} =$$

$$\alpha + \frac{1}{24\pi^2}\int d^4x I_1(x) + \frac{1}{24\pi^2}\int d^4x \int_0^1 dt I_2(t,x)$$

with

$$I_1 = \mathrm{Tr}\{f\partial_\mu f^{-1} f\partial_\mu f^{-1} f\partial_\nu f^{-1} f\partial_\nu f^{-1}\}$$

$$- 5\,\mathrm{Tr}\{f\underbrace{b^+ b}f\underbrace{b^+ b}\} + 4K\,(1 + x^2)^{-2}$$

and

$$I_2 = \varepsilon_{\mu\nu\rho\sigma} \, \mathrm{Tr}\{K^{-1}\partial_t K K^{-1}\partial_\mu K K^{-1}\partial_\nu K K^{-1}\partial_\rho K K^{-1}\partial_\sigma K\}$$

The determination of the integration constant α can be reduced to computing explicitly the 1-instanton case. Those calculations have been done previously [3,17]. The result for the integration constant is

$$\alpha = \{K \frac{2}{3} \sum_{i=1}^{4} e_i \ln M_i - 4\xi'(-1) - \frac{2}{3}\ln 2 + \frac{5}{12}\}.$$

For the case of the 't Hooft solutions analog results were first obtained in the work of Brown and Creamer [18].

SUMMARY AND CONCLUSIONS

The results show that even for a dense instanton gas calculations are still possible to a certain extent. This gives rise to some optimism concerning the future understanding of the QCD instanton gas.

On the other hand an incoherent sum of the pure instanton and anti-instanton contributions runs into difficulties with the cluster theorem and the Θ-vacua cannot be incorporated. This can probably be repaired by taking into account instanton-antiinstanton approximate solutions. Also there are technical problems concerning the computation of correlation functions in a dense instanton gas. For example even for the $\mathbb{C}P^1$, alias O(3) non-linear σ-model, it has so far not been proved rigorously that the dynamically generated finite correlation length of the instanton gas gives rise to the physical mass of the σ-particles which we know to be present in the O(3)σ-model.

ACKNOWLEDGEMENT

This talk is based on work with M. Lüscher.

REFERENCES

1. A. Polyakov, Phys. Lett. 59B, 82 (1975).
2. A. Belavin, A. Polyakov, A. Schwartz, Yu. Tyupkin, Phys. Lett. 59B, 85 (1975).
3. G. 't Hooft, Phys. Rev. D14, 3432 (1976); Phys. Rev. D18, 2199 (1978)
4. R. Crewther, CERN preprint, 1978.
5. E. Witten, Harvard preprint 1979, HUTP-79/AO14; G. Veneziano, CERN preprint 1979
6. V. Fateev, I. Frolov and A. Schwartz, to appear in Nucl.Phys.B.
7. B. Berg and M.Lüscher, DESY preprint 1979/17, to appear in Commun. Math. Phys.
8. A. Belavin, V. Fateev, A. Schwartz and Yu. Tyupkin, Phys. Lett. 83B, 317 (1979)
9. B. Berg and M. Lüscher, DESY preprint 1979, submitted to Nucl. Phys. B
10. M. Lüscher, Phys. Lett. 78B, 465 (1978); A. D'Adda, M. Lüscher and P. Di Vecchia, Nucl. Phys. B146, 63 (1978); V. Golo and A. Perelomov, Phys. Lett. 79B, 112 (1978).
11. J. Fröhlich, Comm. Math. Phys. 47, 233 (1976).
12. M. Atiyah, N. Hitchin, V. Drinfeld and Yu. Manin, Phys. Lett. 65A, 185 (1978).

13. E. Corrigan, D. Fairlie, P. Goddard and S. Templeton, Nucl. Phys. B140, 31 (1978); N. Christ, E. Weinberg and N. Stanton, Phys. Rev. D18, 2013 (1978).
14. C. Bernard, N. Christ, A. Guth and E. Weinberg, Phys. Rev. D16, 2967 (1977).
15. H. Osborn, Nucl. Phys. B140, 45 (1978).
16. E. Corrigan, P. Goddard, H. Osborn and S. Templeton, California Institute of Technology, Calt-68-726. Preprint 1979.
17. S. Chadha, A. D'Adda, P. Di Vecchia and F. Nicodemi, Phys. Lett. 67B, 470 (1977) and Phys. Lett. 72B, 103 (1977).
18. L. Brown and D. Creamer, Phys. Rev. D18, 3695 (1978).

TOPOLOGICAL EXCITATIONS AND QUARK CONFINEMENT

Z.F. Ezawa

Max-Planck-Institut für Physik und Astrophysik

München, Germany

ABSTRACT

It is argued that topological excitations in field variables lead to electric charge confinement in a condensed phase of magnetic (topological) charge. This may occur by way of vacuum tunneling due to instantons or a phase transition due to virtual creations of topological solitons. We propose to study this mechanism by investigating the dual Lagrangian which is a functional Fourier transformation of the original Lagrangian. This is because the perturbative vacuum of the dual Lagrangian is the physical vacuum of the original Lagrangian in the presence of topological excitations. As explicit examples, we analyze the Abelian Higgs model in 1+1 dimensions and the Georgi-Glashow models in 2+1 dimensions as well as 3+1 dimensions. These models are shown to give an ideal realization of the electric quark confinement mechanism conjectured by 't Hooft and Mandelstam.

I. INTRODUCTION

Since topological excitations in field variables such as solitons [1,2] and instantons [3] were discovered, an entirely new insight has been obtained in quantum field theories. In particular, due to the rich topological structure in gauge vacua, a possibility of quark confinement has emerged by way of a non-perturbative effect. I wish to devote my lectures to a study of this problem. I shall illustrate, by taking some simple models, how topological quark confinement will be realized.

First, let us summarize some basic topological properties in field theories. When the Lagrangian theory has a topological current which is absolutely conserved by definition, a field operator $\sigma(x)$ may be constructed so as to create or destroy the topological charge Q at point x. Thus, the operator $\sigma(x)$ is defined by the commutation relation

$$[\sigma(x), Q] = \sigma(x). \tag{1.1}$$

Since a topological charge is a superselection rule, charge, it follows that

$$<0|\sigma(x)|0> = 0, \tag{1.2}$$

where $|0>$ is the perturbative vacuum of the Lagrangian.

If the Lagrangian contains topological solitons, the operator $\sigma(x)$ is a Heisenberg operator which describes these solitons. Namely, $\sigma(x)$ creates a bare solitons at point x from the vacuum $|0>$. Explicit examples of operators $\sigma(x)$ are known for sine-Gordon solitons [4], Nielsen-Olesen vortices [5] and 't-Hooft-Polyakov monopoles [6]

Though the operator $\sigma(x)$ is always well defined in the presence of topological currents, the perturbative vacuum $|0>$ is not necessarily a physical vacuum of the theory. Then, the relation (1.2) may break down for physical vacuum. It seems that there exist two cases in which physical vacuum is different from perturbative vacuum. They are as follows:

i) The Lagrangian contains instantons [3], which induce tunneling among perturbative vacua [7].

ii) The Lagrangian contains solitons but not instantons. Then, the perturbative vacuum is a physical vacuum in the weak coupling regime, but it may be unstable in the strong coupling regime because vacuum fluctuations due to virtual creations of topological solitions will be dominant. Namely, the vacuum will undergo a phase transition.

If the physical vacuum $|0\}$ is different from the perturbative vacuum $|0>$ because of the above reasons, the vacuum $|0\}$ must involve an indefinite number of topological excitations of field variables. Therefore, we expect to have

$$\{0|\sigma(x)|0\} = \text{constant} \neq 0. \tag{1.3}$$

Because the topological current may be identified with the magnetic current in gauge theories, (1.3) implies that the vacuum $|0\}$ is a condensed phase of magnetic charges. Therefore, electric flux is squeezed by the Meissner effect and forms quantized vortices in the physical vacuum. Then, just as magnetic vortices confine magnetic monopoles [8,9], electric vortices confine electric monopoles which are quarks. In this way, we are able to realize electric quark confinement conjectured by Mandelstam [10] and 't Hooft [11].

In order to carry out the above program, we need to determine the physical vacuum together with the canonical field variables acting on it. Namely, we wish to obtain an equivalent Lagrangian whose perturbative vacuum is the physical vacuum $|0\}$. We call such an equivalent Lagrangian the dual Lagrangian. It is rather easy to convince oneself that the dual Lagrangian is constructed by a functional Fourier transformation of the original Lagrangian. Here I mention two basic reasons:

(a) In performing Fourier transformations, we integrate over all field configurations that necessarily involves topological excitations such as instantons and virtual creations of solitons.

(b) In performing Fourier transformations, we interchange the coupling constant g in the original Lagrangian and the coupling constant g^{-1} in the dual Lagrangian.

In general, functional Fourier transformations are extremely complicated. We have to develop a sensible way of approximation.

My lectures are composed as follows. First, in order to illustrate functional Fourier transformation, we analyze the Abelian Higgs model in 1+1 dimensions [12] . The model contains instantons which are Nielsen-Olesen vortices and it gives a prototype of vacuum tunneling [7]. On the basis of the dual Lagrangian, we prove that external charge e is confined for $e/g \neq$ integer. The result confirms the standard analysis [7] . Next, we study the SU(N) Georgi-Glashow model in 2+1 dimensions [13] . The model contains instantons which are 't Hooft-Polyakov monopoles and it gives a prototype of electric quark confinement [14] . On the basis of the dual Lagrangian, we prove that external charge e is confined by quantized electric flux for Ne/g = integer. The result confirms the work of Polyakov [14] in case of N = 2. Moreover, we show that the present scheme gives an ideal realization of the electric confinement mechanism conjectured by 't Hooft [11] . Finally, we analyze the SU(N) Georgi-Glashow model in 3+1 dimensions. The model contains topological solitons which are 't Hooft-Polyakov monopoles. We prove, in the strong coupling limit, that the vacuum $|0\}$ is a condensed phase of magnetic monopoles, and that external charge e is confined by quantized electric flux for Ne/g = integer by the Meissner effect. Hence, the present scheme provides an ideal realization of the electric confinement mechanism conjectured by Mandelstam [10,16] .

II. ELECTRIC CONFINEMENT IN 1+1 DIMENSIONS

In this section we investigate the Abelian Higgs model in 1+1 dimensions, because this is the simplest model that contains instantons. The Lagrangian density is given by

$$L = -\frac{1}{4} F_{\mu\nu} F^{\mu\nu} + \frac{1}{2} |(\partial_\mu - igA_\mu)\Phi|^2 - \frac{\lambda^2}{4}(|\Phi|^2 - v^2)^2 + \frac{e}{2} J\epsilon_{\mu\nu} F^{\mu\nu} , \tag{2.1}$$

where an external field eJ has been explicitly introduced. When we parametrize

$$A_\mu = U_\mu + \frac{1}{g}\partial_\mu \chi \quad \text{and} \quad \Phi = \rho e^{i\chi} \tag{2.2}$$

the field χ decouples from the Lagrangian. Such a field represents a massless unphysical Goldstone mode. Now, let us freeze the Higgs field by letting $\lambda \to \infty$, where $\rho = v$. Then, (2.1) reads

$$L \to -\frac{1}{4} U_{\mu\nu} U^{\mu\nu} + \frac{1}{2} m_v^2 U_\mu U^\mu + \frac{e}{2} J\epsilon_{\mu\nu} U^{\mu\nu} , \tag{2.3}$$

where $m_v = gv$ and $U_{\mu\nu} = \partial_\mu U_\mu - \partial_\nu U_\mu$. Thus, the Lagrangian describes simply a free massive Proca field U_μ. It is easy to see that charge e is not confined.

The above arguments neglect all effects due to instantons. We now consider them by deriving a dual Lagrangian of (2.1). We start with the generating functional

$$Z = \int [d\rho]\,[dU_{\mu\nu}]\,[dU_\mu]\,\delta(U_{\mu\nu} - \partial_\mu U_\nu - \partial_\nu U_\mu)\,\exp\{-\int L\}. \tag{2.4}$$

The Lagrangian (2.1) contains instantons which are Nielsen-Olesen vortices. Their positions are given by Higgs zeros. Each instanton has a topological charge $(g/4\pi)\int d^2x\epsilon_{\mu\nu} U_{\mu\nu} = n =$ integer, and a mass

$$\pi n^2 m_v^2 g^{-2} \log(\lambda/v).$$

Therefore, when we freeze the Higgs field by setting $\rho = v$, we may approximate (2.4) as

$$Z = \int [d(\text{instanton})]\,[dU_{\mu\nu}][dU_\mu]\,\delta(U_{\mu\nu} - \partial_\mu U_\nu - \partial_\nu U_\mu - \epsilon_{\mu\nu} Q) \times \exp\{-\int [\frac{1}{4} U_{\mu\nu}^2 + \frac{1}{2} m_v^2 U_\mu^2 + m - i\frac{e}{2} J\epsilon_{\mu\nu} U_{\mu\nu}]\}. \tag{2.5}$$

where field U_μ now represents quantum fluctuations around instantons, and the instanton configuration is summarized by a topological charge density

$$Q(x) = \frac{2\pi}{g} \Sigma n_q \delta(x-r_q) \tag{2.6}$$

and

$$m(x) = \varepsilon \Sigma n_q^2 \delta(x-r_q), \; \varepsilon = \pi m_v^2 g^{-2} \log(\lambda/v). \tag{2.7}$$

Here, for simplicity, we have assumed vortices to be point-like objects ($m_V \to \infty$). The notation [d(instanton)] stands for an integration over all possible positions of Higgs zeros.

We introduce a dual variable $G_{\mu\nu}$ by

$$\delta(U_{\mu\nu} - \partial_\mu U_\nu + \partial_\nu U_\mu - \epsilon_{\mu\nu} Q)$$
$$= \int [dG_{\mu\nu}] \exp \{i \int G_{\mu\nu}(U_{\mu\nu} - \partial_\mu U_\nu + \partial_\nu U_\mu - \epsilon_{\mu\nu} Q)\}. \tag{2.8}$$

Inserting (2.8) into (2.5), and integrating over $U_{\mu\nu}$ and U_μ, we obtain

$$Z = \int [d(\text{instanton})][dG_{\mu\nu}] \exp\{-\int (m + i\epsilon_{\mu\nu} G_{\mu\nu} Q)\}$$
$$\times \exp\{-\int (G_{\mu\nu} + \frac{e}{2}\epsilon_{\mu\nu} J)^2 - \frac{2}{m_v^2}\int (\partial_\nu G_{\mu\nu})^2\}. \tag{2.9}$$

Let us write

$$G_{\mu\nu} = \frac{1}{2}\epsilon_{\mu\nu} U, \tag{2.10}$$

since $G_{\mu\nu}$ is antisymmetric by definition (2.8). Inserting (2.10) into (2.9) we obtain

$$Z = \int [d(\text{instanton})][dU] \exp\{-\int (m + iQU)\}$$
$$\times \exp\{-\frac{1}{2m_V^2}\int [(\partial_\mu U)^2 + m_V^2 (U + eJ)^2]\}. \tag{2.11}$$

We now integrate over all possible configurations of instantons. We do this calculation by discretizing the space-time into cubic lattice [1,17], where the lattice spacing is $\sim m_V^{-1}$. We make a substitution

$$\exp\{-\int(m+iQU)\} = \exp\{-\varepsilon\Sigma n^2 - \frac{2\pi}{g}\Sigma nU\}, \tag{2.12}$$

where n(x) is an integer-valued field defined on lattice sites. The integration over all possible instanton configurations corresponds to the summation over n(x) independently at each lattice sites. When the vortex mass ε is large enough, we may neglect the terms $|n|\geq 2$ in (2.12). We approximate (2.12) as

$$1 + 2e^{-\varepsilon}\cos\frac{2\pi}{g}U$$

at each lattice site. Hence, (2.11) leads to

$$Z = \int[dU]\exp\{-\frac{1}{2m_v^2}\int[(\partial_\mu U)^2 + m_v^2(U+eJ)^2 - 2m_v^2M^2\cos\frac{2\pi}{g}U]\} \tag{2.13}$$

where $M^2 \sim m_v^2 \exp\{-\pi\frac{m_v^2}{g^2}\log\frac{\lambda}{v}\}$.

We have shown that the dual Lagrangian represents a massive sine-Gordon model;

$$L = \frac{1}{2}(\partial_\mu U)^2 - \frac{m_v^2}{2}(U + \frac{e}{vg}J)^2 - M^2(1 - \cos 2\pi vU), \tag{2.14}$$

where we have rescaled U(x) by the factor m_v. It is easy to see that (2.14) is reduced to (2.3) when M = 0. The periodic potential term results from the instanton effects, which are essentially non-perturbative.

The problem of charge confinement has been solved [18] in models of the type (2.14). Let us review this. We extract the long-range part of the interaction between external charges, by setting $J=\theta(x + R)\,\theta(R - x)$. The computation is simplified when we rewrite (2.14) as

$$L = \frac{1}{2}(\partial_\mu U')^2 - \frac{m_v^2}{2}U'^2 - M^2[1 - \cos(2\pi vU' - \frac{2\pi e}{g}J)], \tag{2.15}$$

where $U' = U + (e/vg)J$. Hence, the interaction energy is given by $2RM^2(1-\cos 2\pi e/g)$. We conclude that charge e is either screened (e/g=integer) or confined ($e/g \neq$ integer).

We have analyzed the 1+1 dimensional Higgs model in physical gauge and illustrated how to derive the dual Lagrangian. However, this is not a good example to illustrate the topological structure which leads to (1.2) and (1.3), though these equations still follow. This is because, in discussing the topological structure of gauge vacua, we need to use the Lorentz gauge by recovering the Goldstone field χ. Here, only the results are mentioned, by referring details elsewhere [12]. The topological current in this model reads

$$J_\mu = \epsilon_{\mu\nu} \partial^\nu \chi \tag{2.16}$$

Since the charge is given by

$$Q = \int dx_1 \, J_0 = \chi(+\infty) - \chi(-\infty), \tag{2.17}$$

the operator $\sigma(x)$ is defined by

$$\sigma(x) = e^{i\tilde{\chi}(x)}, \tag{2.18}$$

where

$$\partial_\mu \tilde{\chi} = \epsilon_{\mu\nu} \partial^\nu \chi \tag{2.19}$$

It is obvious that the perturbative vacuum satisfies (1.2) since $\sigma(x)$ carries a superselection rule [19] . Though it is not trivial to see, a careful analysis shows that the vacuum of the dual Lagrangian (2.14) satisfies (1.3) in the Lorentz gauge [12] . This is because the Lagrangian (2.14) is equivalent to a gauged Fermion system, especially to the massive Schwinger model for $r = 1/\sqrt{\pi}$ in the Lorentz gauge.

III. ELECTRIC CONFINEMENT IN 2+1 DIMENSIONS

In this section we investigate the SU(N)-Georgi-Glashow model in 2+1 dimensions. The model is defined by breaking gauge symmetry SU(N) completely by the Higgs mechanism except for one continuous Abelian symmetry describing electromagnetism. This symmetry is chosen to contain the center Z_N as a subgroup. This model contains instantons which are 't Hooft-Polyakov monopoles[2] . These monopoles have finite masses and their charges assume multiple values of N Dirac units[21] . Hence, a system of instantons is simulated by a system of Dirac monopoles, and characterized by the magnetic source

$$\partial_\mu {}^*F^\mu(x) = \rho(x), \quad ({}^*F^\mu = \frac{1}{2}\epsilon^{\mu\alpha\beta} F_{\alpha\beta}) \tag{3.1}$$

where

$$\rho(x) = 2\pi N \Sigma n_q \delta(x - r_q), \tag{3.2a}$$

and the mass density

$$M(x) = \epsilon \Sigma \, n_q^2 \, \delta(x - r_q). \tag{3.2b}$$

Here, $F_{\alpha\beta}$ is the electromagnetic field and r_q denote the positions of instantons which are given by Higgs zeros. For the sake of simplicity, we have assumed monopole to be point-like objects (i.e., $m_s \to \infty$ and then $m_v \to \infty$).

Let us review briefly the topological structure of the Abelian subgroup generated by the electromagnetism. The system has a topological current

$$J_\mu = \frac{1}{2\pi} \epsilon_{\mu\alpha\beta} F^{\alpha\beta} = \frac{1}{\pi} {}^*F_\mu \tag{3.3}$$

which leads to the magnetic vortex charge

$$Q = \int d^2x J_0 = \frac{1}{2\pi} \oint dx_k A_k \tag{3.4}$$

in the Abelian gauge [21] . Then, the vortex field operator $\sigma(x)$ is constructed so as to create or destroy this charge at point x. Such an operator must satisfy (1.1), and hence is given by [5]

$$\sigma(x) = Z_\sigma \exp\{ ig^{-2} \int d^2y\, f_k(x-y)\, F_{ok}(y)\} , \tag{3.5}$$

where Z_σ is a normalization factor and $f_k(z)$ is a smooth c-number function satisfying

$$\oint dz_k\, f_k(z) = 2\pi. \tag{3.6}$$

We may choose

$$f_k(z) = \epsilon_{kj} \frac{z^j}{r} (1 - e^{-\beta r^2}) \tag{3.7}$$

where the limit $\beta \to \infty$ is taken in order to define the bare vortex operator. The operator (3.5) creates a vortex soliton at point x from the vacuum, provided that the Abelian gauge symmetry is spontaneously broken except for the Z_N symmetry [5]. However, here we wish to keep the Abelian gauge symmetry unbroken. We now prove that the vacuum is a condensed phase of vortices and that the Z_N symmetry is spontaneously broken.

In the following analysis, we treat only the unbroken gauge symmetry explicitly. We may consider, for simplicity, that all other fields are frozen ($m_s \to \infty$ and then $m_v \to \infty$). A careful treatment of these fields shows that their contributions are merely to modify the equation which relates the mass of "photon" to the mass of monopoles [14] . The topological excitations are summarized by the space-time positions of the Higgs zeros. Hence, the generating functional is given by the formula

$$Z = \int [d(\text{instantons})][dF_{\mu\nu}]\ \delta(\partial_\mu {}^*F_\mu - \rho(x))$$
$$\times \exp\{-\frac{1}{4g^2}\int (F_{\mu\nu}F_{\mu\nu} + m)\}, \tag{3.8}$$

where $\rho(x)$ and $m(x)$ are given by (3.2). The notation [d(instantons)] stands for an integration over all possible positions of Higgs zeros.

Rewriting (3.8) as

$$Z = \int [d(\text{instantons})][dF_{\mu\nu}][dB]$$
$$\times \exp\{-\frac{1}{4g^2}\int (F_{\mu\nu}F_{\mu\nu} + m) + i\int B(\partial_\mu {}^*F_\mu - \rho)\} \tag{3.9}$$

we integrate over $F_{\mu\nu}$ and obtain

$$Z = \int [d(\text{instantons})][dB] \exp\{-\frac{g^2}{2}\int (\partial_\mu B)^2 -$$
$$-\int (\frac{m}{2} + i\rho B)\}. \tag{3.10}$$

We now integrate over all possible configurations of instantons. We do this calculation by discretizing the space-time into cubic lattices, where the lattice spacing is $\sim m_V^{-1}$. We may use some techniques familiar in lattice gauge theory [17]. First, we make a substitution

$$\exp\{-\int (\frac{m}{4g^2} + i\rho B)\} = \exp\{-\frac{\varepsilon}{4g^2}\Sigma n(x)^2 - i2\pi N\Sigma n(x)B(x)\} \tag{3.11}$$

where $n(x)$ is an integer-valued field defined on each sites. Here, if the monopole mass ε is large enough, the term $|n| \geqq 2$ are negligible in (3.11). Then, approximating

$$\exp\{-\int (\frac{m}{4g^2} + i\rho B)\} = 1 + 2^{-\varepsilon/4g^2} \cos 2\pi NB,$$

we obtain

$$Z = \int [dB] \exp\{-\frac{g^2}{2}\int [(\partial_\mu B)^2 - M^2 \cos 2\pi NB)]\}, \tag{3.12}$$

where $M^2 \sim M_o^2\, e^{-\epsilon/4g^2}$, M_o being the monopole mass.

We have shown that the dual Lagrangian is given by the sine-Gordon model,

$$L = \frac{g^2}{2}\,[\partial_\mu B \partial^\mu B - M^2(1 - \cos 2\pi NB)] \tag{3.13}$$

together with

$$F_{\mu\nu} = g^2\, \epsilon_{\mu\nu\alpha} \partial^\alpha B \tag{3.14}$$

which follows from (3.9). We call B(x) a magnetic potential. (Note that our terminology for B(x) is not the same as the one used in ref.22). By making use of (3.14), we may express the vortex operator $\sigma(x)$ in terms of the magnetic potential. Since we have

$$\epsilon_{kj}\partial_j f_k(z) = 2\pi\delta^{(2)}(z)$$

in the local limit $\beta \to \infty$ in (3.3), it is easy to rewrite (3.4) as

$$\sigma(x) = e^{-2\pi i B(x)}. \tag{3.15}$$

In this way the Lagrangian density (13) together with (15) describes the Abelian gauge theory embedded in gauge group SU(N).

We proceed to discuss various properties of the system. First, we notice that perturbative treatment is applicable to the usual Lagrangian (4.13) in the strong coupling limit ($g \to \infty$). It is obvious that the vacua of the model must satisfy

$$B(x)\, |n\} = \frac{n}{N}\, |n\}\ , \tag{3.16a}$$

or

$$\sigma(x)|n\} = Z_\sigma e^{-2\pi n i/N}|n\}\ . \tag{3.16b}$$

Therefore, the vacua $|n\}$ are eigenstates of the vortex operator $\sigma(x)$ with eigenvalues being members of the center Z_N of gauge group SU(N). In other words, the Z_N symmetry is spontaneously broken in the strong coupling limit.

It is instructive to notice that the Lagrangian (3.13) together with (3.16) is equivalent to

$$L = \partial_\mu\sigma\partial^\mu\sigma^* + \frac{g^2M^2}{4Z_\sigma^N}(\sigma^N + \sigma^{*N}) - \lambda^2(|\sigma|^2 - Z_\sigma^2)^2, \tag{3.17}$$

where we have set $Z_\sigma = g/2\sqrt{2}\pi$ and the limit $\lambda^2 \to \infty$ is understood. This Lagrangian is precisely the one which 't Hooft has assumed in his paper [11] to discuss the electric confinement mechanism in

2+1 dimensions. However, it is more convenient to use (3.13) rather than (3.17) for this purpose.

We now show that a quantized electric flux is squeezed by the Meissner effect and generates a domain wall which separates two different vacua. We start with the fact that the Lagrangian (3.13) contains a vortex-like soliton which is a band with width $\sim M^{-1}$ in the 2 dimensional space. We consider a static and x_2-independent soliton solution of (3.13). The field equation is

$$\left(\frac{\partial}{\partial x_1} \right)^2 h(x_1) - \pi N M^2 \sin 2\pi N h(x_1) = 0, \tag{3.18}$$

which has a famous sine-Gordon solution with the following asymptotic behaviour;

$$h(x_1 = +\infty) = \frac{1}{N} \text{ and } h(x_1 = -\infty) = 0. \tag{3.19}$$

According to a general prescription[5] , the quantum field operator which generates such a classical soliton is easily constructed and given by

$$W_p(x) = \exp \{ i\, g^2 \int d^2y\, h(x-y)\, \dot{B}(y) N \lambda^{em} \}, \tag{3.20}$$

where λ^{em} is the generator [20] of the electromagnetic subgroup SU(N). Let us consider the state

$$| h(x) \} = W_p(x) | n \} . \tag{3.21}$$

It is trivial to see that

$$\{ h(y) | T(x) | h(y) \} = \exp\{ -2\pi i [h(x-y) + \frac{n}{N}] \}. \tag{3.22}$$

From (3.19), we conclude that the operator $W_p(x)$ generates a domain wall which separates two vacua $|n\}$ and $|n+1\}$. The domain wall is composed of a quantized electric flux, since

$$E_1 = -g\{ h | \partial_2 B | h \} = 0, \quad E_2 = g\{ h | \partial_1 B | h \} = g \partial_1 h,$$

$$H = g \{ h | \partial_0 B | h \} = 0, \tag{3.23a}$$

and

$$\int dx_1\, E_2 = \frac{1}{N} g. \tag{3.23b}$$

The domain wall carries a finite energy per unit length.

It is interesting to take the zero-width limit of the domain wall. This is achieved by replacing $h(x_1)$ with a step function $\theta(x_1)/N$ in (3.20), as is in accord with (3.19). Then, the operator (3.20) reads

$$W_P = \exp \{i \int_P dy_k A_k(y) \lambda^{em}\}, \tag{3.24}$$

where we have used (3.14) and a formal relation $F_{\mu\nu} = \partial_\mu A_\nu - \partial_\nu A_\mu$, though electric potential $A_\mu(x)$ is no longer well defined in the dual Lagrangian. This simply means that the vortex operator (3.20) is related to the non-integrable phase factor (3.24) by a dual transformation. Keeping this notice in mind, we may say that the phase factor (3.24) creates a bare domain wall along the path of integration.

Finally we study the problem of quark confinement. Since the domain wall is composed of a quantized electric flux, it can terminate at electric monopole (quarks) carrying charge $\pm$ g/N. We introduce these quark field operators perturbatively into the previous formalism. It is important to notice that a single quark field cannot be included by itself, simply because it is not single-valued under the operation of the vortex operator. Thus, a quark field must be always accompanied by a domain wall in such a way as

$$\exp \{i \int_x^\infty dz_k A_k(z)\ \lambda^{em}\}\ \ \psi(x).$$

Although this combination can exist topologically, it does not exist dynamically because of an infinite energy carried by the domain wall. Therefore, quarks must be confined by domain walls of finite length stretched among them. For instance, a bare mesonic state is created by the operator

$$M_P = \bar{\psi}(y) \exp \{\ i \int_x^y dz_k A_k(z)\ \lambda^{em}\}\ \ \psi\ (x). \tag{3.25}$$

So far we have only analyzed the "electromagnetic" component of the gauge group explicitly. Thus, the operator (3.25) is invariant under electromagnetic gauge transformation. In order to implement the full gauge invariance, we need to recover all the gauge degrees of freedom. In this case, (3.25) would read

$$M_P = \bar{\psi}(y)\ T \exp \{\ i \int_x^y dz_k A_k^{(a)}\ \ \lambda^{(a)}\}\ \ \psi\ (x). \tag{3.26}$$

where $\lambda^{(a)}$ are the N^2-1 generators of gauge group SU(N). The operator (3.26) is a well-known formula for mesons, and would lead to mesonic strings with width M^{-1} when quantum fluctuations are taken into account.

In conclusion, we have demonstrated that the SU(N) Georgi-Glashow model gives an excellent realization of the 't Hooft picture of quark confinement!

APPENDIX

We wish to discuss a problem of an open domain wall! How can it separate two vacua $|o\rangle$ and $|1\rangle$? The problem is precisely the same as for an open Nielsen-Olesen vortex in superconductor. Just as magnetic monopole needs a Dirac string when described in terms of electric potential A_μ, an electric monopole needs such a string when described in terms of magnetic potential B. First, let us see this in the empty space $|o\rangle$. Setting an electric monopole g/N at the origin, we solve (3.14) together with

$$\partial^k F_{ok} = \frac{g^2}{N} \delta(x),$$

which yields

$$\varepsilon_{ij} \partial^i J^j B(x) = \frac{1}{N} \delta(x).$$

Thus, B(x) is multiple valued. Taking a cut along the x-axis, we obtain

$$B(x) = \frac{1}{2\pi N} \Theta,$$

Θ being the azimuthal angle. The electric flux is isotropic,

$$F_{ok} = \frac{g^2}{2\pi N} \cdot \frac{x^k}{\gamma^2},$$

since it is placed in the empty space. In the condensed phase $|o$ of magnetic charge, such a flux is squeezed into a domain wall as we have discussed before. Therefore, an open domain wall separates two vacua $|o\rangle$ and $|1\rangle$, by forming a closed loop together with an open cut ("Dirac string"). We may as well take a cut along the core of a domain wall. Then, only a single vacuum $|o\}$ is enough to describe this system. Note that in this case there is a discontinuity in the magnetic potential B along the core of the domain wall.

IV. ELECTRIC CONFINEMENT IN 3+1 DIMENSIONS

In this section we investigate the SU(N) Georgi-Glashow model in 3+1 dimensions. The model contains topological solitons which are 't Hooft-Polyakov monopoles. As we have noticed before, their charges assume multiple values of N Dirac units [21]. A system of these monopoles is simulated by a system of Dirac monopoles and characterized by the magnetic current

$$\partial^{\mu *}F_{\mu\nu} = k_\nu \quad , \qquad (^*F_{\mu\nu} = \frac{1}{2}\, \epsilon_{\mu\nu\alpha\beta}F^{\alpha\beta}) \tag{4.1}$$

where

$$k^\mu(x) = 2\pi N\, \Sigma n_q \int d\tau \quad \dot{r}^\mu_q \quad \delta(x - r_q), \tag{4.2a}$$

and the mass density

$$m(x) = \epsilon\, \Sigma n^2_q \int d\tau \sqrt{\dot{r}^\mu_q \dot{r}^\mu_q} \quad \delta(x - r_q). \tag{4.2b}$$

Since the magnetic current (4.1) is conserved absolutely, it defines a topological current in this model. For the sake of simplicity, we have assumed monopoles to be point-like objects ($m_S \to \infty$and then $m_V \to \infty$). Here we emphasize that these monopoles have not been put in by hand. They are topological excitations of fundamental field variables in the Georgi-Glashow model. In particular, the world lines of the Higgs zeros produces the magnetic current (4.1) as parametrized by (4.2a).

Thus, the Georgi-Glashow model allows virtual creations of monopole-antimonopole pairs as vacuum fluctuations. These topological excitations are expected to affect the vacuum structure essentially in the strong coupling limit. Indeed, we shall demonstrate that the Abelian symmetry is dynamically broken and that the vacuum becomes a condensed phase of magnetic monopoles. In other words, 't Hooft-Polyakov monopoles are tachyonic in the strong coupling limit. Hence, our model gives an ideal realization of the electric confinement mechanism conjectured by Mandelstam [10].

Let us start with a brief discussion on a field operator describing magnetic monopoles. Details are found in refs.[10] and [12]. The topological current (1) leads to the topological charge

$$Q = \int d^3x\ k_o.$$

Then, we may construct a monopole field operator T(x) so as to create or destroy this charge. Therefore, such an operator must satisfy (1.1) and hence

$$[\sigma(x),\ A_k(y)\ ] = u_k(x-y) \quad \sigma(x), \tag{4.3}$$

where $A_k(x)$ is an electric potential in the Abelian gauge ($F_{\mu\nu}=\partial_\mu A_\nu - \partial_\nu A_\mu$), while $u_k(z)$ is a c-number function describing a Dirac monopole which carries N Dirac units of charge ($u_k(z) = \frac{N}{2}(1 + \cos\Theta)\ \partial_k\ \varphi$ in the standard spherical coordinate). Note that the Dirac string can be removed by a gauge rotation in the group space SU(N) and that T(x) describes Wu-Yang monopoles in this new gauge [20, 21]. Here, we have only considered a bare monopole, since it has all the topological properties. The 't Hooft-Polyakov monopole is constructed by dressing the Wu-Yang monopole with a coherent cloud of massive gluons. We may write down a monopole operator explicitly,

$$\sigma(x) = Z_\sigma \exp\{ig^{-2} \int d^3y\, u_k(x-y)\, F_{ok}(y)\}, \tag{4.4}$$

where Z_σ is a normalization factor.

Note that the definition (4.4) only refers to a fundamental field in the Georgi-Glashow model, and not to the vacuum of the model. Now, in the weak coupling limit ($g \to o$) we may define the vacuum consistently by

$$\langle o|\, A_\mu(x)\, |o\rangle = o, \tag{4.5a}$$

which implies [5,6]

$$\langle o|\sigma(x)\, |o\rangle = o. \tag{4.5b}$$

Thus, the operator $\sigma(x)$ creates a bare monopole at a point x from the vacuum $|o\rangle$. These monopoles describe physical particles.

We proceed to analyze the strong coupling regime of the SU(N) Georgi-Glashow model. Such a regime corresponds to the weak coupling regime of the dual Lagrangian ($g^{-1} \to o$), since dual transformation interchanges coupling constants g and g^{-1}. In order to obtain the dual Lagrangian, we need to know topological excitations of the SU(N) Georgi-Glashow model. As we have noticed before, they are given by virtual creations of monopole-antimonopole pairs. They trace monopole loops which are world lines of the Higgs zeros parametrized by (4.2a).

In the following analysis, we treat only the unbroken Abelian gauge symmetry explicitly. We may consider, for simplicity, that all other fields are frozen except for their topological properties. A careful treatment of these fields would show that their contributions are merely to modify the equation which relates the mass of "photons" to the mass of monopoles, as is the case in 2+1 dimensions [14] . The essential topological excitations are summarized by the world lines of the Higgs zeros. Hence, the generating functional is given by the formula

$$Z = \int [d(\text{loop})] \int [dF_{\mu\nu}]\, \delta(\partial_\mu {}^*F_{\mu\nu} - k_\nu) \exp\{- \frac{1}{4g^2} \int (F_{\mu\nu}F_{\mu\nu} + m)\} \tag{4.6}$$

where $k_\mu(x)$ and $m(x)$ have been defined by (4.2). The notation [d(loop)] stands for an integration over all possible world lines of the Higgs zeros.

Rewriting (4.6) as

$$Z = \int [d(\text{loop})] \int [dF_{\mu\nu}] \int [dB_\nu] \exp\{ -\frac{1}{4g^2} \int F_{\mu\nu} F_{\mu\nu} + m) + i \int B_\nu (\partial_\mu {}^*F_{\mu\nu} - k_\nu) \} , \tag{4.7}$$

we integrate over $F_{\mu\nu}$ and obtain

$$Z = \int [d(\text{loop})] \int [dB_\nu] \exp\{ -\frac{g^2}{4} \int G_{\mu\nu} G_{\mu\nu} - \int (\frac{m}{4g^2} + i k_\mu B_\mu) \} , \tag{4.8}$$

where $G_{\mu\nu} = \partial_\mu B_\nu - \partial_\nu B_\mu$. We call $B_\mu (x)$ a magnetic gauge potential. Note that our terminology is not the same as the one used in ref.[16].

Now we integrate over all possible configurations of monopole loops. We do this calculation by discretizing the space-time into cubic lattices. We may use some techniques familiar in lattice gauge theory [17]. First we make a substitution

$$\exp \{ - \int (\frac{m}{4g^2} + ik_\mu B_\mu)\} = \exp \{ - \frac{\varepsilon}{4g^2} \Sigma n_\mu^2 - i2\pi N \Sigma n_\mu B_\mu \} , \tag{4.9}$$

where $n_\mu (x)$ is an integer-valued field defined on links, and the summation Σ runs along monopole loops. The integration over all possible monopole loops corresponds to the summation over $n_\mu (x)$ at each links independently, except for the current conservation condition $\partial_\mu n_\mu = 0$ at each sites. Writing this condition as

$$\delta (\partial_\mu n_\mu) = \int [d\chi] \exp \{ - i \Sigma \chi(x) \partial_\mu n_\mu (x) \} \tag{4.10}$$

we perform the summation over n_μ, and obtain

$$\int [d(\text{loop})] \exp\{ - \int (\frac{m}{4g^2} + i k_\mu B_\mu) \} = \sum_{M_\mu} \int [d\chi] \exp \{ - \frac{g^2}{\varepsilon} \Sigma (\partial_\mu \chi - 2\pi N B_\mu + 2\pi M_\mu)^2 \} , \tag{4.11}$$

where we have used the identity

$$\sum_n \exp \{ - an^2 + ibn \} = \sqrt{\frac{\pi}{a}} \sum_m \exp\{ - (b - 2\pi m)^2 / 4a \} .$$

Note that $\chi (x)$ is an angle variable, $2\pi > \chi \geq 0$, because the terms $2\pi\ m_\mu$ imply such a periodicity. Taking the continuum limit of (4.11), and inserting it into (4.8), we obtain

$$Z = \int [dB_\mu] \int [d\chi] \exp \{- g^2 \int [\frac{1}{4} G_{\mu\nu}G_{\mu\nu} + \frac{M^2}{2} (B_\mu - \frac{1}{2\pi N}\partial_\mu\chi)^2\} , \tag{4.12}$$

where M ~ monopole mass.

We have shown that the dual Lagrangian is given by

$$L = g^2 [- \frac{1}{4} G_{\mu\nu}G^{\mu\nu} + \frac{M^2}{2} (B_\mu - \frac{1}{2\pi N} \partial_\mu\chi)^2], \tag{4.13}$$

together with the relation

$$g^2 G_{\mu\nu} = {}^*F_{\mu\nu}, \tag{4.14}$$

which follows from (4.7).

The dual Lagrangian (4.13) may be analyzed perturbatively in the strong limit. Thus, the vacuum is defined consistently by

$$\{0 | B_\mu (x) | 0\} = 0, \quad \{0|\chi(x)|0\} = 0, \tag{4.15}$$

from which follows

$$\{0 | \sigma(x) | 0\} = Z_\sigma , \tag{4.16}$$

where

$$\sigma(x) = Z_\sigma e^{i\chi(x)} . \tag{4.17}$$

Here, we prove that the operator (4.17) is nothing but the monopole operator defined by (4.4). First, we note that $B_\mu (x)$ acquires a mass, and yet that the Lagrangian (4.13) is invariant under a magnetic gauge transformation

$$B_\mu \to B_\mu + \frac{1}{2\pi N} \partial_\mu f$$

$$\chi \to \chi + f. \tag{4.18}$$

These facts imply that $B_\mu (x)$ is composed of two parts, a massive "photon" field $U_\mu(x)$ and a massless gauge field $\chi(x)$; $B_\mu = U_\mu + \frac{1}{2\pi N} \partial_\mu\chi$. Then, because only massless quanta may contribute to the determination of the topological property, we may rewrite (4.14) as

$$\sigma(x) = Z_\sigma \exp \{\frac{i}{2\pi N} \int d^3y\, \epsilon_{kij}\, \partial_i \partial_j\, u_k(x-y)\, \chi(y)\}$$

where we have used (4.14). Now, it is trivial to derive (4.20) since

$$\epsilon_{ijk}\partial_i\partial_j u_k(z) = 2\pi N\,\delta(z).$$

Note that $\chi(x)$ represents an unphysical Goldstone mode associated with a dynamical breakdown of the magnetic gauge symmetry (4.18).

It is instructive to notice that the Lagrangian (4.13) together with (4.19b) is equivalent to

$$L = -\frac{1}{4}G_{\mu\nu}G^{\mu\nu}+\frac{1}{2}|(\partial_\mu - i\frac{2\pi N}{g}B_\mu)\sigma|^2-\lambda(|\sigma|^2-Z_\sigma)^2 \tag{4.19}$$

where we have rescaled B_μ by a factor g and the limit $\lambda \to \infty$ is understood. It is obvious that (4.19) may be regarded as a Higgs Lagrangian where the radial component of τ (x) is frozen. Therefore, this Lagrangian contains Nielsen-Olesen vortices with penetration depth $\sim M^{-1}$. These vortices carry electric flux quantized in the unit of g/N. Hence, they may terminate at electric monopoles which have electric charges ng/N, n = 1,2, ... N-1. Note that field operators introduced in the fundamental representation of SU(N) convey precisely these electric charges. Moreover, according to general arguments [16], the magnetic Abelian symmetry is embedded in the magnetic gauge symmetry $SU(N)/Z_N$. Thus, we can apply the well-known topological analysis to these electric vortices just as in the case of the magnetic vortices [9,20]. Namely, the vortices are characterized by the center Z_N of SU(N). This fact allows us to construct baryonic strings [9]. We conclude that quarks are permanently confined by quantized electric flux stretched among them.

Operators creating mesonic states and baryonic states are given by

$$M_P = \bar{\psi}(x)\ W_P(x,y)\ (y), \tag{4.20a}$$

and

$$B_P = \Pi W_{P_k}(x,y_k)\psi(y_k), \tag{4.20b}$$

respectively, where

$$W_P(x,y) = \exp\{ig^2\int d^3z\ h_i(z)G_{oi}(z)N\lambda^{em}\} \tag{4.21}$$

is a field operator [5] which generates a Nielsen-Olesen vortex stretched from y to x along path P. Here, λ^{em} is the generator of the electromagnetic subgroup in SU(N). The c-number function $h_i(z)$ describes a classical vortex. It is to be determined so as to minimize the energy of the system. When a vortex is long enough and parallel to the z_3-axis, we may approximate

$$h_i(z) = \frac{1}{N}\epsilon_{ij}\frac{z^j}{r}\{\frac{1}{r} - MK_1(rM)\},\ i,j = 1,2, \tag{4.22}$$

away from the endpoints. It is interesting to take the zero-width limit of the vortex tube. This is achieved by making $M\to\infty$ in (4.22). Then, the operator (4.21) reads

$$W_P(x,y) = \exp\{i\int_x^y dz_k A_k(z)\lambda^{em}\}. \tag{4.23}$$

In deriving this, we have used a formal relation

$$g^2 G_{\mu\nu} = \epsilon_{\mu\nu\alpha\beta}\partial^\alpha A^\beta,$$

though electric potential $A_\mu(x)$ is no longer well defined. This simply means that the vortex operator (4.21) is related to the non-integrable phase factor (4.23) by a dual transformation. Keeping this notice in mind, we may say that the phase factor (4.23) creates a bare vortex string which confines quarks in the strong coupling regime.

So far we have only analyzed the electromagnetic component of the gauge group explicitly. Thus, the operator (4.23) transforms properly under the electromagnetic subgroup. In order to implement the full gauge covariance, we need to recover all the gauge degrees of freedom. In this case, (4.23) would read

$$W_P(x,y) = T\exp\{i\int_x^y dz_k A_k^{(a)}\lambda^{(a)}\}, \tag{4.24}$$

where $\lambda^{(a)}$ are the N^2-1 generators of gauge group SU(N). Then, operators (4.20) give well-known formulae for mesons and baryons, and would lead to vortex-strings with width $\sim M^{-1}$ when quantum fluctuations are taken into account.

In conclusion, we have demonstrated that the SU(N) Georgi-Glashow model gives an excellent realization of the Mandelstam picture on quark confinement [10].

V. DISCUSSIONS

Taking some simple models, we have illustrated how topological excitations in field variables lead to electric charge confinement in a condensed phase of magnetic (topological) charge. For this purpose we have derived a Lagrangian equivalent to the original one by way of functional Fourier transformations. We call such a Lagrangian the dual Lagrangian, because field variables in these two Lagrangians are related by dual relations such as (3.14) and (4.14). In the presence of topological excitations, we have argued that the physical vacuum of the original Lagrangian is the perturbative vacuum of the dual Lagrangian.

In performing functional Fourier transformation, we have taken the London limit, $m_S \to \infty$ and then $m_V \to \infty$, for the sake of computational simplicity. However, such an assumption will not be necessary if we make use of a technique of the renormalization group. In our examples, topological excitations are arbitrarily localizable. In an instance of (3.7), the scale of excitations is given by β^{-1}, where β is an arbitrary parameter. In the limit $\beta \to \infty$, topological excitations are point-like objects which we have assumed. The dual Lagrangian contains a mass term which involves the ultraviolet "cut-off" β. It would be possible to write down a renormalization equation with respect to this parameter.

In general, functional transformations are extremely complicated. This is especially so in non-Abelian gauge theories. In these lectures, we have extracted the long-range Abelian subgroup and studied only this component by freezing other degrees of freedom. It is very important to analyze how these non-Abelian components are transformed by functional Fourier transformations. In fact, if we could do this for the pure Yang-Mills theory, we would demonstrate quark confinement in quantum chromodynamics. Finally, we note that the present scheme will give a mathematical framework for the so-called electric-magnetic duality in non-Abelian gauge theories suggested by Mandelstam [16] .

REFERENCES

1. H. B. Nielsen and P. Olesen, Nucl.Phys. B61, 45 (1973).
2. G. 't Hooft, Nucl.Phys. B79, 276 (1974).
A. M. Polyakov, JETP Letters 20, 194 (1974).
3. A. M. Polyakov, Phys.Letters 59B, 82 (1975).
A. A. Belavin et al., Phys.Letters 59B, 85 (1975).
4. S. Mandelstam, Phys.Rev. D11, 3026 (1975).
5. Z. F. Ezawa, Phys.Rev. D18, 2091 (1978); Phys.Letters 82B, 426 (1979).
6. Z. F. Ezawa, Phys.Letters 81B, 325 (1979).
7. C. G. Callan et al., Phys.Lett. 63B 334 (1976); 66B, 375 (1977).
8. Y. Nambu, Phys.Rev. D10, 4262 (1974).
9. S. Mandelstam, Phys.Letters 53B, 476 (1974).
Z. F. Ezawa and H. C. Tze, Phys.Rev. D14, 1006 (1976).
10. S. Mandelstam, Phys.Reports 23, 245 (1976).
11. G. 't Hooft, Nucl.Phys. B138, 1 (1978).
12. Z. F. Ezawa, Nucl.Phys., to be published.
13. Z. F. Ezawa, Phys.Letters, 85B, 87 (1979).
14. A. M. Polyakov, Nucl.Phys. B120, 429 (1977).
15. Z. F. Ezawa, Phys.Letters, to be published.

16. S. Mandelstam, Charge-Monopole Duality and the Phases of Non-Abelian Gauge Theories, November 1978.
17. T. Banks, R. Myerson and J. Kogut, Nucl.Phys. B129, 493 (1977).
R. Savit, Phys.Rev. B17, 1340 (1978).
M. E. Peskin, Ann.Phys. 113, 122 (1978).
18. S. Coleman et al., Ann.Phys. 93, 267 (1975).
19. K. D. Rothe and J. A. Swieca, Phys.Rev. D15, 541 (1977).
Z. F. Ezawa, Nuovo Cimento 51A, 187 (1979).
20. Z. F. Ezawa and H. C. Tze, J.Math. Phys. 17, 2228 (1976).
21. A. Arafune et al., J.Math.Phys. 16, 433 (1975).

INFRA-RED BEHAVIOR OF THE RUNNING COUPLING CONSTANT IN QCD

James S. Ball

Physics Dept. University of Utah

Utah/USA

I. INTRODUCTION

In these lectures I will describe a non-perturbative method of studying the infrared (IR) structure of Quantum Chromodynamics (QCD) [1] and then follow with the description of an investigation of the momentum dependence of vertex functions in gauge theories[2]. It is widely accepted that these IR singularities may cause color to be confined and hence produce quark confinement. In particular it is argued [3] that an "effective potential" in momentum space between quarks is proportional to $g^2(q^2)/q^2$ where $g^2(q^2)$ is the running coupling constant which is singular in the IR i.e. as $q^2 \to 0$. For example if $g^2(q^2) \sim 1/q^2$ as $q^2 \to 0$ we can crudely translate this into a potential growing linearly with distance for large distances.

For these reasons we will focus our attention on calculating the running coupling constant in the IR limit in the case of pure glue (no quarks) as it is expected that the self-coupling of the gluon is the feature that will control the IR singularities of the theory and is in fact the cause of confinement.

A convenient way to study $g^2(q^2)$ is to calculate the gluon propagator in axial gauge, for in this gauge the gluon wave function renormalization constant determines $g^2(q^2)$ directly:

$$g^2(q^2) = Z\,(q^2/\Lambda^2)\,g_0^2 \quad . \tag{1.1}$$

Within the class of axial gauges, Z and hence g is gauge invariant.

A further advantage of the use of the axial gauge is that the form of the Ward identity is the same as that for QED, which will allow us to generalize a simple treatment of the IR signularities in QED to the more complicated QCD case. Our basic approach is the following:

(1) We use the Dyson Eq. to calculate the IR behavior of the gluon propagator $\Delta_{\mu\nu}(q^2)$. This calculation requires $\Delta_{\mu\nu}$ as input, and the bare and dressed gluon vertices.

(2) The system of equations is closed by constructing a dressed gluon vertex which "solves" the Ward identity in the IR limit, and hence expresses the gluon vertex in terms of the propagator.

This procedure produces an integral equation for $\Delta_{\mu\nu}$ which in principle allows one to calculate $\Delta_{\mu\nu}$ and therefore $g^2(q^2)$. The basic assumption required is that the longitudinal parts of the vertices, which can be expressed in terms of Δ through the Ward identities, in fact control the IR singularities of Δ, and that the transverse parts have no effect in the small q region. This property is true in QED and we will assume it to be true in QCD.

II. IR BEHAVIOR OF QED

To demonstrate the simplicity of our method and to illustrate the various steps that are required, we will begin with a discussion of the QED case with spinless electrons. In this case the bare quantities are:

$$S_o(p) = \frac{1}{p^2 - m^2} \qquad \Gamma^o_\mu = (p + p')_\mu \tag{2.1}$$

The Dyson equation for the electron propagator reads

$$S^{-1}(p) = S_o^{-1}(p) + e_o^2 \int d^4k \; D_{\mu\nu}(k) \; S(p+k) \; \Gamma^o_\mu \; \Gamma_\nu \tag{2.2}$$

where Γ_ν is the dressed electron-photon vertex function and $D_{\mu\nu}$ is the photon propagator. We will assume $D_{\mu\nu}$ is known and in the Feynman gauge is

$$D_{\mu\nu} = Z_3 \frac{\delta\mu\nu}{k^2} \tag{2.3}$$

and the physical charge $e^2 = Z_3 e_o^2$. The Ward identity satisfied by Γ_ν is

$$k_\mu \Gamma_\mu = S^{-1}(p+k) - S^{-1}(p). \tag{2.4}$$

Since the construction of a Γ_ν which explicitly solves the Ward identity and is at the same time free of kinematic singularities will be an important ingredient in the QCD calculation, I will go through this process in detail. The vertex must be a vector quantity and hence must have the form

$$\Gamma_\mu = Ap_\mu + Bp'_\mu \qquad (2.5)$$

where $k = p'-p$ and A and B are scalar functions of p^2, p'^2 and q^2. Because of the simple form of (2.5) it is clear that A and B are free of kinematic singularities. The Ward identity gives the following relationship between these functions and the propagator:

$$p \cdot k\, A + p' \cdot k\, B = S^{-1}(p'^2) - S^{-1}(p^2) \qquad (2.6)$$

which can be used to eliminate B

$$\Gamma_\mu = (S^{-1}(p'^2) - S^{-1}(p^2))\frac{p'_\mu}{p' \cdot k} + A\left(p_\mu - \frac{p \cdot k}{p' \cdot k}p'_\mu\right). \qquad (2.7)$$

We have now produced a kinematic singularity in the first term at $p' \cdot k = 0$ which must be cancelled by the second term. Thus A cannot be ignored or set equal to zero to obtain an approximation for Γ_μ.

The required cancellation occurs if

$$\lim_{p' \cdot k \to 0} (p \cdot k)\, A = \lim_{p' \cdot k \to 0} (p'^2 - p^2)\, A = S^{-1}(p'^2) - S^{-1}(p^2). \qquad (2.8)$$

This limit is clearly satisfied by defining

$$A = \frac{S^{-1}(p'^2) - S^{-1}(p^2)}{p'^2 - p^2} + (p' \cdot k)\, A' \qquad (2.9)$$

where A' is now an unconstrained scalar function. The final form of Γ_μ is

$$\Gamma_\mu = \frac{S^{-1}(p'^2) - S^{-1}(p^2)}{p'^2 - p^2}(p + p')_\mu + A'(p' \cdot k\, p_\mu - p \cdot k p'_\mu) \qquad (2.10)$$

To produce a closed equation for S^{-1}, we simply set $A' = 0$, the previously mentioned assumption that the transverse part of the vertex does not affect the IR behavior. The Dyson Eq. after multiplying by S(p) is

$$(p^2 - m^2)\, S(p) = 1 + 4e^2 \int \frac{d^4k}{k^2}\, p^2\, \frac{S(p+k) - S(p)}{(p+k)^2 - p^2}\,. \qquad (2.11)$$

The solution to this equation in the limit $p^2 \to m^2$ can be obtained by using the Laplace transform of S(p) as follows:

$$S(p) = \int_0^\infty ds e^{-(p^2-m^2)\ s + G(s)} \tag{2.12}$$

where we let $p^2 \to m^2$ from above. We can then write

$$(p^2-m^2)\ S(p) = \int_0^\infty ds \left[-\frac{\partial}{\partial s} + \frac{\partial G}{\partial s} \right] e^{-(p^2-m^2)s+G(s)}$$

$$= 1 + \int \frac{\partial G}{\partial s}\ e^{-(p^2-m^2)s+G(s)} ds \tag{2.13}$$

where we have used G(0) = 0, a further condition on G. The r.h.s. of Eq. (2.11) becomes

$$1 + 4e^2 \int ds\ e^{-(p^2-m^2)s+G(s)} \int \frac{d^4k}{k^2}\ \frac{p^2\ e^{-(k^2+2p\cdot k)s}}{k^2+2p\cdot k}\ . \tag{2.14}$$

Consider the integral over k. Nowhere in this integrand does m^2 appear, so nothing special can happen as $p^2 \to m^2$. Thus

$$I(s,p^2) = 4e^2 \int \frac{d^4k}{k^2}\ \frac{e^{-(k^2+2\ p\cdot k)s} - 1}{k^2+2p\cdot k} \tag{2.15a}$$

$$I(s) = I\ (s,p^2)\Big|_{p^2 = m^2}. \tag{2.15b}$$

Combining this, and Eqs. (2.13) and (2.14), we find that

$$\frac{\partial G}{\partial s} = I(s) \quad \text{or } G(s) = \int_0^s I(s')\ ds' \tag{2.16}$$

provides a solution to our integral eq. Clearly the $p^2 \to m^2$ limit of S(p) is given by the large s behavior which is

$$G(s) \longrightarrow s\ \Delta\ m + \frac{\alpha}{\pi}\ \ell n\ s. \tag{2.17}$$

The resulting S is

$$S(p) \sim \left[\frac{1}{p^2 - m^2}\right]^{1 + \frac{\alpha}{\pi}} \tag{2.18}$$

which is the well known QED result. As one can see this method yields in a simple, straightforward manner the IR behavior of the QED propagator. It has the following advantages:

1.) It involves only arguments about the IR region of the integrals.
2.) No special cleverness is required to obtain the desired results.

III. GENERALIZATION TO QCD

In applying the method illustrated in the previous section to QCD, we should try to maximize similarities between QCD and QED as much as possible. In particular, the Ward identity should relate the vertex to the propagator, and no ghosts should be present. This again dictates the use of an axial gauge.

Before plunging into the details of the proposed calculations, let us examine the features that clearly are different in QCD, other than the numbers of indices on tensor quantities. First of all the Dyson Eq. will be a non-linear integral equation as the role of the photon as well as the electron is played by a gluon. The four gluon coupling also provides a basically different structure even though it can be related to the 3-glue vertex through a Ward identity. The gauge direction n provides an additional scalar variable, as well as much additional complication in the tensor decomposition of various quantities. Finally, the large q^2 behavior of QCD is known from asymptotic freedom arguments so that we already have a solution to our equation in this limit

$$g^2(q^2) = \frac{g^2(M^2)}{1-b\, g^2(M^2)\, \ell n\, (\frac{q^2}{M^2})} \qquad (3.1)$$

with which our results must agree. The fact that $b < 0$ prevents continuation of this expression to small q^2, so nothing can be inferred about the IR limit of g^2 from this result.

IV. KINEMATICS OF AXIAL GAUGE QCD

The various quantities that enter the Dyson Eq. must be decomposed into scalar multiples of basic tensor forms before the equations relating the scalar functions can be obtained. In this discussion the color indices will be suppressed in that they appear as over-all factors for the propagator and 3-gluon vertex.

The gluon propagator in the axial gauge must be orthogonal to the gauge vector (n_μ) and symmetric. Thus $\Delta_{\mu\nu}$ must be a linear combination of two possible tensor forms

$$\delta_{\mu\nu} - \frac{n_\mu n_\nu}{n^2}, \qquad N_\mu N_\nu$$

where $N_\mu = n_\mu - q_\mu \frac{n^2}{n\cdot q}$ such that $n\cdot N = 0$. Therefore

$$\Delta_{\mu\nu} = - \frac{B}{q^2}\left(\delta_{\mu\nu} - \frac{n_\mu n_\nu}{n^2}\right) - \frac{C}{q^2 n^2 (1-\gamma)} N_\mu N_\nu \tag{4.1}$$

where

$$\gamma = \frac{n^2 q^2}{(n\cdot q)^2} \quad \text{(addition scalar variable due to } n_\mu \text{)}$$

and B and C are functions of q^2 and γ.

For convenience we provide the relation of our B and C to the quantities defined by Kummer [4] here denoted A_k and B_k.

$$C = q^2 (B_k - A_k) \quad , \quad B = q^2 B_k \tag{4.2}$$

Our B and C are dimensionless. For the special case B=1 and $C = 1-\gamma$ we obtain the bare propagator

$$\Delta^0_{\mu\nu} = - \frac{1}{q^2}\left(\delta_{\mu\nu} - \frac{n_\mu n_\nu}{n^2} + \frac{N_\mu N_\nu}{n^2}\right) \tag{4.3}$$

The gluon self energy "Δ^{-1}" denoted $\Pi_{\mu\nu}$ must be orthogonal to q and again symmetric. If we construct a vector from n_μ and q_μ orthogonal to

$$q, \; Q_\mu = q_\mu - q^2 \frac{n_\mu}{n\cdot q},$$

we can write

$$\Pi_{\mu\nu} = - q^2 \left(\delta_{\mu\nu} - \frac{q_\mu q_\nu}{q^2}\right) b - Q_\mu Q_\nu c. \tag{4.4}$$

The relation between Δ and Π is

$$\Delta_{\mu\lambda} \Pi_{\lambda\nu} = \delta_{\mu\nu} - \frac{q_\mu n_\nu}{n\cdot q} \tag{4.5}$$

which gives

$$b = \frac{1}{B} \qquad c = \frac{1}{1-\gamma}\left(\frac{1}{B} - \frac{\gamma}{A}\right). \tag{4.6}$$

Finally, the bare $\Pi_{\mu\nu}$, denoted $\Pi^0_{\mu\nu}$, is given by

$$\Pi^0_{\mu\nu} = -q^2\left(\delta_{\mu\nu} - \frac{q_\mu q_\nu}{q^2}\right) \tag{4.7}$$

which is independent of n and results for b = 1 and c = 0.

The Ward identities in the axial gauge are

$$q_{1\mu_1}\left(i\,\Gamma^{(q_1,q_2,q_3)}_{\mu_1\mu_2\mu_3}\right) = \Pi_{\mu_2\mu_3}(q_2) - \Pi_{\mu_2\mu_3}(q_3) \tag{4.8}$$

together with cyclic permutations of 1,2,3. The "solution" for Γ which is free of kinematic singularities and satisfies the Ward identity and the requirements of Bose symmetry is the following complicated expression:

$$\begin{aligned} i\,\Gamma^{(q_1,q_2,q_3)}_{\mu_1\mu_2\mu_3} &= \delta_{\mu_1\mu_2}\left[\frac{g_1}{q_1^2}\, q_{1\mu_3} - \frac{g_2}{q_2^2}\, q_{2\mu_3}\right] \\ &+ \frac{\dfrac{g_1}{q_1^2} - \dfrac{g_2}{q_2^2}}{q_1^2 - q_2^2}\left[q_1\cdot q_2\,\delta_{\mu_1\mu_2} - q_{1\mu_2}\, q_{2\mu_1}\right](q_2-q_1)_{\mu_3} \\ &+ \frac{f_2-f_1}{n_3}\,\delta_{\mu_1\mu_2}\, n_{\mu_3} + \left\{\frac{q_{2\mu_3}}{n_1}\left[\frac{f_2}{n_2} + \frac{f_3}{n_3}\right]\right. \\ &\left. - \frac{q_{1\mu_3}}{n_2}\left[\frac{f_1}{n_1} + \frac{f_3}{n_3}\right]\right\} n_{\mu_1}\, n_{\mu_2} \\ &+ n_{\mu_1}\, n_{\mu_2}\, n_{\mu_3}\left\{\frac{f_1}{n_1^2}\left(\frac{q_1\cdot q_2}{n_3} - \frac{q_1\cdot q_3}{n_2}\right)\right\} + \text{cyclic perm.} \end{aligned} \tag{4.9}$$

where to produce a compact notation we have denoted

$g(q_i^2) = g_i$, $f(q_i^2) = f_i$ and $n\cdot q_i = n_i$.

The relation between the functions $g(q^2)$ and $f(q^2)$ and those in in $\Pi_{\mu\nu}$ are $f = q^2 c$ and $g = -q^2(b-c)$.

Here we have assumed that $\frac{1}{n \cdot q}$ factors which also appear in the propagator produce no unacceptable kinematic singularities. We further note that all transverse tensor forms in Γ have the property that they vanish when any of the momenta vanish, just as is the case in QED.

The Dyson equation for $\Pi_{\mu\nu}$ is represented by the diagrams in figure 1. If the Ward identity for Γ_4 is solved in terms of Γ_3, the Dyson Eq. will be a non-linear integral equation for b and c. Each of these are functions of two variables q^2, γ and note that the 4-glue terms involve two loop integrations. Since the two loop integrations are at present too difficult to handle we will not attempt to produce a 4-gluon vertex which satisfies its own Ward identity but will simply state one feature of the bare four-gluon vertex which we will use to avoid the problems associated with the 4-glue coupling. All terms in the bare 4-glue are products of pairs of $\delta_{\mu\nu}$'s where the index of each external line appears once. As a result when $n_\mu \Pi_{\mu\nu}$ is formed all four gluon terms, except the tadpole, will vanish as the bare four gluon coupling always has 3 indices from internal propagators.

The general integral equation which we have obtained is far too difficult to handle directly, so some further assumptions must be made to reduce it to a more manageable form. Let us explore several possibilities.

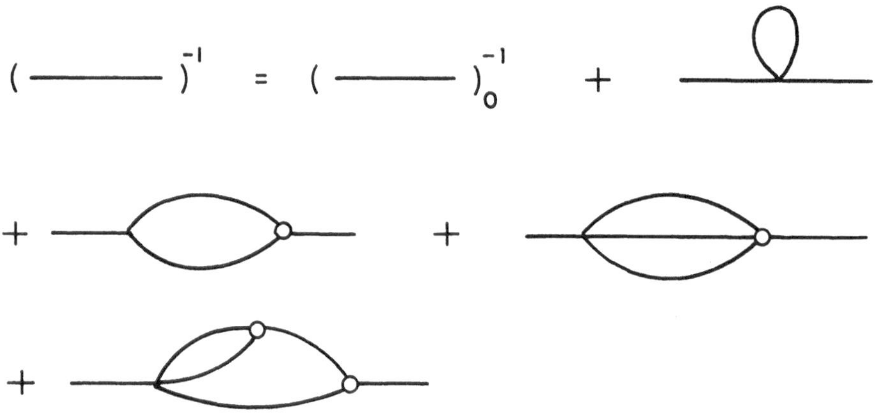

Fig.1 Dyson Equation for QCD; all internal lines are dressed propagators, and open circles are dressed vertices while points are the corresponding bare quantities.

Ignoring the four gluon coupling entirely would certainly simplify the situation. The fact that in perturbation theory Γ_4 is of order g_0^2 while Γ_3 is of order g_0 also lends support to this approximation. On the other hand a Ward identity relates Γ_4 to Γ_3 and a violation of this relation (keeping Γ_3 but $\Gamma_4 = 0$) would certainly cast doubt as to the gauge invariance of the resulting theory. A simple test of this approximation in perturbation theory is the calculation of the UV divergent part of the one loop 3-glue vertex. For the renormalization program to work this one loop term must be proportional to the bare vertex. If the four-glue vertex is ignored, only the triangle graph involving the three-glue vertices survives and the log divergent part is

$$i\,\Gamma^{(2)} \sim 13\ \Gamma^{o} + 9\, n_3 \left[\partial_{\mu_2\mu_3} n_{\mu_1} - \partial_{\mu_3\mu_1} n_{\mu_2}\right] + 9\, n_{\mu_3} \left[n_{\mu_2} q_{3\mu_1} - n_{\mu_1} q_{3\mu_2}\right] + \text{cycl.perm.} \qquad (4.10)$$

However, <u>if</u> the four-glue terms are added, the terms not present in Γ^o (those depending on n) cancel out, and the requirements for renormalization are satisfied once again reminding us of the intimate relationship between gauge invariance and renormalizability.

Another possible assumption is that the most singular part of $\Delta_{\mu\nu}$ in the IR has the same tensor form as the divergent part of Δ in the UV region, namely $\Delta_{\mu\nu} \sim \Delta^{0}_{\mu\nu}$. This assumption roughly corresponds to assuming that the correlation between UV and IR singularities which occurs in perturbation theory is preserved in the complete theory. It is this ansatz we will consider, i.e., $c = 0$ and hence $f = 0$, and it will provide us with a tractable integral equation for the remaining scalar function b. Furthermore, since b is in fact γ independent in the UV we will assume this to be true for the IR region as well. This will result in a non-linear integral eq. for b where b is a function of a single variable q^2. Note however that at least in principle all of these assumptions can be checked.

Since $\Delta_{\mu\nu}$ under our assumption contains only one scalar function, an equation for this function can be obtained by forming

$$\frac{n_\mu \Pi_{\mu\nu} n_\nu}{n^2} = - q^2 b\ (1-\frac{1}{\gamma}) \qquad (4.11)$$

$$= - q^2 (1-\frac{1}{\gamma}) + \int dk\ \ n\cdot\Gamma_o\ \Delta\ \Delta\ \ \Gamma\cdot n + \text{tadpole.}$$

The 4-gluon terms are absent for the reasons previously discussed. Note also that $n_\mu \Pi_{\mu\nu}$ is also independent of the 4-glue terms, because the bare 4-glue vertex always appears on the left side of the diagram in fig. 1. The final form for the

$$\frac{n\cdot\Pi\cdot n}{n^2}$$

equation identifying $b(q^2) = Z^{-1}(q^2)$ is

$$- q^2 (1 - \frac{1}{\gamma}) Z^{-1}(q^2) = - q^2 (1 - \frac{1}{\gamma}) + \int \frac{dk}{k^2 k'^2} (n.k') \times n\cdot(k'-k) \sum_{\sigma\,\sigma'} \tag{4.12}$$

$$\left\{ \frac{Z(k)}{Z(q)} \frac{Z(k')-Z(q)}{k'^2-q^2} (q+k')_\sigma q_\sigma' - Z(k)\,\delta_{\sigma\,\sigma'} - \frac{Z(k')-Z(k)}{k'^2-k^2} (k\cdot k'\,\delta_{\sigma\sigma'} - k'_\sigma k_{\sigma'}) \right\}$$

where $\sum_{\sigma\,\sigma'} = k^2 k'^2 \Delta^o_{\lambda\,\sigma}(k)\, \Delta^o_{\lambda\,\sigma'}(k')$.

The quantity dk stands for

$$i\, g_o^2\, C_A \int \frac{d^4k}{(2\pi)^4}$$

where C_A is the Casimir eigenvalue of the adjoint representation of the color group.

The only other independent eq. we can obtain from $\Pi_{\mu\nu}$ that does not contain the 4-glue terms is

$$\frac{n_\mu \Pi_{\mu\nu} q_\nu}{n\cdot q} = \frac{1}{n\cdot q} \int dk\, n\cdot(k'-k) \Delta_{\lambda\sigma} \Delta_{\lambda\sigma'} \tag{4.13}$$

$$\times \left[\Pi_{\sigma\sigma'}(k') - \Pi_{\sigma\sigma'}(k) \right] = \frac{1}{n\cdot q} \int dk\, n\cdot(k'-k)$$

$$\times \left[Tr(\Delta(k)) - Tr(\Delta(k')) \right]$$

where the Ward identity has been used within the integral. If the integrals are sufficiently convergent the origin can be shifted and we obtain

$$\frac{n.\Pi.q}{n\cdot q} = 2 \int dk \,\mathrm{Tr}\, [\ \Delta\ (k)\ ] \qquad (4.14)$$

which, assuming $\Delta \sim \Delta^{o}$, becomes

$$\sim 2 \int \frac{d^4k}{k^2}\ Z(k)\ \ (2 + \frac{n^2k^2}{(n\cdot k)^2}) = 0. \qquad (4.15)$$

The last step in which an apparently positive definite integral is in fact zero is an example of some of the tricky (non-intuitive) rules in axial gauge QCD calculations. Let us examine Kummer's principle value prescription for the factors $(n\cdot k)^{-1}$ and $(n\cdot k)^{-2}$. Consider the following integral in Euclidean space:

$$\int d^4k\ f(k^2)\ \frac{1}{n\cdot k - n\cdot q} = \frac{4\pi}{n} \int k^2 dk\ f(k^2) \int_{-1}^{1} dz\ \frac{\sqrt{1-z^2}}{z - \frac{n\cdot q}{nk}}. \qquad (4.16)$$

We will define

$$I_1(x) = \int_{-1}^{1} dz\ \frac{\sqrt{1-z^2}}{z-x} \qquad (4.17)$$

which is the basic form required above. This integral can be evaluated using the principle value prescription to obtain

$$I_1\ (x) = \begin{cases} \pi\ (\sqrt{x^2-1}-x) & x > 1 \\ -\ \pi x & -1 < x < 1 \\ -\ \pi\,(x+\sqrt{x^2-1}) & x < -1 \end{cases} \qquad (4.18)$$

If we now obtain I_2, the integral needed for the $(n\cdot k)^2$ denominator, by differentiating I_1,

$$I_2(x) = \int_{-1}^{1} dz\ \frac{\sqrt{1-z^2}}{(z-x)^2} = \frac{\partial I_1}{\partial x}, \qquad (4.19)$$

for x = 0 we find that

$$I_2 = \int_{-1}^{1} dz\ \frac{\sqrt{1-z^2}}{z^2} = -\ \pi \qquad (4.20)$$

in spite of its positive definite integrand. This indicates that some care is required in the treatment of this type of integral. Another example:

$$\int \frac{d^4k}{(n \cdot k)^4} = 0. \tag{4.21}$$

This brings us to another question concerning the integrals which appear in Eq. (4.12). Is it possible to take $q^2 \to 0$ limits before performing the integration? Consider the following typical integral test case:

$$I = \int \frac{d^4k}{k^2 k'^2} \left(2 + \frac{n^2 k^2}{(n \cdot k)^2} \right) \tag{4.22}$$

This integral is UV finite because

$$\int d\,\Omega \left(2 + \frac{n^2\ k^2}{(n \cdot k)^2} \right) = 0 \tag{4.23}$$

and it is IR finite because the integral goes as d^4k/k^2 as $k^2 \to 0$. Since I is dimensionless, $I = I(\gamma)$. If we set $q = 0$ inside the integral we conclude $I = 0$ as the angular integral above vanishes. However, direct integration yields

$$I = 2\pi^2 \left(1 + \ell n \frac{4}{\gamma}\right). \tag{4.24}$$

The reason that the interchange of the limit and the integral fails is that no scale is present other than q, and a major portion of the integral comes from the region $k \sim q$. Thus we see that obtaining the small q limit of Π will not be possible by simply substituting $q = 0$ into the integrand in Eq. (4.12).

We will now return to Eq. (4.12) to see what kind of small q behaviour for Z might be self-consistent. Let us suppose that $Z(q) \to A(\Lambda^2/q^2)^{\alpha}$ as $q^2 \to 0$, for some positive power α, and some constant A. If Z contains no mass scale other than Λ^2, we can scale the q^2 dependence out of the integral in Eq. (4.12) obtaining

$$q^2 \left(\frac{\Lambda^2}{q^2}\right)^{\alpha}$$

times a function of

$$\frac{q^2}{\Lambda^2} \quad \text{and } \gamma.$$

This function, when properly regulated, is analytic in q^2/Λ^2 and we obtain for the integral in (4.12) the expansion

$$q^2\left(\frac{\Lambda^2}{q^2}\right)^{\alpha} F_1(\gamma,\alpha) + q^2\left(\frac{\Lambda^2}{q^2}\right)^{\alpha-1} F_2(\gamma,\alpha) + q^2\left(\frac{\Lambda^2}{q^2}\right)^{\alpha-2} \times F_3(\gamma,\alpha) + \ldots \tag{4.25}$$

as $q^2 \to 0, \gamma$ fixed.

For consistency, the behavior as $q^2 \to 0$ of the entire right hand side of Eq. (4.12) must match the left side and be of order

$$q^2 \, (q^2/\Lambda^2)^{\alpha} \; .$$

This means that all terms more singular than this must vanish. The most singular term on the right hand side is

$$q^2\left(\frac{\Lambda^2}{q^2}\right)^{\alpha} F_1(\gamma,\alpha).$$

Thus we must first require $F_1(\gamma,\alpha) = 0$ for all γ. This provides a constraint on the power α; a constraint, we note, which is independent of $g_0{}^2$ and of which gauge group we discuss. The next most singular term, which is still too singular to match the left hand side, includes the Born term. There must therefore be a term in the expansion (4.25) which cancels the Born term, identically in γ. This strongly suggests that the power α should be an integer.

Suppose, therefore, that $\alpha = 1$. In this case though simple power counting would indicate a logarithmic singular behavior and therefore additional terms in (4.25), these terms are actually absent due to the vanishing of angular integrals. Then (discarding spurious constant terms) the most singular term in (4.25) is q^2F_2, and this must cancel the Born term: thus it must turn out that $F_2 = (1-1/\gamma)$. Finally, the $(q^4/\Lambda^2)F_3$ term in (4.25) must match the left hand side of (4.12), so that $F_3 = - A^{-1}(1 - 1/\gamma)$. Does there really exist a solution in which all of these things happen ?

A complete answer to this question requires actually solving Eq. (4.12). We could imagine doing this by choosing an input Z, calculating the right hand side of (4.12), and thus obtaining an output Z. We would have a solution if the output Z matches the input Z.

Since we must now examine the Dyson equation for a trial input Z, we first decompose the integral appearing in Eq. (4.12) into its most general tensor form. Defining this integral as $I_{\mu\nu}$ we have

$$I_{\mu\nu} = \bar{B}\Pi^{o}_{\mu\nu} + \frac{\bar{C}N_\mu N_\nu}{n^2} + \frac{\bar{D}\; q_\mu q_\nu}{q^2} + \frac{\bar{E}\; n_\mu n_\nu}{n^2} + \frac{F\;(q_\mu n_\nu - q_\nu n_\mu)}{n\cdot q} \tag{4.26}$$

where $\Pi^{o}_{\mu\nu} = -q^2\,(\delta_{\mu\nu} - q_\mu q_\nu/q^2)$ is the bare inverse propagator, $N_\mu = n_\mu - q_\mu n^2/(n\cdot q)$ is a vector perpendicular to n_μ, and $\bar{B}$ through $\bar{F}$ are scalar functions of q^2 and γ. Thus, we find that

$$\frac{n_\mu I_{\mu\nu} n_\nu}{n^2} = -\,q^2\,\bar{B}(1 - 1/\gamma) + \bar{D}/\gamma + \bar{E}\,, \tag{4.27a}$$

$$\frac{n_\mu I_{\mu\nu} q_\nu}{n\cdot q} = \bar{D} + \bar{E} + \bar{F}\,(1-\gamma). \tag{4.27b}$$

Using the Ward identity one can show that the left hand side of (4.27b) vanishes identically provided that $\Delta_{\mu\nu}$ is sufficiently convergent to allow translation of the integration variables. In this case one eliminates $\bar{D}$ to find

$$\frac{n_\mu I_{\mu\nu} n_\nu}{n^2} = (1 - 1/\gamma)\;[-q^2\,\bar{B} + \bar{E} + \bar{F}]. \tag{4.28}$$

Thus, provided the scalar integrals which define $\bar{B}$ through $\bar{F}$ are IR and UV convergent for all γ (in particular for $\gamma = 1$), Eq. (4.28) suggests that the integral in Eq. (4.12) has a zero at $\gamma = 1$, so the factor $(1 - 1/\gamma)$ can be entirely cancelled from Eq. (4.12).

Suppose we simply choose the input Z equal to $A(\Lambda^2/q^2)$, not just as $q^2 \to 0$ but over the entire range of q^2 (A = constant). In this case the above conditions are met and we expect to find the gauge factor $(1 - 1/\gamma)$ in $n_\mu I_{\mu\nu} n_\nu$. After a laborious calculation [5] we in fact obtain for the right hand side of (4.12) the expression

$$-\,(1-1/\gamma)q^2 + \frac{29}{96\pi^2}\,A\,g_o^2\,C_A(1-1/\gamma)q^2 - \frac{22}{96\pi^2}\,g_o^2 A \times C_A(1-1/\gamma)\;1/\gamma\;\frac{q^4}{\Lambda^2} \tag{4.29}$$

That is, the γ dependence of the q^2 term comes out to be precisely the same as that of the Born term. Thus the constant A (essentially the slope of the input Z^{-1}) can be chosen to exactly cancel the Born term, making Z^{-1}_{output} proportional to q^2 as $q^2 \to 0$. Complete consistency would now require the

$$\frac{q^4}{\Lambda^2}$$

term on the right to have the same slope and γ dependence as the input Z^{-1} that appears on the left. It appears that this test fails but actually we cannot at this time check this requirement as we do not know how to calculate the

$$\frac{q^4}{\Lambda^4}$$

term. The problem is that this term depends sensitively on the cutoff procedure. For example, if the cutoff is put

$$\Lambda^2 - aq^2 - b\,\frac{(q\cdot n)^2}{q^2}\,,$$

surely indistinguishable from Λ^2 in the limit $\Lambda^2 \to \infty$, the

$$\frac{q^4}{\Lambda^2}$$

term is changed by an amount proportional to $(a + \frac{b}{\gamma})$. Thus we see that a small modification in the high q^2 region, either in the method of cutoff or in the behavior of Z, can easily produce consistency. These modifications will not change the q^2 term in (4.29) implying that a solution obtained by this method must have the same value of A as that determined above. We therefore conclude that if such a solution exists,

$$Z \sim \frac{1}{q^2}$$

will obtain for any coupling g_0 and for any value of C_A and confinement will hold for any non-Abelian gauge group.

The uncertainties discussed above are clearly removed if one actually solves the integral eq. as this will determine Z for the entire range of q^2 and not just for $q^2 \ll \Lambda^2$. It is certainly not obvious that a solution exists in which Z is independent of γ , though the fact that the factor $(1 - 1/\gamma)$ divides out is certainly encouraging.

There are two assumptions involved in our approach. One is the assumption that the part of Γ not determined by the Ward identity is irrelevant in the IR, as is the case in QED. The second is the ansatz $\Delta \sim \Delta_0$. The first of these is difficult to verify, though a careful analysis of the more general vertex function lends support to it. The second can be proved or disproved through a study of the other independent equation which can be obtained from $\Pi_{\mu\nu}$ by multiplying with $\delta_{\mu\nu}$. This trace equation, currently under investigation, is complicated by the presence of nontrivial 4-gluon terms; an analysis of the 4-gluon vertex analogous to Eq. (4.9) is required.

Any application of a solution of the type we have discussed will involve renormalization. It is clear that if Z is renormalized in the small q^2 region the scale Λ^2 will be replaced by the renormalization mass squared and A will simply divide out of the expression. Any connection between our parameters and those of the high q^2 region disappears because these regions, where Z is known, fail to overlap. A unified approach including the low q^2 behaviour which we have discussed and the asymptotic freedom limit is necessary before anything but the q^2 dependence of IR processes can be calculated.

We are hopeful that modifications of the high q^2 behavior may provide both a consistent solution to our equation and at the same time some natural connection to the UV region. This investigation is now in progress.

V. MOMENTUM DEPENDENCE OF VERTEX FUNCTIONS IN GAUGE THEORIES

In this lecture I would like to report an investigation with T. W. Chiu of the momentum dependence of vertex functions for scalar and spinor electrodynamics. In each case we construct a general tensor form which is free of kinematic singularities and satisfies the Ward identities. The scalar functions are then calculated in perturbation theory to one loop to determine the analytic properties to this order and to see which terms provide the dominate IR behavior. Several questions which we hope to answer are:

1. Transverse components generally have many powers of momentum in the tensor form. This requires scalar functions dimension

   $\frac{1}{q^2}$ and $\frac{1}{q^4}$.

   How are scalars with these dimensions formed in a "natural" way?

2. Is it possible to have kinematic singularities in scalar amplitudes?

3. Are the IR singularities always in the longitudinal terms even off the mass shell?

Finally, we obtain a general expression for the off-shell QED vertex to order α; this contains a single integral and hence the analytic properties are explicit. It is also clear that recurrence formulas for the scalar amplitudes can be obtained for graphs with a particular topology and this might prove to be of use in high order perturbation calculations.

A. Scalar Electrodynamics

The massless scalar electrodynamics vertex which we have already discussed is

$$\Gamma_\mu = \frac{\Delta^{-1}(p') - \Delta^{-1}(p)}{p'^2 - p^2} (p+p')_\mu + A'(p'.kp_\mu - p\cdot kp'_\mu) \qquad (5.1)$$

The first order vertex is given by the diagrams in fig.2. The integrals that appear are:

$$I_0 = \int d^4k \frac{1}{k^2 k'^2 k''^2}$$

$$I_\mu^{(1)} = \int d^4k \frac{k_\mu}{k^2 k'^2 k''^2} \qquad (5.2)$$

$$I_{\mu\nu}^{(2)} = \int d^4k \frac{k_\mu k_\nu}{k^2 k'^2 k''^2}$$

$$I_{\mu\nu\sigma}^{(3)} = \int d^4k \frac{k_\mu k_\nu k_\sigma}{k^2 k'^2 k''^2}$$

where $k' = k-p$, $k'' = k-p'$.

If Feynman parameters are used we obtain integrals of the form

$$\int_0^1 dy \int_0^1 dx \frac{2\ x\ x^\ell\ y^m}{[\Delta^2 - p_1{}^2]} \qquad (5.3)$$

for $\ell = 1, 2, 3$ and $m = 0, 1,2,3$ (only 9 combinations actually occur) where $\Delta^2 = x\,[p^2 y + p'^2(1-y)]$ and $p_{1_\mu} = x\,[\,p_\mu y + p'_\mu\,(1-y)]$.

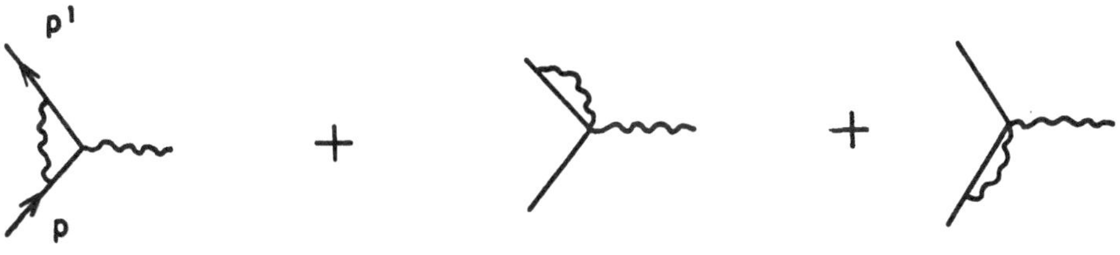

Fig. 2 Feynman graphs for the scalar one loop vertex; wavy lines are photons, solid lines are scalar mesons.

There are relations between these nine integrals but most are not obvious. In fact there is only one independent integral and all the rest can be expressed in terms of elementary functions! This can be seen from the tensor decomposition of these integrals.

$$I_\mu^{(1)} = A\, p_\mu + B\, p'_\mu$$

$$I_A = p\cdot I^{(1)} = p^2 A + p\cdot p' B = \int d^4k \frac{p\cdot k}{k^2 k'^2 k''^2} \qquad (5.4)$$

but $p\cdot k = \frac{1}{2}(-k'^2 + k^2 + p^2)$. Therefore, I_A is a known function plus

$$\frac{p^2}{2} I_o.$$

The other scalar integral $I_B = p'\cdot I^{(1)}$ can also be expressed in terms of I_o and known functions. Thus, $I_\mu^{(1)}$ is given in terms of known functions and I_o. It is clear that this can also be done for $I_{\mu\nu}{}^{(2)}$ and $I_{\mu\nu\sigma}{}^{(3)}$, so all integrals are expressible in terms of I_o and known functions. It is also clear that all integrals could be expressed in terms of I_A or any other scalar integral related to I_o. Note that even in QCD in the Feynman gauge, the one loop vertex will all be expressible in terms of I_o plus known functions. This property is also true for spinor QED though the denominators contain the electron mass.

Our results for massless scalar QED are

$$\Delta^{-1}(p^2) = -\frac{\alpha}{8\pi}\,(5 + 4\,\ell n\, \Lambda^2/p^2)$$

$$A' = -\frac{\alpha}{4\pi}\Big\{4\,\frac{\ell n p'^2 - \ell n\, p^2}{p'^2 - p^2} + (q^2 - p\cdot p')\,\frac{I_B - I_A}{p'^2 - p^2}\Big\} \qquad (5.5)$$

where the symmetry between I_A and I_B makes A' analytic at $p^2 \to p'^2$. Here Λ^2 is the UV cutoff.

If we now express A' in terms of I_A and I_B we obtain

$$A' = -\frac{\alpha}{4\pi}\Big\{\frac{4\ell n(p'^2/p^2)}{p'^2 - p^2} + \frac{p^2 + p'^2 - 4p\cdot p'}{\Delta} \qquad (5.6)$$

$$\times \left[\frac{p \cdot p'}{2} \quad I_0 + \frac{(p^2 + p\cdot p')\ \ell n\ (p^2/q^2) - (p'^2 + p\cdot p')\ \ell n (p'^2/q^2)}{p'^2 - p^2} \right] \}$$

where Δ^2 is the triangle function

$$\Delta^2 = p^2 p'^2 - (p \cdot p')^2 = p^2 q^2 - (p \cdot q)^2 = p'^2 q^2 - (p' \cdot q)^2 \qquad (5.7)$$

The zeros of Δ^2 are at normal thresholds, or in Euclidean space for p,p' colinear. The fact that A' is still analytic is now more difficult to see.

In the colinear limit I_0 can be integrated to yield

$$I_0 = \frac{2\ \ell n\ p^2}{p'(p-p')} - \frac{2\ \ell n\ p'^2}{p(p-p')} - \frac{2\ \ell n (p-p')^2}{pp'} \qquad (5.8)$$

which cancels the other terms in (5.6) and the first term that survives is proportional to Δ^2 hence A' is finite. We note that a singularity of A' might have been considered acceptable in that the tensor form which multiplies A' vanishes in the colinear limit.

The infrared limit of Γ_μ comes entirely from Γ longitudinal

$$\Gamma_\mu^{IR} = (p+p')_\mu \left| 1 - \frac{\alpha}{8\pi} \quad (1 + 4\ \ell n\ \frac{\Lambda^2}{p^2}) \right| \qquad (5.9)$$

Note that the IR limit exists even though the $q^2 \to 0$ limit is not finite unless $p \to p'$.

B. Spinor Electrodynamics

The additional degrees of freedom introduced by having 2 spin - 1/2 particles coupled to the photon considerably complicate the tensor decomposition of the vertex. In this case there are three four-vectors $\gamma_\mu, p_\mu, p'_\mu$ and four types of scalars proportional to $1, \not{p}, \not{p}'$ and $p^\mu p'^\nu \sigma_{\mu\nu}$. The resulting twelve spin amplitudes have been enumerated elsewhere; however, since our goal is producing a vertex which is free of kinematic singularities and that automatically satisfies the Ward identity, we will introduce eight tensors which give no contribution to the Ward identity (the generalization of the second term in Eq. (5.1.)). The remaining four tensors will be completely determined by the Ward identity.

The Ward identity for spinor electrodynamics is

$$q_\mu \Gamma^\mu = S_F^{-1}(p'^2) - S_F^{-1}(p^2) = F(p'^2)\not{p}' - F(p^2)\not{p} + G(p'^2) - G(p^2) \tag{5.10}$$

where F and G are the scalar functions that determine the electron propagator. Note that one of the twelve amplitudes will be identically zero due to the fact that the scalar $p^\mu p'^\nu \sigma_{\mu\nu}$ does not occur on the right of Eq.(5.10). The portion of the vertex which "solves" Eq.(5.10) is

$$\Gamma_o^\mu = \frac{(\not{p} + \not{p}')}{2}(p' + p)^\mu \frac{F(p'^2) - F(p^2)}{p'^2 - p^2} + \frac{F(p'^2) + F(p^2)}{2}\gamma_\mu + \frac{G(p'^2) - G(p^2)}{p'^2 - p^2}(p + p')^\mu . \tag{5.11}$$

Note that the straightforward generalization of the scalar result

$$\Gamma_o^\mu = \frac{S_F^{-1}(p'^2) - S_F^{-1}(p^2)}{p'^2 - p^2}(p + p')^\mu \tag{5.12}$$

has kinematic singularities and is therefore unacceptable.

Here we have constructed a form that is obviously free of kinematic singularities.

The remaining eight tensor forms must satisfy

$$q_\mu T_i^{\ \mu} = 0 \qquad i = 1, 2.....8$$

A set of independent T's which have this property is

$$\begin{aligned}
T_1^{\ \mu} &= Q^\mu = p^\mu (p'\cdot q) - p'^\mu (p\cdot q) \\
T_2^{\ \mu} &= Q^\mu (\not{p} + \not{p}') \\
T_3^{\ \mu} &= q^2 \gamma^\mu - q^\mu \not{q} \\
T_4^{\ \mu} &= Q^\mu p^\lambda p'^\nu \sigma_{\lambda\nu} \\
T_5^{\ \mu} &= \sigma^{\mu\lambda} q_\lambda \\
T_6^\mu &= \gamma^\mu (p'^2 - p^2) - (p + p')^\mu \not{q}
\end{aligned} \tag{5.13}$$

$$T_7{}^{\mu} = \frac{p'^2 - p^2}{2} [\gamma^{\mu} (\not{p} + \not{p}') - p^{\mu} - p'^{\mu}] + (p + p')^{\mu} p^{\nu} p'^{\lambda} \sigma_{\nu\lambda}$$

$$T_8{}^{\mu} = - \gamma^{\mu} p^{\nu} p'^{\lambda} \sigma_{\nu\lambda} + p^{\mu} \not{p}' - p'^{\mu} \not{p} .$$

The complete vertex is then

$$\Gamma^{\mu} = \Gamma_0{}^{\mu} + \sum_{i=1}^{8} A_i T_i{}^{\mu}. \tag{5.14}$$

The only criterion other than simplicity we have for choosing this set rather than some linear combination is the perturbation result. It was found that if instead of T_3 given above we used $Q^{\mu}\not{q}$, a kinematical singularity appeared in A_6 while for the set above all of the A's are separately analytic. Since A_4, A_5 and A_7 are zero to second order in perturbation theory we cannot be sure that higher order calculations might not require particular linear combinations of T_4, T_5 and T_7 rather than the forms given above. The results of the second order calculations are

$$F = \frac{\alpha}{4\pi} [\ell n \frac{\Lambda^2}{m^2} + \frac{3}{2} + \frac{m^2}{p^2} + (\frac{m^4}{p^4} - 1) \ell n \frac{m^2 - p^2}{m^2}]$$

$$G = \frac{\alpha m}{\pi} [(\frac{m^2}{p^2} - 1) \ell n \frac{m^2 - p^2}{m^2} + 1 + \ell n \frac{\Lambda^2}{m^2}] \tag{5.15}$$

$$A_4 = A_5 = A_7 = 0$$

$$A_1 = \frac{\alpha m}{\pi} \frac{J_B - J_A}{p'^2 - p^2} = \frac{\alpha m}{\pi} \frac{1}{\Delta^2} \{ \frac{m^2 + p \cdot p'}{2} J_0 + 2S$$

$$- \frac{(p \cdot p' + p'^2) L' - (p^2 + p \cdot p') L}{p'^2 - p^2} \}$$

$$A_8 = \frac{\alpha}{4\pi} (J_A + J_B - J_0) = - \frac{\alpha q^2}{4\pi\Delta^2} \{ \frac{(m^2 + p \cdot p')}{2} J_0 + 2S \tag{5.16}$$

$$- (p^2 L + p'^2 L') + p \cdot p' (L + L') \}$$

$$A_2 = \frac{\alpha}{4\pi} \frac{J_A - J_B + J_E - J_C}{p'^2 - p^2} = \frac{3}{4} (p \cdot p' + m^2) A_8 - \frac{\alpha}{4\pi\Delta^2} x$$

$$\times \left\{ \frac{(q^2 - 4m^2)}{8} J_o - \frac{1}{2} - \frac{p\cdot p' m^2}{2p^2 p'^2} + \frac{1}{4}(L + L') + \frac{(p^2 + p\cdot p')\left(1 + \frac{m^2}{p^2}\right) L - (p'^2 + p\cdot p')\left(1 + \frac{m^2}{p'^2}\right) L'}{2\,(p'^2 - p^2)} \right\}$$

$$A_6 = \frac{p'^2 - p^2}{2} A_2$$

$$A_3 = -\frac{\alpha}{8\pi}(J_o + 2J_D - J_C - J_E) =$$

$$= \frac{3(m^2 + p\cdot p')}{16m\,\Delta^2} A_1 - \frac{\alpha}{8\pi\Delta^2} \{$$

$$[\Delta^2 - \frac{(p^2 + p'^2 - 2m^2)^2}{8}] J_o - 2(m^2 + p\cdot p')S + \frac{p\cdot p'}{2}$$

$$[(1 - \frac{m^2}{p^2}) L + (1 - \frac{m^2}{p'^2}) L'] + \frac{p^2 - p'^2}{4}(L - L') +$$

$$+ (m^2 + p\cdot p') + \frac{(m^2\, p\cdot p' + p^2 p'^2)}{2\, p^2 p'^2}(p'^2 + p^2)\}$$

The J's are natural scalar integrals that arise in the calculation and can be expressed in terms of the following functions:

$$S = \sqrt{1 - \frac{4m^2}{q^2}} \sinh^{-1}\sqrt{\frac{-q^2}{4m^2}} = \frac{1}{2}\sqrt{1 - \frac{4m^2}{q^2}} \times \ell n \frac{1 + \sqrt{1 - \frac{4m^2}{q^2}}}{1 - \sqrt{1 - \frac{4m^2}{q^2}}}$$

$$L = (1 - \frac{m^2}{p^2})\ \ell n\ (\frac{m^2 - p^2}{m^2}) \qquad (5.17)$$

$$L' = (1 - \frac{m^2}{p'^2})\ \ell n\ (\frac{m^2 - p'^2}{m^2})$$

and

$$J_o = \frac{2}{\pi^2 i} \int d^4k \, \frac{1}{k^2 \, [(k-p)^2 - m^2]\,[(k-p')^2 - m^2]} \, .$$

While the formulas for the A's appear to be quite complicated they are much simpler than any previous results in that they are all expressed in terms of elementary functions and a single scalar integral. We emphasize again that the fact that only one integral appears is well disguised if Feynman parameters are introduced to perform the various vector and tensor integrals. However, if these same integrals are decomposed into scalars the existence of a single integral J_o becomes obvious.

The absence of kinematic singularities at $\Delta^2 = 0$ can again be shown as all integrals can be performed analytically in this limit.

The fact that T_5, the usual magnetic moment coupling, does not appear to this order may seem surprising. However, if we take the mass shell limit of (3.5) we obtain

$$\Gamma_\mu (p,p')\Big|_{\not p = \not p' = m} = \gamma_\mu + \left(\frac{-\alpha}{4\pi}\right) \left\{ (p+p')_\mu \frac{2m}{m^2 + p\cdot p'} S + \gamma_\mu \left[-\frac{1}{2} - \ln \frac{\Lambda^2}{m^2} - 2\,(p\cdot p')\, J_o - 6S \right] \right\} \qquad (5.18)$$

which is in agreement with the usual results, and the magnetic moment term has the proper value.

The final form of the vertex in the IR limit is

$$\Gamma_\mu = \gamma^\mu \left\{ 1 + \frac{\alpha}{4\pi} \left[\ln \frac{\Lambda^2}{m^2} + \frac{3}{2} + \frac{m^2}{p^2} + \left(\frac{m^4}{p^4} - 1\right) \ln \frac{m^2 - p^2}{m^2} \right] \right\} + \frac{\alpha}{2\pi} \left\{ \frac{4mp^\mu}{p^2} \left[1 + \frac{m^2}{p^2} \ln \frac{m^2 - p^2}{m^2} \right] - \frac{\not p\, p^\mu}{p^2} \left[1 + \frac{2m^2}{p^2} + \frac{2m^4}{p^4} \ln \frac{m^2 - p^2}{m^2} \right] \right\} \qquad (5.19)$$

Our conclusions from the analysis of QED are essentially the same as those for scalar electrodynamics. Again the absence of kinematic singularities is much clearer when the A's are expressed in terms of J_A, J_B, J_C, J_D, J_E and J_o rather than in terms of just J_o and elementary functions. Finally, we note the perturbation result

$$A_6 = \frac{p'^2 - p^2}{2} A_2$$

indicates that T_2 could be divided by p'^2-p^2 without producing a kinematic singularity in the vertex. We believe this to be accidental and unlikely to be true for higher order calculations.

C. CONCLUSIONS

Let us now summarize our results.

1.) The form of the vertex that "solves" the Ward identity and is free of kinematic singularities is the form produced in lowest order perturbation theory and the scalar functions have no kinematic zero or poles.

2.) The infrared singularities are contained in the longitudinal parts of the vertex.

3.) The "natural" forms of scalar functions are

$$\frac{F(\ldots.p^2) - F(\ldots.p'^2)}{p^2 - p'^2}$$

or

$$\frac{\text{Symmetric function}}{\Delta^2}$$

though the construction of a symmetric function with zeros at the zeros of Δ^2 cannot be done explicitely.

4.) Only a single function which is not expressible as an elementary function enters into the massless scalar and massive spinor electrodynamics and the same statement holds for the one loop QCD calculation of the 3-glue vertex in the Feynman gauge.

REFERENCES

1. The work described here was done in collaboration with Phil Lucht, University of Utah, F. Zachariasen, Caltech and M. Baker and co-workers at University of Washington. The preliminary description of this program appears in J. S. Ball and F. Zachariasen, Nucl.Phys. B143, 148 (1978).
2. This work was done in collaboration with T. W. Chiu at University of Utah and publication is awaiting the completion of the complete one loop vertex for QCD.
3. J. M. Cornwall and G. Tiktopoulos, Phys.Rev. D15, 2937 (1977).
4. W. Kummer, Acta Phys. Austriaca 41, 315 (1975).

Z(N) GAUGE THEORIES - PHASE TRANSITIONS AND CONFINEMENT

S. Yankielowicz

Tel-Aviv University, Ramat-Aviv, Israel, and

CERN - Geneva

INTRODUCTION

The subject discussed in these lectures is Z(N) lattice gauge theories. The first two questions that come to mind are:

i) Why at all are we interested in Z(N) gauge theories?
ii) Why do we use the lattice formulation?

I'll devote the first part of my lecture to a review discussion on these questions. In particular I would like to indicate the interrelation between a lattice theory and its continuum limit. It is the phase structure of the lattice theory which determines the possible continuum theories one can reach. I'll try to emphasize the characterization of phases in lattice theories in general and gauge theories in particular.

The second part deals with the Hamiltonian approach to the Z(N) gauge theories. I'll introduce the duality transformation and use them as well as strong and weak perturbative calculations to analyze the different phases of the theories in both 2+1 and 3+1 dimensions. A QED like representation of the theories reveals in 3+1 dimensions a three-phase structure for large enough N. Those phases can be characterized as electric confinement, magnetic confinement, and non-confinement. I'll describe these phases in terms of condensation of various topological excitations.

The third part is devoted to the Lagrangian approach to the Z(N) theories which gives us another view-point on the structure of these theories. In particular the role of the topological excitations is evident in a Coulomb gas representation of the theories.

I'll conclude with some remarks on Z(N) gauge theories with matter.

I. WHY Z(N) ?

There are several answers to this question. Perhaps the most ambitious one is to follow 't-Hooft [1] who argued that Z(N), which is the center of SU(N), plays an important role in understanding the problem of confinement. If indeed Z(N) configurations are important in understanding the large distance properties of QCD, then as a first approximation we would like to isolate these configurations and write down an effective Lagrangian or Hamiltonian which describe them. This procedure leads directly to the Z(N) gauge theories.

Whether you like this approach or not, it turns out that the Z(N) gauge theories are very interesting laboratories to check various theoretical ideas. In particular the Z(N) theories give a nice realization of 'tHooft ideas on the type of phases in gauge theories and their characterization. It is the condensation of topological excitation which gives rise to the phase transitions. There is a dual relationship between the confining phase (disordered phase) and the Higgs phase (ordered phase). Nielsen and Olesen [2] were among the first to realize that we do know a physical system which has the property of confinement i.e. the superconductor. If you take a type II superconductor and introduce into it a monopole anti-monopole pair then there will be a linear potential between them due to a magnetic flux tube between the two. Quarks are, however, not monopoles. The way we couple quarks to gauge bosons via the minimal substitution, implies that quarks have electric rather than magnetic charges. It was noted immediately by Mandelstam [3] and 't-Hooft [4] that a situation dual to a superconductor (Higgs) i.e. condensation of magnetic monopoles, will give rise to a linear potential between external quarks via the formation of electric flux tubes (strings).

Another interesting aspect of the Z(N) gauge theories is connected with the limit $N\to\infty$. In this limit we get the PQED (periodic QED) theory. This relation between the Z(N) theories and PQED is important in revealing their phase structure for large N [5,6,7].

Finally let me note that the Z(N) gauge theories are of interest in statistical mechanics being closely related to spin-glass systems [8].

Being particle-physicists we 'll be mainly interested in the Z(N) gauge theories within the framework of the confinement problem.

THE CONFINEMENT PROBLEM

QCD is believed to be the theory which gives the fundamental description of the hadronic world. The short distance behaviour of QCD can be extracted from renormalized perturbation theory due to the fact that the bare coupling is vanishingly small; a property known as asymptotic freedom [9] . An important aspect of asymptotic freedom which I would like to remind you is the following one: Consider two versions of the theory in terms of cut offs "a_0" and "a". For later on it would be advantageous to think of "a_0" and "a" as lattice spacing. But for the time being you may think of $a=1/\Lambda$ with Λ a momentum cut off. In order that the physics of the two formulations be identical their coupling constant must be related

$$g^2(a) = \frac{g_o^2}{1+\frac{cg_o^2}{2\pi}\log(a_o/a)} \xrightarrow[a\to 0]{} \frac{2\pi}{c}\,\frac{1}{\log(a_o/a)} \longrightarrow 0 \qquad (1.1)$$

We learn that the continuum limit ($a\to 0$) of the theory is at zero coupling. Starting with the lattice theory we are in a way far from the continuum limit. After solving the lattice theory we still have to continue the results to zero coupling in order to recover the continuum theory.

While the short distance behaviour of the theory is under control, its long distance global structure is much more difficult to disentangle. To answer questions related to the structure of the vacuum, the low lying spectrum of the theory and quark confinement, we have to develop non perturbative methods to deal with field theories. We can, roughly, divide the methods developed in recent years into

i) The lattice approach [10]
ii) The semi-classical approach [11]

By no means the border line between the two is sharp. One can use the semi-classical approach also in dealing with a lattice theory. Since in these lectures we shall use the lattice approach, a few words about the semi-classical approach are in place.

THE SEMI-CLASSICAL APPROACH

The main idea beyond the semi-classical approach is to identify the most important field configurations which saturate the functional integral. Tacitly, we assume that for the large distance properties of the system not all degrees of freedom are physically relevant at once. In particular, it is assumed that for the question of confinement we can at first neglect the quark field and investigate the gluon sector alone by investigating the behaviour of the Wilson loop [13]

$$<e^{i \oint_c A_\mu dx^\mu}> \equiv \int d[A]\, e^S\, e^{i \oint_c A_\mu dx^\mu} \qquad (1.2)$$

where S is the (Euclidian) action. The gluonic sector by itself is very complicated and one tries to separate out what one believes are the most important gluonic degrees of freedom and investigate them by themselves. In the literature there are various examples of such configurations: instantons [11,12], merons [12], monopoles [3] and finally Z(N) topological configurations which were introduced by 't-Hooft [1]. These are configurations associated with the homotopy group

$$\pi_1 \left(SU(N)/Z(N) \right) = Z(N) \qquad (1.3)$$

THE 't-HOOFT APPROACH [1]

In the 't-Hooft approach field configurations which have non-trivial Z(N) topological charge control the large distance behaviour of QCD and account for quark confinement. This observation of 't-Hooft is based upon the investigation of the algebraic properties of two non-local operators:

i) $\mathfrak{A}(c)$ - Wilson's loop (order parameter)
This operator can be viewed as an operator which measures non-abelian magnetic flux which goes through the curve c. Alternatively, this operator creates or destroys non-abelian electric flux along the curve c.

ii) $\mathfrak{B}(c)$ - 't-Hooft loop (disorder parameter)
This operator is dual to $\mathfrak{A}(c)$. It creates or destroys magnetic flux along the curve c.

The operators $\mathfrak{A}(c)$ and $\mathfrak{B}(c)$ characterize the different phases of a gauge theory. A priori these operators can behave as exp(-area) or exp(-perimeter). The confining phase corresponds to

$$\begin{aligned} \mathfrak{A}(c) &\sim e^{-\text{area } (c)} \\ \mathfrak{B}(c) &\sim e^{-\text{perimeter } (c)} \end{aligned} \qquad (1.4)$$

The two operators satisfy a Z(N) algebra

$$\mathfrak{A}^{+}(c)\,\mathfrak{B}(c')\,\mathfrak{A}(c) = e^{i\frac{2\pi}{N}m}\,\mathfrak{B}(c') \qquad (1.5)$$

where m is the number (mod N) of times the curve c winds around the curve c'. Using this algebra 't-Hooft has excluded the possibility that both operators behave as exp (-area). Hence there are three possible phases that a gauge theory can be in. For large enough N we encounter all of these phaes in the Z(N) gauge theory [5,6,7]. Within the context of the Z(N) theories we'll be able to give an explicit realization to these operators and to check the validity of eq (1.5). Equation (1.4) reveals that the confining phase is characterized by relative abundance of magnetic flux relative to electric flux since the $\mathfrak{A}$(c) operator falls much more rapidly compared to $\mathfrak{B}$(c). Here we see the first hint that the vacuum of the confining phase is a dual transformed state to the Higgs vacuum state (super conductor), where magnetic flux is suppressed (the Meissner effect).

THE LATTICE APPROACH

In discussing the lattice approach I would like first to address the question: WHY THE LATTICE? - There are several answers to this question

i) A field theory is a complicated system with an infinite number of degrees of freedom. The lattice defines for us a cut off version of the field theory. Of course all the difficulties which go under the name of renormalization will have to be eventually faced when we'll discuss the continuum limit.

ii) We are interested in the large distance properties of the system and have to develop non-perturbative methods. The lattice version makes connection with statistical mechanics of spin systems. Hence we can borrow methods used there.

iii) We want to investigate Z(N) gauge theories [14] and as far as I know there exists no continuum pure Z(N) gauge theory. This is related to the fact that Z(N) is a discrete group and we can not perform infinitesimal transformations in the group space.

iv) The Z(N) configurations are classical solutions of the lattice QCD gauge theory [15].

v) The lattice approach provides us with a way to fully maintain the gauge invariance in a non-perturbative way [13]. This is extremely important since we believe that the gauge symmetry is essential in obtaining quark confinement.

It is important to emphasize again that although the lattice theories are of interest by themselves, we shall use them first of all as a device to define cut off field theories. The lattice theory is, therefore, an intermediate step used to analyze field theories. Eventually we would like to consider the continuum limit of the lattice theory. The lattice formulation of a given field theory is not unique. We shall be interested in two types of lattices

i) Space-time symmetric lattice in Euclidean space. This lattice will be of use in the Lagrangian approach;

ii) A lattice in the space directions with time continuous. This lattice will be of use in the Hamiltonian approach.

At this point we can summarize the strategy of the lattice approach

1. Formulate the field theory on a lattice. In doing so one has to be careful to keep all symmetries of the original field theory that one thinks are important to the understanding of the physics.

2. Although the lattice formulation is not unique its critical behaviour (number of phases, critical exponents) is the same.

3. Determine the phase diagram of the lattice theory.

Only if the transition is continuous (second order) one can recover a sensible continuous relativistic field theory at the critical point [16]. Qualitatively we can understand this statement because at the critical point the correlation length is infinite. It is then conceivable that the theory at the critical point loses all memory of the lattice and the continuous space-time symmetries of the field theory are re-established. We learn, therefore, that the lattice phase structure determines the types of continuum theories that can be recovered.

HAMILTONIAN VERSUS LAGRANGIAN FORMULATIONS

I have mentioned the two types of lattice theories we are going to consider. Actually there is a relationship between the two. Starting from the Lagrangian theory formulated on space-time symmetric lattice we obtain a classical statistical mechanics system [16]. Next we can construct the transfer matrix of the theory. In doing it one of the axes is choosen to represent the (Euclidean continued) time direction. Then the transfer matrix is the operator which transfers the system by one step $\Delta\tau$ in the time direction.

This operator T can, therefore, be identified with

$$T = e^{-H \Delta \tau} \tag{1.6}$$

Taking carefully [16,17] the limit $\Delta\tau \to 0$ we can identify and construct the Hamiltonian of the system. If the system under consideration is canonical then the resulting Hamiltonian is just the lattice version of the canonical Hamiltonian. This method is, however, more general and can be applied also in those situations in which we do not have the canonical procedure at our disposal. It is important to note that the Lagrangian approach provides us with a classical statistical mechanics system in $d=D+1$ dimensions. The Hamiltonian approach, on the other hand, provides us with a quantum mechanical system in D space dimensions. The following list can be served as a dictionary between quantities of interest in the two approaches.

Hamiltonian	Lagrangian
Vacuum energy density,	Free energy density,
vacuum expectation value,	average over ensemble
propagator,	correlation function,
mass gap m .	$1/\xi$, ξ-correlation length.

THE IMPORTANCE OF BEING CRITICAL

We have already mentioned the fact that in order to recover a sensible relativistic continuum theory, the lattice theory shall be nearly critical. I would like now to support this observation by some definite examples [16]. Let us first consider a propagator of a scalar theory (continued to Euclidean space). Its large distance behaviour is dominated by $e^{-m|x|}$, where m is the mass gap. On the other hand the correlation function of the statistical mechanics system has the behaviour

$$e^{-|x|/\xi a},$$

where ξ is by definition the correlation length and "a" the lattice spacing. Hence

$$m = {}^{1/}\xi a \tag{1.7}$$

It is clear now that in order to obtain a field theory with small (finite) mass when $a \to 0$, the lattice theory must be nearly critical. As we approach the continuum limit $a \to 0$, $\xi \to \infty$ and there is a chance of recovering an interesting continuum limit.

Next I would like to consider a theory known as compact scalar theory in two Euclidian dimensions. At each point p of the lattice we have an angle variable $0 \leq \phi_p \leq 2\pi$. The functional integral (partition function) is given by

$$Z = \int_0^{2\pi} d\phi_p \, e^{+\frac{1}{g}\sum_{p,\mu}\cos(\phi_p - \phi_{p+\mu})} \tag{1.8}$$

where p and p+μ are two neighboring lattice points, and $\mu = 1,2$ denote the corresponding directions. The propagator is given by the following expresion

$$K(A,B) = \int_0^{2\pi} \pi\phi_p \, e^{+\frac{1}{g}\sum_{p,\mu}\cos(\phi_p - \phi_{p+\mu})} \, e^{i\phi_A} \, e^{-i\phi_B} \tag{1.9}$$

A and B are two points on the lattice.

In order to perform the integrations, we expand the first exponential. Imagine doing first the integration over ϕ_A . It is clear that since

$$\int_0^{2\pi} d\phi \, e^{i\phi} = 0,$$

only a term containing $\cos(\phi_A - \phi_{A+\mu_1})$ will give non zero contribution. Next perform the integration over $\phi_{A+\mu_1}$. In order to get non zero contribution also

$$\cos\,(\phi_{A+\mu_1} - \phi_{A+\mu_1+\mu_2})$$

must be present. Each $\cos\,(\phi_i - \phi_{i+\mu})$ can be represented by a link joining the points i and i+ . It is, then, clear that only terms which can be represented as closed path starting at A and terminating at B give non zero contribution. An example is given in

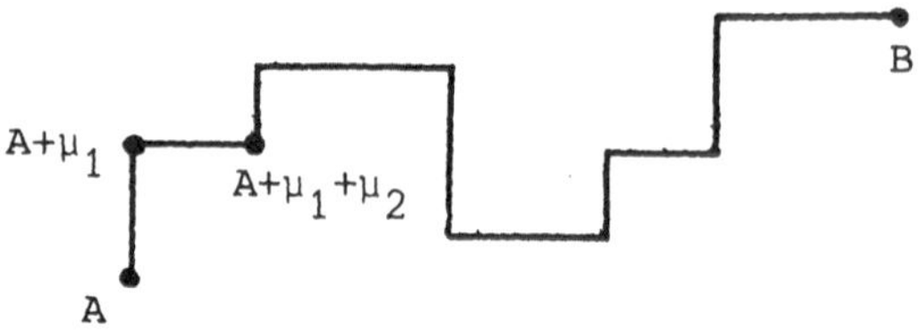

Fig 1 A path contributing to the propagator.

For simplicity let us assume that we are allowed to consider only self avoiding paths (It is possible to show that the non-self avoiding paths do not change the leading large distance behaviour of the propagator). Since each cos is accompanied with a factor of $^1/g$ (g is the coupling constant) we get

$$K(A,B) = \sum_{\substack{\text{self avoiding}\\ \text{path}}} g^{-\text{Length of path (in lattice unit)}} \tag{1.10}$$

Also the recursion formula for the propagator is self evident

$$K(A,B) = g^{-1} \sum_{\nu} K(A,B+\nu) + \delta(A-B) \tag{1.11}$$

where the sum over ν runs over the four points $B+\nu$ which are the neighbors of B . Take the point A to be at the origin and denote by $r \gg 1$ the minimal distance (in lattice units) between A and B . The leading behaviour of the contiunuum propagator is

$$K(B) \simeq e^{-mr} \tag{1.12}$$

Substituting it into equation (1.11) we obtain

$$g^{-1}e^{-mr}\,[e^{-ma\cos\Theta} + e^{ma\cos\Theta} + e^{-ma\sin\Theta} + e^{ma\sin\Theta}] = e^{-mr} \tag{1.13}$$

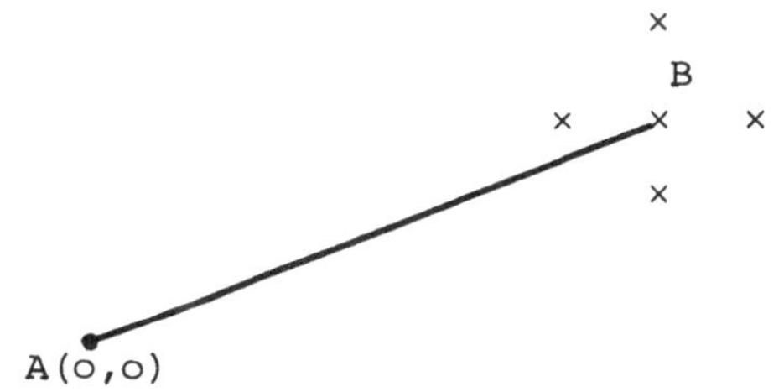

Fig. 2 The minimal distance between the points A and B, which appears in the assymptotic solution for the propagator.

In the continuum limit $a \to o$ eq. (1.13) yields

$$(4+m^2a^2)\ ^1/g = 1 \tag{1.14}$$

In order to have a finite m , $g \to 4$ as $a \to o$; namely a delicate tuning of the coupling constant is needed to ensure a sensible finite mass theory. Actually eq. (1.11) for the propagator can be solved by going into Fourier space

$$K(A,B) = \int_0^1 dk_1\, dk_2\, e^{2\pi i k_1 n}\, e^{2\pi i k_2 P}\, \tilde{K}(k_1 k_2) \tag{1.15}$$

where (n,P) are the coordinates of the point B. A is taken always to be at the origin. Substituting eq. (1.15) into eq. (1.11) gives

$$\tilde{K}(k_1 k_2)[1 - 2g^{-1}(\cos 2\pi k_1 + \cos 2\pi k_2)] = 1 \tag{1.16}$$

Denote

$$x = na, \quad y = Pa$$

$$K(A,B) = \int_0^{2\pi/a} d\left(\frac{2\pi k_1}{a}\right) d\left(\frac{2\ k_2}{a}\right) \frac{e^{i2\pi\left(\frac{k_1}{a}x + \frac{k_2}{a}y\right)}}{1 - 2g^{-1}\left(\cos 2\pi a\frac{k_1}{a} + \cos 2\pi a\frac{k_2}{a}\right)} \tag{1.17}$$

Using dimensionful k_x and k_y

$$K(A,B) = \int_0^\infty dk_x dk_y \frac{e^{i(xk_x + yk_y)}}{1 - 2g^{-1}(\cos(ak_x) + \cos(ak_y))} \tag{1.18}$$

In the continuum limit $a \to o$ the denominator of eq (1.18) gives

$$(1 - 4g^{-1}) + g^{-1}a^2(k_x^2 + k_y^2) \tag{1.19}$$

Hence defining m through

$$1 - 4g^{-1} = a^2 m^2 g^{-1} \tag{1.20}$$

which is the same as eq (1.14), we recover the relativistic rotationally symmetric propagator

$$K(A,B) = \int dk_x\, dk_y \frac{e^{i(xk_x + yk_y)}}{k^2 + m^2} \tag{1.21}$$

PHASE TRANSITIONS AND CONDENSATION OF TOPOLOGICAL EXCITATIONS

The Ising model is a prototype of a model which undergoes a spontaneous symmetry breaking. Let us consider the 2-dimensional Ising model [16,18]

$$Z = \sum_{\{\sigma = \pm 1\}} e^{\beta \Sigma_i \sigma_i \sigma_{i+1}} \qquad \beta = \frac{1}{kT} \tag{1.22}$$

At low T the ground state is doubly degenerate. There are two states, i.e all spin up (+1) and all spin down (-1) which have opposite magnetization. Since going from one state to the other involves turning an infinite number of spins, these states do not mix in any finite order of perturbation theory. We are allowed to choose one of them as our ground state, hence the ground state does not respect the symmetry $(\sigma \to -\sigma)$. The magnetization serves as the order parameter. As the temperature is raised individual spins are allowed to fluctuate. Such fluctuations are responsible for the smooth decrease of the magnetization with increasing T.

There exists, however, another type of fluctuations called domain fluctuations. In such a fluctuation the spins in a whole domain flip as depicted in Fig 3

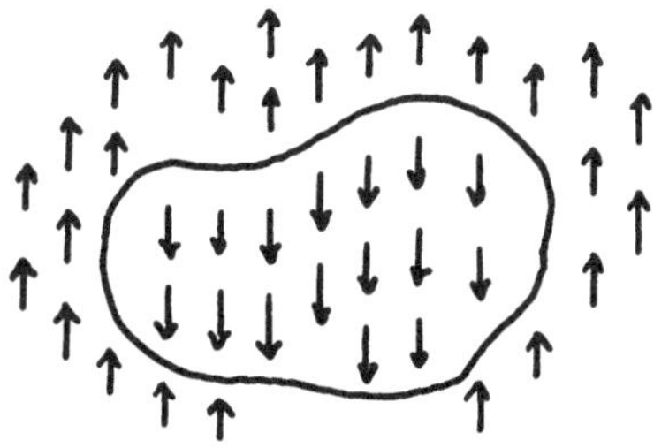

Fig. 3 A domain fluctuation

It is clear from eq (1.22) that to form such a fluctuation we pay in energy only along the boundary of the domain where we have anti-parallel spins.

$$Z\,(\text{domain}) = \exp\,(-2\beta L) \tag{1.23}$$

where 2β is the amount paid for each anti-parallel pair and L is the perimeter of the domain. It is evident that such a domain may appear anywhere on the lattice. Moreover Z(domain), which is related to the probability of its occurrence, just depends on L and not on the particular shape of the domains. Hence

$$Z\,(\text{domain with perimeter } L) = \mu^L \exp\,(-2\beta L) \tag{1.24}$$

with μ a geometrical factor, μ^L counts the number of different shapes with given perimeter L . The factor μ can be estimated using random walk methods; $L \log\mu$ is known in statistical mechanics as the entropy. For small T domains with large L are very unlikely (the energy factor in (1.24) dominates). If T is large the entropy $L \log\mu$ overcomes the energy and large domain walls are abundant. These domain walls can be regarded as the topological excitations which condense at the critical point which this rough consideration estimates to be at

$$\beta_c = \frac{1}{2} \log \mu \tag{1.25}$$

Once the large domain fluctuations are abundant they tend to disorder [19] the system and the magnetization is lost. In dealing with Z(N) gauge theories we'll encounter again the phenomenon that condensation of topological excitations causes a phase transition [5,6,7]. As we have already mentioned we believe that QCD is in a confining phase and that its vacuum can be described in a dual way to that of the Higgs vacuum (superconductor), i.e as a condensate of magnetic monopoles.

For gauge systems there is an important theorem due to S. Elitzur [20] which forbids local symmetry to be broken spontaneously. In particular there is no local order parameter (no magnetization). Only gauge invariant operators can have non zero vacuum expectation values. Such operators are the Wilson and 't-Hooft loops (which are really kind of correlation functions) whose behaviour characterizes the various phases of a gauge theory. The physics beyond Elitzur's theorem is that states which are obtained by the operation of the local symmetry mix in finite order of perturbation theory (unlike the case of a global symmetry), hence the gauge invariant vacuum is always unique.

The last remark concerning lattice gauge theories is that all of them, in particular QCD, confine at large coupling. In this respect the big challenge of lattice QCD is to establish asymptotic freedom as well, as we take the lattice spacing to zero. In order to achieve it there must be no critical point at finite coupling which will forbid the continuation to zero coupling (eq.(1.1)) where asymptotic freedom is recovered [21].

II. HAMILTONIAN FORMULATION OF Z(N) GAUGE THEORIES [22]

We shall consider the following Hamiltonian describing a Z(N) gauge theory

$$H = -\lambda \sum_{\ell} (P_\ell^{+} + P_\ell - 2) - \sum_{P} (Q_{P_1}^{+} Q_{P_2}^{+} Q_{P_3} Q_{P_4} + \text{h.c} - 2) \tag{2.1}$$

The unitary operators P_ℓ and Q_ℓ are associated with the links ℓ of the lattice and obey the Z(N) algebra

$$P_\ell^N = Q_\ell^N = 1 \qquad P_\ell^{+} P_\ell = Q_\ell^{+} Q_\ell = 1 \tag{2.2}$$

$$P_\ell^+ Q_\ell P_\ell = e^{i\delta} Q_\ell \qquad \delta = \frac{2\pi}{N} \tag{2.3}$$

Operators which are associated with different links commute with one another. The second term in eq.(2.1) involves a sum over all plaquettes of the lattice. For each plaquette p one defines the links p_1 and p_4 to be parallel to the unit axes of the lattice. The order 1 to 4 defines a closed loop around the plaquette as shown in Fig 4.

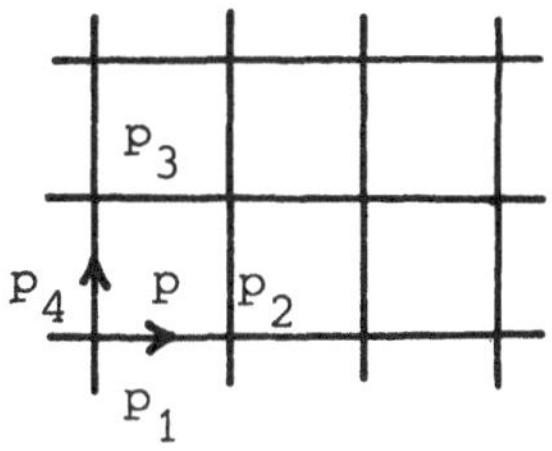

Fig 4. A plaquette notation

An example of a specific representation of the Z(N) algebra is given by

$$P|n\rangle = e^{i\delta n}|n\rangle, \quad Q|n\rangle = |n+1\rangle \qquad n=o,\ldots,N-1 \tag{2.4}$$

The case N=2 is particularly simple and can be written in terms of Pauli matrices

$$H = -\lambda \sum_\ell \sigma_1(\ell) - \sum_p \sigma_3(p_1)\sigma_3(p_2)\sigma_3(p_3)\sigma_3(p_4) \tag{2.5}$$

The Z(N) theories possess a local gauge invariance. At a given point i on the lattice the following operation is a symmetry of the Hamiltonian

$$Q_\ell \rightarrow e^{i\delta} Q_\ell \qquad \ell \in i$$

$$Q_\ell \rightarrow Q_\ell \qquad \ell \notin i$$

$$P_\ell \rightarrow P_\ell \qquad \text{all } \ell \tag{2.6}$$

The symbol $\ell \in i$ means that the link ℓ has i as one of its end points. Each lattice site can be associated with a unitary operator G(i) which commutes with the Hamiltonian and generates this local symmetry

$$[G(i),H] = 0 \tag{2.7}$$

$$G(i) = \prod_{\ell_+ \ni i} P_{\ell_+}^+ \prod_{\ell_- \ni i} P_{\ell_-} \tag{2.8}$$

where the various links ℓ which touch the vertex i are defined to be positive or negative according to whether they are parallel or antiparallel to the basis vectors. The Hilbert space is divided into sectors which are classified by the eigenvalues of all G(i)

$$[G(i)]^N = 1 \Rightarrow G(i) = e^{i\delta n} \qquad n=0,--,N-1 \tag{2.9}$$

The gauge invariant sector has G(i)=1 for all points i

$$G(i)\ |\psi\rangle = |\psi\rangle \tag{2.10}$$

Other sectors where at some point i_0, $G(i_0) = e^{-i\delta n}$ may be interpreted as representing the physics with external charge n(mod N) located at the point i_0, as will become evident from the QED like representation of the models.

QED - LIKE REPRESENTATION OF THE Z(N) GAUGE THEORIES

Let us represent the operators P_ℓ and Q_ℓ in the following way

$$P_\ell = e^{i\delta E_\ell} \qquad Q_\ell = e^{iA_\ell} \tag{2.11}$$

E_ℓ and A_ℓ are dimensionless hermitian operators associated with the link ℓ and, as their name suggests, they are the analogs of the electric field and the vector potential. Formally the Z(N) algebra eq (2.3) is satisfied provided E and A obey the canonical commutation relation

$$[E_\ell, A_m] = i\delta_{\ell,m} \tag{2.12}$$

One should, however, be careful [5] since eq (2.12) cannot strictly hold for finite N when at each link we have a finite norm set of states. This problem was discussed by Schwinger [23]. Inserting eq (2.11) in the definition of G in eq (2.8) we find

$$G(i) = e^{-i\delta\nabla\cdot E} \tag{2.13}$$

where $\nabla\cdot E$ is the obvious lattice definition of the divergence of E

$$\nabla\cdot E(i) = \sum_{\ell_+ \ni i} E_{\ell_+} - \sum_{\ell_- \ni i} E_{\ell_-} \tag{2.14}$$

The identification of the different sectors of Hilbert space with fixed charges at vertices where $G(i) \neq 1$ is the implementation of Gauss' law

$$\nabla\cdot E = \rho \tag{2.15}$$

The analog of the magnetic field B_p is a plaquette variable and is just the lattice version of curl A.

$$B_p = (\nabla \times A)_p = A_{p_1} + A_{p_2} - A_{p_3} - A_{p_4} \qquad (2.16)$$

Using the QED-like representation the Hamiltonian (2.1) can be rewritten in the form

$$H = 2\lambda \sum_{\ell} (1-\cos\delta E_{\ell}) + 2\sum_{p} (1-\cos B_p) \qquad (2.17)$$

A link (plaquette) on which $E_{\ell} \neq 0$ ($B_p \neq 0$) will be said to carry electric (magnetic) flux. These fluxes are defined only modulo N, thus for N=3 three unit electric flux lines can meet at one point of the lattice. This is of course a basic property of a string picture of the structure of baryons. The electric flux lines can be interpreted as strings connecting quarks.

Up to now our discussion was general. I would like now to start to investigate the phase structure of the Z(N) gauge theories and I'll do it separately for the 2+1 dimensions case and the 3+1 dimensions case.

DUALITY TRANSFORMATION IN 2 + 1 DIMENSIONS

The gauge condition eq (2.10) implies that within the gauge invariant sector

$$\left\{ G(i) = 1 \quad \text{for all} \quad i \right\}$$

the original variables of the theory are not independent. We propose to switch to a new set of variables which are independent by solving the gauge condition in an operator language. The new set of variables is defined on the dual-lattice and the transformation to this set is called duality transformation [24]. The dual lattice is obtained from the original lattice by connecting the centers of its plaquettes as shown in fig 5.

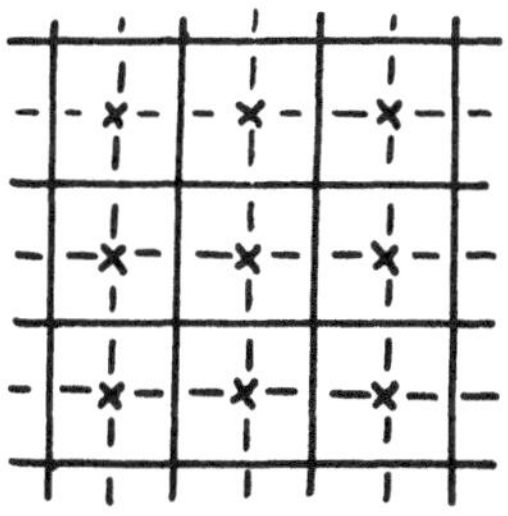

Fig. 5 The lattice and the dual lattice in 2 + 1 dimensions

The geometrical relationship between the lattice and its dual is

<u>lattice</u>	<u>dual-lattice</u>
point	plaquette
link	link
plaquette	point

The dual variables will be defined on the points of the dual lattice

$$R_p = Q_{p_1} Q_{p_2} Q^+_{p_3} Q^+_{p_4}$$
$$P_\ell = S^+_{p'} S_{p''} (\delta_{\ell' p'_1} \delta_{\ell' p''_3} + \delta_{\ell, p'_2} \delta_{\ell, p''_4}) \tag{2.18}$$

The second equation defines the operator S_p in an implicit way. Each operator P_ℓ is expressed as the product of two unitary operators (the S's) which are associated with the two plaquettes (hence points on the dual lattice) touching that link.

$$S^+_p S_p = R^+_p R_p = 1 \qquad S^N_p = R^N_p = 1 \tag{2.19}$$

It is easy to check that in order that P_ℓ and Q_ℓ obey the Z(N) algebra eq (2.3), also R_p and S_p have to obey the same algebra

$$R^+_p S_p R_p = e^{i\delta} S_p \qquad [R_p, S_q] = 0 \quad \text{if } p \neq q \tag{2.20}$$

Moreover it is clear, using eq (2.18) that the gauge invariance condition $G(i) = 1$ is automatically satisfied. Indeed the number of operators in the set (R,S) is half the number of the original variables.

The Hamiltonian in terms of the dual variables is given by

$$H = -\lambda \sum_{<pq>} (S^+_p S_q + \text{h.c.} - 2) - \sum_p (R^+_p + R_p - 2) \tag{2.21}$$

The symbol <pq> denotes nearest neighbors. This representation of the Hamiltonian is true for an infinite lattice. For a finite lattice there are extra terms representing the boundary. The Hamiltonian in eq. (2.21) is recognized as a generalization of an Ising model in a transverse field. This Hamiltonian possesses only global Z(N) symmetry.

To gain some understanding of the physics beyond the duality transformation let us consider eq (2.18). An explicit solution for S_p is

$$S(p) = \prod_{j'=0}^{j} P_x^+(i,j') \tag{2.22}$$

where $p \equiv p(i,j)$ is the point in the dual lattice which is associated with the plaquette P whose lowest left corner is the point (i,j). The product runs over operators which lie on links in the x direction, starting at the boundary of the lattice as depicted in fig 6.

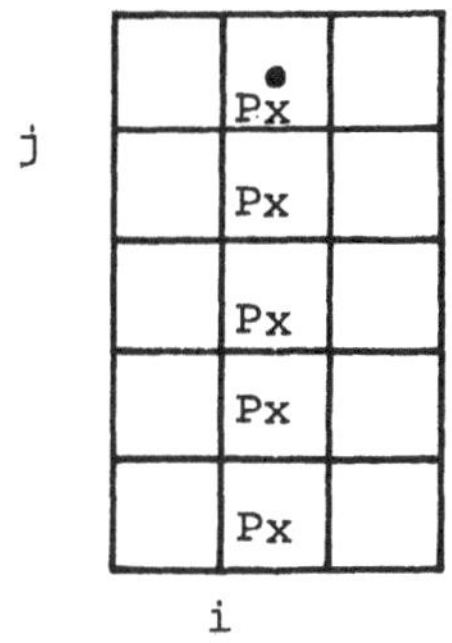

Fig. 6 Solution of the gauge condition in 2 + 1 dimensions

Equation (2.22) is a realization of the relation

$$P_x(i,j) = S^+(i,j)\; S(i,j-1) \tag{2.23}$$

which is contained in eq (2.18). The other part of that set of equations, i.e

$$P_y(i,j) = S^+(i-1,j)\; S(i,j) \tag{2.24}$$

is also satisfied since in the gauge invariant sector one can express the P_y operator in terms of chains of P_x operators

$$P_y(i,j) = \prod_{j'=0}^{j} P_x(i-1,j')P_x^+(i,j') \tag{2.25}$$

Using the QED-like representation eq (2.11) we see that R_p is the Wilson loop operator defined over one plaquette

$$R_p = e^{-i(A_{p_1}+A_{p_2}-A_{p_3}-A_{p_4})} \tag{2.26}$$

The Wilson loop operator for any curve C on the lattice is

$$\mathcal{O}\!\mathcal{L}(C) = e^{-i \oint_C \vec{A}\cdot d\vec{\ell}} \prod_{\ell \in C} \tilde{Q}_1 \qquad (2.27)$$

where $\tilde{Q}_\ell$ equals either Q_ℓ or Q_ℓ^+ according to the relative orientation of the curve C and the link ℓ .

A straight forward algebra (eqs.(2.22),(2.27),(2.3)) yields

$$\mathcal{O}\!\mathcal{L}^+(C) S_p \mathcal{O}\!\mathcal{L}(C) = e^{im\delta} S_p \qquad (2.28)$$

where m is the number of times the curve C winds around the point p of the dual lattice. Equation (2.28) is the 't-Hooft algebra in the 2+1 dimensional case (see eq.(1.5) and the discussion thereafter). S_p plays the role of the monopole field $\phi(x)$ in 't-Hooft's topological $Z(N)$ algebra and serves as the disorder operator. The monopole interpretation can be understood in the following way: the Wilson loop operator can be written

$$\mathcal{O}\!\mathcal{L}(C) = e^{-i \iint_C d\sigma B_z} \qquad (2.29)$$

Equation (2.28) then implies

$$\langle\psi| S_p^+ \mathcal{O}\!\mathcal{L}(C) S_p |\psi\rangle = \langle\psi| \mathcal{O}\!\mathcal{L}(C) |\psi\rangle \times \begin{cases} e^{-im\delta} & \text{if } P\subset C \\ 1 & \text{if } P\not\subset C \end{cases} \qquad (2.30)$$

hence, S_p is an operator which creates one unit ($\delta = \frac{2\pi}{N}$) of magnetic flux flowing through the plaquette p, i.e. it creates a monopole at the point p.

PHASES OF THE Z(N) THEORY IN 2 + 1 DIMENSIONS - ORDER AND DISORDER

Perturbative treatment of the Hamiltonian (2.1) in weak and strong coupling regimes reveals two different phases. We shall describe these phases and the various quantities which characterize them.

1. The large λ phase - confinement

As $\lambda \to \infty$ the vacuum state approaches the state

$$|V\rangle = \prod_\ell |0\rangle_\ell \qquad P_\ell |0\rangle_\ell = 1 \qquad (2.31)$$

Doing perturbation in $1/\lambda$ we obtain

$$E_V = AV + C \qquad (2.32)$$

where V is the "volume" of the lattice (the number of plaquettes in the lattice) and A and B are functions of λ. We would like, next, to define the string state. To do it we study the sector of Hilbert space defined by two charges at distance L.

$$G(i) = \begin{cases} e^{i\delta} & i = (o,o) \\ e^{-i\delta} & i = (L,o) \\ 1 & \text{otherwise} \end{cases} \tag{2.33}$$

The string state is defined to be the ground state of this sector. At $\lambda \to \infty$ it consists of a straight line of electric flux ($E_\ell = 1$) between the two charges. Its energy is

$$E_{string} = AV + TL + B \tag{2.34}$$

where A,T,B are functions of λ which can be calculated in perturbation theory in $^1/\lambda$. The term proportional to L is defined as the string tension [17]

$$T = \lim_{L\to\infty} \frac{E_{string} - E_V}{L} \tag{2.35}$$

A value of λ at which T shows a discontinuous behaviour corresponds to a phase transition. Of particular interest is the point where T vanishes. As long as $T>o$ the energy of the two charges configuration (relative to the vacuum) is proportional to L, i.e. there exists a linear potential. This signifies confinement. The coupling λ_0 at which $T(\lambda_0)=o$ corresponds to a transition from an electric confinement to a non-confinement phase. At this point the vacuum includes arbitrarily long electric flux tubes, since it does not cost anymore in energy proportionally to L. There is a relationship between T and the β function of the theory. The energy of the string state relative to the vacuum is given by

$$\Delta E = L\, T\, ^1/a \tag{2.36}$$

Since ΔE is a physical quantity it should not depend on the unit in which we measure the energy (the lattice spacing a). We have therefore

$$o = a\,\frac{\partial(\Delta E)}{\partial a} = L\left\{-\frac{1}{a}\,T + \frac{\partial T}{\partial\lambda}\frac{\partial\lambda}{\partial a}\right\} \tag{2.37}$$

The β function is defined as usual

$$\beta(\lambda) = a\,\frac{\partial\lambda}{\partial a} \tag{2.38}$$

Using eqs. (2.37), (2.38)

$$\beta(\lambda) = \frac{T/\partial T}{\partial\lambda} \tag{2.39}$$

Hence a zero of T corresponds to a zero of the β function. This derivation is not rigorous in the sense that we disregarded the possible occurrence of anomalous dimensions. Moreover, note that if T jumps from some finite value to zero (corresponding to a first order transition) then a more careful derivation is necessary.

As we have already mentioned, the possible phases of a gauge theory cannot be characterized by means of a local order operator (magnetization). Instead we use the gauge invariant Wilson loop as an order parameter

$$\mathfrak{A}(C) = \langle \prod_{\ell \in C} \tilde{Q}_\ell \rangle_V \tag{2.40}$$

where $\tilde{Q}_\ell$ is equal to Q_ℓ or Q_ℓ^+ according to the orientation of the closed curve C with respect to that of the link ℓ . Using perturbation theory in $1/\lambda$ it is easy to check that the first non zero contribution comes in the order which is equal to the area closed by the curve C

$$\mathfrak{A}(C) \sim \left(\frac{1}{\lambda}\right)^{\text{area}(C)} = e^{-(\log\lambda)\ \text{area}(C)} \qquad \lambda \gg 1 \tag{2.41}$$

Higher orders will change the coefficient in front of area(C). The $e^{-\text{area}}$ behaviour of the Wilson loop is the famous criterion for confinement [25].

The duality transformation allows us to discuss the same theory by using its dual form (eq.(2.21)).The Ising like theory in a transverse field is known to have two phases [26]. The large λ phase is characterized by

$$\langle S_p \rangle \neq 0 \qquad \lambda \gg 1 \tag{2.42}$$

This vacuum expectation value serves as a disorder parameter. Recalling the interpretation of S_p as an operator which creates a magnetic monopole (eq.(2.30)), we learn that the vacuum of the confining phase is a condensate of magnetic monopoles. Starting in the small λ phase we can follow the mass of the monopole as a function of λ (using perturbation theory and Padé approximation). The point λ_c at which this mass goes to zero corresponds to the phase transition into the confining phase.

2. The small λ phase

We can use either the order or the disorder parameter to characterize the small λ phase. It is a straight forward calculation to compute the Wilson loop in perturbation theory. The result is a perimeter law behaviour

$$\mathfrak{A}(C) \sim e^{-b\lambda^2 \text{perimeter } (C)} \qquad \lambda<<1 \qquad (2.43)$$

Thus, the Wilson loop falls at a much slower rate in this phase compared to its behaviour in the large λ phase (eq.(2.41)). The different behaviour of the Wilson loop distinguishes between the two phases.

As for the disorder parameter, equation (2.21) for the dual Hamiltonian reveals that for small λ

$$\langle S_p \rangle = 0 \qquad \lambda<<1 \qquad (2.44)$$

Alternatively in the weak coupling regime we may choose to study the mass of the monopole which is created by the operator S_p. This operator creates a topologically stable excitation in the theory defined by the dual Hamiltonian. The reason is that the Wilson loop operator around the whole lattice commutes with the Hamiltonian. The corresponding quantum number counts the total number of monopoles minus the total number of antimonopoles. We define the mass of the monopole as

$$M_m(\lambda) = E_m(\lambda) - E_{vac}(\lambda) \qquad (2.45)$$

where E_m is the energy of the state which develops with λ out of the state

$$\sum_p S_p|0\rangle \quad \text{at} \quad \lambda = 0.$$

In the small λ phase M_m is positive, hence eq.(2.44) must be satisfied. At the critical point where M_m vanishes, S_p starts developing a vacuum expectation value.

SELF DUALITY OF THE Z(N) GAUGE THEORIES IN 3 + 1 DIMENSIONS

We turn now to the self dual properties of our Hamiltonian (eq.(2.1)) in 3 space dimensions. The dual transformation [27] replaces link variables by plaquette variables and vice versa. Using the QED representation (eq.(2.17)) we note that the link variable is E_ℓ while the plaquette variable is B_p. The physical interpretation of the duality transformation will be the interchange of the electric and magnetic fields. As in the two space demensional case, we define a new set of operators R_p and S_p associated with the plaquettes of the same lattice and obeying the same algebra (eqs. (2.2), (2.3)) as P_ℓ and Q_ℓ. We require the following connection between the two sets of operators

$$R_p^+ = Q_{p_1}^+ Q_{p_2}^+ Q_{p_3} Q_{p_4} \tag{2.46}$$

$$P_\ell^+ = S_{\ell_1}^+ S_{\ell_2}^+ S_{\ell_3} S_{\ell_4} \tag{2.47}$$

The notation is explained in fig 7 which is dual to fig 4.

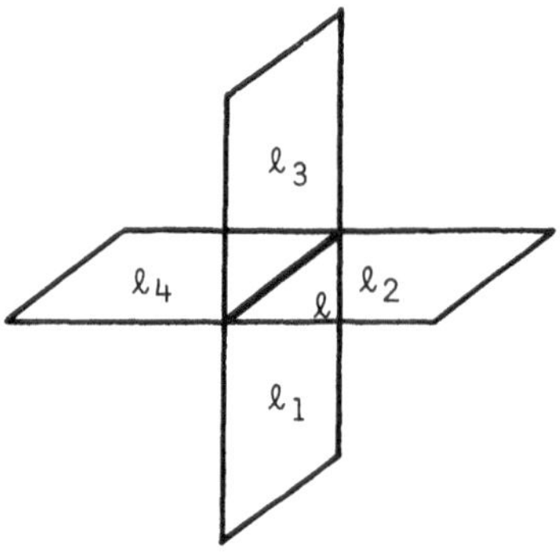

Fig. 7 Notation for the dual transformation in 3 + 1 dimensions

It is clear by inspection that the Z(N) algebra rules of the P_ℓ, Q_ℓ set and the R_p, S_p set are consistent with one another. It is possible to give an explicit solution of the S_p operators in terms of the P_ℓ operators [22] and to check the algebra of the new set directly. The gauge condition $G(i) = 1$ is automatically satisfied by eq. (2.47). This is therefore a solution appropriate for the gauge invariant sector only. Note that unlike the two space dimensional case we have not obtained a set of independent operators. This is indicated by the fact that the number of new variables is the same as before. Equation (2.46) implies that for every cube the product over its enclosing plaquette operators is constrained as follows

$$W(C) = \prod_{p_+ \ni c} R_p^+ \prod_{p_- \ni c} R_p = 1 \tag{2.48}$$

The dual Hamiltonian takes the form

$$H_D = -\lambda \sum_\ell (S_{\ell_1}^+ S_{\ell_2}^+ S_{\ell_3} S_{\ell_4} + h.c. - 2) - \sum_p (R_p^+ + R_p - 2) \tag{2.49}$$

What happens can be easily understood in terms of the electromagnetic variables. Equation (2.49) suggests the identification

$$R_p = e^{i\delta\tilde{E}_p} \qquad S_p = e^{i\tilde{A}_p} \qquad \tilde{B}_\ell = (\nabla x\tilde{A})_\ell \tag{2.50}$$

This leads to the following form of our Hamiltonian

$$H_D = 2\lambda \sum_\ell (i-\cos \tilde{B}_\ell) + 2 \sum_p (1-\cos\delta\tilde{E}_p) \tag{2.51}$$

Comparing to equation (2.17) we find

$$E_\ell \rightarrow \delta^{-1}\tilde{B}_\ell \qquad B_p \rightarrow \delta\tilde{E}_p \tag{2.52}$$

Since by definition $\tilde{B}_\ell$ is divergenceless, we are in the gauge invariant sector. However, in order for $\tilde{E}_p$ to represent the same physics as B_p, it should obey $\nabla\cdot\tilde{E}=o$ condition which is the same as eq (2.48).

In the three space dimensional case the dual transformations lead to a new feature known as self duality. The Hamiltonian H_D of eq (2.49) has similar structure to H of eq.(2.1). Therefore, the physical content of the dual theory $H_D(\lambda)$ is the same as that of $\lambda H(\lambda^{-1})$. Hence, every physical quantity which is linearly dependent on the Hamiltonian must obey the self duality relation

$$F(\lambda) = \lambda F(\lambda^{-1}) \tag{2.53}$$

In particular the phase diagram of this model should exhibit a symmetry under $\lambda \rightarrow \lambda^{-1}$.

As for the two space dimensional case, we can identify order and disorder operators and check their algebra. The order operator is, as before, the Wilson loop $\mathfrak{A}(C)$ (eq.2.40). The definition of the disorder operator $\mathfrak{B}(C)$ is dual to that of $\mathfrak{A}(C)$

$$\mathfrak{B}(C) = \prod_{\ell_D \in C} S_{\ell_D} \tag{2.54}$$

where the curve C is defined now on the dual lattice. Using the Z(N) algebra of our operators we find that $\mathfrak{A}(C)$ and $\mathfrak{B}(C)$ obey the 't-Hooft Z(N) topological algebra (eq.1.5)

$$\mathfrak{A}^+(C)\mathfrak{B}(C')\mathfrak{A}(C) = e^{im\delta}\mathfrak{B}(C') \tag{2.55}$$

where m is the number of times the curve C winds around the curve C'.

PHASES OF THE Z(N) THEORY IN 3 + 1 DIMENSIONS

The self duality of our Hamiltonian in three space dimensions suffices to locate the phase-transition point if it is assumed that the theory has two phases only. The symmetry under $\lambda \to \lambda^{-1}$ implies that in this case the critical point is $\lambda_c = 1$. This is believed to be the case for N = 2 . We shall show that this cannot be the case for all N . For large enough N more than two phases exist. Once the two-phase assumption is relaxed, self duality by itself does not determine the location of the critical point. However, it is still true that if λ_c is a critical point so is also λ_c^{-1}.
The fact that there exist at least two phases can be demonstrated, as in the two space dimensional case, by investigating the behaviour of the order and disorder operators. Using perturbation theory for large and small λ we find

$$\begin{aligned} \mathcal{A}(C) &\sim e^{-\text{area}(C)} \\ \mathcal{B}(C) &\sim e^{-\text{perimeter}(C)} \qquad \lambda \gg 1 \end{aligned} \tag{2.56}$$

$$\begin{aligned} \mathcal{A}(C) &\sim e^{-\text{perimeter}(C)} \\ \mathcal{B}(C) &\sim e^{-\text{area}(C)} \qquad \lambda \ll 1 \end{aligned} \tag{2.57}$$

Hence the large λ phase is the electric confining phase. The two phases are related by the duality transformation. In particular the behaviour of $\mathcal{A}(C)$ and $\mathcal{B}(C)$ in the two phases is complementary. The small λ phase is a Higgs like phase (superconductor) in which we find the phenomenon of magnetic confinement. A physical quantity which we investigate in the strong coupling (large λ) regime is the string tension of the electric flux tube as defined in eqs.(2.33),(2.35). A phase transition occurs at that λ_c where the tension vanishes. Using perturbation theory we calculate the tension as a series in λ^{-1} for very high N, i.e.

$$\delta = \frac{2\pi}{N} \ll 1.$$

The result is

$$T \simeq \frac{1}{2}\lambda\delta^2 - \frac{.333}{\lambda\delta^2} - \frac{.290}{\lambda^3\delta^6} \tag{2.58}$$

Rewriting it in a Padé form one obtains

$$T \simeq \frac{1}{2}\left(\lambda\delta^2 - \frac{1.537}{\lambda\delta^2}\right)\left(1 - \frac{.870}{\lambda^2\delta^4}\right)^{-1} \tag{2.59}$$

which has a zero at $\lambda_c = .031\, N^2$ for $N \gg 2\pi$. The fact that $\lambda_c \sim N^2$

really comes about is because for large N the expansion parameter is $(\lambda\delta^2)^{-1}$ rather than λ^{-1}. For large enough N it is therefore clear that λ_c will become greater than one; hence the self duality ensures us that there exist more than two phases.

The argument for $\lambda_c \sim N^2$ can be made stronger by noticing the connection between the Z(N) theories and PQED. In the limit

$$N \to \infty \qquad \lambda \to \infty \qquad \lambda\delta^2 = e^4 = \text{const.} \tag{2.60}$$

the Hamiltonian (eq. (2.17)) becomes

$$H = \frac{e^4}{2} \sum_{\ell} E_\ell^2 + \sum_p (1-\cos B_p) + e^4 O(\delta^2 \sum_\ell E_\ell^2) \tag{2.61}$$

Neglecting the $\delta^2 E_\ell^2$ and higher terms ($\delta \to o$) we are left with the PQED Hamiltonian

$$\frac{H}{e^2} = \frac{1}{2} (\sum_\ell \bar{E}_\ell^2 + \frac{2}{e^2} \sum_p (1-\cos e\bar{B}_p)) \tag{2.62}$$

$\bar{E}_\ell$ and $\bar{B}_p$ are canonical transformations of the original electric and magnetic fields

$$\bar{E}_\ell = eE_\ell \qquad\qquad \bar{B}_p = e^{-1} B_p \tag{2.63}$$

The PQED theory of eq. (2.62) is known to have a phase transition for a finite value of e . For $e > e_c$ it exhibits confinement of electric charges and formation of electric flux lines. The perturbative equivalence between the high N limit of the Z(N) models and PQED leads to the suggestion that the high λ_c value of Z(N) approaches $e_c^4\delta^{-2}$ as $N \to \infty$. Hence $\lambda_c \sim N^2$. For $e < e_c$ PQED has a non-confining phase. As $e \to o$ it looks more and more like normal QED. Such a non-confining phase follows also in the Z(N) theories throughout the region of constant λ as $N \to \infty$. This can be simply seen by letting $e \to o$ in eq (2.62).

The Z(N) gauge theories are self dual while the PQED theory (eq. (2.62)) is not. There is no contradiction here since PQED is recovered in a particular limit (eq. (2.60)). The dual theory can be obtained by considering the dual limit

$$N \to \infty \qquad \lambda \to o \qquad \frac{\delta^2}{\lambda} \equiv g^4 = \text{const} \tag{2.64}$$

We then find that the theory is perturbatively equivalent to

$$\frac{H}{\lambda g^2} = \frac{1}{2} \left\{ \sum_p \tilde{B}_p^2 + \frac{2}{g^2} \sum_\ell (1-\cos g\tilde{E}_\ell) \right\} \tag{2.65}$$

where the new electric and magnetic fields are

$$\tilde{E}_\ell = g\,\lambda\,E_\ell \qquad \tilde{B}_p = \frac{B_p}{g\,\lambda} \tag{2.66}$$

Equation (2.65) is again a PQED theory but this time the electric and magnetic fields are interchanged. At a new critical point whose value is

$$\lambda_{c'} = \lambda_c^{-1} \sim N^{-2}$$

(recall the self duality of the Z(N) gauge theories) we find a transition between the non confining phase to a magnetic confining one. In this new phase, magnetic flux is confined to flux tubes in much the same way that the electric flux was confined in the large λ region. This is, therefore, a Higgs (superconductor) phase. We could have probed this small λ region directly by using the dual theory and defining the magnetic string tension in an analogous (dual) way to the electric string tension (eqs.(2.33),(2.35). The critical point $\lambda_{c'}$ corresponds to the point at which the magnetic string tension vanishes.

The above considerations lead us to expect that the Z(N)gauge theories will have the phase diagram depicted in Fig 8

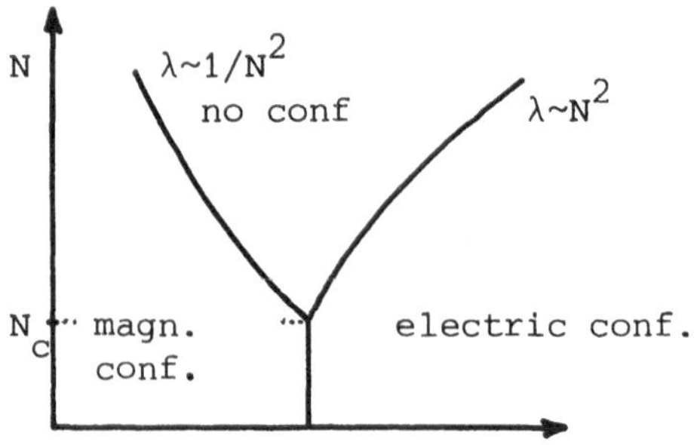

Fig. 8 The phase diagram of the Z(N) gauge theories.

Note that the magnetic confinement phase disappears in the large N limit when PQED is recovered. To understand a little better the origin of the three phases picture for finite N we can start from PQED where the compact U(1) group space is just a circle. The Z(N) gauge theories can be viewed (see also the Lagrangian formulation) as discretization of this circle allowing the angle variable to take only the values

$$\frac{2\pi}{N}\,n\;;\quad n = o,--,N-1.$$

However for discrete spin variables there is always a mass gap at low temperature (coupling) since it costs a finite amount of energy to invert one spin. At low temperature we can ignore entropy considerations, hence we expect a critical temperature at the order of the energy gap i.e. $\sim 1/N^2$. The first phase transition is therefore at $\lambda \sim 1/N^2$ and is associated with the fact that we are dealing with discrete variables. The second phase transition is the one which continuously develops into the critical point of PQED in the $N \to \infty$ limit (eq. (2.60)) and is connected to the periodic (compact) nature of our theories.

To estimate the value of N_c above which a three phases structure develops, we return to the perturbative calculation of the electric tension. Calculation to second order in λ^{-1} gives

$$T = \lambda(1-\cos\delta) - [2\lambda(1-\cos\delta) + \lambda(1-\cos 2\delta)]^{-1} \tag{2.67}$$

This expression vanishes at $\lambda_c = \lambda_c^{(2)}$. Our experience with the Z(N) gauge theories shows us that the fourth order estimate of the zero point is higher,

$$\lambda_c^{(4)} > \lambda_c^{(2)} .$$

Assuming this to be an indication that $\lambda_c > \lambda_c^{(2)}$ we ask at which N we first have $\lambda_c^{(2)} > 1$. The answer is $N_c \lesssim 7$.

CONDENSATION OF TOPOLOGICAL EXCITATIONS AS THE ORIGIN OF THE PHASE TRANSITIONS IN THE Z(N) GAUGE THEORIES

As we have already discussed equations (2.56),(2.57) reveal that the vacuum of the magnetic confining (MC)-phase is a condensate of electric flux tubes [28]. Similarly the vacuum of the electric-confining (EC) phase is a condensate of magnetic flux tubes [29]. In the non-confining (NC) phase both kinds of flux tubes exist in the vacuum and neither is confined.

At this stage it is important to note the relationship between the electric and magnetic couplings (eqs. (2.60), (2.64))

$$eg = \frac{2\pi}{N} . \tag{2.68}$$

This is a Z(N) rather than Dirac quantization condition. We can, however, identify magnetic monopole configurations which do obey Dirac's condition. This is a configuration in which N magnetic flux units emerge from a finite region on the lattice and therefore its total magnetic strength is

$$g_D = Ng = \frac{2\pi}{e} \tag{2.69}$$

This is possible since the flux is conserved only modulo N .

These flux lines have to meet again either on the boundary or in some other region of the lattice which can be identified as an antimonopole. In the MC phase we can observe at most bounded magnetic monopole - antimonopole pairs. In the NC phase, where the magnetic tension vanishes, the pairs can have arbitrarily long strings, i.e. the monopoles are liberated. This corresponds to the first transition. At the second transition point from the NC to EC phase, the magnetic monopoles become massless and the EC vacuum becomes their condensate [30]. Since the theory is self dual we can get a similar description of the phase transitions in terms of electric strings and electric monopoles of charge Ne. The NC phase with a zero mass photon is a phase in which both magnetic and electric monopoles are liberated.

For $N<N_C$ only one phase transition occurs and it is known to be of first order [31]. The intuitive physical picture which has been suggested is the following one [32]: the mass of the monopole becomes negative before the string tension went to zero.Hence, there is only one phase transition in which we go directly from the MC to the EC phase. In this case the tension will drop from some finite value to zero discontinuously causing a first order transition.

III LAGRANGIAN FORMULATION OF Z(N) GAUGE THEORIES

In the Lagrangian approach [33] we start with a partition function (functional integral) for the (Euclidian) theory formulated on a symmetric space time lattice. From now on we shall investigate the Z(N) gauge theories in d=4 dimensions. We start with the partition function of PQED [34]

$$Z_{PQED} = \int_{-\pi}^{\pi} \prod_{i,\mu} d\Theta_\mu(i)\; e^{\frac{1}{g^2}\sum_i \sum_{\mu\nu} \cos\Theta_{\mu\nu}(i)} \tag{3.1}$$

The degrees of freedom are angle variables $\Theta_\mu(i)$ which reside on the links of the lattice. The index i goes over the lattice points, while μ designates the four positive directions which emanate from the point i. Hence $\Theta_\mu(i)$ is the variable associated with the link starting at i and going in the μ direction. We define

$$\Theta_{\mu\nu}(i) = \Theta_\mu(i)+\Theta_\nu(i+e_\mu)-\Theta_\mu(i+e_\nu)-\Theta_\nu(i) \equiv \Delta_\nu\Theta_\mu(i)-\Delta_\mu\Theta_\nu(i) \tag{3.2}$$

to be the lattice curl of Θ. The notation is self-explanatory in fig 9.

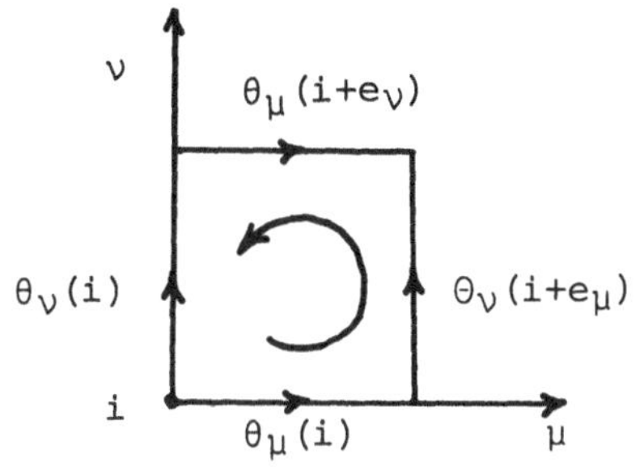

Fig. 9 Notation for the lattice curl.

Note that the action in eq. (3.1) is of the Wilson type. The sum appearing in the action is really a sum over plaquettes. The coupling constant g^2 plays the role of temperature in statistical mechanics.

The theory defined by eq. (3.1) possesses a local U(1) gauge symmetry. The action is invariant under the transformation in which all the variables on links attached to the point i are shifted by the amount $\Theta(i)$ (the sign of the shift depends on wether the link emanates or ends at the point i); i.e.

$$\Theta_\mu(i) \to \Theta_\mu(i) - \Theta(i)$$

$$\Theta_\nu(i) \to \Theta_\nu(i) - \Theta(i) \tag{3.3}$$

Hence $\Theta_{\mu\nu}(i)$ is invariant. To recognize that this is the known gauge transformation we can simultaneously make the transformation at the points i and $i+e_\mu$; then

$$\Theta_\mu(i) \to \Theta_\mu(i) + \Theta(i+e_\mu) - \Theta(i) \equiv \Theta_\mu(i) - \Delta_\mu\Theta(i) \tag{3.4}$$

The Z(N) gauge theory is defined by taking the PQED partition function (eq. (3.1)) and restricting the angle variables to take just N values

$$\Theta_\mu(i) = \frac{2\pi n_\mu(i)}{N} \qquad n_\mu(i) = 0,\text{--},N-1 \tag{3.5}$$

$$Z = \sum_{\{\Theta_\mu(i)\}} e^{\frac{1}{g^2}\sum\limits_{i,\mu\nu} \cos\Theta_{\mu\nu}(i)} \tag{3.6}$$

DUALITY TRANSFORMATION [6,7]

The variables which appear in the partition fuction are just group elements. The main idea beyond the duality transformation is to go from the group itself to the group of the characters (the dual group) by means of a fourier series [27]

$$e^{1/g^2 \cos\Theta_{\mu\nu}} = \sum_{\ell_{\mu\nu}=0}^{N-1} e^{i\ell_{\mu\nu}\Theta_{\mu\nu}} I_{\ell_{\mu\nu}}\left(\frac{1}{g^2}\right) \tag{3.7}$$

where the $\ell_{\mu\nu}$ are integers and the $I_{\ell_{\mu\nu}}$ are N well defined coefficient functions. Instead of continuing with the original theory (eq.(3.1)), we can look at another one by replacing the N parameters $\{I_\ell\}$ with some other N parameters $\{A_\ell\}$. We are implicitly using here what is known in statistical mechanics as universality, i.e. theories which belong to the same universality class have the same phase structure (number of phases, critical indices). The position of the critical points is not universal. The universality class is believed to be determined by very general properties:

i) The dimension of space-time;
ii) The symmetry of the problem;
iii) The type of the interactions (ferromagnetic,anti-ferromagnetic).

The original model (eq.(3.1)) has the property that the free energy (action) has a minimum at $\Theta_{\mu\nu}=2\pi n$ which corresponds to pure gauge configurations. Moreover the action is periodic, which means that on top of the local Z(N) symmetry there is also a symmetry under a shift by an integer multiple of 2π. We shall choose the coefficient $\{A_\ell\}$ in such a way that these properties will be preserved

$$A_\ell = e^{-\frac{1}{2}g^2\ell^2} \tag{3.8}$$

This choice defines the Villan model approximation (periodic Gaussian model)

$$Z_V = \sum_{\{\Theta_\mu(i)\}} \sum_{\{\ell_{\mu\nu}(i)\}} e^{i\sum_{i,\mu\nu} \ell_{\mu\nu}\Theta_{\mu\nu}} e^{-\frac{1}{2}g^2 \sum_{i,\mu\nu} \ell_{\mu\nu}^2} \tag{3.9}$$

Using the Poisson resummation formula it is easy to recast Z_V into the equivalent form

$$Z_V = \sum_{\{\Theta_\mu(i)\}} \sum_{n_{\mu\nu}(i)=-\infty}^{\infty} e^{-\frac{1}{2g^2}\sum_{i,\mu\nu}(\Theta_{\mu\nu}(i)-2\pi n_{\mu\nu}(i))^2} \tag{3.10}$$

with $n_{\mu\nu}$ an integer. This model is obviously in the same universality class as the orginal model (eq.(3.1)). In equation (3.9) Θ_μ appears linearly, hence the summation over $\{\Theta_\mu\}$ can be easily done with the result

$$Z_V = \sum_{\{\ell_{\mu\nu}(i)\}} \Pi(\delta_N(\Delta_\mu \ell_{\mu\nu}))\, e^{-g^2/2 \sum_{i,\mu\nu} \ell^2_{\mu\nu}(i)} \tag{3.11}$$

where δ_N is a δ-function modulo N. Recall that due to the antisymmetry of $\Theta_{\mu\nu}$, $\ell_{\mu\nu}$ is an antisymmetric set of integers as well

$$\ell_{\mu\nu} = -\ell_{\nu\mu} \tag{3.12}$$

The δ_N constraint

$$\Delta_\mu \ell_{\mu\nu} = 0 \pmod N \tag{3.13}$$

can be solved

$$\ell_{\mu\nu} = \varepsilon_{\mu\nu\alpha\beta}\Delta_\alpha \ell_\beta \pmod N \tag{3.14}$$

Note that the new varibale ℓ_β is associated with a link on the dual lattice. Since adding a multiple of N to ℓ_β does not affect $\ell_{\mu\nu}$, we can choose $0 \leq \ell_\beta \leq N-1$ at each dual link. Moreover, ℓ_β is fixed only up to a gauge transformation

$$\ell'_\beta(i) = \ell_\beta(i) + \Delta_\beta \ell(i) \tag{3.15}$$

Both ℓ_β and ℓ'_β lead to the same $\ell_{\mu\nu}$ (eq.(3.14)). The number of gauge transformations is N^V where V is the number of lattice sites. Summing in equation (3.11) on $\{\ell_\alpha\}$ instead of $\{\ell_{\mu\nu}\}$ we multiply the partition function by the irrelevant constant N^V which we shall omit

$$Z_V = \sum_{\{\ell_\alpha(i)\}} \sum_{\{m(i)\}=-\infty}^{\infty} e^{-\frac{g^2}{2} \sum_{i,\mu\nu} (\Delta_\mu \ell_\nu(i) - \Delta_\nu \ell_\mu(i) - Nm(i))^2} \tag{3.16}$$

Defining the dual variable

$$\tilde{\Theta}_\mu = \frac{2\pi}{N} \ell_\mu \qquad \ell_\mu = 0,--,N-1 \tag{3.17}$$

we obtain

$$Z_V = \sum_{\{\tilde{\Theta}_\mu(i)\}} \sum_{\{m(i)\}=-\infty}^{\infty} e^{-\frac{g^2N^2}{8\pi^2} \sum_{i,\mu\nu} (\Delta_\mu \tilde{\Theta}_\nu(i) - \Delta_\nu \tilde{\Theta}_\mu(i) - 2\pi m(i))} \tag{3.18}$$

Comparing eq. (3.18) with the original Villain model eq. (3.10) we learn that the Villain model is self dual with

$$\tilde{g}^2 = \frac{4\pi^2}{g^2 N^2} \tag{3.19}$$

EXISTANCE OF THREE PHASES FOR LARGE ENOUGH N

The self duality eq.(3.19) tells us immediately that if only two phases exist, the critical point should be at

$$\left(g_c^2\right)_N = \frac{2\pi}{N} \tag{3.20}$$

For $g^2 > \frac{2\pi}{N}$ the Wilson loop order parameter will have then an area law behaviour exp $(-b_N A)$. We use A to denote the area of the loop and b_N is a constant depending on g and N. Thus, if $(g_c^2)_{U(1)}$ is the critical point of PQED and if N is large enough so that

$$\frac{2\pi}{N} < (g_c^2)_{U(1)},$$

we shall have in the region

$$\frac{2\pi}{N} < g^2 < (g_c^2)_{U(1)}$$

an area law behaviour. For $N \to \infty$ the Z(N) theories we are looking at, go smoothly into the PQED theory. Since for the PQED theory, as long as $g^2 < (g_c^2)_{U(1)}$, the Wilson loop has a perimeter law behaviour, we face a situation in which as $N \to \infty$ the area law behaviour of the order parameter in the region

$$\frac{2\pi}{N} < g^2 < (g_c^2)_{U(1)}$$

should tend toward a perimeter behaviour. This would say that as we increase N, the system becomes more and more ordered (for fixed g^2, i.e. fixed temperature). This is rather strange, since increasing N means increasing the number of allowed configurations,i.e. the entropy of the system. Having more entropy (for fixed temperature) cannot improve the range of correlation, namely the ordering. The only way out is that our assumption of two phases only, does not hold for large enough N.

A rigorous proof, along this line, for the existence of three phases was given by Elitzur, Pearson and Sigamitsu [6]. The main part of the proof is to establish the inequality

$$(\text{Wilson loop})_{Z(N)} \geqq (\text{Wilson loop})_{U(1)} \quad \text{at given } g^2 \quad (3.21)$$

The physics beyond this inequality is that increasing N (for fixed g^2) means increasing the entropy, hence reducing the order. For $g^2 < (g_c^2)_{U(1)}$ the Wilson loop in PQED obeys a perimeter law behaviour, thus (according to eq.(3.21)) so must also the Wilson loop in the Z(N) gauge theory. The transition form the large g^2 area law behaviour to the small g^2 perimeter behaviour of the Z(N) theory must occur at

$$\left(g_c^2\right)_{Z(N)} \geqq \left(g_c^2\right)_{U(1)} .$$

If

$$N > {}^{2\pi}/(g_c^2)_{U(1)} \qquad \text{then} \quad (g_c^2)_{Z(N)}$$

cannot be the self dual point $\frac{2\pi}{N}$ (eq.(3.19)). By duality there has to exist another critical point

$$\left(g_c^2\right)'_{Z(N)} = \frac{4\pi^2}{N^2} \quad \frac{1}{(g_c^2)_{Z(N)}} \tag{3.22}$$

COULOMB GAS REPRESENTATION [36] OF THE Z(N) GAUGE THEORIES

The Z(N) variables $\{\Theta_\mu\}$ are both compact and discrete. We shall rewrite the theory first in terms of variables which are continuous and compact (angle variables $[0,2\pi]$). Then we shall go to variables $\{A_\mu\}$ which are the usual electrodynamic variables, i.e. continuous and non-compact. We shall start with the PQED partition function (eq.(3.1)). To retrieve the Z(N) gauge theory we shall add a new term to the action

$$e^{h \sum_{i,\mu} \cos(\Delta_\mu \phi(i) - N\Theta_\mu(i))} = \sum_{\{\ell_\mu(i)\}=-\infty}^{\infty} \prod_{i,\mu} I_{\ell_\mu(i)}(h)$$

$$\times\, e^{i \sum_{i,\mu} \ell_\mu(i)(\Delta_\mu \phi(i) - N\Theta_\mu(i))} \tag{3.23}$$

where ℓ_μ runs over the integers and the coefficient functions in the fourier series are Bessel functions. The integration will be over $\{\Theta_\mu\}$ and $\{\phi\}$. Note that $\phi(i)$ is also an angle variable which is defined at each lattice site i . Let me state without entering.

into details that this new term amounts to an introduction into PQED theory of Higgs particle of charge N . Since $\phi(i)$ appears linearly in the action (eq.(3.23)) we can easily perform the integration over $\{\phi(i)\}$ reducing the new term in eq.(3.23) to

$$\sum_{\{\ell_\mu(i)\}} \prod_i \delta(\Delta_\mu \ell_\mu(i))(2\pi h)^{\frac{1}{2}} e^{-h} I_{\ell_\mu(i)}(h)\, e^{-iN\sum_{i,\mu} \ell_\mu(i)\Theta_\mu(i)} \tag{3.24}$$

$$\xrightarrow[h\to\infty]{} \sum_{\{\ell_\mu(i)\}} \prod_i \delta(\Delta_\mu \ell_\mu(i))\, e^{-iN\sum_{i,\mu} \ell_\mu(i)\Theta_\mu(i)}$$

If it were not for the δ function constraint, the sum over $\{\ell_\mu(i)\}$ in eq.(3.24) could be easily done (yielding a δ function) with the result that Θ_μ is essentially a Z(N) variable.

$$\Theta_\mu = \frac{2\pi}{N}\ell_\mu \qquad \ell_\mu \text{ integer} \tag{3.25}$$

The δ function constraint in eq. (3.24) indicates that eq. (3.25) is true up to gauge transformation

$$\Theta_\mu(i) = \frac{2\pi}{N}\ell_\mu(i) + \Delta_\mu\theta(i) \tag{3.26}$$

The added term $\Delta_\mu\Theta(i)$ does not change the expression in eq.(3.24) (integration by part) because of the δ function. To summarize a Z(N) gauge theory can be viewed as the $h\to\infty$ limit of a charge N Higgs theory. This result is actually not surprising. In this limit all the gauge degrees of freedom are frozen apart from those which correspond to the minimum of the potential (eq.(3.23)). The N minima thus obtained correspond to the remaining Z(N) gauge theory. In the Villain approximation we write the partition function (see in particular the discussion which leads to eq.(3.10))

$$Z_V = \int_0^{2\pi} \prod_{i\,\mu} d\Theta_\mu(i) \sum_{n_{\mu\nu}(i)=-\infty}^{\infty} \sum_{\ell_\mu(i)=-\infty}^{\infty} e^{-\frac{1}{2g^2}\sum_{i,\mu\nu}(\Theta_{\mu\nu}(i)-2\pi n_{\mu\nu}(i))^2} \times e^{-iN\sum_{i,\mu}\ell_\mu(i)\Theta_\mu(i)}\, \delta(\Delta_\mu \ell_\mu(i)) \tag{3.27}$$

where $n_{\mu\nu}$ as well as ℓ_μ are integers. Moreover $n_{\mu\nu}$ is antisymmetric $n_{\mu\nu} = -n_{\nu\mu}$. The most general integer antisymmetric tensor can be written as

$$n_{\mu\nu} = (\Delta_\mu n_\nu - \Delta_\nu n_\mu) + \frac{1}{2}\epsilon_{\mu\nu\lambda\kappa} M_{\lambda\kappa} \tag{3.28}$$

with n_μ and $M_{\lambda\kappa}$ integers. The second term on the right hand side of eq.(3.28) ensures that

$$\Delta_\mu \tilde{n}_{\mu\nu} \neq 0 \quad \text{where} \quad \tilde{n}_{\mu\nu} = \frac{1}{2}\epsilon_{\mu\nu\alpha\beta} n_{\alpha\beta}$$

is the dual of $n_{\mu\nu}$. Hence it is to be expected that $M_{\mu\nu}$ will turn out to be connected to magnetic monopoles. Let us define an integer current

$$m_\mu \equiv \frac{1}{2}\Delta_\nu \epsilon_{\mu\nu\lambda\kappa} n_{\lambda\kappa} = \Delta_\nu \tilde{n}_{\mu\nu} \tag{3.29}$$

then

$$\Delta_\nu M_{\mu\nu} = m_\mu \tag{3.30}$$

From equation (3.29) it is clear that m_μ is a conserved current

$$\Delta_\mu m_\mu = 0 \tag{3.31}$$

We can solve eq.(3.30) (for details see refs [6,7,30])

$$M_{\mu\nu} = \hat{n}^\mu (\hat{n}\cdot\Delta)^{-1} m^\nu - (\mu \leftrightarrow \nu) \tag{3.32}$$

where $\hat{n}^\mu$ is a constant unit vector on the lattice. $(\hat{n}\cdot\Delta)^{-1}$ is a line integral

$$(\hat{n}\cdot\Delta)_r \, (\hat{n}\cdot\Delta)_r^{-1} = \delta^4(r) \tag{3.33}$$

At this point we can introduce a non-compact variable A_μ

$$A_\mu(i) = \Theta_\mu(i) - 2\pi n_\mu(i) \tag{3.34}$$

Using eqs.(3.28),(3.32) and (3.34) we can rewrite the partition function eq. (3.27)

$$Z = \int d[A_\mu(i)] \sum_{\{\ell_\mu(i)\}} \sum_{\{m_\mu(i)\}} \delta(\Delta_\mu \ell_\mu(i))\, \delta(\Delta_\mu m_\mu(i))\, \delta(\Delta_\mu A_\mu) \times e^{-\frac{1}{2g^2}\sum_{i,\mu\nu}(F_{\mu\nu} - \epsilon_{\mu\nu\lambda\kappa}\hat{n}_\lambda (\hat{n}\cdot\Delta)^{-1} m_\kappa)^2} \, e^{-iN\sum_{i,\mu}\ell_\mu(i) A_\mu(i)} \tag{3.35}$$

where

$$F_{\mu\nu} = \Delta_\mu A_\nu - \Delta_\nu A_\mu .$$

Note the appearence of the Dirac's string term $\epsilon_{\mu\nu\lambda\kappa}\hat{n}_\lambda (\hat{n}\cdot\Delta)^{-1} m_\kappa$.

We have also introduced a gauge fixing term $\delta(\Delta_\mu A_\mu)$ (Lorentz gauge). The Gaussian integration over A_μ can be carried out with the result (up to irrelevant constant)

$$Z = \sum_{\{\ell_\mu(i)\}\{m_\mu(i)\}} e^{-\frac{1}{2}(Ng)^2 \sum_{i,i'} \ell_\mu(i)\, D(i-i')\ell_\mu(i')} \times e^{-\frac{1}{2}\left(\frac{2\pi}{g}\right) \sum_{i,i'} m_\mu(i)\, D(i-i') m_\mu(i')} \times e^{2\pi i N \sum_{i,i'} m_\mu(i) \epsilon_{\mu\nu\lambda\kappa} \hat{n}_\nu (\hat{n}\cdot\Delta)^{-1} \Delta_\lambda \ell_\kappa} \tag{3.36}$$

where D(i) is the lattice propagator for zero mass photon

$$-\Delta^2 D(i) = -\sum_{\pm\hat{e}} (\delta_{i,\hat{e}} - 2\delta_{i,o})\, D(i) = \delta_{i,o} \tag{3.37}$$

We recognize immediately D(i) as the lattice version of the Coulomb potential. Now the interpretation of eq.(3.36) is quite clear. The partition function is written in terms of a gas of closed ($\Delta_\mu \ell_\mu$=o, $\Delta_\mu m_\mu$=o) electric (ℓ_μ) and magnetic (m_μ) loops. The first (second) term in the action (eq (3.36)) describes the Coulomb interaction of the electric (magnetic) loop. The third term is an interaction term between the electric and the magnetic loops in four Euclidean demensions [37]. It has the form of an interaction between the magnetic current m_μ and the magnetic vector potential generated by the electric current ℓ_μ [37]. It is not too difficult to show that the third term is actually symmetric under the interchange of m_μ and ℓ_μ, hence the theory (eq.(3.36)) is invariant under

$$\ell_\mu \longleftrightarrow m_\mu$$

$$Ng \longleftrightarrow \frac{2\pi}{g} \tag{3.38}$$

which is just the duality transformation (eq.(3.22)) in the Coulomb gas representation. Recall that we have already seen in the Hamiltonian approach (eqs.(2.17),(2.51),(2.52)) that the self duality is just a manifestation of an electric-magnetic symmetry. Note that the charge e running along the electric loop is

$$e = Ng \tag{3.39}$$

while the magnetic charge m is

$$m = {}^{2\pi}/_{g} \tag{3.40}$$

Thus, the Dirac quantization condition is satisfied

$$me = 2\pi N \tag{3.41}$$

THE COULOMB-GAS DESCRIPTION OF THE PHASE TRANSITIONS

We shall not even try to enter into details which can be found in the literature [6,7,30,38], thus let us set aside the interaction term in eq.(3.36) and concentrate on the first (electric) and second (magnetic) terms. For small g it is clear that magnetic charge loops are suppressed. In particular the magnetic self action mass-like term (chemical potential) which is proportional to

$$(\frac{2\pi}{g})^2 \sum_i m_\mu(i) \sim (\frac{2\pi}{g})^2 L$$

(L is the length of the loop), does not allow the occurrence of magnetic loops. A magnetic charge loop can be viewed as a magnetic monopole-antimonopole pair created (e.g. via tunneling) at one point then separating somewhat and reannihilating later on. What we have seen is that at small g such a process is highly improbable. As for the electric charge loop, the action (energy) we have to pay to create such a loop is $\sim(Ng)^2L$ (from the electric self action mass-like term). It seems that for fixed g however small, too large electric loops (large L) are forbidden as well. This is, however, not correct since there is also the entropy factor to be considered. We can essentially repeat the same type of arguments described at the end of the first part of these lecture notes (eqs. (1.22) through (1.25) and the discussion thereafter). The entropy is roughly estimated by the logarithm of the number of loops of length L which pass through a given point to be $\log \mu^L = L \log \mu$. (μ is a geometrical factor - eq.(1.24)). The energy against entropy consideration (see discussion after eq. (1.25)) reveals that a phase transition exists at

$$(Ng_c)^2 L \approx L \log \mu \Longrightarrow g_c^2 \approx \frac{\log \mu}{N^2} \tag{3.42}$$

What happens is that for $g^2 < g_c^2$ the large electric charge loops are not suppressed anymore and there is large charge fluctuation , i.e. the vacuum is a condensate of the electric charge. The small g phase is therefore a Higgs phase. Repeating the same kind of arguments starting at large g we learn from the partition function (eq.3.36)) that the electric charge loops are suppressed. As for the magnetic loops an energy (action) against entropy consideration reveals that a phase transition occurs at

$$\left(\frac{2\pi}{g_c}\right)^2 L \approx L \log \mu \Longrightarrow g_c^2 = \text{finite (indep.of N)} \tag{3.43}$$

For $g^2<g_c^2$ large magnetic loops are suppressed. However, for $g^2>g_c^2$ the vacuum becomes a condensate of the magnetic charge, i.e. we are in the confining phase.

For large enough N it is clear that the two critical points in eqs. (3.42) and (3.43) are different and we get a three phases situation. The intermediate phase is a normal phase with no large fluctuations of either electric or magnetic charge. The gas of both types of current loops is dilute. In this phase a zero mass photon exists.

IV CONCLUDING REMARKS - INTRODUCING MATTER FIELD [39]

These lectures were mainly devoted to the discussion of pure Z(N) gauge theories. It is clear that realistic theories which have any chance of describing nature must eventually include matter fields, i.e. quarks, as well. The reason we investigate first the pure gauge sector is related, as discussed in chapter I, to our belief that if the pure gauge theory confines, then adding matter still results in a physical spectrum being made of color singlets only. It would be, of course, interesting to see how this really works.

The quarks are fermions and belong to the fundamental representation of the gauge group. We would like to keep these properties when we introduce the matter field into the Z(N) gauge theories. It turns out, however, that it is not so easy to deal with fermionic degrees of freedom on the lattice [40]. Therefore at first stage we introduce a scalar Higgs-like field. To give a specific example let us write down the Z(2) case both in the Lagrangian and Hamiltonian formulations.

$$Z = \sum_{\{\sigma_3(\ell)\}} \sum_{\{\tau_3(i)\}} e^{\frac{1}{g^2}\Sigma\sigma_3(P_1)\sigma_3(P_2)\sigma_3(P_3)\sigma_3(P_4)} \times e^{f^2\sum_{<ij>}\tau_3(i)\sigma_3(\ell_{ij})\tau_3(j)} \tag{4.1}$$

where σ_3 and τ_3 are two sets of Pauli matrices representing the gauge and the matter respectively. The σ_3's reside on links while the τ_3's on the vertices. The first sum in the action is over plaquettes and the symbol $<ij>$ denotes nearest neighbors (ℓ_{ij} is the link connecting the points i and j). This theory is invariant under the local gauge transformation at the point i in which:

$$\sigma_3(\ell) \rightarrow -\sigma_3(\ell) \qquad i \in \ell$$

$$\tau_3(i) \rightarrow -\tau_3(i) \tag{4.2}$$

with all other degrees of freedom remain untouched. The symbol $i\in\ell$ means: ℓ is a link attached to the point i. The coupling constants are g^2 and f^2.

The corresponding Hamiltonian is

$$H = -t\sum_{\ell}\sigma_1(\ell) - \frac{1}{t}\sum_{p}\sigma_3(p_1)\sigma_3(p_2)\sigma_3(p_3)\sigma_3(p_4) - $$

$$- \frac{1}{X}\sum_{i}\tau_1(i) - X\sum_{<ij>}\tau_3(i)\sigma_3(\ell_{ij})\tau_3(j) \tag{4.3}$$

Again $\tau_{1,3}$ and $\sigma_{1,3}$ are Pauli matrices defined on vertices and links. As before σ represents the gauge and τ the matter degrees of freedom. The commutation relations are

$$\sigma_1(\ell)\sigma_3(\ell)\sigma_1(\ell) = -\sigma_3(\ell) \tag{4.4}$$

$$\tau_1(i)\tau_3(i)\tau_1(i) = -\tau_3(i)$$

and all other degrees of freedom commute. The generator of gauge transformation at the point i is

$$G(i) = \tau_1(i)\prod_{i\in\ell}\tau_1(\ell) \tag{4.5}$$

It is straightforward to check that

$$[G(i),H] = 0 \tag{4.6}$$

Note that X=o in eq.(4.3) corresponds to the pure gauge theory, while t=o gives a pure Ising (Higgs) theory.

This kind of theories are under current investigation and ref [39] gives a sample of the available literature. What we would like to know first of all is how to characterize the phases of a gauge theory when matter is present. It turns out that if the matter belongs to the fundamental representation, the Wilson loop cannot have any more an area law behavior. The physical reason is that an area law corresponds to a linear potential between a pair of external quark and antiquark. However, if we have a matter in the fundamental representation it can always screen the external charge. The result is a perimeter law behavior. To say the same thing in a slightly different way; as we start to take the two external charges apart, we shall reach a point such that it would be advantegeous to create a pair of dynamical charges out of the vacuum and screen the external charge. The Z(2) gauge theories of eqs. (4.1),(4.3) are good models to investigate this phenomenon [39] since in the Z(2) case we have only one representation; hence the matter is necessarily in the fundamental representation of the gauge group. As it turns out the theory is selfdual in 2+1 dimensions, while in 3+1 dimensions it is dual to a higher gauge theory [41]. The duality property helps us to understand the phase structure of the theory. In particular it has been proven [42] that there is a strip of analyticity in the (t,χ) plane. Hence the confinement (large t) and Higgs (large χ) regions are continuously connected and no phase transition exists between them. Yet the phase diagram is by no means trivial and the reader is referred again to the literature [39].

The whole picture changes when the matter is not in the fundamental representation. In this case external charge in the fundamental representation cannot be screened and the corresponding Wilson loop will have an area law behavior in part of the phase diagram. This situation exists, for example, when we consider a Z(4) gauge theory with a Z(2) matter.

Finally let me return to the fermionic nature of the quarks which we have dropped. The new interesting feature that will enter are the chiral properties of the theory. We would like to write down the theory with zero mass fermionic degrees of freedom in a way that the theory will have a γ_5 symmetry. This is not straightforward but can be done [40]. The important question to investigate then is the interrelation between chiral symmetry breaking and confinement.

ACKNOWLEDGEMENT

I would like to thank Prof. D.Horn with whom I have been working on the Z(N) gauge theories.

It is a pleasure to thank the organizers of the German Summer Institute at Kaiserslautern for their warm hospitality and the stimulating atmosphere they provided us with.

REFERENCES

1. G. 't-Hooft, Nucl. Phys. B138,1 (1978)
   G. 't-Hooft, "A Property of Electric and Magnetic Flux in Non-Abelian Gauge Theories", University of Utrecht preprint (Nucl. Phys. B) (1979)
2. H. B. Nielsen, P. Olesen, Nucl. Phys. B61, 45 (1973)
3. S. Mandelstam, Phys. Rep. 23C, 245 (1976)

4. G. 't-Hooft, "High Energy Physics: Proceedings of the EPS International Conference, Palermo, June 1975", A. Zichichi (ed.), Editrice Compositori, Bologna (1976)
5. D. Horn, M. Weinstein, S. Yankielowicz, Phys. Rev. D19, 3715 (1979)
6. S. Elitzur, R. B. Pearson, J. Shigemitsu, Phys. Rev. D19, 3698 (1979)
7. A. Ukawa, P. Windey, A. Guth, Princeton University preprint (1979)
8. S. F. Edwards, P. Anderson, J. Phys. F5, 965 (1975)
9. D. J. Gross, F. Wilczek, Phys. Rev. Lett. 30, 1343 (1973); H. D. Politzer, Phys. Rev. Lett. 30, 1346 (1973)
10. K. G. Wilson, Phys. Rev. D10, 2445 (1974); J. Kogut, L. Susskind, Phys. Rev. Lett. D11, 395 (1975); S. D. Drell, M. Weinstein, S. Yankielowicz, Phys. Rev. D14, 1627 (1976)
11. A. M. Polyakov, Nucl. Phys. B120, 429 (1977)
12. C. G. Callan, R. F. Dashen, D. J. Gross, Phys. Lett B66, 375 (1977); Phys. Rev. D17, 2717 (1978); Phys. Rev. D19, 1826 (1979)
13. K. G. Wilson, Phys. Rev. D10, 2445 (1974)
14. F. S. J. Wegner, J. Math. Phys. 12, 2259 (1971); R. Balian, J. M. Drouffe, C. Itzykson; Phys. Rev. D11, 2098 (1975); J. M. Drouffe, C. Itzykson, Phys. Rep. C38, 133 (1978); E. Fradkin, L. Susskind, Phys. Rev. D17, 2637 (1978)
15. T. Yoneya, Nucl. Phys. B144, 195 (1978)
16. see e.g. J. Kogut "An Introduction to Lattice Gauge Theory and Spin System" Univ. of Illinois, Urbana preprint (1979) and references therein
17. E. Fradkin, L. Susskind, Phys. Rev. D17, 2637 (1978)
18. R. P. Feynman, "Statistical Mechanics", W.A. Benjamin INC.
19. L. P. Kadanoff, H. Ceva, Phys. Rev. B3, 3918 (1971)
20. S. Elitzur Phys. Rev. D12, 3978 (1975)
21. J. Kogut, R. Pearson, Univ. of Illinois, Urbana preprint (1979)
22. This chapter is based on the work of D. Horn, M. Weinstein, S. Yankielowicz ref.5
23. J. Schwinger, Proc. Nat. Acad. Sciences 46, 570 (1960); "Quantum Kinematics and Dynamics" W.A. Benjamin Inc., N. Y. 1970
24. H. A. Kramers, G. H. Wannier, Phys. Rev. 60, 252 (1941); F. S. J. Wegner, J. Math. Phys. 12, 2259 (1971) For a review on duality transformation see e.g. R. Savit "Duality in Field Theory and Statistical System" Univ. of Michigan preprint (1979),(to be published in Phys. Rep.)
25. K. G. Wilson, Phys. Rev. D10, 2445 (1974); L. Susskind, "17th Scottish Universities Summer School in Physics"(1976)
26. P. Pfeuty, R. J. Elliott, J. Phys. C4, 2370 (1971)
27. For a general review and a complete set of references on duality transformation see R. Savit ref 24
28. M. B. Einhorn, R. Savit, Phys. Rev. D17, 2583 (1978)

29. F. Englert, P. Windey, Proceedings of the XIX International Conference on High Energy Physics, Tokyo 1978; D. Forster, Cornell preprint (1978)

30. T. Banks, J. Kogut, R. Meyerson, Nucl. Phys. B129, 493 (1977) have considered a number of theories (not the Z(N) in which a similar phenomenon takes plase. In particular this is the type of transition which occurs in P.Q.E.D.

31. M. Creutz, L. Jacobs, C. Rebbi, Phys. Rev. Lett. 42, 1390 (1979); BNL-26307 preprint (1979)

32. T. Banks, D. Horn; Tel-Aviv University prepr. TAUP-777 (1979)

33. The autors of refs. 6,7 have used the Lagrangion approach.

34. K. Wilson, Phys. Rev D14, 2445 (1974); A. M. Polyakov, Phys. Lett. 59B, 79 (1975) and ibid. 82 (1975);J. Kogut, L. Susskind, Phys. Rev. D11, 395 (1975); T. Banks, R. Meyerson, J. Kogut ref 30; S. D. Drell, H. Quinn, B. Svetitsky, M. Weinstein, Phys. Rev. D19, 619 (1979)

35. J. Villain, J. Phys. C36, 581 (1975); T. Banks, J. Kogut, R. Meyerson ref 30; A. Casher, Nucl. Phys. B151, 353 (1979)

36. L. P. Kadanoff, J. Phys. All, 1399 (1978); S. Elitzur, R. B. Pearson, J Shigemitsu ref 6; A. Ukawa, P. Windey, A. Gut ref 7

37. J. Schwinger, Phys. Rev. 173, 1536 (1968); "Particles, Sources and Fields" Addison-Wesley Publishing Company (1973)

38. T. Banks, E. Rabinovici, Tel-Aviv University preprint, TAUP 781 (1979)

39. D. Horn, S. Yankielowicz, Tel-Aviv Univ. preprint, TAUP 746 (1979) (to be published in Phys. Lett.); E. Fradkin, S. Shenkar, Phys. Rev. D19, 3682 (1979); M. Green, Nucl. Phys. B144, 473 (1978)

40. J. Kogut, "International Summer Institute for Theoretical Physics" Univ. of Bielefeld (1976); S. D. Drell, M. Weinstein, S. Yankielowicz, Phys. Rev. D14, 1627 (1976); K. G. Wilson, Coral Gable Conference (1976), Cornell preprint CLNS 327 (1976); V. Baluni, J. F. Willemsen, Phys. Rev. D13, 3342 (1976)

41. D. Horn, S. Yankielowicz ref 39; M. Green ref 39;E. Fradkin, S. Shenkar ref 39; M. Creutz BNL-26587 preprint (1979)

42. E. Fradkin, S. Shenkar ref 39; K. Osterwalder, E. Seiler, Annals of Phys. 110, 440 (1978)

THE LATTICE FERMION PROBLEM AND WEAK COUPLING PERTURBATION THEORY

Luuk H. Karsten

Instituut voor Theoretische Fysica
University of Amsterdam
Amsterdam, Holland

ABSTRACT

The problem with the formulation of fermions on a lattice is discussed. The SLAC group has proposed a way of formulating a lattice fermion, which respects continuous chiral invariance. We show, in weak coupling perturbation theory, that a gauge theory with such a fermion gives Lorentz non-covariant results in the continuum limit. A new chiral invariant formulation of a lattice fermion is proposed.

INTRODUCTION

In recent years much attention has been paid to lattice theories. A reason for this is the possibility of handling strong coupling. This is especially relevant for the presumed theory of the hadrons and the strong interactions, QCD, quantum chromo-dynamics.

The lattice implies a denumerable set of degrees of freedom in position space and a (gauge invariant) ultraviolet cut-off in momentum space. This leads to a variety of approximation methods [1,2,3,4], as well as to more mathematical studies [5,6].

A disadvantage of the lattice regularization however is that it explicitly breaks Lorentz invariance. Quantum field theories on a lattice are constructed in such a way, that at the classical level, i.e. without quantum corrections, the continuum limit (lattice distance to zero) is Lorentz covariant. It remains to be shown that the full quantum theory is Lorentz covariant in the continuum limit.

A second question one has to pose oneself studying lattice theories is how to formulate lattice fermions. As we will discuss later the naive lattice transscription of the continuum fermion Lagrangian leads to several mass degenerate fermions instead of the one fermion one wanted to describe. In the literature several methods have been proposed to circumvent this degeneracy. At the classical level these methods are equivalent in the continuum limit. On the lattice they differ widely. Again it remains to be seen what the effect is of quantum corrections.

Some lattice fermion formulations explicitly break continuous chiral invariance, for example Wilson's method [1]. Regarding the importance of chiral invariance in nature one could argue that it is important to be able to formulate lattice gauge theories in a chiral invariant way. Drell, Yankielowicz and Weinstein [3] have proposed a lattice fermion formulation without degeneracy problem and with continuous chiral invariance. This SLAC formulation is non-local: the fermion field in different lattice sites is coupled inversely proportional to the distance between the sites. Classically this non-locality vanishes in the continuum limit. Is this also true when quantum corrections are taken into account?

In the following we will try to get some insight in the effects of the quantum corrections to the continuum limit from weak coupling perturbation theory. We of course do not want to state that we get quantitatively good results with this method for theories like QCD, but we can hope to learn about qualitative properties like Lorentz covariance and locality. In constructive field theory for example one needs the same kind of counter terms as in perturbation theory [7].

Our seminar consists of three parts. First we discuss three lattice fermion formulations: the naive, Wilson's and SLAC's. In the next section the one loop vacuum polarization is calculated in a theory with a SLAC lattice fermion and a vector gauge field. It is found that the result is Lorentz non-covariant and non-local in the continuum limit. This is the reason for proposing a new lattice fermion formulation in the last section.

LATTICE FERMIONS

Consider an (infinite) hypercubic four-dimensional Euclidean lattice. The lattice distance is a, the metric $\delta_{\mu\nu}$. The lattice sites are labelled by n_μ

$$x_\mu = n_\mu a,\ n_\mu = 0, \pm 1, \ldots, \pm \infty,\ \mu = 1,2,3,4.$$

Define fields $\varphi(x)$ on this lattice. The Fourier transform is

$$\tilde{\varphi}(k) = \sum_x \exp(-i\, k\cdot x)\ \varphi(x), \tag{1}$$

$$\sum_x = a^4 \sum_{\{n\}} \tag{2}$$

Because of the finite lattice distance there is a momentum cut-off:

$$|k_\mu| < \frac{\pi}{a} = \Lambda \tag{3}$$

The inverse Fourier transform is

$$\varphi(x) = \int_{-\Lambda}^{+\Lambda} \frac{d^4k}{(2\pi)^4} \exp(i\, k\cdot x)\ \tilde{\varphi}(k) \tag{4}$$

The continuum free fermion action reads

$$I_o = -\sum_\mu \int d^4x\ \bar{\psi}(x) \frac{1}{2i} \gamma_\mu \overleftrightarrow{\partial}_\mu \psi(x) \tag{5}$$

with

$$\{\gamma_\mu , \gamma_\nu\} = -2\,\delta_{\mu\nu} \tag{6}$$

One gets the naive lattice fermion action by replacing the derivative with a difference

$$\partial_\mu \psi \to \frac{1}{a} (\psi(x+a_\mu) - \psi(x)). \tag{7}$$

The naive lattice fermion action is

$$I_o = - \sum_{\mu,x} \bar{\psi}(x) \frac{1}{i} \gamma_\mu \frac{1}{2a} (\psi(x+a_\mu) - \psi(x-a_\mu)) \tag{8}$$

The propagator one obtains from this action is

$$\left(\sum_\mu \gamma_\mu \frac{1}{a} \sin p_\mu a\right)^{-1} \tag{9}$$

In the continuum limit $a \to o$ the propagator is $O(a^o)$ not only for all $p_\mu \approx 0$ but also for one or more components $|p_\mu| \approx \frac{\pi}{a}$. There are sixteen fermions instead of one.

Considerung the free lattice fermion action (8) one can easily see where this degeneracy stems from: the action has extra symmetries. The action is invariant if one multiplies $\psi(x)$ with $\gamma_\mu\gamma_5(\pm 1)$ at alternating sites along one or more space-time directions, e.g.

$$\hat{\psi}(x_1,x_2,x_3,x_4) = \gamma_1\, \gamma_5\, (-1)^{x_1/a}\ \psi(x_1,x_2,x_3,x_4), \tag{10a}$$

$$\hat{\bar{\psi}}(x) = \bar{\psi}(x)\, \gamma_5\, \gamma_1^+\, (-1)^{x_1/a} . \tag{10b}$$

Wilson lifts this degeneracy by adding a symmetry breaking term to the action I_o:

$$- \frac{1}{2a} \sum_{x,\mu} \overline{\psi}(x) \left(\psi(x+a_\mu) + \psi(x-a_\mu) - 2\psi(x) \right).$$

Then the propagator is

$$\left(\frac{1}{a} \sum_\mu \gamma_\mu \sin p_\mu a - \frac{1}{a} \sum_\mu (\cos p_\mu a - 1) \right)^{-1} \tag{12}$$

Because of the $\sum_\mu (\cos p_\mu a \ -1)$ term the degeneracy with $|p_\mu| \approx \pi/a$ is lifted. One could say that the extra particles have got an infinite mass ($O(a^{-1})$).

The extra term (11) however explicitly breaks chiral invariance: one expects mass-counter terms in an interacting theory with this fermion, and one has difficulty implementing local chiral invariance (e.g. when studying weak interactions on a lattice).

Aharonov, Casher and Susskind [2] have proposed a lattice fermion formulation which respects discrete chiral invariance and leaves a two-fold degeneracy, which they interpret as isospin.

A promising method seems the SLAC lattice fermion formulation [3] , which respects continuous chiral invariance and leaves no degeneracy. In this method one replaces the derivative $\partial_\mu \psi$ by the Fourier transform on the lattice of $ik_\mu \tilde{\psi}(k)$

$$\partial_\mu \psi \rightarrow F(i k_\mu \tilde{\psi}). \tag{13}$$

To be more explicit define a function $D_\mu(x)$ on the lattice:

$$D_\mu(x) = \int_{-\Lambda}^{+\Lambda} \frac{d^4k}{(2\pi)^4} i k_\mu \exp(i k \cdot x) , \tag{14}$$

and make the replacement

$$\partial_\mu \psi(x) \rightarrow \sum_y D_\mu(x-y) \psi(y) \tag{15}$$

in the continuum action (5) to get the SLAC lattice fermion action

$$I_o = - \sum_{\mu,x,y} \overline{\psi}(x) \gamma_\mu \frac{1}{i} D_\mu(x-y) \psi(y). \tag{16}$$

(Actually the SLAC group uses a continuous time, we will comment on this at the end of section 3).

This action is manifestly chiral invariant, and has even on the lattice the usual relativistic propagator

$$\left(\sum_\mu \gamma_\mu p_\mu\right)^{-1} , \quad |p_\mu| < \frac{\pi}{a} . \tag{17}$$

There is no degeneracy, the action (16) describes just one fermion. Let us examine more closely $D_\mu(x)$.

$$D_\mu(x) = a^{-4} (-1)^{x_\mu/a} \; 1/x_\mu \quad \text{for } x_\nu = 0, \forall \nu \neq \mu, \; x_\mu \neq 0$$
$$= 0 \quad \text{otherwise.} \tag{18}$$

So there is in (16) a non-local coupling, inversely proportional to the distance between lattice sites. On a classical level this non-locality vanishes in the limit $a \to o$, then

$$D_\mu \; (x) \; \to \partial_\mu \delta^{(4)} \; (x). \tag{19}$$

As we will see in the next section, with quantum fluctuations taken into account this non-locality remains in the continuum limit and leads to Lorentz non-covariant terms.

THE VACUUM POLARIZATION WITH SLAC FERMIONS

The SLAC lattice fermion action

$$I_o = - \sum_{\mu,x,y} \bar{\psi}(x) \; \gamma_\mu \frac{1}{i} \; D_\mu \; (x-y) \; \psi(y) \tag{20}$$

has two global U(1) invariances

$$\hat{\psi} \; (x) = \exp \; (- i \; e \; \chi \;)\psi \; (x) \tag{21a}$$

and

$$\hat{\psi} \; (x) = \exp \; (-ig\gamma_5 \xi) \; \psi(x) \tag{21b}$$

The last invariance is chiral invariance. We make the first invariance local. To this end introduce gauge vector fields V_μ (x), transforming as

$$\hat{V}_\mu(x) = V_\mu(x) + \frac{1}{a} \; (\; \chi(x+a_\mu) \; - \; \chi(x) \;) \tag{22}$$

(Compare with the continuum transformation law

$$\hat{V}_\mu \; (x) = V_\mu \; (x) + \partial_\mu \; \chi(x) \;).$$

The action (20) contains bilocal terms $\bar{\psi}(x) \; \psi(y)$. In continuum theory these terms are made gauge-invariant by completing them with a Schwinger line-integral:

$$\hat{\psi}(x) \; \exp \; (\; i \int_x^y dz \; e \; V_\mu(z))\psi(y).$$

On the lattice one defines a line sum, e.g.

$$\sum_{z_4=x_4}^{y_4} \varphi(z) = a \sum_{\ell=m}^{n-1} \varphi(\vec{x},\ell a) \text{ if } x_4<y_4$$

$$= - a \sum_{\ell=n}^{m-1} \varphi(\vec{x},\ell a) \text{ if } x_4>y_4, \tag{23}$$

with $x_4 = ma$, $y_4 = na$ and $z_4 = \ell a$. Notice that in the action (20) only terms with $x_\mu = y_\mu$ for all but one μ occur, because of the definition of $D_\mu\ (x - y)$.

A gauge invariant extension of I_0 is

$$I = - \sum_{\mu,x,y} \bar{\psi}(x)\gamma_\mu \frac{1}{i} D_\mu\ (x-y) \exp\ \{ ie \sum_{z_\mu=x_\mu}^{y_\mu} V_\mu(z)\}\ \psi(y) \tag{24}$$

(for $a \to o$: $I \to I_{\text{continuum}} = -\sum_\mu \int d^4x\ \bar{\psi}(x)\gamma_\mu \frac{1}{i} (\partial_\mu + ieV_\mu(x)\ )\ \psi(x)$, because $D_\mu\ (x) \to \partial_\mu\ \delta(x)$).

One then adds to I a gauge field action, an Higgs action for the fermion mass m and gauge fixing terms in the usual way.

From the action (24) one derives Feynman rules for a weak coupling perturbation expansion by Fourier transformation.

First a comment on momentum conservation. In continuum theory one finds momentum conservation at the vertices from integrals like

$$\int d^4x \exp\ (i\ (p+q+k)\cdot x) = (2\pi)^4 \delta^{(4)}(p+q+k).$$

The lattice analogue is

$$\sum_x \exp\ (i(p+q+k)\cdot x) = (2\pi)^4 \delta^{(4)}(p+q+k; \text{modulo } 2\pi/a).$$

On the lattice there is momentum conservation modulo $2\pi/a$. One can easily convince oneself, that the Feynman rules can be formulated <u>with</u> exact momentum conservation, if the propagators and vertex functions are made periodic $(2\pi/a)$.

We will need the following propagator and vertices:

$$\left(m + \sum_\mu \gamma_\mu P_\mu(p)\right)^{-1} \tag{25a}$$

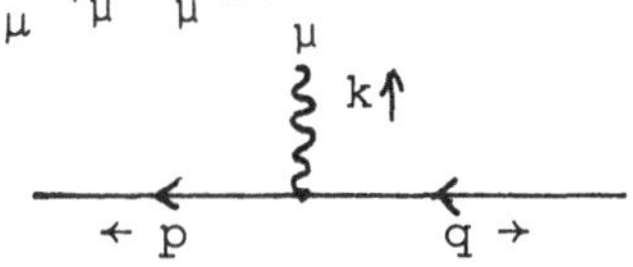

$$e\gamma_\mu \frac{P_\mu(p) + P_\mu(q)}{S_\mu(k)}, \quad p + q + k = o \tag{25b}$$

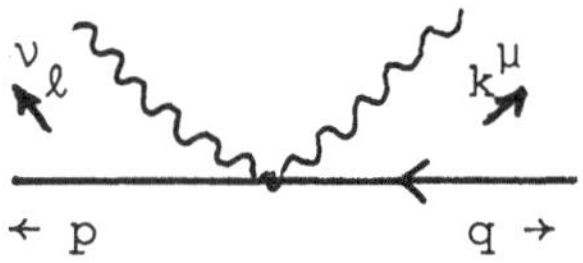

$$-e^2 \delta_{\mu\nu}\gamma_\mu \frac{P_\mu(p) - P_\mu(p+k) - P_\mu(p+\ell) - P_\mu(q)}{S_\mu(k)\, S_\nu(\ell)}, \tag{25c}$$

$$p + q + k + \ell = o.$$

We have defined the saw tooth function $P_\mu(p)$

$$P_\mu(p) = p_\mu - 2\,n\,\Lambda, \text{ for } (2\,n-1)\Lambda < p_\mu < (2\,n+1)\Lambda \tag{26}$$

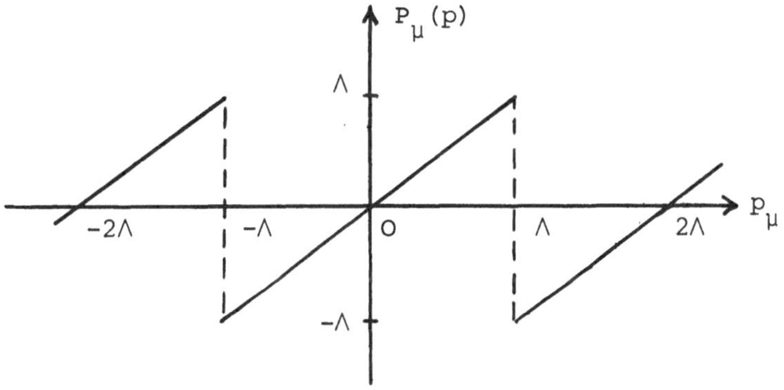

Fig. 1 The saw tooth function.

The function $S_\mu(k)$,

$$S_\mu(k) = \frac{2}{a} \sin \frac{1}{2} k_\mu a, \tag{27}$$

occurs in lattice Ward identities instead of k_μ, as one can, at the tree level, easily see from the Feynman rules (25). More details can be found in refs. [8,10].

With these Feynman rules we are now able to calculate the vacuum polarization $\langle 0| T(A_\mu A_\nu)|0\rangle$ with SLAC lattice fermions at the one loop level, $I_{\mu\nu}(p)$ [9]. Two diagrams contribute

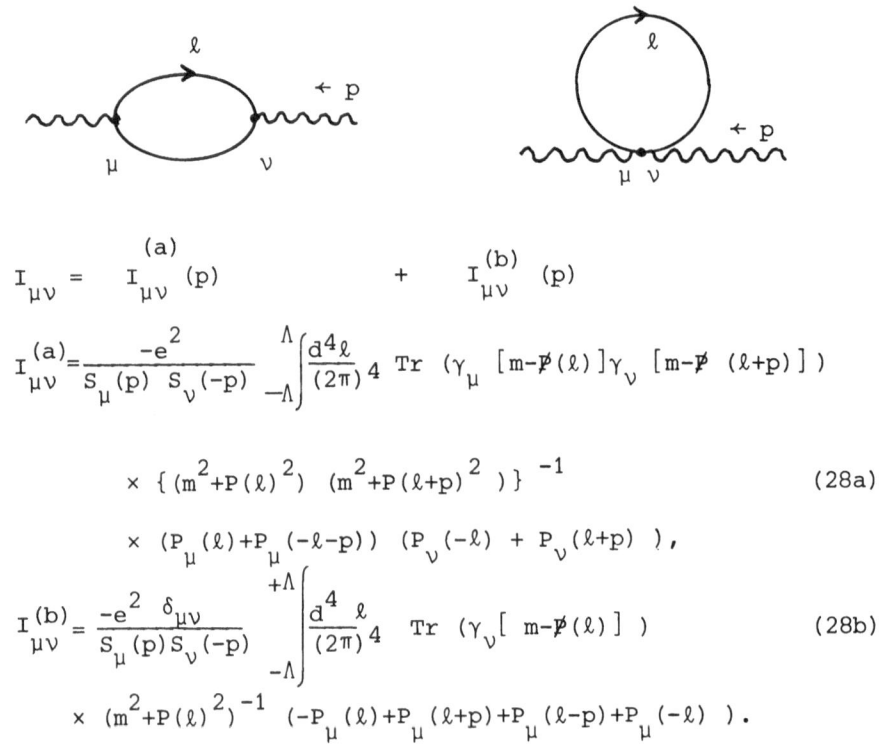

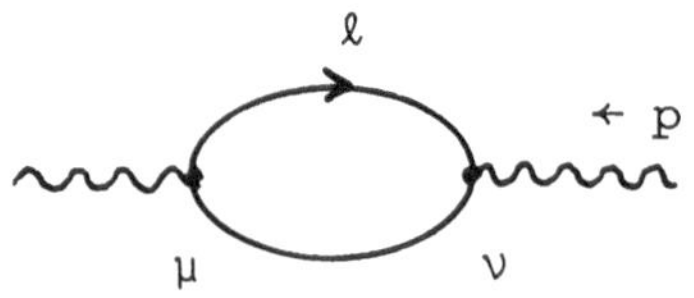

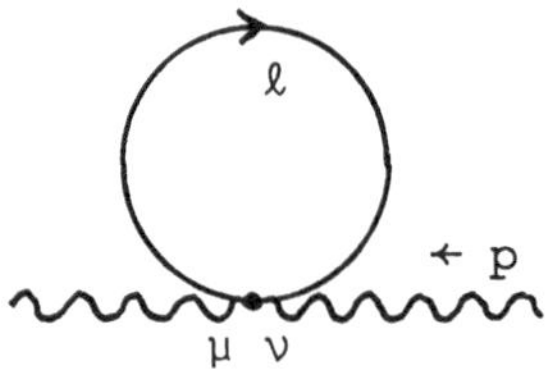

$$I_{\mu\nu} = I^{(a)}_{\mu\nu}(p) + I^{(b)}_{\mu\nu}(p)$$

$$I^{(a)}_{\mu\nu} = \frac{-e^2}{S_\mu(p)\,S_\nu(-p)} \int_{-\Lambda}^{\Lambda} \frac{d^4\ell}{(2\pi)^4}\, \mathrm{Tr}\,(\gamma_\mu\,[m-\not{P}(\ell)]\gamma_\nu\,[m-\not{P}(\ell+p)])$$

$$\times \{(m^2+P(\ell)^2)\,(m^2+P(\ell+p)^2)\}^{-1} \tag{28a}$$

$$\times\,(P_\mu(\ell)+P_\mu(-\ell-p))\,(P_\nu(-\ell)+P_\nu(\ell+p)),$$

$$I^{(b)}_{\mu\nu} = \frac{-e^2\,\delta_{\mu\nu}}{S_\mu(p)S_\nu(-p)} \int_{-\Lambda}^{+\Lambda} \frac{d^4\ell}{(2\pi)^4}\, \mathrm{Tr}\,(\gamma_\nu[m-\not{P}(\ell)]) \tag{28b}$$

$$\times\,(m^2+P(\ell)^2)^{-1}\,(-P_\mu(\ell)+P_\mu(\ell+p)+P_\mu(\ell-p)+P_\mu(-\ell)).$$

The Ward identity is

$$\sum_\mu S_\mu(p)\,I_{\mu\nu}(p) = 0. \tag{29}$$

The verification is straightforward: one has to use the periodicity of the integrands of the convergent integrals in (28).

Evaluating the integrals gives the result:

$$I_{\mu\nu} = -8e^2\left(\frac{1}{2\pi}\right)^4 [\{\delta_{\mu\nu}\,\frac{\sum_\rho p_\rho}{p_\nu} - 1\}\,\{4\Lambda^2\,e_2 + 4e_3(\tfrac{7}{3}p^2 - 2m^2)\}$$

$$+\{\frac{1}{12}\,\delta_{\mu\nu}\,p_\nu \sum_\rho p_\rho - \frac{1}{24}\,(p_\mu^2+p_\nu^2)\}\;4e_2$$

$$+\{\delta_{\mu\nu}\,(\sum_\rho p_\rho + \frac{p^2}{p_\nu}) - p_\mu - p_\nu\}\Lambda\,(c_2 - 4e_2 + 16\,e_3)$$

$$+ \{\delta_{\mu\nu}(\frac{\Sigma_\rho p_\rho^3}{p_\nu} + p_\nu \Sigma_\rho p_\rho) - (p_\nu^2 + p_\mu^2)\} \quad (-44\, e_3 + 104 e_4)$$

$$+ \{-\delta_{\mu\nu}\, p^2 + p_\mu p_\nu\} c\, (p^2, m^2)\;] + O\, (\Lambda^{-1}) \tag{30}$$

with $c_2 = \pi^2/4$; $e_2 = \pi/3\sqrt{3}$; $e_3 = 4\pi/3\sqrt{3} + \frac{1}{2}$; $e_4 = \frac{1}{12}(\pi/\sqrt{3}+\frac{7}{24})$.

This is the result for $p_\sigma > 0, \forall\sigma$; for general p_σ one finds for example

$$\delta_{\mu\nu} \frac{|p_\mu| \Sigma_\rho |p_\rho|}{p_\mu\, p_\nu} - \frac{|p_\mu|\, |p_\nu|}{p_\mu\, p_\nu}$$

in the first line of (30). The last line of (30) contains the standard result and is Lorentz covariant [11] .

The other terms are <u>Lorentz non-covariant</u> and <u>non-local</u> (i.e. non-polynomial in p, singular at $p_\sigma = 0$). This is a disaster for the theory. These terms are only covariant under the lattice symmetry group: rotations over 90° and reflections. They constitute remnants of the lattice in the continuum limit.

Before further discussing the result we will first give some details of the calculation and trace the cause of the Lorentz non-covariant terms. The first term in (30) can most easily be calculated. Let us look where a tensor structure $\delta_{\mu\nu}\, p_\rho/p_\nu$, $\rho \neq \mu = \nu$, can come from.

Evaluating the trace in $I_{\mu\nu}^{(b)}$ shows that $I_{\mu\nu}^{(b)}$ does not depend on p_ρ , $\rho \neq \mu=\nu$. $I_{\mu\mu}^{(a)}$ however does depend on $p_\rho, \rho \neq \mu$. Consider

$$\tilde{I}_{\mu\mu}^{(a)} = \frac{(2\pi)^4\, S_\mu(p)\, S_\mu(-p)}{-e^2}\, I_{\mu\mu}^{(a)} \; .$$

Using scaled variables, combining propagators and evaluating the trace gives:

$$\tilde{I}_{\mu\mu} = -\Lambda^4 \int_{-1}^{+1} d^4A \int_0^1 dx\, [\hat{p}_\mu - 2\, (h^+ + h^-)_\mu]^2$$

$$\times [-4\hat{m}^2 + 4\, (A+(1-x)\hat{p}-2h^+)_\mu\, (A-x\hat{p}+2h^-)_\mu +$$

$$-4\, \Sigma_{\rho\neq\mu} (A-x\hat{p}+2h^-)_\rho\, (A+(1-x)\hat{p}-2h^+)_\rho]$$

$$\times\; [\, A^2 + \varepsilon + \Sigma_\lambda N_\lambda\,]^{-2} \tag{31}$$

with $\hat{p} = p/\Lambda$ etc.,

$$N_\lambda = 4 \{ (A - x\hat{p} + 1)_\lambda \, h^-_\lambda - (A - (1-x)\hat{p} - 1)_\lambda \, h^+_\lambda \} , \tag{32}$$

$$\varepsilon = x\,(1-x)\hat{p}^2 + \hat{m}^2 . \tag{33}$$

The hat functions $h^\pm(A)$ are defined by

$$P_\mu(\,\Lambda(A+(1-x)\hat{p})\,) = \Lambda\,(A+(1-x)\hat{p}-2h^+)_\mu \tag{34a}$$

and

$$P_\mu(\,\Lambda(A-x\hat{p}+2h^-)\,) = \Lambda(A-x\hat{p}+2h^-)_\mu \, . \tag{34b}$$

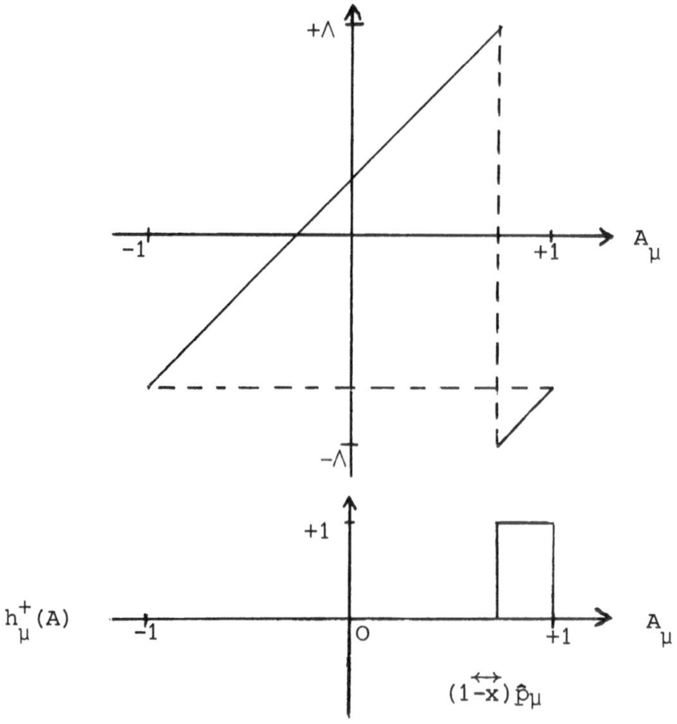

Fig.2 The definition of the hat functions

One gets a factor of p_ρ and the highest power in Λ if one takes from the squared factor in (31) the square of the hat functions and from the following factor the ρ-term:

$$-\Lambda^4 \int_{-1}^{+1} d^4A \int_0^1 dx \;\; 4\,(h^+ + h^-)_\mu \; (-4)\; 2A_\rho (h^+ - h^-)_\rho \; [A^2+\varepsilon]^{-2} =$$

$$= -\,32\,\Lambda^4 (\hat{p}_\mu \hat{p}_\rho \int_{-1}^{+1} dA_\sigma dA_\tau \; [2+A_\sigma^2+A_\tau^2]^{-2} + O(\Lambda^{-3})\;) =$$

$$= -32\,\Lambda^2\,(\pi/3\sqrt{3})\; p_\mu p_\rho + O(\Lambda) \quad (\mu,\rho,\sigma,\tau \text{ all different}).$$

This result is compatible with (30). More details can be found in [10] .

So we saw that the non-local derivative led to non-analytic propagators and vertex factors, and this gave non-local Lorentz non-covariant results.

How to get rid of these unwanted terms? Counter terms are not attractive, furthermore many of them are needed, probably infinitely many (because the multi-vector meson vertex functions are as divergent as the two point function).

We want to make a remark. Originally the SLAC group formulated their fermion in latticized space with time continuous. Then a similar calculation as above gives Lorentz non-covariant space-space components for $I_{\mu\nu}$, which is also unacceptable.

In conclusion we think that one has to look for a lattice fermion formulation different from that of the SLAC group if one wants to have a satisfactory chiral invariant lattice gauge theory. We will propose such a formulation in the next section.

A NEW FORMULATION OF LATTICE FERMIONS

In the previous section we found that the Lorentz non-covariant terms in the vacuum polarization with SLAC fermions were singular at p = o. Because the propagators and vertex factors were not analytic in the momenta such terms were not forbidden. To avoid such terms one should start with a theory, that has analytic vertex factors and propagators, like Wilson's formulation of fermions. (Lattice QED with Wilson fermions is Lorentz covariant for $a \to o$ in weak coupling perturbation theory [12]). We will try to formulate Wilson's fermions chiral invariantly.

The Wilson free fermion action reads

$$\begin{aligned} I_o = &- \sum_{x,\mu} \bar{\psi}(x)\, \frac{1}{i}\gamma_\mu \frac{1}{2a}\left(\psi(x+a_\mu) - \psi(x-a_\mu) \right) \\ &+ m \sum_x \bar{\psi}(x)\, \psi(x) \\ &- r \sum_{x,\mu} \bar{\psi}(x)\, \frac{1}{2a}\left(\psi(x+a_\mu) + \psi(x-a_\mu) - 2\psi(x)\right) \end{aligned} \tag{35}$$

(We have multiplied the extra term with r to keep track of it more easily).

Let us generate the extra r-term, just like the mass term by spontaneous symmetry breaking. We introduce a Higgs field $\varphi(x)$, transforming under chiral transformations as

$$\hat{\varphi}(x) = \exp(-2\,i\,g\,\xi)\,\varphi(x) \tag{36}$$

$$(\hat{\psi}(x) = \exp(-i\,g\,\gamma_5\,\xi)\;\psi(x))$$

Give φ a (real) vacuum expectation value F and couple φ to the fermion field ψ in the following way:

$$G\sum_{x,\mu}\varphi(x)\;\overline{\psi}(x)\;(1-\gamma_5)\{\frac{1}{4}\psi(x) - \frac{R}{2a}(\psi(x+a_\mu)+\psi(x-a_\mu)-2\psi(x))\}$$

$$+\text{" c.c.".} \tag{37}$$

With the identifications $m = 2\,FG$ and $r = 2\,FGR$ we have now indeed generated the extra Wilson term in (35). However, the $\varphi - \psi$ coupling has effective scale dimension 5 : in the limit $a \to 0$ the R-term becomes

$$-\frac{1}{2}\;GRa\int d^4x\;\{\overline{\psi}\;(1-\gamma_5)\;(\Box\psi)\varphi\}\quad + \text{c.c.}\;. \tag{38}$$

One can lower the effective scale dimension to 4 by multiplying the R-term in (37) with a . Then the free fermion propagator is

$$\{\frac{1}{a}\sum_\mu\gamma_\mu\;\sin p_\mu a + r\sum_\mu(1-\cos p_\mu a) + m\}^{-1} \tag{39}$$

We seem to have the fermion doubling back again: for $|p_\mu|a \simeq \pi$ the propagator is $O(a^0)$. The difference with the naive lattice fermion formulation is that the various fermions now have different masses, namely m, m+2r, m+4r, m+6r, and m+8r.

m+8r	1		+g
m+6r	4		-g
m+4r	6	fermions	+g
m+2r	4		-g
m	1		+g
	16		

It is straightforward to couple these fermions to a vector and an axial vector gauge field. One-loop calculations then give Lorentz covariant results, and the usual axial anomaly is absent, because 8 of the 16 particles have chiral charge +g and 8 have -g. (They all have electrical charge +e).

As far as we know fermions do not occur in nature in 16-plets, furthermore this multiplet structure is a lattice artifact, depend-dent on the lattice shape. In contrast to the naive lattice fermion formulation we now have a handle on the masses of the 15 extra fermions: by choosing r high enough we can make these masses so large that they escape detection with the present day accelerators.

Alternatively we can remove the extra fermions entirely from the spectrum by taking the limit $r \to \infty$. A certain number of vertex functions will then become infinite. These infinities should be cancelled by (r dependent) counter terms. The feasibility, to all orders of perturbation theory, of this approach remains to be proved.

The work described above was performed in collaboration with Dr. J. Smit.

REFERENCES

1. K. G. Wilson, Phys.Rev. D10, 2445 (1974); A. M. Polyakov, Phys.Lett. 59B, 82 (1975).
2. Y. Aharonov, A. Casher and L. Susskind, Phys.Rev. D5, 988 (1972); J. Kogut and L. Susskind, Phys.Rev. D11, 395 (1975); L. Susskind, Phys.Rev. D16, 3031 (1977).
3. S. D. Drell, M. Weinstein and S. Yankielowicz, Phys.Rev. D14, 487, 1627 (1976); Phys.Rev. D16, 1769 (1977); S. D. Drell, H. R. Quinn, B. Svetitsky and M. Weinstein, Phys.Rev. D19, 619 (1979).
4. J. Kogut, D. K. Sinclair and L. Susskind, Nucl.Phys. B114, 199 (1976); T. Banks et al., Phys.Rev. D15, 111 (1977); H. Bergknoff, Phys.Rev. D19, 1543 (1979).
5. K. Osterwalder and E. Seiler, Ann.of Physics 110, 440 (1978).
6. J. Fröhlich, F. Guerra, lectures at this Summer Institute.
7. Constructive Quantum Field Theory, ed. G. Velo and A. S. Wightman, Vol. 25, Lecture Notes in Physics (Springer, New York, 1973).
8. L. H. Karsten and J. Smit, Nucl.Phys. B144, 536 (1978).
9. L. H. Karsten and J. Smit, Phys.Lett. B85, 100 (1979).
10. L. H. Karsten, Thesis (Amsterdam, October 1979).
11. R. P. Feynman, Phys.Rev. 76, 769 (1949).
12. H. S. Sharatchandra, Phys.Rev. D18, 2042 (1978).

GREEN FUNCTIONS, DETERMINANTS AND INDUCED ACTIONS IN GAUGE THEORIES

K. D. Rothe and B. Schroer

Institut für Theoretische Physik

Freie Universität Berlin, Berlin

INTRODUCTION

In a theory of quarks and gluons for which the Lagrangian depends bilinearly on the matter fields, important physical properties only become exposed after integration over the matter fields. The conventional integration rules for this integration are well-known [1]:

a) replace the contraction functions of the matter fields by the external field dependent euclidean Green function G e.g.

$$i\not{D}G(x,y;A) = -\delta(x-y)$$

b) Compute the functional determinant

$$\Gamma[A] = -\ln\frac{\not{D}}{\not{\partial}}$$

The functional integral for correlation functions of ψ can then be written as

$$<\psi(x_1)\ldots \quad> = \text{field independent fact.}\int d[A_\mu]e^{-S_{ind}(x_1\ldots)}$$

where

$$S_{ind}(x_1\ldots) = S_0 + \Gamma + \Gamma_{source}(x_1\ldots)$$

$\Gamma_{source}(x_1\ldots)$ = external field dependent part of Green's function written in exponential form

The explicit form of S_{ind} is known for abelian and nonabelian gauge theories in two space-time dimensions. The construction of S_{ind} [2,3,4] and the discussion of its physical consequences will be the main topic of these lectures. The above formalism has to be modified due to the presence of topology in the A_μ-configuration.

The system described by S_{ind} shares many properties with a Higgs gauge Lagrangian. The induced equations of motion for d=2 are of the London form: $(D^2-m^2)F^a_{\mu\nu}$ = source-terms. The finiteness of the induced action leads to the quantization of topological charge, a property which the "bare" action S_0 often does not share [2].

The effective action for abelian d=2 theories is particularly simple. This theory screens the colour of quarks if they are massless and confines them if they are massive [5] apart from a discrete set of values of the vacuum Θ-angle for which massive "bleached" quarks come out [6]. This very rich "confinement versus screening" picture [7] which goes beyond the Wilson loop picture, is sometimes attributed to the fact that two-dimensional gluons without quarks have no physical degrees of freedom of their own so that the introduction of quarks can drastically change the medium [8]. The confinement versus screening picture holds however also for models in which the gluons are coming from a CP^n σ-model [9]. Here the crucial property is the chiral invariance of the quark Lagrangian. If quarks are massless or are described by a chiral invariant quadrilinear coupling (Chiral Gross-Neveu model) one can construct bleached quark states, whereas massive quarks, even if the mass is spontaneously generated (ordinary Gross-Neveu model), lead to confinement [9] apart from exotic (TP invariant!) θ angles. Although these properties are most conveniently exposed by "bosonization" of fermions, the only known systematic approach which has any chance to unravel such features in QCD_2 and QCD_4 is the method of the induced action.

The first part gives a mathematical discussion of Green functions and determinants. We use the one-point compactified euclidean space and construct fermion determinants in the ζ-regularization approach based on the De Witt-Seeley-Gilkey asymptotic expansions. The axial anomaly relation which we derive in this formalism provides enough information for constructing two-dimensional determinants.

The second part contains model discussions for generalized abelian models and QCD_2. The induced instantons of the abelian torus-model reappear in QCD_2 as quasiclassical extrema in gauge invariant correlations for long distances. The quasiclassical (non-fluctuating) part of the effective potential in QCD_2 suggests a similar distinction between screening in the massless and confinement in the massive case as in the abelian models.

I. REPRESENTATION OF EUCLIDEAN GREEN FUNCTIONS AND FUNCTIONAL DETERMINANTS

In the method of functional integration one is often led to consider conformally covariant elliptic differential equations. Consider for example a trilinear interaction between a matter field density and a "gluon" field. The functional integration will give rise to an external (gluon) field-dependent conformally covariant differential operator acting on the matter field. This operator has to be inverted (field dependent Green function) and its determinant must be computed. Sometimes the Lagrangian contains self interactions, for example in the general Thirring model or in σ-models. In that case there exist mathematical tricks for introducing auxiliary fields which allow to reformulate the problem in terms of the above language of matter-density-gluon interactions [10].

The prototype of such an elliptic differential operator in an external field is the euclidean Dirac-operator in a (matrix valued) vector-potential:

$$i\not{D} \equiv i\gamma_\mu(\partial_\mu - igA_\mu) \tag{1}$$

Many other operators can be reduced to this situation.

Classically this operator is known to be conformally covariant. Under a conformal mapping [11]

$$ds^2 \longrightarrow ds'^2 = \Omega\, ds^2 \tag{2}$$

$$i\not{D} \longrightarrow i\hat{\not{D}} = \Omega^{-\frac{1}{4}(d+1)}\, i\not{D}\,\Omega^{\frac{1}{4}(d-1)} \tag{3}$$

This is most easily seen by demanding hermiticity of the Ansatz

$$i\hat{\not{D}} = \Omega^a\, i\,\not{D}\,\Omega^b$$

with respect to the inner product.

$$(\phi,\psi) = \int d^dx\, \hat{\phi}^+ \sqrt{g}\hat{\psi} \tag{4}$$

Usually one takes for Ω the stereographic mapping

$$\Omega = \left(\frac{2R^2}{R^2+x^2}\right)^2$$

since for certain A_μ configurations it allows to use angular momentum algebra [12].

For this case:

$$\sqrt{g} = \left(\frac{2R^2}{R^2+x^2}\right)^{d} \tag{5}$$

This metric leads to a one point compactification of the euclidean space: R^d_c. The change of the inner product implied by

$$\psi = \left(\frac{2R^2}{R^2+x^2}\right)^{\frac{d}{4}} \hat{\psi} \tag{6}$$

allows to express the spectrum of $i\not{D}$ in (6) in terms of an "associated" Dirac equation: [13]

$$i\not{D}\psi_\ell = E_\ell \frac{2R^2}{R^2+x^2} \psi_\ell \tag{7}$$

$$(\phi,\psi) = \int d^dx\phi^+ \left(\frac{2R^2}{R^2+x^2}\right)^{\frac{d}{2}} \psi. \tag{8}$$

We take a an off-diagonal realization of hermitean γ's with

$$\gamma_5 = \begin{pmatrix} 1 & 0 \\ 0 & -1 \end{pmatrix}, \quad i\not{D} = \begin{pmatrix} 0 & L \\ L^+ & 0 \end{pmatrix} \tag{9}$$

The eigenvalues lie symmetrically around zero

$$\psi_\ell \to \gamma_5\psi_\ell, \quad E_\ell \to -E_\ell \tag{10}$$

The classical Greens function:

$$i\not{D}G(x,y;A) = -\delta(x-y) \tag{11}$$

has the following model representation

$$G(x,y;A) = -\sum_\ell \frac{u_\ell(x)u^+_\ell(y)}{RE_\ell} \tag{12}$$

where u_ℓ denotes the orthonormalized eigenfunctions of the associated Dirac equation (7):

$$(u_\ell, u_\kappa) = R\delta_{\ell\kappa} \tag{13}$$

Note that the inner product contains a quantity of dimension (in mass units) = -1 because our spinors have dimension $\frac{d-1}{2}$. In order to define determinants we pass from E_ℓ to dimensionless eigenvalues

$$\lambda_\ell = RE_\ell \tag{14}$$

The concept of dimensionless eigenvalues is only defined within a common scale factor since what one calls the $(\mathrm{Vol})^{1/d}$ or the diameter of the compact manifold $\mathfrak{M}$ is a matter of convention.

A prerequisite for the discrete representation (12) of G is the requirement that A_μ lives on R_c^d (with patches and transition functions as specified by the fibre bundle discription of gauge connections). This means that the topological numbers which can be formed from A_μ, i.e. the "winding number", is integer valued. This implies physically speaking a fall-off property of $G(x,y)$ as $s \to \infty$ or $y \to \infty$ such that it can be decomposed in terms of the R_c^d complete set u_ℓ.

In the case of zero modes in (10) there is a clash between the requirement that G can be viewed as a distribution on R_c^d and its defining relation (11). Maintaining the first property, the defining relation (11) must be relaxed:

$$i\not{D}G' = -\delta(x-y) + P_o(x,y) \tag{15}$$

P_o = projector on zero modes

$$G' = -\sum_\ell{}' \frac{u_\ell(x)\, u_\ell^+(y)}{\lambda_\ell} \tag{16}$$

' = omission of zero modes.

The G' depends on the metric in R_c^d (i.e. the radius R); in using G' in correlation functions, the R dependence must be absorbable into the renormalization constants.

Besides then "classical" quantities one needs one-loop objects as the fermion determinants.

Assume now that D is any positive semi-definite selfadjoint elliptic operator on a compact manifold $\mathfrak{M}$. The formal definition of its determinant

$$\text{"det D"} = \prod_{\lambda_i > 0} \lambda_i \quad \text{or}$$

$$\text{"}\Gamma\text{"} = -\ln \text{"detD"} = -\sum{}' \ln\lambda_i$$

may be converted into an existing analytic expression with the help of the function:

$$\zeta(s) = \sum_i{}' \frac{1}{(\lambda_i)^s} \tag{17}$$

$$\Gamma_\zeta = - \frac{d}{ds} \zeta(s)\Big|_{s=0} \tag{18}$$

In the presence of zero modes Γ_ζ and $\det D = e^{-\Gamma_\zeta}$ are often called Γ' and $\det' D$, whereas the prime is omitted in the absence of zero modes. The functional integration (see later section)

$$Z = \int d[\psi]d[\bar{\psi}]\; e^{-\int \bar{\psi} i \not{\partial} \psi} \tag{19}$$

using the Grassmann rules gives

$$Z \sim \prod_{\lambda_i > 0} \lambda_i^2 = c^{-\Gamma_\zeta} \tag{20}$$

with

$$\Gamma_\zeta = - \frac{1}{2} \ln (i\hat{\not{\partial}})^2 = - \frac{1}{2} \frac{d}{ds} \zeta_{(i\hat{\not{\partial}})^2}(s)\Big|_{s = 0} \tag{21}$$

In the case of a complex Bose field the integration is Gaussian and one obtains

$$Z \sim \prod \frac{1}{\lambda_i} = e^{\Gamma_\zeta} \tag{22}$$

$$\Gamma_\zeta = - \frac{d}{ds} \zeta_{\hat{D}^2}(s)\Big|_{s = 0}$$

Because of the aforementioned scale ambiguity in defining the λ_ℓ and the property

$$-\zeta'(0) \xrightarrow[\text{scale tr.}]{} - \zeta'(0) + \zeta(0) \ln M \tag{23}$$

it is reasonable, to make this ambiguity manifest by defining

$$\Gamma_M = - \zeta'(0) + \zeta(0) \ln M$$

The meromorphy properties of $\zeta(s)$ become visible with the help of the heat equation (proper time) kernel to D

$$h(t;x,y) = \sum_i e^{-E_i t}\, \hat{u}_i(x)\, \hat{u}_i^{+}(y)\, \frac{1}{L^{\deg D}}$$

$$(\hat{u}_i, \hat{u}_j) = L^{\deg D}\, \delta_{ij}\,, \quad L = \text{size of } \mathfrak{M} \tag{25}$$

$$(\frac{\partial}{\partial t} + D)h = 0, \quad h(0;x,y) = \frac{1}{\sqrt{g}}\, \delta(x,y)$$

$$\int h(0,x,y)\, d(\text{vol}) = 1$$

The diagonal part of h has a well-known asymptotics which was first treated systematically by Seeley [14]. For second degree operators, deg D=2, one finds:

$$h(t;x,x) \xrightarrow[t\to o]{} \frac{1}{(4\pi t)^{d/2}} [a_o + a_1 t + a_2 t^2 + \ldots] \tag{26}$$

Before we indicate the computation of the matrix valued expansion coefficients a_i let us try to understand the meromorphy of ζ. One defines a matrix-valued ζ function by Mellin-transformation

$$\zeta(s;x) = \frac{1}{\Gamma(s)} \int_0^\infty dt\, t^{s-1} [h(t;x,x) - \sum_i \hat{u}_i^{(o)}(x)\hat{u}_i^{(o)+}(x)] \tag{27}$$

Insertion of the definition (27) yields

$$\zeta(s;x) = \sum_i{}' \frac{1}{\lambda_i} \hat{u}_i(x)\, \hat{u}_i(x)^+$$

where the subtraction of the zero mode contribution in (29) was necessary in order to obtain Σ'. Inserting the asymptotic expansion (28) one sees that the negative t powers are responsible for poles in s; the rest will be an analytic function in s. The singular part

$$\frac{1}{\Gamma(s)} \frac{1}{(4\pi)^{d/2}} \int_0^\varepsilon dt\, t^{s-1-\frac{d}{2}} [a_o + a_1 t + a_2 t^2 + \ldots$$

$$- t^{\frac{d}{2}} \Sigma\, \hat{u}_i^{(o)}(x)\, \hat{u}_i^{(o)}(x)^+] \tag{28}$$

behaves near $s \sim d/2 - k > o$

$$\frac{\alpha_k}{s - \frac{d}{2} + k} \times \left(\lim_{s\to\frac{d}{2}+k} \frac{\varepsilon^{s - \frac{d}{2} + k}}{(4\pi)^{d/2}\Gamma(s)} \right)$$

The last factor is 1 and therefore the residuum of the pole at

$$s = \frac{d}{2} - k > 0$$

is

$$\operatorname*{Res}_{s = \frac{d}{2} - k} \zeta(s,x) = \frac{1}{\Gamma(\frac{d}{2} - k)} \frac{1}{(4\pi)^{d/2}} a_k \tag{29}$$

In particular the value at s=o is finite:

$$\zeta(o,x) = \frac{a_{d/2}}{(4\pi)^{d/2}} - \sum_i \hat{u}_i^{(o)}(x)\hat{u}_i^{(o)}(x)^+ \tag{30}$$

All physically interesting operators are of the form:

$$D = -D_\rho D^\rho + X \tag{31}$$

D_ρ = covariant derivative (including gauge fields)

X = matrix-valued function

For example the square of the Dirac operator leads to

$$X = -\frac{1}{4}R - \frac{i}{2}[\gamma_\mu,\gamma_\nu]\, i\, F^{\mu\nu} \tag{32}$$

R = scalar curvature, F = gauge field-strength matrix. The computational result for the first three Seeley-DeWitt coefficients is [15,16]

$$a_o = 1$$

$$a_1 = \frac{R}{6} + X \tag{33}$$

$$a_2 = \frac{1}{2}\left(\frac{R}{6} + X\right)^2 + \frac{1}{30}R_{;\mu}{}^{\mu} - \frac{1}{180}R_{\mu\nu}R^{\mu\nu}$$

$$+ \frac{1}{180}R_{\mu\nu\sigma\tau}R^{\mu\nu\sigma\tau} - \frac{1}{12}Y_{\nu\tau;}{}^{\nu\tau} + \frac{1}{12}Y_{\nu\tau}Y^{\nu\tau} + \frac{1}{6}X_{;\tau}{}^{\tau}$$

Here the usual notation for the Riemann and Ricci tensors is used; the semicolon denotes covariant derivative and

$$Y_{\lambda\nu} = [D_\lambda, D_\nu] \tag{34}$$

a_3 has been computed by Gilkey [16]. It is a very lengthy expression and we are not going to need it. The $\zeta(s)$ function is obtained by integration:

$$\zeta(s) = \int d^d x \sqrt{g}\, \mathrm{tr}\, \zeta(s,x), \tag{35}$$

tr = trace over matrix indices.

The Gilky method is completely rigorous, but the Dewitt procedure is easier to understand for physicists. Therefore we will briefly indicate DeWitts approach.[17] He starts from the Ansatz

$$h(t;x,x') = \frac{\Delta^{1/2}}{(4\pi t)^{d/2}} e^{-\frac{\sigma}{2t}} \sum_{\ell=0}^{\infty} a_\ell(x,x')\, t^\ell \tag{36}$$

$$\sigma(x,x') = \frac{1}{2}\,(\text{geodetic distance})^2 = \text{"geodesic interval."} \tag{37}$$

$$\Delta(x,x') = \frac{1}{\sqrt{g}} \det \sigma_{;\mu\nu} \frac{1}{\sqrt{g'}} \tag{38}$$

$$= \text{Van Vleck determinant}$$

$;\mu\nu'$ = covariant differentiation with respect to x_μ, x'_ν respectively.

The form of the Ansatz is geared to the form of the free field heat kernel. σ and Δ are very natural quantities in Riemannian geometry. The local limits of their covariant derivative are known geometrical objects which can be iteratively determined. The local limits are indicated by square brackets. For example:

$$[\Delta] \equiv \Delta(x,x) = 1 \tag{39a}$$

$$[\sigma_{;}{}^{\mu}] = 0, \; [\sigma_{;}{}^{\mu}{}_{\nu}] = g^{\mu}{}_{\nu}, \; [\sigma_{;}{}^{\mu}{}_{\nu\mu}] = 0 \tag{39b}$$

$$[\Delta^{1/2}{}_{;\mu}{}^{\mu}] = \frac{R}{6} \tag{40a}$$

$$[\Delta^{1/2}{}_{;\mu}{}^{\mu}{}_{\nu}{}^{\nu}] = \frac{1}{5} R_{;\mu}{}^{\mu} + \frac{1}{36} R^2 - \frac{1}{30} R_{\mu\nu} R^{\mu\nu}$$

$$+ \frac{1}{30} R_{\mu\nu\sigma\tau} R^{\mu\nu\sigma\tau} \tag{40b}$$

The insertion of the De Witt Ansatz into the heat equation leads to the following recursive system:

$$t^{-2} : \frac{1}{2} \sigma_{;\mu} \sigma^{;\mu} - \sigma = 0 \tag{41a}$$

$$t^{-1} : \frac{1}{2} (\Delta^{1/2} a_0 \sigma^{;\mu})_{;\mu} - 2\Delta^{1/2} a_0 = 0 \tag{41b}$$

$$t^{\ell} : \sigma_{;\mu}\, a_{\ell+1;}{}^{\mu} + (\ell+1) a_\ell = - \Delta^{-\frac{1}{2}} D\, (\Delta^{\frac{1}{2}} a_\ell) \tag{41c}$$

The first equation is an identity for the geodetic interval. With the help of the relation (known from Riemannian geometry)

$$\Delta^{-1} (\Delta \sigma^{;\mu})_{\mu} = d \tag{42}$$

the second equation simplifies to

$$a_{o;\mu} \sigma^{;\mu} = 0$$

i.e. the parallel transport equation along the geodesic. Using the coincidence condition:

$$[a_o] = 1$$

which follows from the initial condition of the heat equation kernel, one obtains

$$a_o = P \exp [- i \int_{geod.} d\xi_\mu \omega^\mu] \tag{43}$$

ω^μ = connection matrix (metric + gauge).

The coincidence limit for a_1 can be read off from (41c) for $\ell = o$ (use relation (42))

$$a_1 = [(\Delta^{\frac{1}{2}} a_o)_{;\mu}{}^{\mu}] + X \tag{44}$$

and similarly:

$$2[a_2] = [(\Delta^{1/2} a_1)_{;\mu}{}^{\mu}] + X [a_1] . \tag{45}$$

So we shall need the coincidence limits of derivation of $\Delta^{\frac{1}{2}}$, a_o and a_1. We already know those of Δ (40) (which follow from differentiating (42) and higher derivative coincidence limits of σ).

The differentiation of the parallel transport equation for a_o gives

$$\sigma_{;}{}^{\mu} a_{o;\mu\nu\lambda} + \sigma_{;\mu}{}^{\mu}{}_{\lambda} a_{o;\mu\nu} + \sigma_{;}{}^{\mu}{}_{\nu} a_{o;\mu\lambda} + \sigma_{;}{}^{\mu}{}_{\nu\lambda} a_{;\mu} = 0$$

e.g. to the coincidence relation

$$[a_{o;\mu\nu}] = - [a_{o;\nu\mu}]$$

and hence:

$$[a_{o;\mu\nu}] = \frac{1}{2} [D_\nu, D_\mu][a_o] = \frac{1}{2} [D_\nu, D_\mu] \equiv \frac{1}{2} Y_{\nu\mu} . \tag{46}$$

A tedious 4th order differentiation of the parallel transport equation gives

$$[a_{o;\nu}{}^{\nu}{}_{\tau}{}^{\tau}] = -\frac{1}{2} Y_{\nu\tau;}{}^{\nu\tau} + \frac{1}{2} Y_{\nu\tau} Y^{\nu\tau} \, .$$

Two fold differentiation of the a_1 equation yields

$$[a_{1;\mu}{}^{\mu}] = \frac{1}{3} [\Delta^{-\frac{1}{2}}{}_{;\tau}{}^{\tau}][\Delta^{\frac{1}{2}}{}_{;\mu}{}^{\mu}] + \frac{1}{3} [\Delta^{\frac{1}{2}}{}_{;\mu}{}^{\mu}{}_{\tau}{}^{\tau}]$$

$$+ \frac{1}{3} [a_{o;\mu}{}^{\mu}{}_{\tau}{}^{\tau}] + \frac{1}{3} X_{;\tau}{}^{\tau}$$

This leads to the results mentioned before (33). The computation of a_3 would be a major task in this formalism. Fortunately we do not need this quantity

The matrices $Y_{\mu\nu}$ and X for $s = \frac{1}{2}$ are:

$$Y_{\mu\nu} = [D_\mu, D_\nu] = - ie F_{\mu\nu} + \frac{1}{2}\frac{1}{4} [\gamma_\alpha, \gamma_\beta] R^{\alpha\beta}{}_{\mu\nu} \qquad (47)$$

$$X = -\frac{1}{4} [\gamma_\mu, \gamma_\nu] Y^{\mu\nu} = -\frac{ie}{4} [\gamma_\mu, \gamma_\nu] F^{\mu\nu}$$

$$- \frac{1}{8} \gamma_\mu \gamma_\nu \gamma_\alpha \gamma_\beta R^{\mu\nu\alpha\beta}$$

$$= -\frac{ie}{4} [\gamma_\mu, \gamma_\nu] F^{\mu\nu} - \frac{R}{4} \qquad (48)$$

The coefficients a_1, a_2 may be computed for any spin. They are very important in the discussion of shielding versus amplification [18] (Diamagnetism or Paramagnetism).[19] Let us return to the construction of the determinant. The ζ-function definition together with the scale ambiguity lead us to Γ_M. We will now discuss the relation of Γ_M with a more field theoretic definition of Γ: the Pauli-Villars regularized determinant $\Gamma_{P.V.}$. Let M, $i = 1 \ldots \nu$, be large regulator "masses" (in our case dimensionless) and choose e_i such that

$$\sum_{i=1}^{\nu} c_i = -1 \, , \quad \Sigma c_i M_i^\ell = 0 \, , \quad \ell = 1, \ldots, \nu-1$$

Consider the Pauli-Villars regularized ζ-function:

$$\zeta_{pv}(s) = \zeta(s) + \Sigma c_i \zeta_{M_i}(s)$$

$$\zeta_{M_i}(s) = \sum_{\ell}{}' \frac{1}{(\lambda_\ell + M_i)^s} \tag{49}$$

Clearly the minimal number ν of PV parameters is related to the polynomial bound for the density of the asymptotic spectral values. The minimal ν is the one for which $\zeta_{pv}(s)$ and its derivative exist near $s = o$ without analytic continuation. Instead of investigating asymptotic spectral densities it is more convenient to consider the asymptotic behaviour for small t. Eq. (49) may be rewritten as

$$-\zeta'_{pv}(0) = -\lim_{s\to o} \frac{d}{ds} \frac{1}{\Gamma(s)} \int_0^\infty dt\, t^{s-1} (1 + \Sigma c_i e^{-M_i t})\ \mathrm{Tr}\ (e^{-tD} - P_o) \tag{50}$$

Where Tr (with capital T) denotes the trace with respect to the internal degrees of freedom and space-time.

The singularities for small t will cancel for $\nu = [\frac{d}{2}]$. Restricting to an even dimension all the inverse powers in t are integer and a short computation [11] reveals:

$$\begin{aligned} -\zeta'_{pv}(o) &= -\zeta'(o) + \sum_i (\alpha_{\frac{1}{2}d} - p)\, c_i \ln M_i \\ &+ \sum_{l>o}^{d/2} \sum_i \frac{(-1)^\ell}{\ell!}\, \alpha_{\frac{d}{2}-\ell}\, c_i M_i^\ell \ln M_i + O(M_i^{-2} \ln M_i) \end{aligned} \tag{51}$$

Here

$$\alpha_{\frac{d}{2}-\ell} = \int d^2x \sqrt{g}\ a_{\frac{d}{2}-\ell}(x) \tag{52}$$

$$p = \mathrm{Tr}\ P_o .$$

We therefore define:

$$\Gamma_M = -\zeta'(o) + (\alpha_{\frac{d}{2}} - p) \ln M + \sum_\ell \sum_i \frac{(-1)^\ell}{\ell!} \alpha_{\frac{d}{2}-\ell}\, c_i M_i \ln M_i$$

$$\ln M = \sum_i c_i \ln M_i$$

The coefficients are the residues of ζ and

$$\alpha_{\frac{d}{2}-p} = \zeta(o) \tag{54}$$

A further simplification may be obtained in gauge theories. We define the determinant to be used in functional integration as:

$$\Gamma = \Gamma[A] - \Gamma_M [A=0] \qquad (55)$$

i.e. we divide by the free field determinant. In this difference we loose all power-like regularization terms for d=2,4. The reason is that a_0 is field independent and

$$\mathrm{tr}\ a_1 \Big|_0^A = 0 \qquad (56)$$

as can be seen directly from (33).

Gauge determinants in d=2,4 contain therefore only the Hawking [20] scale parameter M as a regularization parameter.

From experience with renormalization theory one knows that this parameter is used in the process of ultraviolett (wave function and coupling constant) renormalization. But gauge theories in d=2 are superrenormalizable. Looking at the coefficient of lnM one realizes that here the dependence on the regularization is related to zero modes. In fact one knows that in such theories the functional integrand contains terms of the form

$$\int d^2x' D(x-x')\epsilon_{\mu\nu}F_{\mu\nu}(x')$$

which for nontrivial winding contain an infrared regularization. The lnM dependence in Γ is used to compensate this dependence on the infrared parameter and to obtain the known, regularization independent, correlation functions.

The determinant is the starting point for the introduction of various induced composite fields. For fermionic determinants in gauge theories one defines

$$-gj_\mu = \frac{\delta\Gamma}{\delta A_\mu} \qquad (57)$$

$$-g_w j_{\mu_5} = \frac{\delta\Gamma}{\delta W_\mu}\Big|_{W_\mu=0} \quad , \ W_\mu \text{ coupled to pseudo-current} \qquad (58)$$

$$\Theta_{\mu\nu} =: -2\,\frac{\delta\Gamma}{\delta g_{\mu\nu}} \qquad (59)$$

In flat space gauge theories one very often works with R_c^d using a spherical metric of radius R. In that case

$$R\frac{\delta\Gamma}{\delta R} = R\int d^dx\,\theta_{\mu\nu}(x)\frac{\delta g_{\mu\nu}}{\delta R}$$

$$= \int d^dx \frac{2R^2}{R^2+x^2}\frac{x^2}{R^2}\,\Theta_\mu^{\ \mu}(x) \qquad (60)$$

where the trace of the energy-momentum tensor is

$$\Theta_\kappa^{\ \kappa} = -2\left\{-\frac{d}{ds}\Sigma'\frac{s}{(\lambda_i^2)^{s+1}}\,2\lambda_i\frac{\delta\lambda i}{\delta g^{\kappa\nu}}g^{\nu\kappa}\Big|_{s=0}\right.$$

$$\left.+\ \ln M\,\Sigma'\frac{s}{(\lambda_i^2)^{s+1}}\,2\lambda i\frac{\delta\lambda i}{\delta g^{\kappa\nu}}g^{\nu\kappa}\Big|_{s=0}\right\}_0^1 \qquad (61)$$

Using the first variation:

$$\frac{\delta\lambda i}{\delta g^{\mu\nu}} = \frac{i}{2}\sqrt{g}\,\hat{u}_i^{\dagger}\gamma_\mu\overleftrightarrow{D}_\nu\hat{u}_i$$

the result is

$$\Theta_\kappa^{\ \kappa} = -2\left\{-\lim_{s\to 0}\Sigma'\frac{1}{(\lambda_i^2)^{s+1}}\cdot 2\lambda_i^2\,\hat{u}_i^{\dagger}\hat{u}_i\right\}_0^A$$

$$= 2\left[\zeta_A(0,x) - \zeta_{A=O}(0,x)\right]\sqrt{g} \qquad (62a)$$

therefore:

$$\Theta_\kappa^{\ \kappa} = \text{zero mode contr.}\ ,\quad d=2$$

$$\Theta_\kappa^{\ \kappa} = 2\frac{\sqrt{g}}{8\pi^2}\left\{\mathrm{tr}\,\frac{1}{2}X^2 + \frac{1}{12}Y_{\mu\nu}Y^{\mu\nu}\right\}_0^A + \text{zero mode contr.},d = 4 \qquad (62b)$$

The simple R-dependence and the triviality of the coefficient of lnM for d=2 can therefore be understood in terms of the absence of an anomaly in $\Theta_\mu^{\ \mu}$.

A third interesting induced composite is the pseudocurrent. Since the lnM-dependent term in Γ can be represented as:

$$\zeta(0) = \frac{1}{2}\int d^dx\,\Theta^\mu_{\ \mu} \qquad (63)$$

and since the dependence of $\Theta_{\mu\nu}$ on an axial potential is quadratic, the functional differentiation of ζ with respect to W_μ will lead to a vanishing result after setting $W_\mu = 0$. Therefore $j_{\mu 5}$ will be independent of lnM:

$$j_{5\mu} = \left(-\frac{1}{g_W}\frac{\delta\Gamma}{\delta W_\mu}\right)_{W=O} = -\sqrt{g}\,\Sigma'\left(\frac{\lambda i}{(\lambda_i^2)^{s+1}}\frac{1}{g_W}\frac{\delta\lambda i}{\delta W_\mu(x)}\right)_{W_\mu=O}$$

$$= \sqrt{g}\,\Sigma'\frac{\lambda i}{(\lambda_i^2)^{s+1}}\hat{u}_i^+\gamma_\mu i\gamma_5\hat{u}_i \tag{64}$$

where from now on λ_i and u_i are referring to the W_μ=o zero situation. Taking the divergence (observe that ∂^μ commuted with $\sqrt{g}$ will be converted into the covariant derivative in R_c^d !) one finally obtains:

$$\partial^\mu j_{5\mu} = -2\sqrt{g}\,\mathrm{tr}\zeta(0,x)\gamma_5$$

$$= -\sqrt{g}\frac{2}{(4\pi)^{d/2}}\,\mathrm{tr}\,a_{\frac{1}{2}d}\gamma_5 + 2\sqrt{g}\,\Sigma\,\hat{u}_i^{(0)+}(x)\gamma_5\hat{u}_i^{(0)}(x) \tag{65}$$

With:

$$\mathrm{tr}\,a_1\gamma_5 = e\,\varepsilon^{\mu\nu}F_{\mu\nu} \qquad \text{for} \quad d = 2 \tag{66a}$$

$$\mathrm{tr}\,a_2\gamma_5 = -e\,\mathrm{tr}\,F_{\mu\nu}\tilde{F}^{\mu\nu} + \frac{1}{48}NR_{\rho\sigma\mu\nu}R_{\alpha\beta}{}^{\rho\sigma}\varepsilon^{\alpha\beta\mu\nu} \tag{66b}$$

the equation is the axial current version [21] of the Atiyah-Singer relation:

$$\partial^\mu j_{5\mu} = 2\begin{cases} -\dfrac{e}{4\pi}\,\mathrm{tr}\,F + (n_+(x)-n_-(x)\;, \quad d=2 & (67a)\\[2ex] \dfrac{e}{8\pi^2}\,\mathrm{tr}\,F\wedge F + \dfrac{1}{8\pi^2}\dfrac{N}{24}R^a{}_b\wedge R^b{}_a + (n_+(x)-n_-(x)) \\ \qquad\qquad d = 4 & (67b)\end{cases}$$

$$n_+(x) - n_-(x) = \sqrt{g}\,\Sigma\,\hat{u}_i^{(0)+}(x)\gamma_5\hat{u}_i^{(0)}(x) \tag{68}$$

$$F = \frac{1}{2}F_{\mu\nu}\,dx^\mu\wedge dx^\nu\;, \quad R_b^a = \frac{1}{2}R^a{}_{b\mu\nu}$$

The integration over R_C^d gives a vanishing left hand side and hence the global Atiyah-Singer index theorem. Note that the Pontryagin-density

$$\rho_1 = \frac{1}{8\pi^2} R \wedge R$$

vanishes for a sphere.

II. GENERIC CONSTRUCTION OF DETERMINANTS

The method can be immediatly applied to colour carrying currents and pseudo-currents. For QCD_2 we obtain

$$D_\mu j^a_{5\mu} = -\frac{1}{2\pi} \varepsilon_{\mu\nu} F^a_{\mu\nu} + 2 \sum_i u_i^{(o)\dagger} \gamma_5 \frac{\lambda^a}{2} u_i^{(o)} \tag{1}$$

The pseudo-current $j^a_{5\mu}$ is not gauge invariant and hence this relation does not have a direct physical implication. Nevertheless it serves as an important mathematical tool to derive the generic form of the fermion determinant. This method is the same for QED_2 and QCD_2:
With

$$j^a_{5\mu} = \varepsilon_{\mu\nu} j^a_\nu = -\varepsilon_{\mu\nu} \frac{\delta\Gamma}{\delta A^a_\nu} \tag{2}$$

the anomaly equation can be integrated [4]:

$$j^a_\mu(x) = \frac{e}{4\pi} \int d^2y \; \tilde{K}^{ab}_\mu(x,y) \; \varepsilon_{\lambda\rho} F^b_{\lambda\rho}(y) + \text{zero mode contr.} \tag{3}$$

with

$$D^{ab}_\mu \tilde{K}^{ac}_\mu = 0 \tag{4a}$$

$$\tilde{D}^{ab}_\mu \tilde{K}^{bc}_\mu = \delta^{ac} \delta^2(x-y) \tag{4b}$$

In terms of euclidean $\pm$ $(z,\bar{z})$ components these two equations read:

$$\begin{pmatrix} 0 & -D_- \\ D_+ & 0 \end{pmatrix} \begin{pmatrix} 0 & K_- \\ -K_+ & 0 \end{pmatrix} = - \mathbb{1} \, \delta(x-y) \tag{5}$$

So the current functional can be written in closed form by employing the adjoint spinor Green function. It is easy but somewhat lengthy to show that this result is obtained from Feynman integrals if one performs all the one loop integrals:

Fig. 1

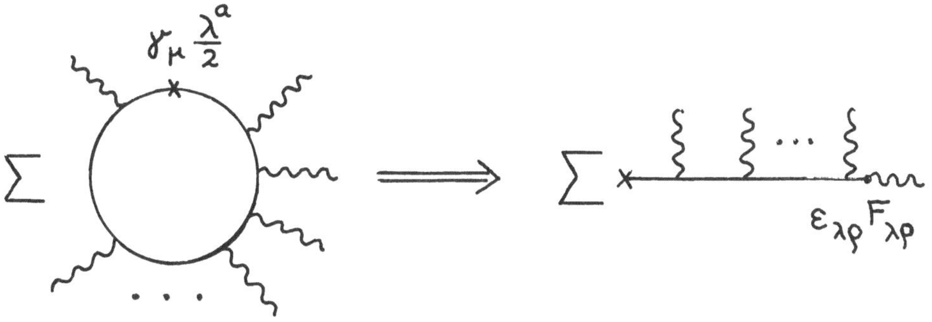

vertices: λ - coupling vertices: f-couplings

Restricting the gauge configuration to the torus:

$$\underset{\sim}{A}_\mu = \sum_{i_D=1}^{N-1} \frac{\lambda^{i_D}}{2} A_\mu^{i_D} \tag{6}$$

one obtains the previously known result:

$$\tilde{K}_\mu = \partial_\mu \tilde{D}, \quad D(\xi) = -\frac{1}{4\pi} \ln \mu^2 \xi^2 \tag{7}$$

$$j_\mu^{i_D}(x) = \frac{g}{4\pi} \int d^2x' \tilde{\partial}_\mu D(x-x') \varepsilon_{\lambda\rho} F_{\lambda\rho}^{i}(x') + \text{zero mode contr.}$$

Let us for a moment ignore the zero mode contribution. The "Schwinger part" of Γ (i.e. the part not involving zero modes) is

$$\frac{\delta \Gamma s}{\delta A_\mu^a} = -g j_\mu^a, \quad \Gamma_s = -\text{i.f.} \int_0^g dg' \int d^2x \, j_{5\mu}(x,g') A_\mu(x') \tag{8}$$

Here i.f. denotes an infrared-finite part operation. Two remarks on this result for Γ_s are in order:

1) The variable coupling constant representation is a solution of the functional differential equation if and only if the integrability condition:

$$\frac{\delta}{\delta A_\nu^b(y)} j_{5\mu}^a(x) = \frac{\delta}{\delta A_\mu^a(x)} j_{5\nu}^b(y)$$

is satisfied. The above solution fulfills this condition.

2) The infrared-operation i.f. is most easily performed by a partial integration. For example in the abelian U(1) case one obtains the well-known gauge invariant infrared-finite expression

$$\Gamma_s = \frac{e^2}{8\pi} \int d^2x \int d^2x' \varepsilon_{\mu\nu} F_{\mu\nu}(x) \, D(x-x') \varepsilon_{\rho\sigma} F_{\rho\sigma}(x') \qquad (9)$$

In order to integrate the zero mode part $\Delta\Gamma$

$$\tilde{D}_\mu^{ab} \frac{\delta \Delta\Gamma}{\delta A_\mu^b} = 2 \sum_i u_i^{(o)+} \gamma_5 \frac{\lambda^a}{2} u_i^{(o)} \qquad (10)$$

let us first consider the abelian (U(1)) case. The zero mode wave function are easily obtained from the wave functionals which appear in the factorized form of the Green's function [4]

$$G(x,y) = \begin{pmatrix} 0 & h_-(x) \, G_-(x-y) h_-^{-1}(y) \\ -h_+(x) \, G_+(x-y) \, h_+^{-1}(y) & 0 \end{pmatrix} \qquad (11a)$$

$$h_{\pm}(x) = e^{H_{\pm}(x)} , \quad H_{\pm}(x) = -ig \int d^2y \, D(x-y) \left(\partial_\mu A_\mu(y) \mp \frac{i}{2} \varepsilon_{\mu\nu} F_{\mu\nu}(y) \right) \qquad (11b)$$

If the winding number:

$$n = \frac{g}{4\pi} \int d^2x \, \varepsilon_{\mu\nu} F_{\mu\nu}(x)$$

is positive, the h_+ function behaves as

$$h_+(x) \sim e^{-\frac{n}{2} \ln \mu^2 x^2} \qquad (12)$$

and there are n normalizable (in the R_c^2 metric) solutions

$$\psi_{\ell_+}(x) = R^{-\ell + \frac{1}{2}} z^{\ell-1} \, h_+(x) \; , \quad \ell = 1, \ldots, n, \quad z = x_1 + i x_2$$

Define the overlap matrix:

$$\left(N_+^{-1} \right)_{\ell' \ell} = \int d^2x \frac{2R}{R^2 + x^2} \psi_{\ell_+}^{+}(x) \psi_{\ell'_+}(x) \qquad (13)$$

In case of n negative we define a N_-^{-1} in an analogous manner. In our case one of the matrices will be trivial for a definite winding.

$$\text{Theorem:} \quad \Delta\Gamma = \mathrm{tr}\,\ln N_+ + \mathrm{tr}\,\ln N_- \tag{14}$$

The proof follows from:

$$\tilde{\partial}_\mu \frac{\delta}{\delta A_\mu(x)} \mathrm{tr}\ln N_{\pm} = \pm 2g \frac{R}{R^2+x^2} \mathrm{tr} N_{\pm} \hat{n}_{\pm}(x) \tag{15}$$

$$[\hat{n}_{\pm}(x)]_{\ell'\ell} = \psi^+_{\ell_{\pm}}(x)\, \psi_{\ell'_{\pm}}(x)$$

which in turn is a result of the Ward identity:

$$\partial^y_\mu \frac{\delta h_{\pm}(x)}{\delta A^c_\mu(y)} = \pm g\delta^2(x-y) h_{\pm}(x) \tag{16}$$

The $\Delta\Gamma$ for QCD_2 is derived in an entirely analogous fashion. First one establishes the factorization of the Greens function. The wave functionals $h_{\pm}$ which in this case are a NxN matrix are so called harmonic "spinors"

$$\begin{pmatrix} 0 & -D_- \\ D_+ & 0 \end{pmatrix} \begin{pmatrix} 0 & h_- \\ h_+ & 0 \end{pmatrix} = 0 \,, \quad h_{\pm}(A_\mu = 0) = 1 \tag{17}$$

$$h_{\pm}(x) = 1 + \sum_1^\infty (-ig)^n \int d^2y_1 \ldots d^2y_n G_{\pm}(x-y_1)\ldots G_{\pm}(y_{n-1}-y_n) \times A_{\mp}(y_1)\ldots A_{\mp}(y_n) \tag{18}$$

Again it is evident that G satisfies the defining relation of the euclidean Greens function. The $h_{\pm}$ can also be obtained by a resummation of the Feynman series for G. One then selects those columns of $h_{\pm}$ which are normalizable in R^2_c and multiplies them with the maximally possible "angular momentum" factors z resp. $\bar{z}$ in order to define $N_{\pm}$. The relevant Ward identities

$$D^{ac}_\mu(y) \frac{\delta h_{\pm}(x)}{\delta A^c_\mu(y)} = \pm g\delta^2(x-y) \frac{\lambda^a}{2} h_{\pm}(x) \tag{19}$$

are easily established. From this we obtain our main result:

Theorem:
$$\Gamma = \Gamma_s + \Delta\Gamma + (n_1 + n_-) \ln M \tag{20}$$

The wave functionals $h_{\pm}$ (A_μ, x) have a number of remarkable properties. They behave like semiinfinite line integrals under gauge transformations

$$U(g) h_{\pm}(A^g_\mu, x) = h_{\pm}(A_\mu, x) \tag{21a}$$

where U belongs to the fundamental representation. At the same time thay are Lorentz scalars

$$h_{\pm}(\Lambda^\nu_\mu A_\nu, \Lambda x) = h_{\pm}(A_\mu, x) \tag{21b}$$

Their A_μ flux is not concentrated on a line but spreads out. Some remarks on four dimensional QCD_4 factorization and determinants are in order. In this case the Dirac operator

$$i\not{D} = \begin{pmatrix} 0 & T^+ \\ T & 0 \end{pmatrix}, \quad T^+ = i\,\sigma_\mu D_\mu \tag{22}$$

after its chiral decomposition still depends on Pauli-matrices. The wave functionals h would therefore depend on σ_μ. The noncommutativity of the σ's would prevent strict factorization. This can only be expected after making some sort of partial wave decomposition. The Green function on the ADHM instantons [22] have a factorizing appearance:

$$S = \begin{pmatrix} 0 & S_- \\ S_-^+ & 0 \end{pmatrix} \tag{23}$$

$$S_- = i\,\sigma_\mu D_\mu G(x,y) \tag{24}$$

$$G(x;y) = \frac{V^+(x) V(y)}{4\pi^2 |x-y|^2} \tag{25}$$

However the Greens function property

$$D^2_\mu G = -\delta^2(x-y) \tag{26}$$

is only satisfied due to a very subtle interplay of properties of the $(n+2k)\times n$ matrices with the free Green function:

$$[D_\mu^2 V^+(x) + 2(D_\mu V^+)\partial_\mu]\frac{1}{4\pi(x-y)^2} \tag{27}$$

Therefore there is no strict factorization as in QCD_2. Related to this lack of factorization is the inutility of the heat equation method to this problem. Since one only knows the Greens function to use the (mathematically less rigorous) point split method of Schwinger to obtain the current

$$j_\mu(x) = \lim_{y\to x} [P \exp ig \int_x^y d\xi_\mu A_\mu \; \mathrm{tr} G(x,y)\frac{\lambda^a}{2}] \tag{28}$$

where the limit $y \to x$ indicates a directional (Vierbein) averaging. The variation of Γ under changes of the instanton parameter may than be written as [23]

$$\delta\Gamma = -\int j_\mu^a \; \delta A_\mu^a \tag{29}$$

δA_μ^a and j_μ^a are known functions on the instanton manifold. The problem of quadrature for the fundamental representation was solved recently by Berg and Lüscher [24].

Our generic method based on the anomaly equation works for the CP^{n-1} determinant [25]. This model is formulated in terms of the action

$$S = \frac{1}{2}\int d^2x \, \overline{D_\mu \phi} \, D_\mu \phi \quad , \quad \bar{\phi}\phi = 1 \tag{30}$$

$$D_\mu = \partial_\mu - \bar{\phi}\partial_\mu\phi \quad , \quad \phi = n \text{ comp. vector}$$

The Lagrangian multiplier method leads to the equation of motion. The finiteness of the action implies the quantization of the topological charge

$$Q = \frac{1}{2\pi}\int d^2x \; i \; \varepsilon_{\mu\nu} \, \overline{D_\mu \phi} \, D_\mu \phi = \frac{1}{2\pi}\int d^2x \; \varepsilon_{\mu\nu} \, \partial_\mu A_\nu$$

$$A_\nu = i\bar{\phi}\partial_\mu \phi$$

The inequality $S \geq \pi|Q|$ leads to minima

$$D_\mu \phi = \pm i \, \varepsilon_{\mu\nu} \, D_\nu \phi$$

or $D_{\mp}\phi = 0$

Upon introduction of homogeneous coordinates these instanton equations are transformed into Cauchy-Riemann equations. The global solutions are of the form

$$\phi^c_\alpha \frac{p_\alpha(x)}{|p(x)|}$$

where p_α are polynomials of degree k (instantons number) in $\bar{z}=x-iy$ which for different α have no common factor

$$p_\alpha(\bar{z}) = c_\alpha \prod_{j=1}^{k} (\bar{z} - a^j_\alpha) \tag{31}$$

$c_n = 1$ (without loss of generality)

Introducing $D_\pm$ instead of D_μ and using

$$Q = k = \frac{1}{4\pi} \int d^2x (\overline{D_+\phi} D_+\phi - \overline{D_-\phi} D_-\phi)$$

we obtain:

$$S = \frac{1}{2} \int d^2x\, \overline{D_-\phi} D_-\phi + 2\pi k \tag{32}$$

Decomposing ϕ into the classical instanton and the fluctuating part η

$$\phi = (1 - |\eta|^2)\phi^c + \eta\,, \quad D_-\phi^c = 0\,, \quad \phi\eta = 0 \tag{33}$$

we obtain:

$$D_-\phi_\alpha = (\delta_{\alpha\beta} - \phi^c_\alpha \bar{\phi}^c_\beta)\, D^c_-\eta_\beta + O(|\eta|^2) \tag{34}$$

We now define an auxilary Dirac like problem

$$i\not{D}_{\alpha\beta}\hat{\eta}_\beta = (\delta_{\alpha\beta} - \phi_\alpha \bar{\phi}^c_\beta) \begin{pmatrix} 0 & -D_- \\ D_+ & 0 \end{pmatrix} \begin{pmatrix} \hat{\eta}^\beta_- \\ \hat{\eta}^\beta_+ \end{pmatrix} \tag{35}$$

and the subsidary condition: $\bar{\phi}^c\hat{\eta} = 0$. The differential operator commutes with this condition. The $A_\pm$ potential in this Dirac operator is

$$iA_\mp = \pm\, \partial_\mp \ln|\rho|^2, \quad A_\mu = -\, \varepsilon_{\mu\nu}\partial_\nu \ln|\rho|^2 \tag{36}$$

Let us forget for a moment that the function ϕ^c which defines the subsidary condition is related to the vector potential A_μ

in $i\not{D}$. Then we are dealing with a (generic) Dirac-problem in which the spinors have a U(n) flavour index which is constrained by the function ϕ^c. Instead of computing the determinant of fluctuations perpendicular to ϕ^c we may consider the unconstrained determinant divided by the determinant of the Dirac operator in ϕ^c direction [26]. This operator is of the form

$$i\tilde{\not{D}} = i\gamma_\mu(\partial_\mu - i\tilde{A}_\mu) \tag{37}$$

$$\tilde{A}_\mu = A_\mu + (A_\mu - i\bar{\phi}^c\partial_\mu\phi^c)$$

The fluctuation determinant is therefore

$$\Gamma_{CP^{n-1}} = (n-1)\Gamma_s - \Gamma_s(A_\mu - i\bar{\phi}^c\partial_\mu\phi^c) + \Delta\Gamma + \text{const} \tag{38}$$

The zero mode contribution $\Delta\Gamma$ is again a trace ln of an overlap matrix formed from those $\hat{\eta}^{(\ell)}_{o\alpha}$ which are perpendicular to ϕ^c. For the actual (non-generic) case the second term vanishes since:

$$A_\mu = i\bar{\phi}^c\partial_\mu\phi^c \tag{39}$$

The basic CP^{n-1} fields are complex bose fields; the determinant for the fluctuations is

$$\det_{CP^{n-1}} = e^{\Gamma}$$

Berg und Lüscher [25] have shown how this determinant can be used for the computation of an (infrared-finite) dense instanton gas.

Finnally we would like to comment on attempts to compute fermion determinants without using the R_c^2 formulation. Patrascious [27] tried to obtain the determinant by starting with a quantization box of radius R. However the Dirac equation is unfortunately not leading to a selfadjoint problem if we choose

$$\psi(R) = 0$$

i.e. Dirichlet boundary conditions. One could of course try to find other consistent local boundary conditions (e.g. the MIT bag boundary conditions). But according to a theorem of Atiyah and Bott [28] local boundary conditions become incompatible if there is nontrivial topology in the A_μ ; they lead to "topological obstructions".

A useful non local boundary condition which lives in complete harmony with topology is the so called "spectral" boundary condition [29]. One writes the Dirac operator in polar coordinates

$$i\not{D} = -i\gamma_r(-\partial_r + B) \tag{40}$$

$$\gamma_r = \begin{pmatrix} 0 & ie^{-i\phi} \\ -ie^{i\phi} & 0 \end{pmatrix}, \quad B = \frac{i}{r}\begin{pmatrix} -1 & 0 \\ 0 & 1 \end{pmatrix}\left(\partial_\phi - iA_\phi\right) \tag{41}$$

For example, the radially symmetric A_μ

$$A_\mu = -\frac{\alpha}{1+x^2}\,\varepsilon_{\mu\nu}x_\nu$$

leads to

$$A_\phi = \frac{\alpha r^2}{1+x^2} \tag{42}$$

One then proceeds to compute the spectrum of the "little" Dirac operator on the boundary circle

$$B\psi_\ell = \lambda_\ell\psi_\ell \tag{43}$$

It is very important to use anticyclic boundary conditions:

$$\psi(2\pi) = -\psi(o) \tag{44}$$

The deeper reason for this is the observation of Milnor [30] that the "spin-structure" obtained from restricting the topologically spin-structure on the disk to the circle becomes nontrivial i.e. it is of the Moebius type. In contrast to even dimensional Dirac equations which have a spectrum symmetric around the origin (γ_5 invariance), the odd dimensional equation leads to an asymmetric spectrum. A measure for its asymmetry is the famous η invariant

$$\eta(s) = \Sigma\ \mathrm{sgn}\lambda_i\ |\lambda_i|^{-s} \tag{45}$$

In our special example the spectrum (for the upper component) is

$$\lambda_n = n + \frac{1}{2} - \frac{\alpha R^2}{1+R^2} \tag{46}$$

the $\frac{1}{2}$ originates from the antiperiodicity. For this spectrum

$\eta(o)$ is easily computed to be

$$\eta(o) = 1 - 2\left(\frac{1}{2} - \frac{\alpha R^2}{1+R^2}\right) \tag{47}$$

here we assumed

$$0 < \frac{1}{2} - \frac{\alpha R^2}{1+R^2} < 1 \tag{48}$$

Note that the little Dirac equation develops zero modes if $\alpha R^2/(1+R^2)$ = half integer. Let h denote the number of zero modes ($h=o$ or 1 in our model). The spectral boundary condition for the big Dirac equations in the disk is:

$$\psi_+|_R \in \mathcal{H}_{<o}\,, \quad \psi_-|_R \in \mathcal{H}_{\geqslant o}$$

$\mathcal{H}_{<o}$: negative energy space of little Dirac operator

$\mathcal{H}_{\geqslant o}$: non-negative energy space of little Dirac operator

$\psi_{\pm}$ are positive (negative) chirality solutions.

It is easy to check that this boundary condition leads to self-adjointness

$$\int_{disc} \psi^+ \, i\not{D}\phi = \int_{disc} (i\not{D}\psi)^+\phi$$

The index of the boundary value problem may now be expressed in terms of the η invariant as

$$\text{index } i\not{D} = \frac{1}{4\pi}\int_{disc} \varepsilon_{\mu\nu} F_{\mu\nu} - \frac{\eta(o)+h}{2} \tag{49}$$

In our case $h = o$,

$$\frac{1}{4\pi}\int_{disc} \varepsilon_{\mu\nu} F_{\mu\nu} = \frac{\alpha R^2}{1+R^2}$$

and we therefore obtain:

$$\text{index } i\not{D} = \frac{\alpha R^2}{1+R^2} - \frac{\alpha R^2}{1+R^2} = 0 \tag{50}$$

Suppose now that α increases so that the inequality (48) will be violated on the lefthand side say

$$-1 < \frac{1}{2} - \frac{R^2}{1+R^2} \leqslant 0 \tag{51}$$

In case of the equality $\eta(o) = o$, $h = 1$ and therefore the index still vanishes

$$\text{index } i\not{D} = \frac{1}{2} - \frac{1}{2} = 0 \tag{52}$$

If in (51) the inequality holds, we obtain for $\eta(o)$ and h :

$$\eta(o) = -1 + 2\left|\frac{1}{2} - \frac{\alpha R^2}{1+R^2}\right| = -1 - 2\left(\frac{1}{2} - \frac{\alpha R^2}{1+R^2}\right)$$

$$h = 0$$

and hence:

$$\text{index } i\not{D} = 1 \tag{53}$$

Therefore, as the Bohm-Aharanov flux inside the disc surpasses the value $\frac{1}{2}$

$$\frac{1}{4\pi}\int_D \varepsilon_{\mu\nu} F_{\mu\nu} > \frac{1}{2}$$

the index jumps from zero to one which is a consequence of the jump in $\frac{\eta+h}{2}$.

One easily checks that a further increase in α will not change this index as long as

$$\frac{\alpha R^2}{1+R^2} \leqslant \frac{3}{2}$$

Passing through $^3/2$ will lead to the next jump, increasing the index to 2. Everytime the flux goes through a halfinteger value, the index will increase by one unit. This may be seen directly from the spectral boundary conditions. The formal solution for the homogeneous Dirac equation are:

$$\begin{pmatrix} z^1 (1+r^2)^{-\alpha/2} \\ 0 \end{pmatrix} , \quad \begin{pmatrix} 0 \\ \bar{z}^1 (1+r^2)^{\frac{\alpha}{2}} \end{pmatrix} \tag{54}$$

In order to restrict them to the circle, we have to adjust the frame, i.e. perform a rotation by ϕ which means multiplication with

$$e^{i\gamma_5 \frac{\phi}{2}}$$

Clearly we will obtain eigenstates of the little Dirac operator. The negative chirality eigenstates will never lie in the nonnegative part of the "little" spectrum. The positive chirality eigenstates will be admissable if the flux $\alpha R^2/(1+R^2)$ surpasses half-integer values: first the state with ℓ = o for

$\alpha R^2/(1+R^2) > \frac{1}{2}$, then the second state ℓ = 1 for

$\alpha R^2/(1+R^2) > \frac{3}{2}$ etc.

In computing the "Schwinger part" Γ_S of the determinant one must interprete the fermion propagators in  as being those of the spectral boundary value problems.

Note that only if the total flux is integer one is able to get rid of the spherical compactification. We expect that for fractional (rational) fluxes one may get rid of the boundary only at the expense of obtaining a more complicated manifold (many-sheeted Riemann sphere?)

The general message to be learned from this simple excercise is that usual (local) "box"-boundary conditions are not always physical. In certain cases involving Dirac spinors they are plainly inconsistent.

Box boundary conditions have been applied to study fluctuation problems in supersymmetric theories [31]. They lead to a contradiction of results established with other methods [32]. Those investigations ought to be rechecked in the light of our remarks [33].

III. TWO-DIMENSIONAL QED

1. Some topological aspects reconsidered

In this lecture we shall illustrate the general results obtained in the previous lectures for QED_2 and examine their consequences.

Because of the two-dimensionality of space-time, the Coulomb gauge is pathological. As was shown by Lowenstein and Swieca [34] the Coulomb gauge correlation functions do not exist as tempered distributions. Hence following these authors and J. Schwinger [35] we shall discuss QED_2 in the Lorentz gauge. Following the by now standard "gauge fixing" procedure of Fadeev and Popov we then have for the euclidean generating functional of fermionic correlation functions,

$$Z[\bar{\eta},\eta] = \mathcal{N}^{-1} \int d[A_T] d[\psi] d[\bar{\psi}] e^{-S_0[A] - \int d^2z \bar{\psi} i \not{D} \psi} \times e^{\int d^2z (\bar{\eta}\psi + \bar{\psi}\eta)} \tag{1a}$$

where $d[A_T]$ is short hand for

$$d[A]\ \delta(\partial_\mu A_\mu)$$

D_μ is the usual covariant derivative,

$$D_\mu = \partial_\mu - ieA_\mu \tag{1b}$$

and S_0 is the (euclidean) action associated with the free electromagnetic field:

$$S_0[A] = \tfrac{1}{4} \int d^2z\ F_{\mu\nu}(z) F_{\mu\nu}(z) \tag{1c}$$

$$F_{\mu\nu} = \partial_\mu A_\nu - \partial_\nu A_\mu \tag{1d}$$

η and $\bar{\eta}$ are Grassman valued external sources. $\mathcal{N}$ is the usual normalization constant $\mathcal{N} = \langle 0|0\rangle$.
As is well known, there are not Fadeev-Popov ghosts in this gauge.

Expanding the integrand of (1) in powers of the coupling constant we generate the familiar perturbation expansion for the euclidean correlation functions upon differentiating with respect to the external sources, and setting these sources equal to zero after differentiation. This is nothing new. There exists however a feature of the model which would not show up in such an expansion and is thus of non-perturbative nature: the existence of an infinite set of degenerate ground states. These ground states are known to carry a chiral and fermion-number selection rule [34]. Thus tunneling is expected for operators carrying chirality and fermion number. Because of the well known [36] connection between this vacuum structure and topology, we expect a non-trivial topology of the gauge-field configuration to be responsible for these non-perturbative effects. The detailed study of these effects is the objective of this lecture; their topological origin provides the link with the **preceding** lectures.

The clue to these non-perturbative effects is the existence of zero modes of the euclidean Dirac equation. Their number $N = n_+ + n_-$ is related to the Chern number n of the gauge field via the Atiyah-Singer index theorem [29]

$$n_+ - n_- = n \qquad (2)$$

where $n_\pm$ is the dimensionality of the null-space of $\not{D}$ of positive and negative chirality, respectively, and

$$\int \frac{d^2x}{4\pi} e\, \varepsilon_{\mu\nu} F_{\mu\nu}(x) = n \qquad (3)$$

Relation (4) follows from the U(1)-version of eq. (I.67b),

$$\partial_\mu j_{5\mu}(x) = -\frac{e}{2\pi} \varepsilon_{\mu\nu} F_{\mu\nu}(x) + 2\left(n_+(x) - n_-(x)\right) \qquad (4)$$

upon integrating over all space and noting that the current (I.57)

$$j_\mu(x) = \sum_i{}' \frac{\bar{u}_i(x)\gamma_\mu u_i(x)}{\lambda_i}$$

tends to zero asymptotically. Relation (2) so far only states that the Dirac operator (1b) in an external field with integral Chern number will have zero modes; it does not say how many.

Let us pause for a moment and reflect on another crucial ingredient which leads to eq.(2); it is the assumption that the Dirac operator $\not{D}$ has a one-point compactification. This presupposes of course, that the field configuration in question "lives" on the corresponding compact manifold $\mathfrak{M}$. A necessary requirement for this to be true is, that

$$A_\mu(x) \xrightarrow[x\to\infty]{} g(x)\partial_\mu g^{-1}(x) \;, \quad g(x) \in U(1)_G \qquad (5)$$

Let us illustrate by a simple example, why this is important. We choose $\mathfrak{M}$ to be the manifold obtained by stereographic projection of R_2 onto a sphere of radius R . This corresponds to the choice (I.5) for the metric.

Consider the field configuration (vortex)

$$A_\mu^{[n]}(x) = \frac{n}{e} \frac{1}{R^2+x^2} \varepsilon_{\lambda\mu} x_\lambda \tag{6}$$

It has the property

$$A_\mu^{[n]}(x) \xrightarrow[x\to\infty]{} \frac{n}{e} \varepsilon_{\lambda\mu} \frac{x_\lambda}{x^2} = -\frac{1}{e} \partial_\mu \Lambda(x)$$

with

$$\Lambda(x) = n\,\text{arc ctn}\, \frac{x_2}{x_1}$$

Identify euclidean infinity with a circle of radius $r \to \infty$. Parametrize $x_{1,2}$ on this circle by r and an angle θ:

$$x_1 = r \sin\theta \;, \quad x_2 = r\cos\theta$$

Then

$$\Lambda(r,\theta) = n\,\theta$$

Hence for n integer, A_μ fullfills the condition (5); in fact, it corresponds to the Chern number n:

$$e\int \frac{d^2x}{2\pi} \varepsilon_{\lambda\mu}\partial_\lambda A_\mu^{[n]}(x) = n\int_{S_R^1} \frac{d\Omega}{2\pi} = n$$

Let us violate condition (5) by replacing n by a non-integer constant ν:

$$A_\mu(x) = \frac{\nu}{e} \frac{1}{R^2+x^2} \varepsilon_{\lambda\mu} x_\lambda$$

The corresponding field on the stereographic sphere of radius R is given by [13]

$$\hat{A}_\mu = \frac{1}{2} \left(\frac{R^2+x^2}{R^2}\right) A_\mu - \frac{x_\mu x_\nu}{R^2} A_\nu = \frac{\nu}{2e} \varepsilon_{\lambda\mu} \frac{x_\lambda}{R^2} \;, \quad \mu = 1,2$$

$$\hat{A}_3 = -\frac{x_\mu}{R} A_\mu$$

Let us rewrite $\hat{A}_\mu$ in terms of the coordinates on that sphere by identifying the northpole with euclidean infinity:

$$x_1 = \frac{R \sin\Theta \cos\phi}{1 - \cos\Theta} \quad , \quad x_2 = \frac{R \sin\Theta \sin\phi}{1 - \cos\Theta}$$

Defining a unit vector tangent to the stereographic sphere, pointing in the azimuthal direction

$$\underset{\sim}{e}_\phi = (-\sin\phi, \cos\phi)$$

we have then,

$$\underset{\sim}{\hat{A}} = \frac{\nu}{2eR} \frac{\sin\Theta}{1-\cos\Theta} \underset{\sim}{e}_\phi \quad , \quad \hat{A}_3 = 0$$

We see that $\hat{A}_\mu$ only exists in a patch excluding the north pole $\theta = 0$. We could however equally well map infinity onto the southpole. In that case we obtain

$$\underset{\sim}{\hat{A}}' = \frac{-\nu}{2eR} \frac{\sin\Theta}{1+\cos\Theta} \underset{\sim}{e}_\phi \quad , \quad \hat{A}_3' = 0$$

$\hat{A}_\mu'$ lives on the stereographic sphere excluding the south pole. $\hat{A}_\mu$ and A_μ thus live on two different patches defined by

(I) $\quad 0 < \Theta < 2\pi$

(II) $\quad -\pi < \Theta < \pi$

In the regions of overlap they are related by

$$\underset{\sim}{\hat{A}}' = \underset{\sim}{\hat{A}} + \frac{\nu}{eR \sin\Theta} \hat{e}_\phi \tag{7}$$

We now recall that the gradient operator has the following form:

$$\underset{\sim}{\nabla} \equiv \underset{\sim}{e}_r \frac{\partial}{\partial r} + \underset{\sim}{e}_\Theta \frac{1}{r}\frac{\partial}{\partial\Theta} + \underset{\sim}{e}_\phi \frac{1}{r\sin\Theta}\frac{\partial}{\partial\phi}$$

Hence we may write (7) in the form

$$\underset{\sim}{\hat{A}}' = \underset{\sim}{\hat{A}} + \frac{1}{e}\underset{\sim}{\nabla}\Lambda$$

with

$$\Lambda = \nu\phi$$

$\hat{A}_\mu'$ and $\hat{A}_\mu$ thus differ by a bonafide gauge transformation in the overlap region if, and only if $\nu = n =$ integer; if this is the case, the corresponding gauge invariant quantities exist globally on the stereographic sphere.

After this brief digression, which was supposed to illustrate the importance of integral winding in our considerations, we shall now sharpen the affirmation (4) by stating a

Vanishing Theorem for QED_2

Affirmation:

For an external field configuration $A_\mu^{[\pm n]}$ with winding $\pm n$ there exist precisely n zero modes of chirality ± 1 respectively.

We prove this theorem by explicitely constructing the "zero-energy" eigenfunctions of eq.(I.7) For this purpose it is sufficient to construct any complete set of such eigenfunctions of $\not{D}$; they need not be orthogonal. One readily verifies that

$$\psi_{\ell_+}^{(o)}(x) = \left(\frac{1}{2\pi R}\right)^{\frac{1}{2}} \left(\frac{x_1+ix_2}{R}\right)^{l-1} \begin{pmatrix} h_-(x) \\ o \end{pmatrix} \tag{8a}$$

$$\psi_{\ell_-}^{(o)}(x) = \left(\frac{1}{2\pi R}\right)^{\frac{1}{2}} \left(\frac{x_1-ix_2}{R}\right)^{l-1} \begin{pmatrix} o \\ h_+(x) \end{pmatrix} \tag{8b}$$

with

$$h_{\mp}(x) = e^{ie\int d^2z D(x-z)\left(\partial_\mu A_\mu(z) \pm i\,\varepsilon_{\mu\nu}\partial_\mu A_\nu(z)\right)}$$

$$= e^{ie\int d^2z D(x-z)\partial_{\mp} A_{\pm}(z)} \tag{9}$$

$$\Box D(z) = -\delta^2(z)\ , \quad D(z) = -\frac{1}{4\pi}\ln\mu^2 z^2 \tag{10}$$

are all solutions of the Dirac equation (7) for a zero eigenvalue. For $x \to \infty$ we have

$$\psi_{l_\pm}^{(o)}(x) \xrightarrow[x \to \infty]{} \left(\frac{1}{2\pi R}\right)^{\frac{1}{2}} \left(\frac{1}{R}\right)^{l-1} e^{\pm i\Theta} (\mu|x|)^{l-1\mp n} \tag{11}$$

where $\Theta = \text{arc ctn}\,\frac{x_2}{x_1}$ and n is the "winding number" of the configuration in question. For $n > o$ ($n < o$) there exist therefore precisely $|n|$ zero-modes with positive (negative) chirality which are normalizable with respect to measure (5). This proves our assertion.

2. THE NEW FEYNMAN RULES

The zero mode problem is a non-perturbative phenomenon. It leads to a modification of the Matthews-Salam rules whenever winding configurations are involved. To obtain them consider the generating functional (1a) with the A_μ-integration restricted to field configurations $A_\mu^{[n]}$ with Chern number n . We denote the corresponding generating functional by $Z^{[n]}[\eta,\bar{\eta}]$.

Expand ψ and $\bar{\psi}$ in terms of the orthonormalized eingenfunctions (I.13) with respect to the measure (5):

$$\psi_\alpha(x) = \sum_i b_i u_i(x)_\alpha \,, \quad \bar{\psi}_\alpha(x) = \sum_i \bar{b}_i u_i^*(x)_\alpha \tag{12}$$

$$i \not{D} u_i(x) = \frac{2R^2}{R^2+x^2} E_i \, u_i(x) \tag{13}$$

$$\int d^2x \, \frac{2R^2}{R^2+x^2} \, \bar{u}_i(x) \, u_j(x) = R \, \delta_{ij} \tag{14}$$

The completeness of the u_i's follows from the observation that there exists a one-to-one correspondence with the eigenfunctions $\hat{u}_i$ in (I.25) of the Dirac operator $\hat{\not{D}}$ on the stereographic sphere of radius R . Note that as the result of the orthonormalization, the eigenfunctions $u_i(x)$ will depend non-trivially on R. Note also that in euclidean space

$$\bar{u}_i = u_i^\dagger, \quad \bar{\psi} = \psi^\dagger \ , \text{ etc.}$$

Substituting (12) into (1a) and using (14) we obtain

$$Z^{[n]}[\eta,\bar{\eta}] = \mathcal{N}^{-1} \int d[A_T^{[n]}] \left(\prod_j db_j(o) d\bar{b}_j(o) \right) \left(\prod_i{}' \, db_i d\bar{b}_i \right) \times$$

$$\times \, e^{-S_0[A] - \sum_i{}' \bar{b}_i b_i \lambda_i^2} \tag{15}$$

$$\times \, e^{\sum_i (\bar{b}_i u_i[\bar{\eta}] + u_i[\eta] \, b_i)}$$

where

$$\lambda_i \equiv R \, E_i \tag{16}$$

and

$$\bar{u}_i[\eta] = \int d^2z \bar{u}_i(z)\eta(z), \quad u_i[\bar{\eta}] = \int d^2z \bar{\eta}(z) u_i(z) \tag{17}$$

Here the substitution

$$d[\psi]d[\bar{\psi}] = \prod_i db_i d\bar{b}_i$$

was made and the zero mode contributions have been explicitely separated out. Using the standard Grassman rules for fermion integration

$$\int d\bar{b}db \begin{pmatrix} 1 \\ b \\ \bar{b} \\ \bar{b}b \end{pmatrix} = \begin{pmatrix} 0 \\ 0 \\ 0 \\ 1 \end{pmatrix}, \quad b^2 = \bar{b}^2 = 1 \tag{18}$$

we obtain, after dividing through by det $i\hat{\partial}$, the familiar result

$$Z^{[n]}[\eta,\bar{\eta}] = \mathcal{N}\int d[A_T^{[n]}]\prod d\bar{b}_i^{(0)} db_i^{(0)} e^{\sum_i \left(\bar{b}_i^{(0)} u_i^{(0)}[\bar{\eta}] + u_i^{(0)}[\eta] b_i^{(0)}\right)}$$

$$\times\ e^{-S_o[A]-\Gamma[A]_e - \iint d^2z d^2z' \bar{\eta}(z) G'(z,z';A)\eta(z')} \tag{19}$$

where $e^{-\Gamma}$ is as defined in section I

$$\exp(-\Gamma) = \prod_i{}' \lambda_i^2[A]/\prod_i \lambda_i^2[0]$$

and $G'(z,z';A)$ is the "pseudo Green's function" (I.16)

$$G'(x,y;A) = \Sigma' \frac{u_i(x)u_i^\dagger(y)}{\lambda_i} \tag{20}$$

satisfying eq. (I.15). The $u_i^{(0)}[\eta]$ are defined as in (17) with $u_i(x) \longrightarrow u_i^{(0)}(x)$.

Note that the normalizability of the u_i's on the two-dimensional stereographic sphere implies, that $G'(x,y;A)$ goes to zero asymptotically as x and (or) y go to infinity. The remaining integration over the zero-mode degrees of freedom can also be performed using the Grassman rules (18):

$$Z^{[n]}[\eta,\bar{\eta}] = \mathcal{N}\int d[A_T^{[n]}] W_n[A;\eta,\bar{\eta}]\ e^{-S_o[A]}\ e^{-\Gamma[A]}$$

$$\times\ e^{-\iint d^2z d^2z' \bar{\eta}(z) G'(z,z';A)\eta(z')} \tag{21}$$

where $W_n[A;\eta,\bar\eta]$ is the "generating wave functional"

$$W_n = \frac{1}{(n!)^2}\left(\sum_i \varepsilon_{i_1\ldots i_n} u^{(o)}_{i_1}[\bar\eta]\ldots u^{(o)}_{i_n}[\bar\eta]\right)$$

$$\left(\sum_j \varepsilon_{j_1\ldots j_n} \bar u^{(o)}_{j_1}[\eta]\ldots \bar u^{(o)}_{j_n}[\eta]\right) \tag{22}$$

In the absence of zeromodes (trivial winding) we recover the familiar Matthews-Salam rules: $W_n[A;\eta,\bar\eta]$ in (21) is to be replaced by one, G' is replaced by the true external field Greens-function G and Γ has the usual graphical interpretation in terms of one fermion-loop graphs. In particular for QED_2, it is entirely given in terms of the lowest order fermion loop, eq.(9) all higher order one loop graphs being zero.

By integrating the anomaly equation (4), one obtains the U(1)-version of the result (II, 20,14)

$$\Gamma[A] = \Gamma_s[A] + \mathrm{tr}\,\ell n\, N[A] + c \tag{23}$$

where $\Gamma_s[A]$ is the Schwinger result (II.9) and $N[A]$ is the matrix defined by

$$N = \begin{pmatrix} N_+ & 0 \\ 0 & N_- \end{pmatrix} \tag{24a}$$

with

$$\left(N_{\pm}^{-1}\right)_{\ell'\ell} \equiv \int d^2x \,\frac{2R}{R^2+x^2}\, \psi^{(o)}_{\ell'\pm}(x)\, \psi^{(o)+}_{\ell\pm}(x) \tag{24b}$$

Because of the vanishing theorem, $N_+(N_-)$ is zero for configurations with negative (positive) chirality.

Several comments concerning the result (23) are in order:

a) The normalization matrix (24) is unique only up to a multiplicative constant reflecting our freedom in choosing the normalization of the zero-energy wavefunctions (8). This arbitrariness can however be absorbed into the integration constant c.

b) The integration constant c can only depend on the Chern-class to which A_μ belongs, and not on a detailed representative within this class. Hence we may calculate it by explicetely computing $\Pi\lambda_i^2$ for a conveniently chosen representative within each class. This has been done in ref. [3] for the vortex (6). It involves computation of the eigenvalues λ_i and calculation of the infinite product $\Pi\lambda_i^2$ using some (ζ-function, Pauli-Villars etc.) regularization. As mentioned already in sect. I. the ambiguity thus introduced by the choice of regularization will be of the form

$$[\zeta_A(o) - \zeta_o(o)]\ln M^2 = -n\ln M^2$$

which is independent of the details of the configuration. Comparing the result of this computation with the one obtained from (23) for the vortices (6) one finds

$$c = -n^2\ln\mu R - n\ln M^2 \tag{25}$$

c thus remains an arbitrary constant since μ and M are arbitrary. However,

c) as the result of a remarkable cancellation to be discussed in sect. III.4, the correlation functions are found to be unique up to the multiplicative constant

$$Z_n = \mathcal{N}^{-1}\left(\frac{M}{\mu R}\right)^{2n}$$

(which also includes the dependence of Γ_S and the wave function on the infrared regulator mass μ).The remaining arbitrariness is entirely compatible with the superrenormalizability of QED_2 since it merely reflects the fact that the normalization of the infinite set of degenerate ground states is not specified within our functional approach. Z_n may be determined by observing that the infinite vacuum degeneracy implies a breakdown of the linked-cluster property:

$$<o|Q^{[-2n]}(\xi_1\ldots\xi_n)\psi_{\alpha_1}(x_1)\ldots\psi_{\alpha_n}(x_n)\bar{\psi}_{\beta_1}(y_1)\ldots\bar{\psi}_{\beta_n}(y_n)|o>$$

$$\xrightarrow[\substack{\xi_i\to\infty \\ \xi_i - \xi_j \text{ finite}}]{}$$

$$<o|Q^{[-2n]}(\xi_1\ldots\xi n)|n><n|\psi_{\alpha_1}(x_1)\ldots\psi_{\alpha_n}(x_n)\bar{\psi}_{\beta_1}(y_1)\ldots\bar{\psi}_{\beta_n}(y_n)|o>$$

Here care should be taken that $Q^{[-2n]}$ is a gauge invariant operator in order to avoid the excitation of unphysical, zero-norm states. This cluster violation has been used [37] to derive Feynman-path representations for the tunneling correlation functions in QED_2. The details of carrying out the limit are much too involved to be presented here. The functional representation for the tunneling amplitudes obtained in the way are identical with those calculated from the generating functional if we choose $\frac{M}{\mu R} = 1$.

3. CALCULATION OF $G'(x,y;A)$

Making use of the completeness relation

$$\sum_i u_i(x)_\alpha u_i^+(y)_\beta = \frac{R^2+x^2}{2R} \delta^2(x-y)\delta_{\alpha\beta}$$

we see that $G'(x,y;A)$, defined by (20), satisfies the equation

$$i\not{D}G'(x,y;A) = \delta^2(x-y) - \frac{2R}{R^2+x^2} \sum_i u_i^{(o)}(x) u_i^{(o)+}(y) \qquad (26)$$

where the second term is just the projector P_O on the zero modes introduced earlier. It is convenient to write,

$$G' = G + g \qquad (27)$$

where

$$i\not{D}G(x,y;A) = -\delta^2(x-y) \qquad (28)$$

Then

$$i\not{D}g(x,y;A) = -\frac{2R}{R^2+x^2} \sum_i u_i^{(o)}(x) u_i^{(o)}(y)^+ \qquad (29)$$

The solution to (28) is of the form (II.11) with $h_{\pm}(x)$ given by (9).

With the aid of G we now easily construct the solution to eq. (29):

$$g(x,y;A) = g_o(x,y;A) - \sum_i \int d^2z \frac{2R}{R^2+z^2} G(x,z;A)\, u_i^{(o)}(z) u_i^{(o)}(y) \qquad (30)$$

where g_O is a solution to the homogeneous Dirac equation:

$$i\not{D}g_o(x,y;A) = 0$$

Note that G' carries the same chiral selection rules as G_O. g_O is chosen such that $G'(x,y;A)$ satisfies the required "boundary condition" of falling off at infinity in both variables, x and y.

Evaluation of the right hand side involves the orthonormal zero modes $u_i^{(o)}$. It is thus clear that $G'(x,y;A)$ unlike $G(x,y;A)$ cannot be represented in a simple closed form for generic field configurations. It is nevertheless instructive to compare G' with G for the particular vortex (6) with n=1, where all integrations may be performed. Choosing R=1,

$$A_\mu^{[1]} = \frac{1}{e}\frac{1}{1+x^2}\varepsilon_{\lambda\mu}x_\lambda$$

we obtain from (II.11a) and (9)

$$G(x,y;A^{[1]}) = -\frac{1}{2\pi}\begin{pmatrix} 0 & \left(\frac{1+y^2}{1+x^2}\right)^{\frac{1}{2}} \frac{x_- - y_-}{(x-y)^2} \\ -\left(\frac{1+x^2}{1+y^2}\right)^{\frac{1}{2}} \frac{x_+ - y_+}{(x-y)^2} & 0 \end{pmatrix} \tag{31}$$

with $x_\pm = x_1 \pm ix_2$. Thus $G(x,y;A^{[1]})$ does not fall off asymptotically in x and y, as expected. For n=1 we have only one normalizable zero mode. From (8) we obtain, after normalization,

$$u^{(o)}(x) = \frac{1}{\sqrt{2\pi}}\begin{pmatrix} \frac{1}{\sqrt{1+x^2}} \\ 0 \end{pmatrix}$$

corresponding to positive chirality. Hence eqn.(29) becomes

$$i\not{\partial}g(x,y;A^{[1]}) = -\frac{1}{\pi}\begin{pmatrix} \frac{1}{(1+x^2)^{\frac{3}{2}}(1+y^2)^{\frac{1}{2}}} & 0 \\ 0 & 0 \end{pmatrix}$$

Choosing

$$g_o(x,y;A^{[1]}) = -\frac{1}{2\pi}\begin{pmatrix} 0 & \frac{y_-}{(1+x^2)^{\frac{1}{2}}(1+y^2)^{\frac{1}{2}}} \\ 0 & 0 \end{pmatrix}$$

we obtain from (30)

$$g(x,y;A^{[1]}) = -\frac{1}{2\pi}\begin{pmatrix} 0 & \dfrac{y_-}{(1+x^2)^{\frac{1}{2}}(1+y^2)^{\frac{1}{2}}} \\ \dfrac{x_+}{(1+x^2)^{\frac{1}{2}}(1+y^2)^{\frac{1}{2}}} & 0 \end{pmatrix}$$

Combining this result with (31) in (27), we finally have

$$G'(x,y;A^{[1]}) =$$

$$= -\frac{1}{2\pi}\begin{pmatrix} 0 & \left(\dfrac{1+y^2}{1+x^2}\right)^{\frac{1}{2}}\dfrac{x_- - y_-}{(x-y)^2} + \dfrac{y_-}{(1+x^2)^{\frac{1}{2}}(1+y^2)^{\frac{1}{2}}} \\ -\left(\dfrac{1+x^2}{1+y^2}\right)^{\frac{1}{2}}\dfrac{x_+ - y_+}{(x-y)^2} + \dfrac{x_+}{(1+x^2)^{\frac{1}{2}}(1+y^2)^{\frac{1}{2}}} & 0 \end{pmatrix}$$

$G'(x,y;A^{[1]})$ is seen to fall asymptotically in both x and y, as expected.

4. A REMARKABLE CANCELLATION

Consider the "generating wave functional" $W_n[A;\eta,\bar{\eta}]$ introduced in (22). All zero modes $u_i^{(o)}$ participating in W_n carry either positive or negative chirality. Expanding the $u_i{}^{(o)}$ in terms of the $\psi_\ell{}^{(o)}$

$$u_i^{(o)}(x) = \sum_\ell a_{i\ell}\psi_\ell^{(o)}(x)$$

and noting the identity

$$\sum_i \varepsilon_{i_1 \ldots i_n} a_{i_1 \ell_1} \ldots a_{i_n \ell_n} = \varepsilon_{\ell_1 \ldots \ell_n} \det a$$

we obtain

$$W_n[A;\eta,\bar{\eta}] = \det N[A]\mathcal{W}_n[A;\eta,\bar{\eta}] \tag{32}$$

where $N[A]$ is the normalization matrix (24) and

$$\mathcal{W}_n = \frac{1}{(n!)^2}\left(\sum_i \varepsilon_{i_1 \ldots i_n}\psi_{i_1}^{(o)}[\bar{\eta}]\ldots\psi_{i_n}^{(o)}[\bar{\eta}]\right)$$

$$\left(\sum_j \varepsilon_{j_1 \ldots j_n}\bar{\psi}_{j_1}^{(o)}[\eta]\ldots\bar{\psi}_{j_n}^{(o)}[\eta]\right) \tag{33}$$

with

$$\psi_i^{(o)}[\eta] = \int d^2z\bar{\eta}(z)\psi_i^{(o)}(z)$$

$$\bar{\psi}_i^{(o)}[\eta] = \int d^2z\bar{\psi}_i^{(o)}(z)\eta(z) \tag{34}$$

Nothing that

$$\det N[A] = e^{\mathrm{trln}\, N\,[A]}$$

we see that this factor in (32) cancels the corresponding factor of $e^{-\Gamma}$ in (21), leaving us with the generating functional

$$Z^{[n]}[\eta,\bar{\eta}] = \mathcal{N}_n \int d[A_T^{[n]}]\mathcal{W}_n[A;\eta,\bar{\eta}]\, e^{-S_o[A] - \Gamma_s[A]}$$

$$\times\, e^{-\iint d^2z d^2z'\bar{\eta}(z)\ G'(z,z';A)\eta(z')} \tag{35}$$

where $\mathcal{N}_n = \mathcal{N}^{-1}e^{-c}$ with c given by (25). Note that this cancellation is not model specific. For QED_2 it means that correlation functions involving functional integration over winding fields are again given in terms of Gaussian integrals, and hence are exactly computable. This was expected since the exact operator solutions to the model are completely known.

5. THE NOTION OF INDUCED ACTION

In this section we introduce a notion, which will play a crucial role in our later considerations.

In the previous section we made the important observation that the non-polynomial and non-local functional dependence of Γ introduced through the zero modes actually cancelled in the generating functional (21), leaving us with (35).

From (35) we obtain the 2n-point correlation function of chirality 2n,

$$<n|\psi_1(x_1)\ldots\psi_1(x_n)\bar{\psi}_1(y_1)\ldots\bar{\psi}_1(y_n)|0> =$$

$$= \mathcal{N}_n\int d[A_T^{[n]}]\,\mathcal{W}_n[A;x_1\ldots x_n;y_1\ldots y_n]\; e^{-S_o[A]\;\Gamma_s[A]} \qquad (36)$$

where

$$\mathcal{W}_n[A;x_1\ldots x_n;y_1\ldots y_n] = \left\{\sum_\ell \varepsilon_{\ell_1\ldots\ell_n}\psi^{(o)}_{\ell_{1_+}}(x_{\ell_1})\ldots\psi^{(o)}_{\ell_{n_+}}(x_{\ell_n})\right\}$$

$$\left\{\sum_\ell \varepsilon_{\ell_1\ldots\ell_n}\bar{\psi}^{(o)}_{\ell_{1_+}}(y_{\ell_1})\ldots\bar{\psi}^{(o)}_{\ell_{n_+}}(y_{\ell_n})\right\} \qquad (37)$$

Replacing the zero-energy wave functions $\psi^{(o)}_{\ell_+}$ by (8a), we see that $\mathcal{W}_n$ has the structure

$$\mathcal{W}_n[A^{[n]};x_1\ldots y_n] = \left(\frac{1}{2\pi R}\right)^n \mathcal{L}_n(x_1\ldots y_n)\; e^{-\Gamma_{source}} \qquad (38)$$

with (in the Lorentz gauge)

$$\Gamma_{source} = e\sum_{i=1}^{n}\int d^2z\left[D(x_i-z) + D(y_i-z)\right]\varepsilon_{\mu\nu}\partial_\mu A_\nu(z) \qquad (39)$$

and $\mathcal{L}_n$ an "angular momentum" factor which can conveniently be written in the form

$$\mathcal{L}_n(x_1\ldots y_n) = \prod_{j<k}\left(\frac{x_j-x_k}{R}\right)_+\left(\frac{y_j-y_k}{R}\right)_- \qquad (40)$$

Let us look for "universal"- that is correlation function independent- instanton configurations. They correspond to solutions of

$$\frac{\delta S_o[A]}{\delta A_\mu(z)} = -\frac{\delta\Gamma_s[A]}{\delta A_\mu(z)} \equiv e\, j_\mu(z;A) \qquad (41)$$

In the Lorentz gauge , (9) reduces to

$$\Gamma_s[A] = \frac{e^2}{\pi}\int d^2z A_\mu(z)A_\mu(z) \qquad (42)$$

so that

$$j_\mu(z;A) = -\frac{e}{\pi} A_\mu(z) \tag{43}$$

Moreover, in this gauge

$$S_0[A] = -\int d^2z A_\mu(z) \Box A_\mu(z) \tag{44}$$

so that eqn. (41) becomes

$$\left(\Box - \frac{e^2}{\pi}\right) A_\mu(z) = 0 \tag{45}$$

This euclidean differential equation has only the trivial solution, as is most easily seen by taking its Fourier transform. Hence, unlike the case of QCD4, there exist no universal instantons in QED2. There exist however "induced" - correlation function dependent - instantons; they are induced by the wave functionals $\mathcal{W}_n$. For the correlation function (36)

$$j_\mu(z;A)_{\text{source}} = -\frac{1}{e}\frac{\delta\Gamma\text{source}}{\delta A_\mu(z)} = \sum_{i=1}^{n} \tilde{\partial}_\mu^z (D(x_i-z)+D(y_i-z))$$

Note that this induced current is independent of A_μ . Regarding it as an additional source in the effective action, eqn. (45) is replaced by

$$\left(\Box - \frac{e^2}{\pi}\right) A_\mu(z)_{cl} = -e\sum_{i=1}^{n}\tilde{\partial}_\mu^z\left(D(x_i-z) + D(y_i-z)\right) \tag{46 }$$

The solution is

$$A_\mu(z)_{cl} = \frac{\pi}{e}\sum_{i=1}^{n}\tilde{\partial}_\mu^z\left(\mathcal{D}(x_i-z) + \mathcal{D}(y_i-z)\right) \tag{47}$$

where

$$\mathcal{D}(z) = D(z) - \Delta(z;\frac{e^2}{\pi})$$

$$\left(\Box - \frac{e^2}{\pi}\right)\Delta(z;\frac{e^2}{\pi}) = -\delta^2(z) \tag{48}$$

For $z \to \infty$

$$A(z)_{c\ell} \xrightarrow[z\to\infty]{} \frac{n}{e}\varepsilon_{\nu\mu}\frac{z_\nu}{z^2}$$

corresponding to a winding number equal to n, as expected.

It is easy to see that the 2n-point correlation function (36) will vanish for configurations with winding $m \neq n$.

The reason is that
a) for $m<n$ G'- which carries the usual chiral selection rule - will come into play;
b) for $m>n$ there exist too many zero modes.

The notion of induced instantons will play a central role in our discussion of the long range behaviour of fermionic correlation functions in QCD_2. However, before turning to QCD_2 we need to make a brief digression on abelian generalization of QED_2.

6. GENERALIZATION TO THE TORUS OF SU(N)

The above considerations can be easily generalized to the maximal abelian sugroup $SU(N)_D$ - the torus of SU(N) - by making the substitutions

$$S_o[A] \longrightarrow \sum_{i_D} \int d^2z \, \frac{1}{4} F^{i_D}_{\mu\nu} F^{i_D}_{\mu\nu} \tag{49a}$$

$$i\not\partial \longrightarrow i\not\partial + g \sum_{i_D} \frac{\lambda^{i_D}}{2} \not A^{i_D} \tag{49b}$$

$$h_{\mp}(x) \longrightarrow e^{\, ig \sum \frac{\lambda^{i_D}}{2} \int d^2z D(x-z) \left(\partial_\mu A_\mu^{i_D}(z) \pm i\varepsilon_{\mu\nu} \partial_\mu A_\nu^{i_D}(z) \right)} \tag{49c}$$

where the sum only extends over the "diagonal" components. Also in the torus model there exist zero-modes with a structure similar to those of QED_2. It is thus tempting to apply once again the "machinery" developed in the preceding lectures to this case. This would lead us again to a generating functional of the form (35) with

$$\Gamma_s[A] = \frac{g^2}{16\pi} \sum_{i_D} \iint d^2z d^2z' \, \varepsilon_{\mu\lambda} F^{i_D}_{\mu\lambda}(z) D(z-z') \varepsilon_{\nu\rho} F^{i_D}_{\nu\rho}(z') \tag{50}$$

(Note that $e^2 \rightarrow g^2/2$; the factor 1/2 arises from

$$\mathrm{tr} \left(\frac{\lambda^i}{2} \frac{\lambda^j}{2} \right) = \frac{1}{2} \delta ij \, .)$$

The winding number of a given configuration will now be defined by

$$\nu_c = \sum_{i_D} \frac{\lambda_{cc}^{i_D}}{2} \int \frac{d^2z}{4\pi} \varepsilon_{\mu\nu} \partial_\mu A_\nu^{i_D}(z) \tag{51}$$

The zero-modes corresponding to winding ν_c will behave asymptotically like

$$\psi_{\ell_\pm}^{(o)}(x)^a \sim e^{\mp\left(\nu_a \mp (\ell-1)\right)\ell n(\mu|x|)} \tag{52}$$

Let us consider the expectation value:

$$\left\langle J_+^a(x) \right\rangle_A \tag{53}$$

where $J_+^a(x)$ is the chirality=2 density

$$J_+^a(x) = \bar\psi^a(x) \frac{1+\gamma_5}{2} \psi^a(x) = \bar\psi_1^a(x)\psi_1^a(x) \tag{54}$$

Let us suppose that the whole "machinery" we have developed so far is also applicable to the torus model. In that case there would be associated with the expectation value (53) the induced instantons,

$$A_\mu^{i_D}(z)_{c\ell} = \frac{2\pi}{g} \lambda_{aa}^{i_D} \tilde\partial_\mu^{\,z}\left(D(z) - \Delta(z;\frac{g^2}{2\pi})\right) \tag{55}$$

These configurations correspond to fractional winding

$$\nu_c = \sum_{i_D} \frac{\lambda_{cc}^{i_D}}{2} \lambda_{aa}^{i_D} = \left(\delta_{ca} - \frac{1}{N}\right) \tag{56}$$

so that the application of the formalism developed in the preceding lectures becomes highly questionable if no boundary is introduced. Indeed, it is easy to see that a naive application of this formalism leads to wrong results: From (52) we see that there will exist precisely one normalizable zero mode of chirality + 1 if the scalar product (15) is maintained. This is just the correct number to yield

$$\langle J_+^a(x)\rangle_{A_{c\ell}} \neq 0$$

which would imply vacuum tunneling giving rise to a non-vanishing expectation value $\langle J_+^a(x)\rangle \neq 0$ in the "true" vacuum. This is in contradiction with the result obtained from the exact operator solutions constructed in ref. [6], where the soliton operator

$$S_\alpha(x) = e^{i\sqrt{\frac{\pi}{N}}\left(\gamma^5_{\alpha\alpha}\phi(x) + \int_{x^1}^{\infty} dy^1 \partial_0 \phi\right)}$$

$$\Box \phi = 0 \tag{57}$$

carries a U(1)-chiral selection rule and prevents $J_+^a(x)$ from developing a vacuum expectation value.

On the other hand we also know from the operator solutions [6] that $J_+{}^a$ will develop a vacuum expectation value upon adding an electromagnetic interaction, as a result of "spurionization" of the soliton operator. This can also be understood on the functional level: The "wave function" (49c) is now multiplied by the U(1) factor (9) and Γ_s in (50) is replaced by

$$\Gamma_s[A] = \frac{1}{2}\int d^2z \left(\frac{g^2}{2\pi} \sum_{i_D} A_\mu^{i_D}(z) A_\mu^{i_D}(z) + N\frac{e^2}{\pi} A_\mu(z) A_\mu(z)\right) \tag{58}$$

The induced instanton associated with the electromagnetic part of the effective action is found to be

$$A_\mu(z) = \frac{2\pi}{N}\left[D(x-z) - \Delta(x-z;\frac{Ne^2}{\pi})\right] \tag{59}$$

This configuration has winding 1/N. Its contribution cancels the 1/N term in (56) and leaves us with a total winding number = 1. Correspondingly there is again precisely one positive chirality zero-mode, and our formalism, which is now applicable, tells us that J_+ develops a vacuum expectation value, in agreement with the operator results.

The same remarks apply to higher-point correlation functions of J_+. They show that the $SU(N)_D$-theory, unlike the $U(1)xSU(N)_D$ case, falls in general outside the framework established in these lectures. This should serve as a warning against a naive application of our formalism. There nevertheless exist also $SU(N)_D$-correlation functions which lead to induced instanton configurations with integral winding, and are thus treatable by our methods: this is for instance the case for correlation functions of the color-determinant of ψ_α^a,

$$\det[\psi_\alpha] = \sum_P \varepsilon a_1 \ldots a_N \, \psi_\alpha^{a_1} \ldots \psi_\alpha^{a_N} \tag{60}$$

which plays the role of the fermion field in QED_2. In particular one might consider

$$<\det[\psi_1]\ \det\ [\bar{\psi}_1]>_A \qquad (61)$$

Because of the tracelassness of the SU(N) generators, the corresponding induced instanton carries zero winding. Hence there exist no normalizable zero modes and the correlation function (61) vanishes as a result of the chiral U(1) selection rule carried by G. This agrees with the result obtained from the aforementioned operator solution [6] where the soliton operator (57) carries this selection rule. As we have remarked already this soliton operator S spurionizes if an electromagnetic interaction is added, and the correlation function (61) no longer vanishes. This is again easily understood in the functional language: The addition of an electromagnetic interaction now leads to induced instanton configurations with total winding equal to one. Hence we have presisely one normalizable zero mode for each field operator ψ^a_α in (61) and the familiar " vacuum seizing"mechanism witnessed previously is again operative.

After these cautionary remarks we now turn to a discussion of QCD_2, where the torus configurations will be shown to play a central role.

IV. TWO-DIMENSIONAL QCD

1. INTRODUCTORY REMARKS

It has been argued [38] that in QCD_2 the SU(N)-gauge group may be broken down to the maximal abelian subgroup - the torus - of SU(N). This would reflect a spontaneous breakdown of local gauge invariance, a notion which is inadmissable, by definition. This does not exclude however the possiblility, that the pure torus may play an important role in QCD_2 for some limiting situations. In fact, it has been suggested on the basis of the axial current anomaly in the operator-current $j^a_{5\mu}$ that an intrinsic Higgs mechanism similar to that of the torus-model may be responsible for giving a mass to the gauge-fields A^a_μ. This would resemble the situation encountered in the pure torus theory, and raises the question, whether torus configurations could play the leading role in some limiting situations. This presupposes, in the first place, that a consistent picture can be developed in which suitable chosen torus configurations indeed represent local extrema of the effective action. The examination of this question is the object of this lecture. As we shall see, a no-go theorem can be formulated in this respect for non-gauge invariant correlation functions.

Zero modes also exist in QCD_2; examples may explicitely be constructed in terms of torus configurations. We shall however limit ourselves to correlation functions for which no such zero-modes occur, i.e., where the usual Feynman path representation applies.

2. LONG RANGE BEHAVIOUR OF CORRELATION FUNCTIONS

Consider the QCD_2 generating functional for correlation functions involving the fermion fields. Following the procedure of Fadeev-Popov we have in the Lorentz-gauge, after fermion integration,

$$Z[\eta,\bar{\eta}] = \int d[A]\delta(\partial_\mu A^a_\mu)\Delta g[A]\; e^{-S_o[A]-\Gamma[A]}\; e^{-\iint d^2z d^2z' \bar{\eta}(z)G(z,z';A)\eta(z')} \tag{1}$$

where $G(z,z';A)$ and $\Gamma[A]$ have been given in (II.11,18,8),and $\Delta_g(A)$ is the usual Fadeev-Popov determinant

$$\Delta_g[A] = \det(\Box - gT\cdot A_\mu\partial_\mu) \;,\; T_a{}^c{}_b = f_{acb}$$

representing the contribution of the ghosts.

For pure torus configurations $\Delta_g[A]$ reduces to the A_μ-independent expression, $\Delta_g[A] = \det(\Box)$ and (1) reduces to the corresponding generating functional of the pure torus theory in this case.

In order to avoid unnecessary complications, we shall illustrate our considerations for the gauge-invariant correlation function $\langle o|J_-(x)J_+(y)|o\rangle$ where $J_\pm$ are defined as in (54)

$$J_\pm(x) = \bar{\psi}(x)\,\frac{1\pm\gamma^5}{2}\,\psi(x) \tag{2}$$

summation over the internal color degrees of freedom being understood. From (1) and (II.11) we obtain

$$\langle o|J_-(x)J_+(y)|o\rangle = \int d[A_T]\Delta_g[A]e^{-S_o[A]-\Gamma[A]}\; H(x,y)G_-(x-y)\;G_+(y-x) \tag{3}$$

where

$$H(x,y) = \mathrm{tr}\,(h_+^{-1}(x)h_-(x)h_-^{-1}(y)h_+(y)) \tag{4}$$

with (see eq. (II.8))

$$h_{\pm}(x) = 1+\sum_{n=1}^{\infty}(ig)^n\int d^2z_1\ldots d^2z_n G_{\pm}(x-z_1)G_{\pm}(z_1-z_2)\ldots G_{\pm}(z_{n-1}-z_n)$$

$$\times\ [A_{\mp}(z_1)\ldots A_{\mp}(z_n)] \qquad (5)$$

where

$$A_\mu(z) = \sum_c \frac{\lambda^c}{2} A^c_\mu(z),\ A_{\pm} = A_1 \pm iA_2$$

As in the case of QED_2, there do not exist "universal" instantons lying along the torus of SU(N). Hence we are again **led to define** an "induced"action,

$$\Gamma_{source} = -\ \ln H(x,y) \qquad (6)$$

in terms of which

$$\langle o|J_-(x)J_+(y)|o\rangle \sim \int d[A_T]\Delta_g[A]e^{-S_{ind}[A;x,y]}G_-(x-y)G_+(y-x) \qquad (7)$$

with

$$S_{ind} = S_0 + \Gamma + \Gamma_{source} \qquad (8)$$

In the spirit of semi-classical approximation, we develop $S_{ind}[A]$ around some suitably chosen classical configuration by writing

$$A^a_\mu(x) = A^a_\mu(x)_{c\ell} + \alpha^a_\mu(x)$$

Then

$$S_{ind}[A] = S_{ind}[A_{c\ell}] - \qquad (9)$$

$$- g\int d^2z\alpha^a_\mu(z)[\mathfrak{D}^{ab}_\lambda(z)_{c\ell}F^b_{\lambda\mu}(z)_{c\ell}+gj^a_\mu(z;A_{c\ell})_{ind}]+S_{fluc}[A]$$

where

$$\mathfrak{D}^{ab}_\mu = \delta^{ab}\partial_\mu + g\ f_{abc}\ A^c_\mu \qquad (10)$$

and

$$j^a_\mu(z;A)_{ind} = j^a_\mu(z;A) + j^a_\mu(z;A)_{source}$$

with $j^a_\mu(z;A)$ given by (II.3), and

$$gj^a_\mu(z;A)_{source} = \frac{1}{H(x,y)} \frac{\delta H(x,y)}{\delta A^a_\mu(z)} \tag{11}$$

Note, that it follows form (II.20) and the corresponding equations for $h^{-1}_{\pm}(x)$,

$$\tilde{\mathfrak{D}}^{ac}_\mu(z) \frac{\delta h^{-1}_{\pm}(x)}{\delta A^c_\mu(z)} = \mp \frac{g}{2} \delta^2(x-z) h^{-1}_{\pm}(x) \frac{\lambda^a}{2}$$

$$\mathfrak{D}^{ac}_\mu(z) \frac{\delta h^{-1}_{\pm}(x)}{\delta A^c_\mu(z)} = i\frac{g}{2} \delta^2(x-z) h^{-1}_{\pm}(x) \frac{\lambda^a}{2}$$

that $j^a_\mu(x,A)_{source}$ is covariantly conserved

$$\mathfrak{D}^{ac}_\mu j^c_\mu(z;A)_{source} = 0$$

We now make the crucial observation that on torus configurations, $j^a_\mu(x;A)$source, just as $j^a_\mu(x;A)$ - has components only along the torus direction. This is a consequence of the trace over the color-degrees of freedom in (3) and is therefore not true for correlation functions transforming non-trivially under global 'rotations' of SU(N). This observation is the essence of our "no-go" theorem for non-gauge invariant correlation functions. For gauge invariant ones it means that the existence of pure torus configurations corresponding to extrema of S_{ind} are a priori not excluded. Such torus configurations would have to be solutions of the equation

$$\Box A^a_\mu(z)_{tor} = -gj^a_\mu(z;A_{tor}) - gj^a_\mu(z;A_{tor})_{source} \tag{12}$$

Now, integration of eq. (II.4) yields [6]

$$K^{ab}_{\pm}(x,y) = \delta^{ab} G_{\pm}(x-y) + \sum_n (-ig)^n \int d^2z_1 \ldots d^2z_n$$

$$G_{\pm}(x-z_1) G_{\pm}(z_1-z_2) \ldots G_{\pm}(z_{n-1}-z_n) G_{\pm}(z_n-y) \, \mathrm{tr}\left\{\frac{\lambda^a}{2} [A_{\mp}(z_1)[A_{\mp}(z_2) \ldots\right.$$

$$\left. \ldots [A_{\mp}(z_n), \frac{\lambda^b}{2}]\right\} \tag{13}$$

From here one learns

$$K^{aD}_{\pm}(x,y) \Big|_{A_{tor}} = \delta^{aD} G_{\pm}(x-y)$$

where "D" (diagonal) stands for any one of the N-1 torus directions. Thus, one has from (13)

$$j^a_\mu(z;A_{tor}) = -\frac{g}{2\pi} A^D_\mu(z) \delta^{aD}$$

Since $j^a_\mu(z;A_{tor})_{source}$ only has components along the torus directions, we may compute it by functionally differentiating $\Gamma_{source}[A_{tor}]$, a quantity which is easily computed from (6). Indeed for configurations along the torus directions, the series (5) is easily summed by using the identity

$$\sum_{perm} G_{\pm}(z-z_{i_1})\ldots G_{\pm}(z_{i_{n-1}}-z_{i_n}) = \prod_1^n G_{\pm}(x-z_i)$$

valid in two dimensions. The result is just the already familiar torus expression

$$h_{\pm}(x)_{tor} = e^{\pm g \sum_D \frac{\lambda^D}{2} \int d^2z D(x-z)\tilde{\partial}_\mu A^D_\mu(z)} \tag{14}$$

In order to simplify the discussion to follow, we restrict ourselves to the group SU(2). We then have from (4) and (14)

$$H(x,y) = 2\cosh\phi(x,y;A^3) \tag{15}$$

where

$$\phi(x,y;A) = -g\int d^2z(D(x-z) - D(y-z))\tilde{\partial}_\mu A^3_\mu(z) \tag{16}$$

or from (11)

$$gj^3_\mu(z;A^3)_{source} = g\tilde{\partial}^z_\mu(D(x-z)-D(y-z))\tanh\phi(x,y;A^3) \tag{17}$$

Combining (17) with (14), eq. (12) becomes

$$\left(\Box - \frac{g^2}{2\pi}\right)A^3_\mu(z)_{c\ell} = - g\tilde{\partial}^z_\mu\left(D(x-z)-D(y-z)\right)\tanh\phi(x,y;A^3_{c\ell}) \tag{18}$$

Note that the solutions always occur in conjugate pairs $\pm A^3_\mu$. We want to learn something about the asymptotic behaviour of $A^3_\mu(z)_{c\ell}$. To this end we need to study the corresponding behaviour of $\phi(x,y;A^3_{c\ell})$. Recalling the definition (16), we may readily convert the differential equation (18) into an algebraic equation for ϕ:

$$\phi = 4\pi\,(\mathcal{D}(o) - \mathcal{D}(x-y))\tanh\phi \tag{19}$$

where $\mathcal{D}(z)$ is defined as in (48) with $e^2 \to \frac{g^2}{2}$.
The transcendental algebraic equation (19) has non-trivial solutions for all x,y such that

$$4\pi\,(\mathcal{D}(o) - \mathcal{D}(x-y)) > 1 \tag{20}$$

Since

$$\mathcal{D}(z) \underset{z\to\infty}{\sim} - \frac{1}{4\pi}\,\ell n\mu^2 z^2$$

condition (20) will be satisfied for sufficiently large separations x-y.

Note that ϕ depends only on the difference x-y. It is thus useful to define a "running" coupling constant by the relation

$$\bar{g}(x-y) = g\,|\tanh\phi(x,y;A^3_{c\ell})|$$

In terms of this running coupling constant the solution to eqn.(18) becomes

$$A^3_\mu(z)_{c\ell} = \pm\frac{2\pi}{g}\left(\frac{\bar{g}(x-y)}{g}\right)\tilde{\partial}_\mu{}^z(\mathcal{D}(x-z)-\mathcal{D}(y-z)) \qquad (21)$$

From (19) one learns that g is an infrared fixed point of $\bar{g}(x-y)$

$$\bar{g}(x-y) \xrightarrow[x-y\to\infty]{} g$$

so that

$$A^3_\mu(z)_{c\ell} \sim \pm\frac{2\pi}{g}\tilde{\partial}^z_\mu(\mathcal{D}(x-z)-\mathcal{D}(y-z)) \qquad (22)$$

The right hand side of (22) are just the "induced instantons" of the SU(2)-torus theory. Hence

$$S_{ind}[A^3_{c\ell}] \xrightarrow[x-y\to\infty]{} S^{tor}_{ind}[A^3_{c\ell}]$$

Moreover, recall the $\Delta g[A_{tor}] = \det\Box$; hence, if the contribution due to the quantum fluctuations around $A^3_\mu(z)_{c\ell}$ becomes independent of x and y as $x-y\to\infty$, we would have arrived at the important statement, that in the limit of large separations $x-y\to\infty$,the correlation function (3) - and presumably also other gauge invariant fermionic correlation functions - approaches the correlation of the pure torus theory. The following observation suggests that this is indeed so: If we make the change

$$z \to z + \frac{x+y}{2}$$

in integration variable the individual fluctuation terms become manifestly translational invariant and formally tend to an x,y - independent limit as $x-y\to\infty$. Assuming this property to carry over to the correlation function we are lead to conclude that the long range behaviour of the gauge invariant fermionic correlation functions in QCD_2 resembles that of the torus theory. Since the massless torus theory is exactly soluble [6], this would tell us something about the lowest lying bosonic states in QCD_2. In particular it would tell us that the bosonic spectrum of QCD_2 starts at $m^2=g^2/2\pi$.

3. THE WILSON LOOP INTEGRAL

Another quantity of great physical interest is the vacuum expectation value of the Wilson-operator

$$\Omega = \mathrm{tr}\, P e^{-iQ \oint_C dz_\mu A_\mu(z)} \tag{23}$$

where the trace is taken with respect to the internal (SU(N)) degrees of freedom and P denotes path ordering. It is convenient to parametrize the path along the closed loop C by a parameter ranging between 0 and 1. The path-ordering operation is then defined by

$$\Omega = P e^{-iQ \oint dz_\mu A_\mu(z)} = P e^{-iQ \int d\tau \left(\frac{dz_\mu(\tau)}{d\tau}\right) A_\mu(z(\tau))}$$

$$= 1 + \sum_{n=1}^{\infty} (-iQ)^n \int_0^1 d\tau_1 \int_0^{\tau_1} d\tau_2 \ldots \int_0^{\tau_{n-1}} d\tau_n \left(A_{\mu_1}(z(\tau_1)) \frac{dz_{\mu_1}(\tau_1)}{d\tau_1}\right) \ldots$$

$$\ldots \left(A_{\mu_n}(z(\tau_n)) \frac{dz_{\mu_n}(\tau_n)}{d\tau_n}\right) \tag{24}$$

We shall be interested in computing

$$\langle\Omega\rangle = \int d[A_T]\Delta_g[A]\Omega[A] e^{-S_0[A] - \Gamma[A]} = \int d[A_T]\Delta_g[A] e^{-S_{ind}[A]} \tag{25a}$$

where

$$S_{ind}[A] = S_0[A] + \Gamma[A] + \Gamma_\Omega[A] \tag{25b}$$

$$\Gamma_\Omega[A] = - \ln\Omega[A] \tag{26}$$

We specialize to the contour shown in Fig. (2).

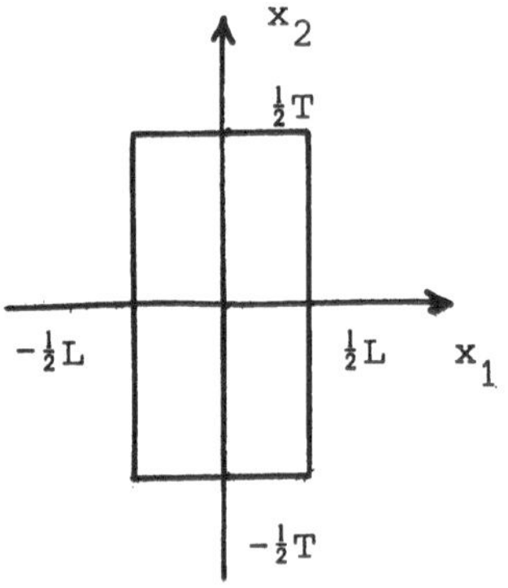

Fig. 2

We may then write

$$\oint dz_\mu A_\mu(z) = \int d^2z A_\mu(z) J_\mu(z) \tag{27}$$

where

$$J_\mu(z) = -\varepsilon_{\mu\nu}\partial_\nu \Phi(z) \tag{28a}$$

with

$$\Phi(z) = \theta(\tfrac{L}{2} - z_1)\theta(\tfrac{L}{2} + z_1)\theta(\tfrac{T}{2} - z_2)\theta(\tfrac{T}{2} + z_2) \tag{28b}$$

Define the "Wilson current" by

$$-Qj^a_\mu(z;A)_\Omega = \frac{\delta\Gamma_\Omega[A]}{A^a_\mu(z)} = -\frac{1}{\Omega[A]}\frac{\delta\Omega[A]}{\delta A^a_\mu(z)} \tag{29}$$

From the gauge invariance of Ω follows that $j^a_\mu(z;A)_\Omega$ is covariantly-conserved. We look for configurations extremizing S_{ind}. They correspond to solutions of

$$\mathcal{D}^{ac}_\lambda[A]F^c_{\lambda\mu}(z)_{c\ell} + gj^a_\mu(z;A_{c\ell}) + Qj^a_\mu(z;A_{c\ell})_\Omega = 0 \tag{30}$$

We now make again the crucial observation that on torus configurations $j^a_\mu(z;A_{tor})_\Omega$ also points in torus-direction. This follows from its definition (29) upon differentiating (24) with respect to A_μ and taking the trace; we may thus look for non-trivial torus-solutions of eqn. (30). It also means that $j^a_\mu(z;A_{tor})_\Omega$ may be calculated by functional differentiations of $\Omega[A_{tor}]$ along the torus. Since for torus configurations pathordering in (24) becomes the idendity operation, $j^a_\mu(z;A_{tor})_\Omega$ is easily calculated. In order to simplify the discussion, we restrict ourselves again to SU(2). In that case

$$\Omega[A^3] = 2\cosh\phi[A^3] \tag{31}$$

where

$$\phi[A^3] = -i\frac{Q}{2}\int d^2z J_\mu(z)A^3_\mu(z) \tag{32}$$

Hence, from (29) we have

$$j^3_\mu(z;A^3)_\Omega = -i\frac{Q}{2} J_\mu(z)\tanh\phi[A^3] \tag{33}$$

For torus configurations, eq. (30) thus becomes,

$$(\Box - m^2)A^3_\mu(z)_{c\ell} = i\frac{Q}{2} J_\mu(z)\tanh\phi[A^3_{c\ell}] \tag{34}$$

where $m^2 = g^2/2\pi$.

It is easy to see that eq. (34) has only the trivial solution for real ϕ . Indeed, recalling (32) we have from (34)

$$\int d^2z A^3_\mu(z)_{c\ell}\left(\Box - m^2\right) A^3_\mu(z)_{c\ell} = -\phi_{c\ell}\tanh\phi_{c\ell} \tag{35}$$

Since $A^3(z)_c$ is pure imaginary for real ϕ , the left hand side of (35) is positive, whereas the right hand side is negative. This proves our assertion.

Eqn. (34) has however oscillatory solutions for ϕ pure imaginary. Should we expand S_{ind} around them? A brief degression to the pure SU(2)-torus model reveals that such an expansion cannot be stable in the limit where $T \to \infty$. To see this observe that the above procedure could equally well be applied to the torus-model. With ϕ pure imaginary we would find $\langle\Omega\rangle \to$ const. as $T \to \infty$ if we neglect the fluctuations. On the other hand, $\langle\Omega\rangle$ may be exactly calculated in the torus model by evaluating each term in

$$\langle\Omega\rangle_{SU(2)\text{-torus}} = \langle \sum_{\pm} e^{\mp iQ \oint dz_\mu A^3_\mu(z)} \rangle_{\text{torus}}$$

separately. The gaussian integration yields

$$\langle\Omega\rangle_{SU(2)\text{-torus}} \underset{T\to\infty}{\sim} e^{-TV_Q(L)} \tag{36}$$

where

$$V_Q(L) = \frac{Q^2}{4m^2}\left(1 - e^{-mL}\right) \tag{37}$$

This is the exact result.

Hence we conclude that the fluctuations around the oscillatory solutions must give the dominant contribution for $T\to\infty$. On the other hand, repeat both calculations for imaginary charges $Q = -ig$. In this case there exist solutions to (34) for ϕ real and we obtain for both calculations, if we neglect the fluctuations,

$$\left\langle \mathrm{tr}\, e^{-q\oint dz_\mu A_\mu(z)} \right\rangle_{\text{torus}} \underset{T\to\infty}{\sim} e^{+TV_q(L)} \tag{38}$$

This shows that this time the fluctuations do not contribute in the limit $T \to \infty$ except for an irrelevant normalization constant. Observe that from (38) we recover again (36) upon continuing q back to iQ.

The above digression strongly suggests that for large T the expectation value (25a) may be calculated by a) first continuing in Q to the (negative) imaginary axis, b) then expanding the functional integral around the solutions to

$$(\Box - m^2) A^3_\mu(z)_{c\ell} = \frac{q}{2} J_\mu(z) \tanh\hat{\phi}[A^3_{c\ell}] \tag{39}$$

$$\hat{\phi}\,[A^3] = -\frac{q}{2}\int d^2z J_\mu(z) A^3_\mu(z) \tag{40}$$

neglecting fluctuations, and c) continuing the result back to g=iQ.

Eqn.(39) may be formally integrated to yield a transcendental algebraic equation for A^3. Recalling the definition of J (z) one obtains,

$$\begin{aligned} A^3_1(z)_{c\ell} &= -\frac{q}{2m}[e^{-m|z_2 - \frac{T}{2}|} - e^{-m|z_2 + \frac{T}{2}|}]\tanh\hat{\phi}[A^3_{ce}] \\ A^3_2(z)_{c\ell} &= \frac{q}{2m}[e^{-m|z_1 - \frac{L}{2}|} - e^{-m|z_1 + \frac{L}{2}|}]\tanh\hat{\phi}[A^3_{ce}] \end{aligned} \tag{41}$$

for z within the square of Fig. (2), and $A^3_\mu = 0$ otherwise. Substituting these results into (40), we obtain a transcendental equation for $\hat{\phi}$:

$$\hat{\phi}_{c\ell} = \frac{q^2}{2m}\left\{L[1 - e^{-mT}] + T[1 - e^{-mL}]\right\}\tanh\hat{\phi}_{c\ell} \tag{42}$$

from where we deduce that

$$\hat{\phi}_{c\ell} \xrightarrow[T\to\infty]{} TV_q(L) \quad , \quad \text{or} \quad \tanh\hat{\phi}_{c\ell} \xrightarrow[T\to\infty]{} \pm 1 \tag{43}$$

There exist two solutions $\pm A^3_\mu(z)_{c\ell}$ to eqns.(41). They both contribute equally to the effective action. In fact, from the equation of motion (39) and (43) we deduce,

$$S_0[\pm A^3_{c\ell}] + \Gamma[\pm A^3_{c\ell}] = \frac{1}{2}\hat{\phi}_{c\ell} \tanh\hat{\phi}_{c\ell} \xrightarrow[T\to\infty]{} \frac{1}{2}|\hat{\phi}_{c\ell}|$$

and from (26) and (31) one has (in obvious notation).

$$\Gamma_{\hat{\Omega}}[\pm A^3_c] \to -|\hat{\phi}_{c\ell}|$$

Combining everything we find,

$$\langle\hat{\Omega}\rangle \xrightarrow[T\to\infty]{} 2e^{TV_q(L)}\,\hat{\Omega}_{\text{fluct.}} \quad (T\to\infty)$$

The above digression on the SU(2)-torus model suggests that $\hat{\Omega}_{\text{fluct}}$ approaches a (possibly L-dependent) constant as $T \to \infty$. This seems very reasonable because of the exponential fall off in T of the T-dependent component of $A_\mu(z)_{c\ell}$. This observation, supplemented with the assumption of analyticity of (25a) in the neighbourhood of $Q = o$ then leads to

$$\langle\Omega\rangle \xrightarrow[T\to\infty]{} \text{const (L)} \cdot e^{-TV_Q(L)}$$

The potential $V_Q(L)$ is depicted in Fig. (3). It is identical with the potential seen by two external 'probes' with charge Ω separated by a distance L, as obtained in ref. [7] for QED_2, following a different method. As discussed in that reference, the absence of a linear rise with separation L reflects screening - rather than "confinement"- of the two test charges by the induced vacuum polarisation. This is characteristic of the theory of massless fermions, and is an exact feature of QED_2 as already discussed in ref. [7]. The considerations of this section suggest that it is also a characteristic feature of massless QCD_2.

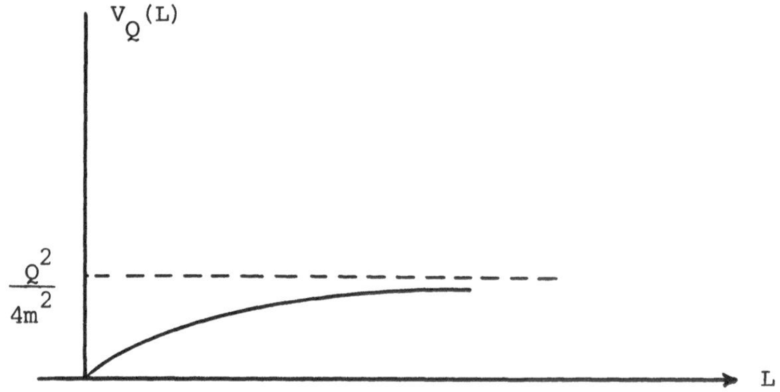

Fig. 3

ACKNOWLEDGEMENT:

We are very grateful for enlightening discussions with N.K. Nielsen, J.A. Swieca and M. Hortaçsu. We thank the organizers of the school for having succeeded to create a scientifically stimulating atmosphere.

REFERENCES AND FOOTNOTES

1. A. Salam and P.T. Matthews,Phys. Rev. 90, 690 (1953).
2. The notion of induced action was introuduced in connection with the functional representation of correlation functions in the Schwinger model: N.K. Nielsen and B. Schroer, Nucl. Phys. B120, 62 (1977); N.K. Nielsen and B. Schroer, Phys. Lett. 66B, 373 (1977).
3. M. Hortaçsu, K.D. Rothe and B. Schroer, FUB/HEP March 79, to be published in Phys. Rev. D (Nov. 79).
4. N.K. Nielsen, K.D. Rothe and B. Schroer, FUB/HEP May 79, to be published in Nucl. Phys. B. See also C. Sorensen and G.H. Thomas, ANL-HEP-PR-79-13; The method of these authors does not yield however the zero mode contribution to the determinant.
5. V. Kurak, B. Schroer and J.A. Swieca, Nucl. Phys. B134, 61 (1978); S. Coleman, Ann. Phys. N.Y. 101, 239 (1976). See also J.A. Swieca's lectures at this school.
6. L.V. Belvedere, K.D. Rothe, B. Schroer and J.A. Swieca, Nucl. Phys. B153, 112 (1979).
7. H.J. Rothe, K.D. Rothe and J.A. Swieca, Phys. Rev. D19, 3020 (1979).
8. This argument was used by Coleman in a private discussion with one of the authoers (B.S.)
9. M. Lüscher, Phys. Lett. 78B, 465 (1978). A. D'Adda, P. Di Vecchia and M. Lüscher, Nucl.Phys. B146, 63 (1978).
10. M. Karowski and P. Weisz, Nucl. Phys. B139, 455 (1978).
11. A.S. Schwarz, Comm. Math. Phys. 64, 233 (1979).
12. R. Jackiw and C. Rebbi, Phys. Rev. D14, 517 (1976).
13. N.K. Nielsen and B. Schroer, Nucl. Phys. B127, 493 (1977).
14. R.R. Seeley, Proc. Symp. Pure and Appl. Math. 10, 288 (1967).
15. B. DeWitt, Dynamical Theory of Groups and Fields, Gordon and Breach 1965.
16. P.B. Gilkey, Proc. Symp. Pure Math. 27 (1973), and J. Diff. Geom. 10, 601 (1975).
17. See also an unpublished paper by N.K. Nielsen, Nordita 78/24.
18. See the lectures of H. Leutwyler at this school.
19. H. Hogreve, R. Schrader and R. Seiler, Nucl. Phys. B142, 525, (1978).
20. S. Hawking, Comm. Math. Phys. 55, 133 (1977).
21. B. Schroer, Acta Austriaca, Suppl. XIX, 155 (1978).
22. M.F. Atiyah, N.J. Hitchin, V.C. Drinfeld and Yu.I. Manin, Phys. Lett. 65A, 185 (1978); E.F. Corrigan, D.B. Farlie, P. Goddard and S. Templeton, Nucl. Phys. B140, 31 (1978). N.H. Christ, E.J. Weinberg and N.K. Stanton, Phys. Rev. D18, 2013 (1978).
23. L.S. Brown and D.B. Creamer, Phys. Rev. D18, 3695 (1978).
24. B. Berg and M. Lüscher, DESY 79/40.
25. B. Berg and M. Lüscher, DESY 79/17.

26. This simplification for the computation of the determinant of the constrained Dirac operator in terms of the unconstrained determinant divided by the Dirac operator parallel to the constraint was suggested to us by M.F. Atiyah.
27. A. Patrascioiu, Phys. Rev. D20, 491 (1979).
28. M.F. Atiyah and R. Bott., "The Index Theorem for Manifolds with Boundary", Differential Analysis (Bombay Colloquium) Oxford, 1964.
29. M.F. Atiyah, V.K. Patodi and I.M. Singer, Math. Proc. Camb. Phil. Soc. 77, 43 (1975); Math. Proc. Camb. Phil. Soc. 78, 405 (1975).
30. J. Milnor, Enseignement Mathematique 9, 198 (1963).
31. J.F. Schonfeld, Minnesota preprint 1979.
32. P. D'Adda, R. Horsley and P. Di Vecchia, Phys. Lett. 76B, 298 (1978).
33. We thank N.K. Nielsen for bringing this controversy to our attention.
34. J. Lowenstein and J.A. Swieca, Ann. of Phys. N.Y. 68, 172 (1971).
35. J. Schwinger, Phys. Rev. 128, 2425 (1962); Theoretical Physics (IAEA Vienna, 1963) 88.
36. K.D. Rothe and J.A. Swieca, Phys. Rev. D15, 541 (1977).
37. K.D. Rothe and J.A. Swieca, Ann. of Phys. N.Y. 117, 382 (1979).
38. Subsequent to our paper, ref.(6), P. Mitra and Probir Roy reconsidered the torus solutions interpreting them as QCD_2 solutions in the broken phase (Bombay preprints I-III to appear in Phys. Rev. D); see also Patrascioiu, Phys. Rev. D15, 3592 (1977). Their interpretation is wrong.

THE BOOTSTRAP PROGRAM FOR 1+1 DIMENSIONAL FIELD THEORETIC MODELS WITH SOLITON BEHAVIOUR

M. Karowski

Institut für Theoretische Physik
Freie Universität Berlin
D-1000 Berlin 33, Arnimallee 3

ABSTRACT

A review is given of the present status of the bootstrap program for 1+1 dimensional field theoretic models with soliton behaviour. From the existence of infinitely many conservation laws, unitarity, crossing, "minimality", internal symmetries, and some assumptions on the one particle spectrum first the exact S-matrix is derived explicitely. We repeat the procedure for the massive Thirring-model (alias Sine-Gordon) S-matrix starting with two fermion (alias soliton) scattering and then deriving the boundstate (alias breather) S-matrix. The same complete S-matrix is obtained by starting with the determination of the boson (breather) S-matrix and then deriving the boson-soliton and soliton-soliton scattering from the assumption that the soliton is a soliton-boson bound state. By means of the same procedure we obtain the Gross-Neveu S-matrix including the kink scattering. Then for some models generalized form factors and Green's functions for the Z(2)-Ising model in the scaling limit are calculated.

I. THE S-MATRIX

For some field theoretic models we calculate exactly S-matrix elements [1] like

$$^{out}\langle p_1' \ldots p_n' \;\; p_1 \ldots p_n \rangle^{in}$$

for the scattering of any number and kinds of particles appearing in the models.

1. Introduction: The Models

i) The Sine-Gordon (SG) model defined by the relativistic wave equation

$$\Box \phi + \frac{\alpha}{\beta} \sin \beta \phi = 0 \tag{1}$$

is on the classical level completely integrable by means of the inverse scattering method [2] . There are solutions called soliton (antisoliton)

$$\phi_{sol}(x,t) = \phi_{antisol}(-x,t) = \frac{4}{\beta} \, tg^{-1} \, e^{\sqrt{\alpha} x} \tag{2}$$

and breathers which may be considered as soliton-antisoliton bound states. On the semiclassical level there is a sequence of breathers b_k with the masses [3]

$$m_k = 2m \sin \frac{k\pi}{2\lambda} \quad , \; k = 1,2,\ldots<\lambda \tag{3}$$

where m is the soliton mass and λ a parameter related to the coupling constant:

$$\lambda = \frac{8\pi}{\beta^2} - 1 \tag{4}$$

The soliton is a coherent state

$$|sol\rangle = \exp \int dx \, \phi \, (x) \, i\overleftrightarrow{\partial}_t \, \phi_{sol}(x) | \, o\rangle \tag{5}$$

or

$$\langle sol|\phi(x) \, |sol\rangle = \phi_{sol}(x)$$

which means that the soliton is an eigenstate of the annihilation operator

$$\phi^{(+)}(x) \, |sol\rangle = |sol\rangle \, \phi^{(+)}_{sol}(x) \quad . \tag{6}$$

On the quantum level the fundamental boson corresponding to the field $\phi(x)$ is the lowest breather b_1. Eq. (6) means that the soliton is a soliton-breather bound state.

ii) The massive Thirring (MT) model defined by the Lagrangian

$$L = \bar{\psi}(i\partial\!\!\!/ - m)\psi - \frac{g}{2} (\bar{\psi}\gamma^{\mu}\psi)^2 \tag{7}$$

describes the interaction of fermions f and $\bar{f}$. For $g > o$ there exist fermion-antifermion bound states b_k.

The famous equivalence of the quantum Sine-Gordon and the massive Thirring model due to Coleman [4] says: identify the SG-solitons and breathers with the MT-fermions and bound states, respectively, relate the SG-field to the MT-current by

$$\beta\, \varepsilon^{\mu\nu} \partial_\nu \phi = -\, 2\pi\, \bar{\psi}\gamma^\mu \psi$$

and the coupling constants by

$$\gamma = \frac{8\pi}{\beta^2} - 1 = 1 + \frac{2g}{\pi} \tag{8}$$

iii) The Gross-Neveu (GN) model [5] defined by the Lagrangian

$$L = \sum_{i=1}^{N} \bar{\psi}_i\, i\partial\!\!\!/\, \psi_i + \frac{g^2}{2} \left(\sum_{i=1}^{N} \bar{\psi}_i\, \psi_i\right)^2$$

describes an O(N)-symmetric interaction of N(even) selfconjugate fermions f_i with a dynamically generated mass m_1. There are $2^{N/2}$ kinks k_α which are O(N)-isospinors [6] . The kink-kink bound states $(i_1 \ldots i_j)_k$ (j=0,2,...,k for even k, j=1,3 for odd k) which are antisymmetric tensors have the spectrum eq.(3) with $\lambda = \frac{N}{2} - 1$ [7] . The lowest bound states (k=1) are the fundamental fermions f_i, the higher ones may also (as for the SG model) be considered as bound states of k fundamental fermions.

iv) There are further models to which the methods used in this paper can be applied, e.g. the O(N)-nonlinear σ-model [8] , a chiral SU(n) model [9] , a Z(N)-Ising model in scaling limit $T \to T_c$ [10].

2. Infinitely many conservation laws

The models presented in the Introduction are in some sense simple because they have soliton behaviour, this means they possess infinitely many conservation laws.

2.1 The classical case [11]

Let us take the SG-model as an example. There exist an infinite set of local conservation laws

$$\partial_\mu J_k^{\ \mu}(x) = 0 \quad , \quad k = 1,3,5,\ldots \tag{9}$$

This can be seen as follows: Define recursively as a formal power series

$$\sigma(x,\varepsilon) = \sum_{k=1}^{\infty} \varepsilon^k \sigma_k(x) = \left[1 - \varepsilon\partial_+ - \frac{\varepsilon}{2}\, \sigma \partial_+ \varphi\right]^{-1} \frac{\varepsilon}{2}\, \partial_+ \varphi$$

and

$$(J+, J-) = \sum_{k=1}^{\infty} \varepsilon^k (J_{k+}, J_{k-}) = (\sigma\partial_+\varphi, \varepsilon(\sigma \sin\varphi - \cos\varphi))$$

where

$$\partial_\pm = \partial_0 \pm \partial_1,\quad J_\pm = J_0 \pm J_1 .$$

Then for any solution of the SG-equation $\Box\varphi$ sin+φ=0 eq.(9) is fulfilled and the quantities

$$\begin{aligned} Q_k &= \int dx\, J_k^o \\ &= \int dx\, [i\partial_o\varphi\partial^k + \varphi + O(\varphi^3)] \end{aligned} \tag{10}$$

are conserved, $\dot{Q}=0$. The contribution $O(\varphi^3)$ vanishes on asymptotic states. If we consider the classical theory as the tree approximation of the quantum theory we obtain

$$Q_k |\, p_1,\ldots,p_n >^{in} = \sum_{i=1}^{n} (p_+)^k \, |p_1,\ldots,p_n >^{in} \tag{11}$$

for n=1,3,5,....(for n=even Q_n vanishes).

Since the S-matrix commutes with Q and eq.(11) holds for any k for a scattering process the set of incoming and outgoing momenta are equal.

$$\{p_1,\ldots,p_n\}^{in} = \{p_1',\ldots,p_n',\}^{out} \tag{12}$$

This means absence of particle production and only momentum exchange. For some models there exist also non local conservation laws [12] with similar consequences.

2.2 Quantization

The BPHZ quantization[13] of the conservation laws (9) apparently produce anomalies which cancel after redefinition of the currents J_k^μ [14]. The nonlocal charges are also conserved in the quantized models [15] . In summary we can say the infinitely many conservation laws survive quantization.

2.3 Consequences for scattering

Since eqs.(11) and (12) are also true for the quantum models we have absence of particle production and only momentum exchange between particles with the same mass. Furthermore one can show [16] that the n-particle S-matrix factorizes into a product of two-particle S-matrices

$$S^{(n)}(p_1,\ldots,p_n) = \prod_{i<j} S^{(2)}(p_i,p_j) \tag{13}$$

For non-vanishing reflection the factors on the r.h.s. do not commute and we have to specify the order. For n=3 and

$p_1^1 > p_2^1 > p_3^1$

there are two possible orders (with $S^{(2)}(p_i,p_j)=S_{ij}$)

$$S^{(3)} = S_{12}\, S_{13}\, S_{23} = S_{23}\, S_{13}\, S_{12} \tag{14}$$

This so called factorization equation gives relations between the two-particle scattering amplitudes which allow to calculate the S-matrix exactly [1] .

3. Construction of the S-matrix

3.1 General procedure

We calculate the S-matrix from the assumption:

i) Factorization
ii) Qualitative knowledge of the one-particle spectrum; i.e. we assume e.g. for the MT-model the existence of a coupling region with no bound states, for the SG-model the existence of a (b_1-b_1) bound state and the soliton to be a (soliton-b_1) bound state.
iii) "Minimality", i.e. absence of redundant poles and zeros in the physical sheet of the transmission amplitudes (this is true for potential scattering [17])
iv) Unitarity, crossing, internal symmetry (O(N),U(N),Z(N)), e.g. for the MT-model: $f,\bar{f}$ = fundamental representation of U(1); for the SG-model: b_1 = self conjugate isosinglett.
v) Asymptotically good behaviour for $p_1p_2\to\infty$

$$S^{(2)} = o\left(\exp \frac{p_1p_2}{m^2}\right).$$

By these assumptions the S-matrix is "uniquely" determined. "Uniquely" means we have

for $U(1)\cong O(2)$ a one-parametric set of solutions [1],
for O(N) (N>2) two solutions, [1][18],
for U(N) N >1 five solutions [18],
for Z(N) one solution [10].

A sketch of the proof: [1] .

i) The factorization equation (14) implies e.g. for the MT-model fermion-antifermion amplitudes

$$h(\alpha+\beta) = h(i\pi+\alpha)\, h(\beta) + h(\alpha)\, h(i\pi-\beta) \tag{15}$$

where $h(\theta)$ is the ratio of transmission and reflection amplitudes and θ the rapidity difference defined by $p_1p_2=m^2\text{ch}\theta$. The solution of eq.(15) is

$$h(\theta) = \frac{\text{sh } \lambda\theta}{\text{sh } \lambda i\pi} \tag{16}$$

where λ is a free parameter.

ii) Unitarity, crossing, "minimality" and eq.(16) imply uniquely the two particle S-matrix.

3.2 Bound State Scattering [19][20]

Let b be a bound state with mass m_b of two particles α and β, then the two particle S-matrix has a pole at θ=ia where a is given by

$$m_b^2 = m_\alpha^2 + m_\beta^2 + 2m_\alpha m_\beta \cos a \tag{17}$$

The S-matrix for the scattering of the bound state b with a particle γ is obtained by considering the three-particle S-matrix for the scattering of $\alpha(p_1)$, $\beta(p_2)$, $\gamma(p_3)$ at the pole θ_{12}=ia. Factorization and unitarity imply [20]

$$S_{1+2,3} = |R_{12}|^{-\frac{1}{2}} S_{23} S_{13} |R_{12}|^{\frac{1}{2}} \tag{18}$$

where

$$R_{12} = \underset{\theta = ia}{\text{Res}} S_{12}$$

3.3 Examples

i) Massive Thirring

All states of two particles with momenta p_1 and p_2 are $ff, f\bar{f}, \bar{f}f, \bar{f}\bar{f}, fb_1, \ldots, b_1b_1, \ldots$. Since reflection can only appear for two different particles with the same mass, the two-particle S-matrix has the form

$$S_{12} = \begin{pmatrix} u & & & & & & \\ & t & r & & & & \\ & r & t & & & & \\ & & & u & & & \\ & & & & S_{fb_1} & & \\ & & & & & \ddots & \\ & & & & & & S_{b_1b_1} \\ & & & & & & & \ddots \end{pmatrix} \tag{19}$$

For g< o there are no bound states. The U(1)-S-matrix obtained from factorization, unitarity, crossing and minimality in 3.1 has no pole in the physical sheet $0 < \mathrm{Im}\,\theta < \pi$ for $0<\lambda<1$. Thus we propose first in this coupling region

$$(u,t,r)=\left(1,\frac{\mathrm{sh}\lambda\theta}{\mathrm{sh}\,\lambda\,(i\pi-\theta)},\frac{\mathrm{sh}\lambda i\pi}{\mathrm{sh}\,\lambda\,(i\pi-\theta)}\right)\exp\int_0^\infty \frac{dx}{x}\,\frac{\frac{x}{2}(1-\frac{1}{\lambda})}{\mathrm{sh}\,\frac{x}{2\lambda}\,\mathrm{ch}\,\frac{x}{2}}$$

$$\times\ \mathrm{sh}\,x\,\frac{\theta}{i\pi} \tag{20}$$

For $\lambda>1$ the $f\bar{f}$-transmission amplitude $t\,(\theta)$ has poles at

$$\theta = i\pi\,(1-\frac{k}{\lambda})\ ,\ k = 1,2,\ldots<\lambda \tag{21}$$

These poles correspond to $f\bar{f}$-bound states b_k with masses

$$m_k = 2\,m\,\sin\frac{\pi k}{2\lambda} \tag{22}$$

This agrees with the WKB-spectrum eq.(3) for the SG-model if we identify

$$\lambda = \frac{8\pi}{\beta^2} - 1 = 1 + \frac{2g}{\pi}$$

The amplitudes given by eq.(20) have first been proposed by Zamolodchikov [21] by different arguments. They have been checked in perturbation theory at $g\to 0$ and $\beta\to 0$ [21][22] .

The scattering of fermions and bound states b_k can be calculated by means of eq. (18) [19] . E.g. for b_1 we obtain

$$S_{fb_1}(\theta,\lambda) = \frac{\mathrm{sh}\,\theta + i\,\sin\frac{\pi}{2}\,(1+\frac{1}{\lambda})}{\mathrm{sh}\,\theta - i\,\sin\frac{\pi}{2}\,(1+\frac{1}{\lambda})} \tag{23}$$

The pole at $\theta = \frac{1}{2}\,i\pi\,(1+\frac{1}{\lambda})$ corresponds to the bound state $f = (fb_1)$ which was anticipated by eq.(6).

The scattering of two bound states can be derived from eqs. (18) and (23). The result is e.g. for b_1-b_1:

$$S_{b_1b_1}(\theta,\lambda) = \frac{\mathrm{sh}\,\theta + i\,\sin\frac{\pi}{\lambda}}{\mathrm{sh}\,\theta - i\,\sin\frac{\pi}{\lambda}} \tag{24}$$

The pole at $\theta = i\,\frac{\pi}{\lambda}$ corresponds to the bound state $b_2=(b_1b_1)$. (The generalization of this formula is $b_{i+j+k}=(b_i+b_j+b_k+\ldots.)$).

The amplitude $S_{b_1b_1}$ can be checked in SG-perturbation theory at $\beta \to 0$.

ii) Sine-Gordon Model

In the last section we have derived the complete S-matrix of the MT-model alias the SG-model by starting with the soliton S-matrix and then calculating the breather scattering. Now we will do it the other way round without using directly the factorization equation (15) but the property (6) that the soliton is a bound state of itself and a breather.

For the breather-breather S-matrix absence of particle production, unitarity, crossing, "minimality" and the existence of a $(b_1 b_1)$-bound state imply for the scattering of a selfconjugate isosinglett b_1 uniquely the S-matrix (24). The existence of a two particle bound state $b_2=(b_1 b_1)$ with mass $m_2=m_1 \sin\frac{\pi}{\lambda} \sin\frac{\pi}{2\lambda}$ and factorization imply the existence of a k-particle bound state

$$b_k=(b_1 \ldots b_1)= \ldots = (b_{i1} \ldots b_{ij})=(b_{k-1} b_1) \text{ with } i_1+\ldots+i_j=k$$

and the spectrum (3) [23] . This can be seen inductively as follows: The S-matrix of b_{k-1} and b_1 is given by eq.(18) to be

$$S_{b_{k-1}b_1}(p_1+\ldots+p_{k-1},p_k)=S_{b_1b_1}(p_1,p_k)\ldots S_{b_1b_1}(p_{k-1},p_k) \quad (25)$$

where (for suitable ordered momenta) the rapidity differences are

$$\theta_{jj+1}= i\,\frac{\pi}{\lambda} \;(j=1,\ldots,k-2).$$

The bound state pole of $S_{b_{k-1}b_1}$ is again at $\theta_{k-1,k}=i\frac{\pi}{\lambda}$.

Therefore the mass of b_k is

$$m_k=m_1\left|\sum_{j=1}^{k-1} e^{i\frac{\pi}{\lambda}j}\right|= m_1 \frac{\sin\frac{\pi k}{2\lambda}}{\sin\frac{\pi}{2\lambda}} \quad (26)$$

which agrees with eq.(3). Note that there are two cases: a) there exist kinks with mass

$$m = m_1/2 \sin\frac{\pi}{2\lambda}$$

then $k=1,2,..<\lambda$ as for the massive Thirring model and the Gross-Neveu model or b) there exist no kinks then $k=1,2,..<2\lambda$ as for the chiral SU(N) model [9] and a Z(N)-Ising model [10].

For the scattering of a soliton and a breather the property that the soliton is a bound state of itself and a breather together with eq.(18) and unitarity implies

$$S_{fb_1}(\Theta) = S_{fb_1}(\Theta+i\pi\tfrac{1}{\lambda})\, S_{b_1b_1}(\Theta-i\pi(\tfrac{1}{2}-\tfrac{1}{2\lambda}))$$

$$= S_{b_1b_1}(\Theta+i\pi(\tfrac{1}{2}-\tfrac{1}{2\lambda}))\, S_{fb_1}(\Theta - i\pi\tfrac{1}{\lambda}). \tag{27}$$

These equations have the unique "minimal" solution (23). For the soliton-soliton scattering the same argument leads to

$$u(\Theta) = u(\Theta+i\pi\tfrac{1}{\lambda})\, S_{fb_1}(\Theta-i\pi(\tfrac{1}{2}-\tfrac{1}{2\lambda}))$$
$$= S_{fb_1}(\Theta+i\pi(\tfrac{1}{2}-\tfrac{1}{2\lambda}))\, u(\Theta-i\pi\tfrac{1}{\lambda}). \tag{28}$$

The "minimal" solution (with u(o)=1) is obtained as follows:

$$u(\Theta)=\exp \frac{1}{2\pi i}\int_C \frac{dz}{sh(z-\lambda\Theta)}\,\frac{sh\ \Theta}{sh\ z}\,\ell n\ u(\tfrac{z}{\lambda}) \tag{29}$$

$$=\exp \frac{1}{2\pi i}\int_{-\infty}^{\infty} \frac{dz}{ch(z-\lambda\Theta)}\,\frac{sh\ \lambda\Theta}{ch\ z}\,\ell n\,\frac{u(\frac{z}{\lambda}-i\pi\frac{1}{2\lambda})}{u(\frac{z}{\lambda}+i\pi\frac{1}{2\lambda})}$$

where C encloses the strip $-\frac{\pi}{2}\leq\ \text{Im}\ z\leq\frac{\pi}{2}$.

Using eq.(28) we can express u(o) in terms of $S_{fb_1}(\Theta)$. The result agrees with eq.(20) which can be understood as another proof of Coleman's equivalence.

iii) The Gross-Neveu Model

The same approach as in ii) can be used to calculate the fermion-kink and kink-kink scattering for the GN-model. The direct method using factorization is complicated because of group theoretical problems[6]. From factorization, unitarity, crossing, "minimality" and the existence of a (ff)-bound state in the O(N)-isoscalar and antisymmetric tensor channel one obtains for the scattering of fundamental fermions [24]

$$f_i(p_1)+f_j(p_2) \longrightarrow f_k(p_1)+f_\ell(p_2)$$

$${}_{k\ell}S_{ij}(\Theta)= (\delta_{ik}\cdot\delta_{j\ell} - \frac{N-2}{2\pi i}(i\pi-\Theta)\,\delta_{i\ell}\delta_{jk} - \frac{N-2}{2\pi i}\,\Theta\,\delta_{ij}\,\delta_{k\ell})\ \times \tag{30}$$

$$S_{b_1b_1}(\Theta,\lambda=\frac{N}{2}-1)\ \exp\int_0^\infty \frac{dx}{x}\ \frac{e^{-\frac{2x}{N-2}}-1}{\mathrm{ch}\,\frac{x}{2}} \times \mathrm{ch}\,x\,(\frac{\Theta}{i\pi}-\frac{1}{2}) \tag{30}$$

where $S_{b_1b_1}(\Theta,\lambda)$ is given by eq.(24). Analogous to the SG-case we obtain for fermion-kink scattering+)

$$f_i(p_1) + k_\alpha(p_2) \to f_j(p_1) + k_\beta(p_2)$$

$${}_{j\beta}S_{i\alpha}(\Theta) = -\,(\delta_{ij}\ \delta_{\alpha\beta} - \frac{1}{N}\gamma^j_\beta\ \gamma^i_\alpha + \frac{2(N-2)\Theta - i\pi N}{2(N-2)\Theta + i\pi N}\ \frac{1}{N}\gamma^j_\beta\ \gamma^i_\alpha)\ \times$$
$$\times\ S_{fb_1}(\Theta,\gamma=\frac{N}{2}-1)\ \exp\int_0^\infty \frac{dx}{x}\ \frac{e^{-x(1+\frac{1}{N-2})}}{\mathrm{ch}\,\frac{x}{2}}\ \mathrm{sh}\,x\,\frac{\Theta}{i\pi} \tag{31}$$

where $S_{fb_1}(\Theta,\lambda)$ is given by eq.(23) and the $\gamma_\alpha^{\ i}$ are the usual γ matrices of O(N), satisfying

$$\gamma^i\ \gamma^j + \gamma^j\ \gamma^i = 2\ \delta_{ij}$$

Again analogous to the SG-case we obtain for kink-kink scattering+)

$$k_\alpha\ (p_1) + k_\beta\ (p_2) \to k_\gamma\ (p_1) + k_\delta\ (p_2)$$

$${}_{\gamma\delta}S_{\alpha\beta}\ (\Theta) = \sum_{n=0}^{N} \frac{1}{n!2^{N/2}}\ \sigma^{(n)}_{\delta\alpha} \otimes \sigma^{(n)}_{\gamma\beta}\ u_n(\Theta) \tag{32}$$

where $u_n(\Theta) = (-1)^{\frac{N}{2}}\ u_{N-n}(\Theta) = u(\Theta,\lambda=\frac{N}{2}-1)\times$

$$\times\ \exp\int_0^\infty \frac{dx}{x}\ \frac{e^{-x(1+\frac{1}{N-2})}}{\mathrm{ch}\,\frac{x}{2}\ \mathrm{sh}\,x\,\frac{2}{N-2}}\ (\mathrm{sh}\,\frac{x}{2}\ (1+\frac{2}{N-2}) - 2\ e^{x\frac{n-1}{N-2}}\mathrm{ch}\,\frac{x}{2}\ \mathrm{sh}\,x\,\frac{n}{N-2})$$
$$\times\ \mathrm{sh}\,x\ (\frac{\Theta}{i\pi}-1) \tag{32a}$$

for n even $\leq N/2$

$$u_n(\Theta) = -\ (-1)^{\frac{N}{2}}\ u_{N-n}(\Theta) = u(\Theta,\lambda=\frac{N}{2}-1)\ \times$$

+) More details will be published elsewhere [24] .

$$\times \exp \int_0^\infty \frac{dx}{x} \frac{e^{-x(1+\frac{1}{N-2})}}{\mathrm{ch}\frac{x}{2}\,\mathrm{sh}x\frac{2}{N-2}} \left(\mathrm{sh}\frac{x}{2}\left(1-\frac{2}{N-2}\right) - 2\,e^{x\frac{n}{N-2}}\mathrm{ch}\frac{x}{2}\,\mathrm{sh}x\frac{n-1}{N-2}\right)$$

$$\times \quad \mathrm{sh}x\left(\frac{\theta}{i} - 1\right) \tag{32b}$$

for n odd $\leq N/2$

and

$$\sigma^{(n)}_{\delta\alpha} \otimes \sigma^{(n)}_{\gamma\beta} = \frac{1}{n!} \sum_{\pi\in S_n} (-1)^{\pi} \left(\gamma^{\pi(i_1)}\cdots\gamma^{\pi(i_n)}\right)_{\delta\alpha} \frac{1}{n!} \sum_{\pi'\in S_n} (-1)^{\pi'} \left(\gamma^{\pi'(i_1)}\cdots\gamma^{\pi'(i_n)}\right)_{\gamma\beta}$$

The SG-amplitude $u(\theta,\lambda)$ is given by eq. (20).

II. THE FIELD

We will calculate "generalized" form-factors

$${}^{out}\langle p_1'\ldots |O(x)| p_1\ldots\rangle^{in}$$

for local operators and Green's functions

$$\langle T\,\phi(x_1)\ldots\phi(x_n)\rangle.$$

These problems are up to now only solved for special simple cases.

1. Introduction

The form factor problem is simple for the Z(N)-Ising model mentioned in chap. I.1. It is completely solved for N=2. We will give a short introduction to these models.

Z(N)-symmetry means that the order variable fulfils

$$\sigma^N = 1 \quad \text{or} \quad \sigma^+ = \sigma^{N-1}.$$

If b_1 is the fundamental particle to the field $\sigma(x)$ and there exists a two particle bound state $b_2=(b_1b_1)$ then $\sigma^+=\sigma^{N-1}$ implies that the antiparticle is a N-1-particle bound state

$$\bar{b}_1 = (b_1\ldots b_1)$$

This fact together with factorization, unitarity, crossing and "minimality" leads to the S-matrix [10]

$$S_{b_1 b_1}(\theta) = \frac{\mathrm{sh}\,\frac{1}{2}\,(\theta + \frac{2\pi i}{N})}{\mathrm{sh}\,\frac{1}{2}\,(\theta - \frac{2\pi i}{N})} \tag{33}$$

and the spectrum for particles $b_k = \bar{b}_{N-k}$

$$m_k = m_1 \frac{\sin\frac{k\pi}{N}}{\sin\frac{\pi}{N}} \tag{34}$$

For the Z(2)-Ising model we get a self conjugate isosinglet $b_1 = \bar{b}_1 = b$ and

$$S_{bb} = -1. \tag{35}$$

2. Watson's Theorem

We derive a set of equations for matrix elements of local operators which follow from general principles of quantum field theory, maximal analyticity and the S-matrix factorization. For simplicity we first consider the case where we have only one kind of boson in the model and a hermitean operator $\sigma(x)$. If we define (with $p_1 p_2 = m^2 \mathrm{ch}\,\theta$)

$$\langle o \,|\, O(o) \,|\, p_1 p_2 \rangle^{in} = F(\theta) \tag{36}$$

it follows from

CTP-invariance: $\langle o | O(o) | p_1 p_2 \rangle^{out} = F(-\theta)$

unitarity and factorization: $F(\theta) = \langle o | O(o) | p_1 p_2 \rangle^{out} S(\theta)$

and crossing: $\langle p_1 | O(o) | p_2 \rangle = F(i\pi - \theta)$.

Hence we have Watson's equations

$$F(\theta) = F(-\theta)\, S(\theta).$$
$$F(i\pi - \theta) = F(i\pi + \theta). \tag{37}$$

The corresponding equations for the general case where α_i denote different kinds of particles read

$$^{out}\langle \alpha_1(p_1) \ldots \alpha_m(p_m) | O(o) | \ldots \alpha_n(p_n) \rangle^{in} =$$

$$= F_{\alpha_1 \ldots \alpha_n}(\theta_{ij}, i\pi - \theta_{rs}, \theta_{k\ell}) =$$

$$= {}_{\alpha_1 \ldots \alpha_m} S^{(m)}_{\alpha_1' \ldots \alpha_m'} (\theta_{ij}) \, F_{\alpha_1' \ldots \alpha_n'} (-\theta_{ij}, i\pi + \theta_{rs}, -\theta_{k\ell})$$
$$\times \, {}_{\alpha_{m+1}' \ldots \alpha_n'} S^{(n-m)}_{\alpha_{m+1} \ldots \alpha_n} (\theta k_\ell) \tag{38}$$

where $1 \leq i < j \leq m$, $1 \leq r \leq m < s \leq n$, $m < k < \ell \leq n$ and $0 \leq m \leq n$. Solutions of these equations are only known for simple cases.

3. Construction of Form Factors

The problem constructing form factors has been solved for models with arbitrary number of kinds of particles for the case n=2. This means usual form factors, i.e. one particle expectation values of a local operator, are known exactly for e.g. the massive Thirring [25], the nonlinear σ-, and the Gross-Neveu model [26]. Furthermore arbitrary n-particle matrix elements are known for the Z(2)-Ising model in scaling limit with one kind of particles and $S^{(2)}=-1$ [27]. There are two distinct problems:

3.1 The "Matrix Problem"

The set of matrix equations (38) has not been solved for the general case. For n=2 we diagonalize $S^{(2)}$ and get for each eigenvalue the simple equations (37). Theorem [26] :

A function $F(\theta)$ fulfilling eqs.(37) is uniquely (up to a normalization) determined by the positions of the poles (and zeros) at $\theta = ia_k$ in the physical strip $0 \leq \text{Im}\,\theta \leq \pi$:

$$F(\theta) = K(\theta) F^{min}(\theta) \tag{39}$$

$$\text{where } K(\theta) = \text{const}\Big[\prod_k \text{sh}\,\frac{1}{2}(\theta - ia_k)\ \text{sh}\,\frac{1}{2}(\theta + ia_k)\Big]^{-1} \tag{39a}$$

and

$$F^{min}(\theta) = \exp \frac{1}{4\pi i} \int_{-\infty}^{\infty} \frac{dz}{\text{sh}\,\frac{1}{2}(z-\theta)} \, \frac{\text{ch}\,\frac{1}{2}\theta}{\text{ch}\,\frac{1}{2}z} \, \ell n\, S(z). \tag{39b}$$

For vanishing reflection $S^{(n)}$ is diagonal and a solution of the general equations (38) is:

$$F_{\alpha_1 \ldots \alpha_n}(\theta_{12}, \ldots) = K_{\alpha_1 \ldots \alpha_n}(\theta_{12}, \ldots) \prod_{i<j} F^{min}_{\alpha_i \alpha_j}(\theta_{ij}) \tag{40}$$

where K is a solution of eqs.(38) with S=1.

3.2 The "Pole Problem"

The poles in the physical region of the function

$$F_{\alpha_1 \cdots \alpha_n}(\theta_{12}\cdots) = \langle o|O(o)|\alpha_1(p_1)\ldots\alpha_n(p_n)\rangle^{in}$$

which are contained in the function K are determined by one-particle states in all subchannels $\alpha_{i1}\ldots\alpha_{ij}$. If we make the minimality assumption that there are no further so called redundant poles and no zeros in the physical region of F , we can make proposals for exact form factors, which can then be checked by perturbation theory.

4. Examples

4.1 Sine-Gordon model

The electromagnetic S.G. soliton form factor is defined by

$$\langle f(p_1)|j^\mu(o)|f(p_2)\rangle = \bar{u}(p_1)\gamma^\mu u(p_2)\, F(i\pi-\theta)$$

where $p_1p_2=m^2\text{ch}\theta$. Since there are no bound states for $g< o$, we propose in this region F to be the minimal solution of Watson's equations (37) with the negative C-parity S-matrix eigenvalue t-r given by eq.(20). For $g > o$ we take the analytic continuation [25]:

$$F(i\pi-\theta) = \frac{\text{ch}\,\frac{1}{2}\,\theta}{\text{ch}\,\frac{1}{2}\,\lambda\theta}\,\exp\int_0^\infty \frac{dx}{x}\,\frac{\text{sh}\,\frac{x}{2}(1-\frac{1}{\lambda})}{\text{sh}\,\frac{x}{2\lambda}\text{ch}\,\frac{x}{2}}\,\frac{\sin^2\frac{x\theta}{2\pi}}{\text{sh}x}\,. \tag{41}$$

This formula was checked [25] in perturbation theory $g\to o$ and in the classical limit $\beta \to 0$ for which one gets the classical soliton solution eq.(2).

4.2 Z(2)-Ising model

For the Z(2)-Ising model in the scaling limit the problem of calculating arbitrary matrix elements of the order variable

$$\langle o|\,\sigma(o)\,|\,p_1\ldots p_n\rangle^{in} = F^{(n)}(\theta_{12},\ldots\theta_{n-1,1}) \tag{42}$$

has been solved completely [27]. The minimal solution of Watson's equations with $S^{(2)}= -1$ is

$$F^{min}(\theta) = -\,i\,\text{sh}\,\frac{\theta}{2}. \tag{43}$$

Since there is no reflection we can use formula (40)

$$F^{(n)}(\theta_{12},\ldots) = K^{(n)}(\theta_{12}\ldots) \prod_{i<j} F^{min}(\theta_{ij}). \tag{44}$$

The function K contains the poles from one-particle states in all subchannels $p_{i1}+\ldots+p_{ij}$. For n=3 we obtain

$$K^{(3)} = \frac{\text{const}}{(p_1+p_2+p_3)^2-m^2} = \frac{Z^{(3)}}{\text{ch}\,\frac{\theta_{12}}{2}\,\text{ch}\,\frac{\theta_{13}}{2}\,\text{ch}\,\frac{\theta_{23}}{2}}. \tag{45}$$

The constant $Z^{(3)}$ is determined to be $Z^{(3)}=2$ by the requirement that $F^{(3)}$ reproduces the correct S-matrix $S^{(2)}=-1$. For $n\geq 5$ the arguments are just slightly more involved. Again the functions $K^{(n)}$ must have poles at $(p_i+p_j+p_k)^2 = m^2$ where any set of particles i,j,k are combined to give a one-particle intermediate state. However, the residues of "higher poles" where more than three particles build up a one-particle state must vanish, since the presence of such poles would be in contradiction to the absence of particle production. Hence the minimality hypothesis implies that $K^{(n)}$ has the form

$$K^{(n)} = \frac{\text{const}\left[\prod_{i<j}(p_i+p_j)^2\right]^{\frac{n-3}{2}}}{\prod_{i<j<k}\left[(p_i+p_j+p_k)^2-m^2\right]} = \frac{Z^{(n)}}{\prod_{i<j}\text{ch}\,\frac{\theta_{ij}}{2}} \tag{46}$$

The numerator in eq.(46) is necessary to modify the severe singularity of the denominator at $p_i+p_j=0$ in the matrix element (42) into simple poles. The constant $Z^{(n)}$ can again be determined to be $Z=2^{n-1/2}$ by crossing and taking the off shell leg $\sigma(x)$ in eq.(42) on shell. Finally we obtain

$$\langle o|\,\sigma(o)\,|p_1\ldots p_n\rangle^{in} = (2\,i)^{\frac{n-1}{2}} \prod_{i<j} \text{th}\,\frac{\theta_{ij}}{2}. \tag{47}$$

By crossing only the connected part of the matrix element is given by analytic continuation. The complete matrix element can be determined as follows e.g. for n=3:

For the transition $\langle o|\sigma(o)|p_1p_2p_3\rangle^{in} \to \langle p_1|\sigma(o)|p_2p_3\rangle^{in}$

we have to replace $\theta_{1i} \longrightarrow i\pi - \theta_{1i}$

this means $\text{th}\,\frac{\theta_{1i}}{2} \longrightarrow P(\text{th}\frac{\theta_{1i}}{2})^{-1} + f\,\delta(p_1-p_i)$

where P is the principal part and the δ-function part is determined by the disconnected part of the matrix element. It can be calculated by considering the asymptotic LSZ-limit $t\to\pm\infty$:

$$t \to +\infty \quad \langle p_1 | a^{out}(p) \, | p_2 p_3 \rangle^{in} = (\delta_{p_1 p_2} \, \delta_{p p_3} + 2 \leftrightarrow 3) \, S^{(2)}$$

$$t \to -\infty \quad \langle p_1 | a^{in} (p) | \, p_2 p_3 \rangle^{in} = \delta_{p_1 p_2} \, \delta_{p p_3} + 2 \leftrightarrow 3$$

We find that f is proportional to $1+S^{(2)}$. Since for the Z(2)-Ising model $S^{(2)}=-1$ there is only the principal part term. Hence

$$^{out}\langle p_1 \ldots p_m | \sigma(o) | \ldots p_n \rangle^{in} = (2\,i)^{\frac{n-1}{2}} \prod_{i<j\leq m} \text{th}\, \frac{\theta ij}{2}$$

$$\prod_{r\leq m<s} P(\text{th}\, \frac{\theta rs}{2})^{-1} \prod_{m<k<\ell} \text{th}\, \frac{\theta k\ell}{2} \tag{48}$$

is the complete matrix element of the order variable.

5. Green's Functions

Till now the problem of constructing Green's functions has only been solved for the simplest field theory with soliton behaviour, the Z(2)-Ising model in the scaling limit $T \to T_c$ [27]. Having deduced all matrix elements of the σ-field, we can immediately write down expressions for the Green's functions. For example the two point function is

$$\tau(p) = \int d^2 x \, e^{ipx} \langle T \, \sigma(x) \, \sigma(o) \rangle = i \int d\varkappa^2 \, \frac{\rho \, (\varkappa^2)}{p^2 - \varkappa^2 + i\varepsilon}$$

with the spectral function ρ given by

$$\rho \, (p^2) = \sum_{n \text{ odd}} \frac{1}{n!} \int \frac{d\theta_1}{4\pi} \ldots \int \frac{d\theta_n}{4\pi} \, 2\pi \, \delta^{(2)} (p - \Sigma p_i) \prod_{1\leq i<j\leq n} \text{th}^2 \, \frac{\theta_{ij}}{2} \, .$$

There are similar formulae for the n-point Green's functions which are all in agreement with those derived in Ref. [28]. This agreement supports the validity of the minimality hypothesis and gives us some confidence in the feasibility of the bootstrap program. The determination of the correlation functions for one of the more involved soliton field theories remains, however, a challenging open problem.

ACKNOWLEDGEMENT

This paper is based on collaborations at the Institut für Theoretische Physik, FU-Berlin with B. Berg, V. Kurak, B. Schroer, R. Seiler, H.J. Thun, T.T. Truong and P. Weisz which I would like to thank for many discussions.

REFERENCES

1. M. Karowski, H. J. Thun, T. T. Truong, P. Weisz, Phys.Lett. 67B, 321 (1977);
A.B. and Al. B. Zamolodchikov, Nucl.Phys. B133, 525 (1978); Phys.Lett. 72B, 481 (1978); R. Shankar and E. Witten, Phys.Rev. D17, 2134 (1978); B. Berg and P. Weisz, Nucl.Phys. B146, 205 (1978); V. Kurak and J. A. Swieca, Phys.Lett. 82B, 289 (1979); R. Köberle and J. A. Swieca, Factorizable Z(N) Models, S. Carlos preprint (1979).
2. M. S. Ablowitz, O. J. Kaup, A. C. Newell, N. Segur, Phys.Rev. Lett. 31, 125 (1973); L. A. Takhtadzhyun and L. D. Faddeev, Theor.Math.Phys. 21, 1046 (1975).
3. R. Dashen, B. Hasslacher and A. Neveu, Phys.Rev.D10, 4114, 4130, 4138, (1974), D11, 3424 (1975).
4. S. Coleman, Phys.Rev. D11, 2088 (1975); S. Mandelstam, Phys. Rev. D11, 3027 (1975); R. Seiler, D. Uhlenbrock, Ann.Phys. 105, 81 (1977); B. Schroer, T. T. Truong, Phys.Rev. D15, 1684 (1977).
5. D. Gross and A. Neveu, Phys.Rev. D10, 3235 (1974).
6. R. Shankar and E. Witten, Nucl.Phys. B141, 349 (1978).
7. R. Dashen, B. Hasslacher and A. Neveu, Phys.Rev. D12, 2443 (1975).
8. A. B. and Al. B. Zamolodchikov, c.f. ref. [1] .
9. B. Berg and W. Weisz, c.f. ref. [1] ; V. Kurak and J. A. Swieca, c.f. ref. [1] .
10. R. Köberle and J. A. Swieca, c.f. ref. [1] .
11. M. D. Kruskal and D. Wiley, American Mathematical Society, Summer Seminar on Nonlinear Wave Motion, ed. A. C. Newell, Potsdam, N.Y. July 1972; B. Berg, M. Karowski and H. J. Thun, Phys.Lett. 62B, 187 (1976), 64B, 286 (1976); B. Yoon, Phys.Rev. D13, 3440 (1976); R. Flume, D. K. Mitter, N. Papanicolaou, Phys.Lett. 64B, 289 (1976); P. P. Kulish, E. R. Nissimov, Pisma v JETP 24, 247 (1976).
12. K. Pohlmeyer, Comm.Math.Phys. 46, 207 (1976); M. Lüscher, K. Pohlmeyer, Nucl.Phys. B137, 48 (1978); H. Eichenherr, M. Forger, preprint Freiburg (Germany) THEP 79/2; E. Brezin, C. Itzykson, J. Zinn-Justin, J. B. Zuber, preprint Saclay (France) DPh-T/79/1.

13. W. Zimmermann, Ann.Phys. 77, 536 (1973); M. Gomes, J. H. Lowenstein, Phys.Rev. D7, 550 (1973).
14. R. Flume, Phys.Lett. 62B, 93 (1976) and corrigendum; B. Berg, M. Karowski, H. J. Thun, Phys.Lett. 62B, 633 (1976); Nuovo Cimento 38A, 11 (1977); R. Flume, S. Meyer, Nuovo Cimento Lett. 18, 236 (1977); E. R. Nissimov, Bulg. Journ. Phys.L. 113 (1977); J. H. Lowenstein, E. R. Speer, Commun.Math.Phys. 63, 97 (1978); preprint, New York, NYU/TR3/7.
15. M. Lüscher, Nucl.Phys. B135, 1 (1978).
16. P. P. Kulisch, Theor.Math.Phys. 26, 198 (1976); D. Iagolnitzer, Phys.Rev. D18, 1275 (1978).
17. B. Berg, M. Karowski, W. Theis, H. J. Thun, Phys.Rev. D17, 1172 (1978).
18. B. Berg, M. Karowski, V. Kurak, P. Weisz, Nucl.Phys. B134, 125 (1978).
19. M. Karowski, H. J. Thun, Nucl.Phys. B130, 295 (1977); A. B. Zamolodchikov, Moscow preprint ITEP 12/1977.
20. M. Karowski, Nucl.Phys. B153, 244 (1979).
21. A. B. Zamolodchikov, Commun.Math.Phys. 55, 183 (1977).
22. P. Weisz, Nucl.Phys. B122, 1 (1977).
23. B. Schroer, T. Truong, P. Weisz, Phys.Lett. 63B, 422 (1976).
24. M. Karowski, H. H. Thun, to be published.
25. P. Weisz, Phys.Lett. 67B, 179 (1977); A. B. Zamolodchikov, Moscow preprint ITEP 112 (1977).
26. M. Karowski, P. Weisz, Nucl.Phys. B139, 455 (1978).
27. B. Berg, M. Karowski, P. Weisz, Phys.Rev. D19, 2477 (1979).
28. R. Z. Bariev, Phys.Lett. 55A, 456 (1976); B. M. McCoy, C. A. Tracy, T. T. Wu, Phys.Rev.Lett. 38, 793 (1977); M. Sato, T. Miwa, M. Jimbo, Proc.Japan Acad. 53A, 6 (1977).

SCREENING AND CONFINEMENT IN SOLUBLE MODELS

J.A. Swieca

Dep.de Fisica, Universidade Federal de São Carlos,
São Carlos, S.P., Brazil and Dep.de Fisica, PUC,
Rio de Janeiro, Brazil

Features of screening and confinement are discussed in the Schwinger and chiral Gross-Neveu models.

I. INTRODUCTION

In these lectures I want to report on work done during the last year [1,2,3] which is directly related with screening and/or confining properties of model field theories.

The first lecture (section II) will be devoted to an analysis of the behaviour of static potential between quarks in a supposedly confining gauge theory (the Schwinger model).

We will see that as a result of vacuum polarization effects the growing potential characteristic of confining theory saturates after a critical distance corresponding to the break up of the original hadron into pairs [1,4,5] . For sufficiently light quarks the behaviour of these potentials suggest screening rather than confinement of the colour degrees of freedom. This is in agreement with the fact that operator solutions of the Schwinger model with flavour do allow for states [2,6] whose only interpretation is that of a bleached quark i.e. one which has lost its colour but not the fundamental flavour.

In the second lecture (section III) we will deal with screening aspects of a non-gauge theory: the chiral Gross-Neveu model [7] . Here as first stressed by Witten [8] one has a dynamically generated fermion mass without spontaneous breakdown of the chiral symmetry.

This apparent paradox is resolved by realizing that the chirality (and the charge) of the physical fermions are screened. Both those quantum numbers are carried by a free massless ("gauge") excitation which completely decouples from what we would call physical states. The interpolating field for the physical fermions obeys generalized statistics [9] and satisfies a remarkable algebraic identity which means that anti-particles are bound states of particles [3] . This property together with the expected S-matrix factorizability of this model [10,11,12] allows one to write down its exact S-matrix [3, 13] .

It is a pleasure to thank my friends and collaborators, L.V. Belvedere, R. Koberle, K. D. Rothe, H. Rothe, V. Kurak and B. Schroer for their help and encouragement.

II. SCREENING VERSUS CONFINEMENT

Recent theoretical work [14,15] has taught us that we can expect confinement of quarks via a mechanism of flux tube formation dual to that known to occur in a superconductor [16] . This rising potential implied by those flux tubes [17] is tested by the Wilson loop criterion usually applied to the pure gauge theory, quarks being considered as external probes. One may well expect that if the quarks are sufficiently heavy, freezing the vacuum in this way will not affect the confining features of the theory since in this case there should be a critical length L_C ($L_C \sim$ hadronic size) below which pair production is negligible. For distances larger than L_C however, polarization effects will take over and destroy the rising potential picture. By staying in two dimensional space-time where the existence of flux tubes is a direct consequence of Gauss law one may get a more detailed understanding of the influence of vacuum polarization effects on the confining potential.

It should be clear that in two dimensions colour confinement i.e. the absence of coloured states in the physical space is automatically guaranteed.*) Indeed consider the dipole state

$$|D\rangle = \psi(\underset{\sim}{x})\ P\ e^{ig \int_{\underset{\sim}{x}}^{\underset{\sim}{y}} \underset{\sim}{A}(\underset{\sim}{z}) d\underset{\sim}{z}} \psi^*(\underset{\sim}{y})\,|0\rangle \qquad \text{(II.1)}$$

Since in the temporal gauge the electric field is the canonical momentum conjugate to A it is obvious that the expectation value of the Hamiltonian

*) Notice that this argument does not depend on the dimensionality of space. In higher dimensions the real dynamical problem related to flux tube formation is to find out whether (II.1) is the lowest energy state compatible with Gauss law.

$$H = \frac{1}{2}\int (\underset{\sim}{E}^2 + \ldots) dx_1$$

grows linearly with the separation $|\underset{\sim}{x} - \underset{\sim}{y}|$. This means that one cannot isolate states carrying colour. Colour confinement however could mean that either the colour of the quarks has been screened or that the quarks have been permanently bound into hadrons (confinement proper). Those rather different mechanisms for the absence of coloured states will correspond to a drastically different behaviour of the quark potential as we shall now illustrate in the Schwinger model.

We start from the Hamiltonian for the massive Schwinger model in the presence of two external charges $\pm Q$ located at $\pm L$ in bosonized form [18]

$$H = \frac{1}{2}\int dx_1 \{\dot{\Sigma}^2 + (\nabla\Sigma)^2 + \mu^2(\Sigma-\phi)^2 + \frac{M^2}{2\pi}[1-\cos(2\sqrt{\pi}\Sigma)]\} \qquad \text{(II.2)}$$

where

$$\mu = \frac{e}{\sqrt{\pi}}\ ; \ M \sim \text{quark mass;} \quad \Sigma = \text{quantized electric field;}$$

and

$$\phi_L(x^1) = \sqrt{\pi}\ \frac{Q}{e}\ \theta(L+x^1)\theta(L-x^1) \qquad \text{(II.3)}$$

is the external electric field of the test charges. Here we have chosen the cosmological angle θ [19,20] equal to zero. The potential between the external charges including all polarization effects is given by

$$H(L) = \min_{\psi} \langle\psi|H|\psi\rangle \qquad \text{(II.4)}$$

The behaviour of $H(L)$ for large L was found in Ref. [18] to approach a constant for values of Q which are integer multiples of e. This is clearly an effect of the dominating role of vacuum polarization for large separations. Here we wish to obtain a more detailed picture of the behaviour of $H(L)$ for all L's. Since we do not hope to solve the quantum mechanical problem exactly we will replace it by the corresponding classical one:

$$H(L) = \min \int dt\{\frac{1}{2}\dot{q}^2 - V_L(q,t)\} \qquad \text{(II.5)}$$

where we have replaced x^1 by t and Σ by q, and

$$V_L(q,t) = -\frac{\mu^2}{2}(q-\phi_L(t))^2 - \frac{M^2}{4\pi}[1-\cos(2\sqrt{\pi}q)] \qquad \text{(II.6)}$$

To obtain H(L) eq.(II.5) we have to consider the notion of a particle in the potential (II.4) subject to the boundary conditions that for $t \to \pm\infty$ it is at the origin to ensure finiteness of H(L). During its motion the particle changes suddenly at $t=-L$ from a potential V_0 to another potential V_Q keeping its kinetic energy, and reverses its motion at t=o. V_0 and V_Q are given by

$$V_0 = -\frac{\mu^2}{2}(q)^2 - \frac{M^2}{4\pi}\,[1-\cos(2\sqrt{\pi}q)] \tag{II.7}$$

$$V_Q = -\mu^2\,(q-\sqrt{\pi}\,\frac{Q}{e}) - \frac{M^2}{4\pi}\,[1-\cos(2\sqrt{\pi}q)]$$

In computing H(L) using this mechanical analogy it is convenient to use the relation

$$\frac{dH(L)}{dL} = -2\varepsilon(L) = -2\mu q(-L)Q + Q^2 \tag{II.8}$$

where $\varepsilon(L)$ is the energy of the particle when it moves on the potential V_Q. Eq.(II.8) follows from the minimal action principle since the boundary conditions are L independent. The total time the particle spends on V_Q is given by

$$2L = 2\int_{q(-L)}^{q(o)} \frac{dt}{\sqrt{\varepsilon - V_Q(q)}} \tag{II.9}$$

From (II.9) one finds $\varepsilon(L)$ as an explicit function of L and therefore from (II.8)

$$H(L) = \int_o^L -2\varepsilon(L')dL' \tag{II.10}$$

For the massless Schwinger model one expects screening to set in immediately [21] since pair production is energetically cheap and physical states are collective excitations analogous to plasmons. The potentials V_0 and V_Q for this case are shown in Fig. 1 . Solving (II.9) for $\varepsilon(L)$ one finds in this case

$$\varepsilon(L) = -\frac{1}{2}Q^2e^{-2\mu L}$$

so that one arrives with (II.10) at the well known result

$$H(L) = \frac{Q^2}{2\mu}\,(1-e^{-2\mu L}) \tag{II.11}$$

The linearly growing part of this potential corresponding to the region where pair production is negligible occurs only for $L \ll 1/\mu$ whereas the size of the plasmon is of the order of $\frac{1}{\mu}$ definitely forbidding the interpretation of the plasmon as a quark-antiquark bound state. Equation (II.10) is typically a screening potential. Notice that the fact that this potential saturates for an arbitrary value of Q is a consequence of the equal height of the maxima of V_0 and V_q (Fig.1) in our mechanical analog model.

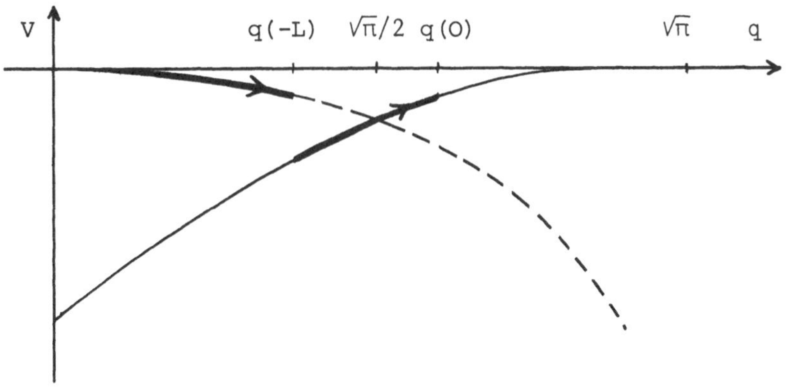

Fig.1 Potentials V_0 (dotted line) and V_Q (solid line) for M=O. Heavier line with arrow represents particle motion.

As soon as a mass is given to the fermion this equality of the heights of the maxima will happen only for Q=ne. The fact that for $M \neq o$ H(L) grows indefinitely for Q not an integer multiple of e is sometimes taken as an indication for confinement. However, one should probe the system with the charges occurring in the theory. For Q=ne and

$$\frac{M^2}{\mu^2} \ll 1$$

one gets the same qualitative behaviour as in (II.11) indicating again screening and not confinement.

Let us turn now to the opposite situation

$$\frac{M^2}{\mu^2} \gg 1$$

(Fig.2 corresponds roughly to $\frac{M^2}{\mu^2} = 10$ and Q=e). For $M \sim \mu$ a novel feature appears which is the presence of secondary maxima. As a result there are 3 classes of motions satisfying the boundary conditions. In class one the particle jumps from V_0 to V_Q at a point between O and

$$\frac{\sqrt{\pi}\,\mu^2}{2M^2}$$

with a turning point that approaches the local maximum of V_Q when $L \to \infty$. This gives rise to a linearly rising branch of $H(L)$, $0 \leq L \leq \infty$

$$H_I(L) = \left(1-\frac{\mu^2}{M^2}\right) M\mu^2 L + \frac{\pi\mu^4}{2M^3}\left(1-e^{-2ML}\right) \tag{II.12}$$

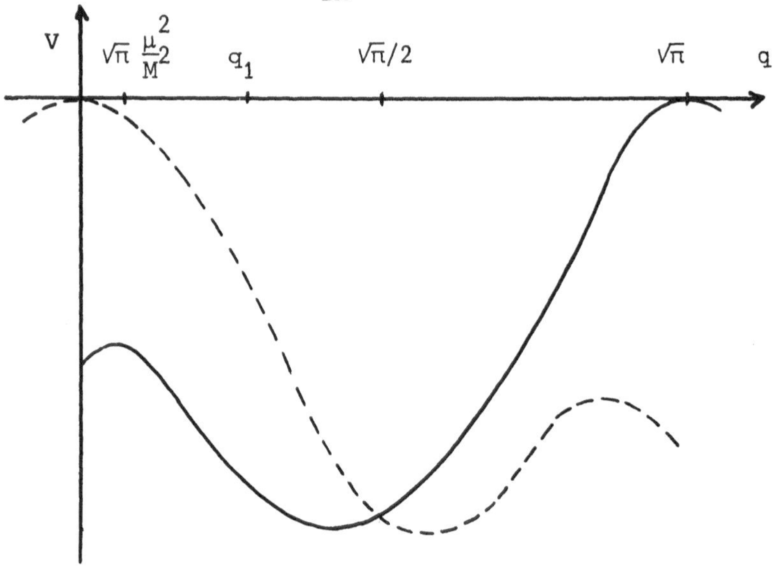

Fig.2 Potentials V_O (dotted line) and V_Q (solid line) for $M>e$, $\theta=0$.

The second and third class of solutions exist only for

$$\frac{1}{M} \ln \frac{M^2}{\mu^2} \lesssim L < \infty$$

with the jumping point $q(-L)$ lying in the ranges

$$\left[\frac{\sqrt{\pi}\mu^2}{2M^2}, q\right] \quad \text{and} \quad \left[q, \frac{\sqrt{\pi}}{2}\right]$$

respectively. From (II.8) one realized that $H_{II}(L) \geq H_{III}(L)$, and up to corrections of order μ/M

$$H_{III}(L) = \frac{4\pi M}{\pi}\left(1-2e^{-2ML}\right) \tag{II.13}$$

$$L > \frac{1}{M}$$

It is clear that the true potential $H(L)$ is obtained by going smoothly from branch I to branch III (see Fig. 3) at the intersection point

$$L = L_c \simeq \frac{4\pi}{e^2}$$

thus guaranteeing for all L the minimum of the action (II.5). As is clearly seen from Fig. 3 for $M \gg \mu$ there exists a rather big region within which the quark potential is linearly rising. In this regime one can indeed speak of quark confinement since the size of a quark-antiquark bound state can be estimated to be of order of

$$\left(\frac{e}{M}\right)^{4/3} L_c \ll L_c .$$

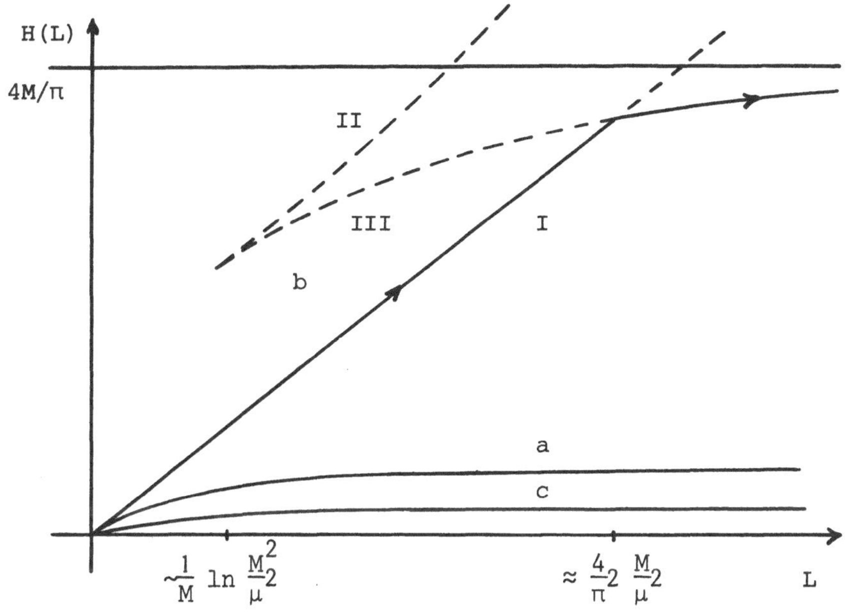

Fig. 3 Interparticle potential H(L) in the three cases a), b) and c). Dotted lines represent metastable states of the polarization cloud.

Neutralization of colour therefore happens via hadron formation with M setting the scale of hadronic masses. For $L > L_c$ also here polarization effects become important and the whole potential picture breaks down via hadron production. The Yukawa behaviour of H(L) for $L > L_c$ corresponds to a residual interaction between colour neutral hadrons.

Decreasing now the ratio $\frac{M}{\mu}$ for $M \sim \mu$ the secondary maxima of V_Q disappear corresponding to a saturation of branch I and eventually the three branches merge into one leading us again to the screening potential (II.11).

It is also instructive to discuss what happens if one introduces a cosmological angle Θ into the Hamiltonian (II.2). For a generic Θ the qualitative picture of screening vs. confinement as a function of $\frac{M}{\mu}$ remains unchanged. For $\Theta = \pi$ and $M > \mu$ a new phenomenon comes into play since due to a spontaneous symmetry breakdown kink states make their appearance [2,6] . Indeed, as is clearly seen from Fig.4 there is in this case an additional degeneracy between the maxima of V_0 and V_Q.

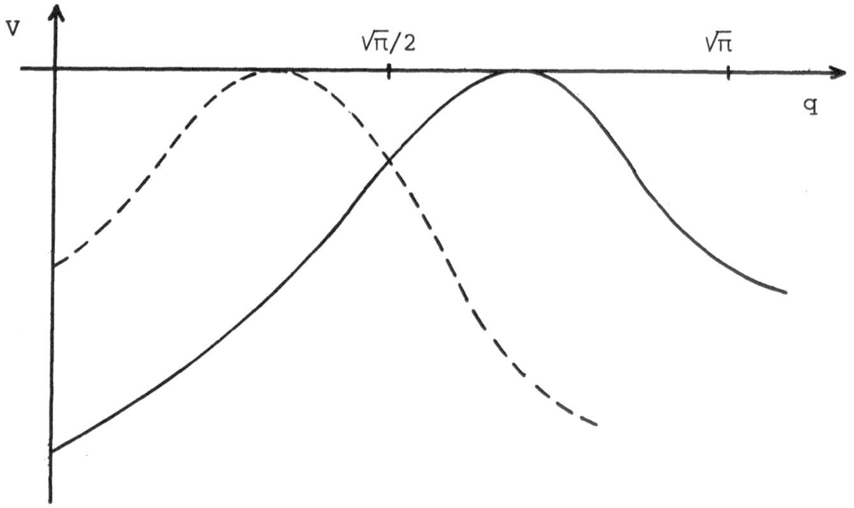

Fig.4 Potentials V_0 (dotted line) and V_Q (solid line) for $M > e$, $\Theta = \pi$.

Since those maxima are separated by a distance of only $\sqrt{\pi}\,\frac{\mu^2}{M^2}$ the motions in the equivalent mechanical problem minimizing H(L) will take place between them. In this case just as in the massless Schwinger model one finds

$$H(L) = \frac{\pi}{2}\,\frac{\mu^4}{M^3}\,(1-e^{-2ML}) \tag{II.14}$$

This very shallow potential corresponds to screening effects even more violent than those occurring in the massless Schwinger model: the kink states appear as bleached quarks.

It is interesting to compare the picture we developed based on the quark potential with what is known from operator solutions of the Schwinger model [2,6,19] . In the massless Schwinger model the physical (gauge invariant) quark field is a coherent state of plasmons:

$$\hat{\psi} = e^{i\sqrt{\pi}\Sigma}$$

By introducing flavour into the model $\hat{\psi}$ will carry fundamental flavour. This can only mean that the charge of the quark has been screened. The addition of a fermion mass gives infinite energy to $\hat{\psi}$ |o> eliminating such a bleached quark from the physical spectrum. For $\Theta = \pi$ however, the state K $\hat{\psi}$|o> with K a kink operator [22] will have finite energy and carry fundamental flavour: the bleached quark has reappeared.

What lessons can we draw from our mechanical exercise concerning a realistic confining theory?

There is no doubt that the saturating property of the inter-quark potential will come out automatically whenever the quark vacuum is unfrozen [5] .

A deeper probing of the theory is also required if one wants to be sure that not only colored states are absent but that the colour carriers (quarks) bind themselves into hadrons according to the confining picture phenomenologists employ. The loop criterion applied to the pure gauge theory is therefore a necessary but not sufficient condition for a reasonable confining theory.

III. SCREENING IN THE CHIRAL GROSS-NEVEU MODEL

In this lecture we will examine screening effects which are completely unrelated to any confining mechanism providing thus a counterpoint for our discussion of the preceding section. The model characterized by the Lagrangian

$$L = i \sum_{i=1}^{N} \bar{\psi}_i \not{\partial} \psi_i + \frac{g^2}{2N} [(\sum_1^N \bar{\psi}_i\psi_i)^2 - (\sum_1^N \bar{\psi}_i\gamma^5\psi_i)^2] \tag{III.1}$$

would like [7] to have a spontaneous chiral symmetry breakdown which is however inhibited by the strong massless fluctuations which forbid any spontaneous breakdown of a continuous symmetry in two dimensions [23] . Nevertheless as in the more often studied X-Y model [24,25] it has a phase with short range correlations [8] with a dynamically generated fermion mass. On the other hand there are many indications coming from classical, semi-classical and perturbative computations [10-12, 26] that it belongs to the privileged class of models with factorizable S-matrices [27-30] which amounts to its exact solubility, at least at the S-matrix level.

As in the previous section it is convenient to use the bosonized form of the model. In terms of bose variables [31]

$$\psi_i(x) = K_i \sqrt{\frac{m}{2\pi}}\, e^{i\frac{\pi}{4}\gamma^5} \exp\{ - i\sqrt{\frac{\pi}{N}}[\gamma^5\phi(x) + \tag{III.2a}$$

$$+ \int_x^{\infty} dz^1 \, \phi(z)] - i\sqrt{\pi}\, [\gamma^5 \hat{\phi}_i(x) + \int_x^{\infty} dz^1 \, \hat{\phi}_i(z)]\}$$

$$i = 1, \ldots N \qquad \text{(III.2a)}$$

$$\sum_{i=1}^{N} \hat{\phi}_i(x) = 0 \qquad \text{(III.2b)}$$

where K_i is a Klein factor needed to ensure the anticommutation relations for the different SU(N) components. $\phi(x)$ is the pseudo-potential of the U(1) current and $\hat{\phi}_i(x)$ are the pseudo-potentials of the diagonal SU(N) currents. The numerical factors in the exponential ensure that ψ_i is a fermion field with the same short distance singularities as the free field in accordance with the asymptotic freedom of the Lagrangean (III.1).

As a consequence of the U(1)xU(1) symmetry ϕ is a massless canonical free field implying that $\psi(x)$ describes infraparticles [32] , and has long ranged correlation functions. In order to set up a proper scattering theory for this model and exhibit its real particle content one introduces the field $\hat{\psi}$ defined as

$$\hat{\psi}_i(x) = K_i \sqrt{\frac{m}{2\pi}}\, e^{i\frac{\pi}{4}\gamma^5} \exp\{-i\sqrt{\pi}[\gamma^5\hat{\phi}_i(x) + \int_x^{\infty} dz'\hat{\phi}_i(z)]\} \qquad \text{(III.3)}$$

thus "gauging" away the infrared cloud surrounding the fermions. In this process ψ becomes a bona-fide interpolating field for the model and the charge and pseudo charge of the theory are screened. Notice the formal analogy of going from ψ to $\hat{\psi}$ in the massless Schwinger model.

Carrying neither charge nor pseudo-charge $\hat{\psi}$ transforms according to SU(N) and therefore it is not surprising it satisfies (cf. Eq.III.3) the following algebraic identity

$$\hat{\psi}_i^{*} = \frac{1}{(N-1)!} (\prod_e Ke)\, \varepsilon_{ij_1j_2\ldots j_{n-1}} \hat{\psi}_{j_1} \ldots \hat{\psi}_{j_{n-1}} \qquad \text{(III.4)}$$

where a suitable normal product prescription for the r.h.s. of (III.4) is understood. In a theory with particle content eq.(III.4) means that antiparticles should be identified with bound states of N-1 particles. It also follows from (III.4) that $\hat{\psi}$ is neither a bose nor fermi field but rather carries "spin"

$$s - \frac{1}{2}(1-1/N).$$

Therefore one expects an appropriate collision limit to reveal asymptotic states with generalized statistics [9].

In order to set up the scattering theory appropriate to the field $\hat{\psi}$ we find it convenient to introduce an auxiliary field ψ', in terms of which a conventional scattering theory can be carried through:

$$\psi'_i(x) = e^{i\sqrt{\frac{\pi}{N}}\,[\gamma^5 A(x)+B(x)]}\,\psi_i(x) \qquad \text{(III.5)}$$

where A and B are two independent auxiliary free massless fields quantized with a metric opposite to the one of ϕ. Their role as can be seen from (III.2) and (III.5) is to compensate the infrared structure of ψ without affecting its statistics. In this way the U(1)xU(1) charges are spurionized by a mechanism analogous to the one occurring in the Schwinger model [19] , with the essential difference that here the spurions carry non-trivial "spin". It is thus clear that the Greens function of ψ' should have the structure of a conventional massive fermi theory.

Following Araki-Haag [33] we have

$$\psi'(vt,t) \underset{t\to\pm\infty}{\longrightarrow} \frac{1}{\sqrt{|t|}}\,[e^{-im\gamma t}\, a_{\substack{out\\in}}(m\gamma v) + e^{im\gamma t}\, b^{+}_{\substack{out\\in}}(m\gamma v)] \qquad \text{(III.6)}$$

with a and b^+ the usual annilitation and creation operator for fermions resp. antifermions and $\gamma = (1-v^2)^{-1/2}$.

For $\hat{\psi}$ we envisage a similar asymptotic limit

$$\psi(vt,t) \underset{t\to\pm\infty}{\longrightarrow} \frac{1}{\sqrt{|t|}}\,[e^{-im\gamma t}\, \hat{a}_{\substack{out\\in}}(m\gamma v) + e^{im\gamma t}\, \hat{b}^{+}_{\substack{out\\in}}(m\gamma v)] \qquad \text{(III.7)}$$

From the equal time commutation relations for $\hat{\psi}$

$$\hat{\psi}(x,t)\hat{\psi}(y,t) = e^{2\pi i S\varepsilon(x-y)}\,\hat{\psi}(y,t)\hat{\psi}(x,t) \qquad \text{(III.8)}$$

we formally derive using the limit (III.7) that

$$\hat{a}^{+}_{\substack{out\\in}}(p)\,\hat{a}^{+}_{\substack{out\\in}}(p') = e^{\pm 2\pi i S\varepsilon(p-p')}\,\hat{a}^{+}_{\substack{out\\in}}(p')\,\hat{a}^{+}_{\substack{out\\in}}(p) \qquad \text{(III.9)}$$

and the same for b^+. The fields ψ' and $\hat{\psi}$ differing only by a spurion should describe the same asymptotic states with a different asignment of statistics [9] . Therefore one should be able to express $\hat{a},\hat{b}$ in terms of a,b and vice-versa:

$$\hat{a}^+_{in}(p) = a^+_{in}(p)e^{2\pi i(S-\frac{1}{2})\int_p^\infty N_{in}(p')dp'} \tag{III.10a}$$

where N_{in} is the number operator. From PT the corresponding expression for the out operators is

$$\hat{a}^+_{out}(p) = a^+_{out}(p)e^{2\pi i(S-\frac{1}{2})\int_{-\infty}^p N_{out}(p')dp'} \tag{III.10b}$$

Equations (III.9) generalize expressions obtained in the investigation of the scaling limit of the Ising model [34] . They allow us to relate S-matrices corresponding to different statistics assignment for the particles of a theory. It is in terms of the field $\hat{\psi}$ that the idea that antiparticles are bound states of particles can be consistently formulated. On the other hand the whole analytic factorizable S-matrix program can be carried out using ψ' as an interpolating field. From eqs. (III.10) we can establish the following identity between the S-matrix elements computed with either ψ' or $\hat{\psi}$ as interpolating fields:

$$\begin{aligned} &<0|a_{out}(p_1')\ldots a_{out}(p_n')a^+_{in}(p_1)\ldots a^+_{in}(p_n)|0> = \\ &<0|\hat{a}_{out}(p_1')\ldots \hat{a}_{out}(p_n')\hat{a}^+_{in}(p_1)\ldots \hat{a}^+_{in}(p_n)|0> \end{aligned} \tag{III.11}$$

where $p_1 > p_2 > \ldots p_n$, $p_1' > p_2' > \ldots p_n'$. A different ordering of the operators will introduce non-analytic phase factors in the S-matrix elements of $\hat{\psi}$.

In order to obtain the exact S-matrix of this model we notice that using ψ' as an interpolating field one may view our problem as having a U(n) symmetry and therefore it should belong to the general classification of [35] . On the other hand from (III.11) and (III.4) one should identify antiparticles with bound states of particles effectively reducing the symmetry of the S-matrix of SU(n). Such an identification requires:

a) There should be a bound state in the scattering amplitude of n-1 → n-1 particles in the $\{\bar{N}\}$ channel with the same mass as the original particle.

b) The scattering of this bound state with the original particle should coincide with the particle-antiparticle scattering. An immediate consequence is that the reflection amplitude must vanish and the S-matrix must belong to class II of the classification [35] . Furthermore from a) the antisymmetric particle-particle amplitude must have a bound state with mass

$$m_2 = m \frac{\sin \frac{2\pi}{N}}{\sin \frac{\pi}{N}} \tag{III.12}$$

inducing [36] the following bound state spectrum

$$m_n = m \frac{\sin \frac{n\pi}{N}}{\sin \frac{\pi}{N}} \qquad n = 1 \ldots N-1 \tag{III.14}$$

implying

$$m_{n-1} = m \tag{III.15}$$

From the above it now follows [3] that the S-matrix elements of this model are given by

$$<\delta'(\theta_2)\gamma'(\theta_1)|S|\gamma(\theta_1)\delta(\theta_2)> = u_1(\varphi)\delta_{\gamma\gamma'}\delta_{\delta\delta'}+u_2(\varphi)\delta_{\gamma\delta'}\delta_{\delta\gamma'} \tag{III.16a}$$

$$< \bar{\delta}'(\theta_2)\gamma'(\theta_1)|S|\gamma(\theta_1)\bar{\delta}(\theta_2)> =t_1(\varphi)\delta_{\gamma\gamma'}\delta_{\delta\delta'}+ t_2(\varphi)\delta_{\gamma\delta}\delta_{\gamma'\delta'} \tag{III.16b}$$

where $\varphi = (\theta_1 - \theta_2)/i\pi$ and

$$u_1(\varphi) = \frac{\Gamma\,(1-\frac{\varphi}{2}\,\Gamma(\frac{\varphi}{2}-\frac{1}{N})}{\Gamma\,(1-\frac{\varphi}{2}-\frac{1}{N})\Gamma(\frac{\varphi}{2})}\ , \quad u_2(\varphi) = -\frac{2}{N\varphi}\,u_1(\varphi) \tag{III.17a}$$

$$t_1(\varphi) = u_1(1-\varphi), \quad t_2(\varphi) = u_2(1-\varphi) \tag{III.17b}$$

Let us illustrate in detail the procedure by means of which anti-particles are identified with bound states of particles in the case of N=3. From the scattering amplitude of 3 particles with two of them projected into the $\{\bar{3}\}$ channel one finds from the residuum of this amplitude at the pole $\theta = \frac{2}{3}i\pi$ the bound state-particle amplitude

$$<\bar{\Sigma}'(\theta_2)\gamma'(\theta_1)|S|\gamma(\theta_1)\bar{\Sigma}(\theta_2)> =-(\delta_{\Sigma'\Sigma}\delta_{\gamma'\gamma}T_1+\delta_{\Sigma'\gamma'}\delta_{\Sigma\gamma}T_2) \tag{III.18}$$

with T_1, T_2 given in terms of u_1, u_2.

Using now eq.(III.11)

$$<\bar{\Sigma}'\gamma'|S|\gamma\bar{\Sigma}> = <\gamma'\bar{\Sigma}'|S|\gamma\bar{\Sigma}> = <\gamma'\bar{\Sigma}'|\hat{S}|\gamma\bar{\Sigma}> \qquad \text{(III.19)}$$

where $\hat{S}$ denotes the S-matrix computed using $\hat{\psi}$ as interpolating field. On the other hand for the particle-antiparticle amplitude

$$<\bar{\delta}'\gamma'|S|\gamma\bar{\delta}> = - <\gamma'\bar{\delta}'|S|\gamma\bar{\delta}> = - <\gamma'\bar{\delta}'|\hat{S}|\gamma\bar{\delta}> \qquad \text{(III.20)}$$

so that

$$<\gamma'\bar{\delta}'|\hat{S}|\gamma\bar{\delta}> = -(\delta_{\delta'\delta}\delta_{\gamma'\gamma}t_1 + \delta_{\gamma'\delta'}\delta_{\gamma\delta}t_2) \qquad \text{(III.21a)}$$

$$<\gamma'\bar{\Sigma}'|\hat{S}|\gamma\bar{\Sigma}> = -(\delta_{\Sigma'\Sigma}\delta_{\gamma'\gamma}T_1 + \delta_{\gamma'\Sigma'}\delta_{\gamma\Sigma}T_2) \qquad \text{(III.21b)}$$

Up to the Klein factor we see from (III.4) that the two amplitudes in Eq.(III.21) should be identical. Since the Klein factor has eigenvalues ± 1 and on the other hand it belongs to the center of SU(N) it is identically 1 for all odd N's. In the case of SU(3) therefore we should have

$$t_i = T_i \quad i = 1,2 \qquad \text{(III.22)}$$

which is satisfied by explicit computation [3] . In this way we consistently identify the bound state $\bar{\Sigma}$ with the antiparticle $\bar{\delta}$.

To illustrate the role played by the Klein factor in Eq.(III.4) let us consider the case N=2. Although in this case there is no bound state implied by the algebraic identity we can nevertheless make the necessary identification which will lead us from a U(2) invariant S-matrix to an SU(2) invariant one.

From

$$\hat{\psi}^*_\alpha = e^{2\pi i Q_3} \varepsilon_{\alpha\beta}\hat{\psi}_\beta \qquad \text{(III.23)}$$

where Q_3 is the third component of the isospin one finds

$$\hat{b}^+_\alpha = e^{2\pi i Q_3} \varepsilon_{\alpha\beta}\hat{a}^+_\beta \qquad \text{(III.24)}$$

From Eq.(III.24) we get

$$\begin{aligned}&<\bar{\gamma}(\theta_1)\delta(\theta_2)|\hat{S}|\hat{\alpha}(\theta_1)\beta(\theta_2)> = \\ &= - \varepsilon_{\gamma\gamma'}\varepsilon_{\alpha\alpha'}<\gamma'(\theta_1)\delta(\theta_2)|\hat{S}|\alpha'(\theta_1)\beta(\theta_2)>\end{aligned} \qquad \text{(III.25)}$$

where the minus sign in Eq.(III.25) arises from the Klein factor. Eq.(III.25) in conjunction with Eq.(III.11) gives

$$t_1 = -(u_1+u_2),\quad t_2 = u_2 \tag{III.26}$$

which is again satisfied by the formulae (III.17) for N=2.

Finally one can compare the proposed exact S-matrix with the result of a 1/N expansion [3]. The original Lagrangean is plagued with infrared problems [12] which have their origin in a pseudo Goldstone boson and a consequent lack of a particle structure. By employing the field ψ' we should be removing the infrared cloud. This means using the Lagrangean

$$L = i\sum_1^N \bar{\psi}'_i \not\partial \psi'_i + \frac{1}{2N} g^2 [(\sum_1^N \bar{\psi}'_i \psi'_i)^2 - (\sum_1^N \psi'_i \gamma^5 \psi'_i)^2] \tag{III.27}$$

$$- \frac{1}{2}(\partial_\mu A)^2 - \frac{1}{2}(\partial_\mu B)^2 + \bar{\psi}' [\frac{\alpha}{\sqrt{N}} \gamma^5 \not\partial A - \frac{\beta}{\sqrt{N}} \not\partial B] \psi'$$

where the minus sign in the kinetic energy terms of A and B is due to their indefinite metric quantization and α,β are the unrenormalized coupling constants. Following standard procedures [7] one finds that the renormalized values $\sqrt{\pi}$ in (III.5) correspond to $\alpha \to \infty, \beta = \sqrt{\pi}$. In that limit the Goldstone pole disappears. To lowest order in 1/N the S-matrix coincides with one that has been obtained by other methods by Abdalla, Berg and Weisz [13] reproducing to that order the proposed exact S-matrix.

REFERENCES

1. H. J. Rothe, K. D. Rothe and J. A. Swieca, Phys.Rev. D19, 3020 (1979).
2. L. V. Belvedere, K. D. Rothe, B. Schroer and J. A. Swieca, Nucl.Phys. B153, 112 (1979).
3. V. Kurak and J. A. Swieca, Antiparticles as Bound States of Particles in the Factorized S-Matrix Framework, Phys.Letters to appear; R. Köberle, V. Kurak and J. A. Swieca, Phys.Rev. D20, 897 (1979).
4. L. van Hove, Remarks on the Quark Confinement Problem, CERN preprint, 1978.
5. H. G. Dosch and V. F. Müller, Vacuum Polarization Effects in Lattice Gauge Theories, Heidelberg preprint 1979.
6. V. Kurak, B. Schroer and J. A. Swieca, Nucl.Phys. B134, 61 (1978); S. Coleman, Ann.of Phys. N.Y., 101, 239 (1976).

7. D. Gross and A. Neveu, Phys.Rev. D10, 3235 (1974).
8. E. Witten, Harvard preprint HIUTP-78/A027.
9. J. A. Swieca, Fortschr.der Physik 25, 303 (1977).
10. A. Neveu and N. Papanicolau, Comm.Math.Phys. 58, 31 (1978).
11. S. Shei, Phys.Rev. D14, 535 (1976).
12. B. Berg and P. Weisz, Nucl.Phys. B146, 205 (1978).
13. E. Abdalla, B. Berg and P. Weisz, preprint DESY 79/04.
14. C. G. Callan, R. Dashen and D. Gross, Phys.Rev. D17, 2717 (1978); L. P. Kadanoff, Rev.Mod.Phys. 49, 267 (1977); A. M. Polyakov, Nucl.Phys. B120, 429 (1977).
15. S. Mandelstam, in "Extended Systems in Field Theory", J. L. Gervais and A. Neveu (eds.), Physics Reports 23C No. 3 (1976); G. 't Hooft, Nucl.Phys. B138, 1 (1978); E. Fradkin and L. Susskind, Phys.Rev. D17, 3637 (1978); F. Englert, Cargèse Lectures 1977.
16. Y. Nambu, Phys.Rev. D10, 4262 (1974).
17. K. Wilson, Phys.Rev. D10, 2445 (1974); K. Kogut and L. Susskind, Phys.Rev. D11, 395 (1975).
18. S. Coleman, R. Jackiw and L. Susskind, Ann.Phys. N.Y. 93, 267 (1975).
19. J. H. Lowenstein and J. A. Swieca, Ann.Phys.N.Y. 68, 172 (1971).
20. S. Coleman, Ann.Phys. N.Y. 101, 239 (1976).
21. H. J. Rothe, K. D. Rothe and I. O. Stamatescu, Ann.Phys. N.Y. 105, 63 (1975); A. Casher, J. Kogut and L. Susskind, Phys.Rev. D10, 732 (1974).
22. B. Schroer and J. A. Swieca, Nucl.Phys. B121, 505 (1977).
23. S. Coleman, Commun.Math.Phys. 31, 259 (1973); H. Ezawa and J. A. Swieca, Comm.Math.Phys. 5, 330 (1967).
24. V. L. Berezinski, Sov.Phys.JETP 32, 493 (1970); ibid 34, 610 (1971).
25. J. M. Kosterlitz and D. J. Thouless, J.Phys. C6, 1181 (1973).
26. S. Meyer and R. Trostl, Phys.Lett. 79B, 429 (1978).
27. A. B. Zamolodchikov, Comm.Math.Phys. 55, 183 (1977).
28. M. Karowski, H. J. Thun, T. T. Truong and P. Weisz, Phys.Lett. 67B, 321 (1977); M. Karowski and H. J. Thun, Nucl.Phys. B130, 295 (1977).
29. A. B. Zamolodchikov and Al. B. Zamolodchikov, Nucl. Phys. B133, 525 (1977); Phys.Lett. 72B, 481 (1978).
30. R. Shankar and E. Witten, Nucl.Phys. B141, 349 (1978), Phys.Rev. D17, 2134 (1978).
31. S. Mandelstam, Phys.Rev. D11, 3026 (1975); M. B. Halpern, Phys.Rev. D12, 1684 (1975).
32. B. Schroer, Fortschr.der Physik 11, 1 (1963).
33. H. Araki and R. Haag, Comm.Math.Phys. 4, 77 (1967).
34. M. Sato, T. Miwa and M. Jimbo, RIMS preprint 207 (1976); B. Schroer and T. T. Truong, Nucl.Phys. B144, 80 (1978).
35. B. Berg, M. Karowski, V. Kurak and P. Weisz, Nucl.Phys. B134, 125 (1978).
36. B. Schroer, T. T. Truong and P. Weisz, Phys.Lett. 63B, 422 (1976).

ISING FIELD THEORY

T. T. Truong

Institut für Theoretische Physik
Freie Universität Berlin
Berlin, W.-Germany

ABSTRACT

We review the construction of quantum field operators associated with the two dimensional Ising model in the scaling limit.

Throughout the years the two dimensional Ising model has been a source of invaluable information and has provided deep forays into the understanding of the nature of interacting infinite systems. Here we would like to review briefly the recent developments pertaining to this exactly solvable system.

As it is well known the key to this solubility is the existence of an underlying free fermion structure originally discovered by L. Onsager and B. Kaufman [1], and subsequently put into a greatly simplified form by T.D. Schultz,et al.[2]. Various intensive quantities were then computed exactly in the thermodynamic limit such as free energy per spin, spontaneous magnetization...

But the exact calculation of the correlation functions has only appeared recently in the works of B. McCoy, C. A. Tracy and T. T. Wu [3], R. Z. Bariev [4], D. B. Abraham [5] due to the enormous complications of the mathematical structure of these functions. The availability of these functions raises the interesting question whether there exists a relativistic quantum field theory defined by the analytic continuation into the Minkowski region of the n-point correlation functions in the scaling limit. To put it differently, can one only construct a relativistic quantum field operator such that its n-point Wightman functions continued in the Euclidean region are precisely those obtained in the two dimensional Ising model?

The first quantum field operator theory was proposed in 1976 by M. Sato, T. Miwa and M. Jimbo [6] . They have observed that within the framework of Onsager-Kaufman, the original dynamical variables entering in the formulation of the row to row transfer matrix at a lattice site m on a row

$$s_m = I \otimes \dots \otimes \begin{pmatrix} 1 & 0 \\ 0 & -1 \end{pmatrix} \otimes \dots \otimes I$$

$$m^{th} \text{position} \qquad (1)$$

$$C_m = I \otimes \dots \otimes \begin{pmatrix} 0 & 1 \\ 1 & 0 \end{pmatrix} \otimes \dots \otimes I$$

once converted into fermion variables p_m, q_m, through a Jordan-Wigner transformation:

$$p_m = C_0 C_1 \dots C_{m-1} s_m \qquad p_0 = s_0$$

$$q_m = C_0 C_1 \dots C_m \, s_m \qquad (2)$$

or conversely

$$s_m = p_m \, t_m \qquad C_m = q_m p_m$$

with

$$t_m = q_{m-1} p_{m-1} \dots q_0 p_0 \qquad (3)$$

have "abnormal" commutation relations with these fermion variables:

$$t_m \, p_{m'} = \begin{cases} -p_{m'} \, t_m & m' < m-1 \\ p_{m'} \, t_m & m' > m \end{cases}$$

$$t_m \, q_{m'} \quad \begin{cases} -q_{m'} \, t_m & m' < m-1 \\ q_{m'} \, t_m & m' > m \end{cases} \qquad (4)$$

s_m has similar commutation relations with the opposite sign on the r.h.s. of eq.(4). To make contact with critical phenomena theory, we shall denote s_m/t_m by $\sigma(m)/\mu(m)$ known as order/disorder parameter [7] . It is known that the transfer matrix is diagonalized by a new set of fermion variables $\psi(\theta_\ell),\psi^+(\theta_\ell)$ $\ell = 1,2,\ldots,M$ through a Bogoliubov transformation which turns out to have two choices of parametrization corresponding to regimes above or below a critical temperature T_c, thereby yielding two sets of fermion variables in the thermodynamic limit $\hat{\psi}_{\gtrless}(\theta)$, $\hat{\psi}^+_{\gtrless}(\theta)$ (θ is now a continuous variable in the interval $(-\pi,\pi)$). Heuristically through the kinematical equations (4) $\mu(m)$ can be viewed as defining "singular" rotation of angle π/o (mod 2π) if the argument of μ is left/right of the fermion argument. Sato [6] have succeeded in computing an explicit form for both μ and σ in the basis of the fermion degrees of freedom $\hat{\psi}_{\gtrless}(\theta),\hat{\psi}^+_{\gtrless}(\theta)$. For example for $T>T_c$ at a lattice site (m,n) they are the $\ast\ast$ ordered expressions:

$$\mu_>(m,n) = {}^\ast_\ast \exp M_1(m,n){}^\ast_\ast \tag{5}$$

$$\sigma_>(m,n) = {}^\ast_\ast \hat{\psi}_0(m,n) \exp M_1(m,n){}^\ast_\ast$$

with

$$\hat{\psi}_0(m,n) = \int_{-\pi}^{\pi} \frac{d\theta}{2\pi}\frac{1}{b_>(\theta)}\{e^{-im\theta-n\gamma(\theta)}\hat{\psi}^+(\theta)+e^{im\theta+n\gamma(\theta)}\hat{\psi}(\theta)\}$$

$$M_1(m,n) = \frac{1}{2}\int\!\!\int_{-\pi}^{\pi}\frac{d\theta}{2\pi}\frac{d\theta'}{2\pi}(\hat{\psi}^+(\theta)\hat{\psi}(\theta))\begin{pmatrix}\hat{R}^{--} & \hat{R}^{-+}\\ \hat{R}^{+-} & \hat{R}^{++}\end{pmatrix}\begin{pmatrix}\hat{\psi}^+(\theta')\\ \hat{\psi}(\theta')\end{pmatrix}$$

and

$$\hat{R}^{\varepsilon\varepsilon'}=\{\varepsilon\left|\frac{|b_>(\theta)|}{|b_>(\theta')|}\right|-\varepsilon'\frac{b_>(\theta)}{b_>(\theta)}\}\frac{e^{i(m-1)(\varepsilon\theta+\varepsilon'\theta')+n(\varepsilon\gamma(\theta)-\varepsilon'\gamma(\theta'))}}{1-e^{i(\varepsilon\theta+\varepsilon'\theta'-io)}}$$

$$\varepsilon,\varepsilon' = \pm 1.$$

The factors $b_>(\theta)$ and $\gamma(\theta)$ are defined by the diagonalization conditions of the transfer matrix:

$$\cosh\gamma(\theta) = \cosh 2K_1 \cosh 2K_2^* - \sinh 2K_1 \sinh 2K_2^* \cos\theta$$

$$b_>(\theta) = \sqrt{(1-\alpha_1 e^{i\theta})(1-\alpha_2^{-1}e^{i\theta})}$$

$$\alpha_1(\Theta) = \tanh K_1 \tanh K_2^* \qquad \alpha_2(\Theta) = \tanh^{-1} K_1 \tanh K_2^*$$

and the usual energy coupling constants K_1 and K_2 of the two dimensional Ising model.

Below T_C the existence of a degeneracy of the ground state necessitates a recomputation of μ and σ (with $b_<(\Theta)$ now) given choosen periodic boundary conditions for the original fermion variables. Otherwise it is necessary to introduce a Z_2-"spurion" degree of freedom in order to account for the mutation of the pair $(\mu_>, \sigma_>)$ into $(\sigma_<, \mu_<)$ as one crosses through the critical temperature, and to insure the survival of the "duality" algebra [8] .

Mathematically speaking (μ, σ) are elements of the Clifford group of the fermion algebra. Their composition law allows the computation of the n-point correlation functions [6] obtained in ref.3),4),5), but we shall not elaborate on these involved technical points.

So far we have discussed the theory in the thermodynamic limit, we would like to consider now the scaling limit. To do so we let $T \to T_C$, shrink the lattice constant $a \to o$, but let $(m,n) \to \infty$ such that $x_1 = ma$ and $x_0 = ina$ remain fixed. Then the fermion pair $(\hat{\psi}^+(\Theta), \hat{\psi}(\Theta))$ goes over into a relativistic free Majorana field as expected and the $\mu_>, \sigma_>$ operators reach a limiting form:

$$\begin{aligned} \mu_>(x) &= {}^*_*\exp M_1(x)^*_* \\ \sigma_>(x) &= {}^*_*\hat{\psi}_o(x) \exp M_1(x)^*_* \end{aligned} \qquad (6)$$

where

$$\hat{\psi}_o(x) = \int_{\infty}^{\infty} d\Theta_p \{c_p\, e^{-ipx} - c_p^+\, e^{ipx}\}$$

$$M_1(x) = \frac{1}{2}\iint d\Theta_p\, d\Theta_q \{ \left[-\frac{1}{2\pi} \text{P.V.}\coth \frac{\Theta_p - \Theta_q}{2} - 2\pi\delta_{pq} \right] e^{i(p-q)x} c_p^+ c_q^+$$
$$\left[-\frac{1}{2\pi} \tanh \frac{\Theta_p - \Theta_q}{2}\, e^{i(p-q)x} \right] c_p^+ c_q^+ - \text{h.c.}\}$$

with $p = (m\cosh\Theta_p, m\sinh\Theta_p)$ and $\{c_p^+, c_q\} = \delta(\Theta_p - \Theta_q)$.

The operators $(\mu_>, \sigma_>)$ generate precisely the n-point correlation functions of the Ising model in the scaling limit. Lastly let us quote one more result: this construction of μ and σ gives also an explicit solution to the Riemann-Hilbert problem, which in turn is closely related to the monodromy preserving deformation theory of differential equations [9].

However, it is to be noted that the expressions of μ and σ are not written with a local ordering of a function of any local field operator of the Majorana field. To circumvent this one may try to consider two non-interacting Majorana fields (resp. two non-interacting Ising lattices) and built the theory on a Dirac free fermion field. The idea is not new, since it was suggested by R. Ferrell [10] when he studied the two dimensional Ising model in the scaling limit directly from a Hamiltonian approach. Moreover since we are eventually or ultimately interested in the n-point functions of the more general massive Thirring model [11] (resp.the eight vertex model on a lattice, known to be equivalent to two interacting Ising lattices [12])the doubling could provide a natural checking point for a special value of the coupling constant.

We expect of course a continuum version of eq.(4) to be valid say for $T > T_c$:

$$\mu_D(y)\ \psi_D(x) = \begin{cases} -\ \psi_D(x)\ \mu_D(y) & x_1 < y_1 \\ \psi_D(x)\ \mu_D(y) & x_1 > y_1 \end{cases} \tag{7}$$

where D refers to the Dirac field, $\sigma_D(x)$ now obeys the same commutation as $\mu_D(x)$ but is odd under charge conjugation and has a vanishing vacuum expectation value.

It turns out that both μ_D and σ_D can be expressed as a local ordered expression of simple functions of a local field operator of the Dirac theory, namely the pseudo-potential $\varphi(x)$ defined by:

$$\overset{*}{\underset{*}{}}\psi_D^+(x)\ i\ \gamma^5\gamma^\mu\psi_D(x)\overset{*}{\underset{*}{}} = \frac{1}{\sqrt{\pi}}\partial^\mu\varphi(x) \tag{8}$$

and the boundary

$$\lim_{x_1 \to \infty} \varphi(x) = \sqrt{\pi}Q$$

Q being the total charge operator [13] : It is clear that $\varphi(x)$ is bilinear in the fermion-antifermion degrees of freedom. We have the following [14] :

$$\mu_D(x) = \vdots\ \cos\ \sqrt{\pi}\varphi(x)\vdots$$

$$\sigma_D(x) = \vdots\ \sin\ \sqrt{\pi}\varphi(x)\vdots \tag{9}$$

The local ordering $\vdots\quad\vdots$ is defined in ref.[13]and it is easy to check that the commutation relations (7) are just the consequence of the non-locality of $\varphi(x)$ with respect to the Dirac field $\psi_D(x)$. Algebraically the pair $\mu_D(x)$ and $\sigma_D(x)$ obeys composition laws of the Baker-Hausdorff-Campbell type which help to generate n-point functions that are squares of the n-point functions of the simple Ising model.

Some physical features emerge from this formulation besides the manifest locality and covariance.

As the temperature crosses its critical value, the mass of the fermion field reverses its sign, the corresponding spinor field is now multiplied by γ^5 and thanks to Mandelstam's formula [15] $\varphi(x)$ goes to $\varphi(x) + \sqrt{\pi}/2$ which implies in turn an interchange of sin $\leftrightarrow$ cos functions and thereby an interchange of the operators $\mu_D \leftrightarrow \sigma_D$ (modulo constant phases). This fact can be viewed as an operator implementation of the Kramers-Wannier duality transformation.

The doubling procedure allows us to study simply the field theoretical scale invariant zero mass limit in simple terms [14]. If one makes the following rescaling

$$\mu_D(x) \to m^{1/4}\mu_D(x) \qquad\qquad \sigma_D(x) \to m^{1/4}\sigma_D(x)$$

then the n-point functions show in the Euclidean region a short distance behavior in agreement with the naive zero mass limiting picture and the Γ-selection rule of Kadanoff-Ceva [7] on a line (in particular for functions involving both $\mu_D(x)$ and $\sigma_D(x)$).

In conclusion, let us mention a few facts which should even more emphasize the constructive role of the two dimensional Ising model in theoretical model-making. In the context of the "bootstrap" program it has been shown recently that the operator $\sigma(x)$ can be reconstructed from its S-matrix [16] thus serve as guide line to any operator reconstruction one may hope to achieve in the same spirit. Moreover it has been found that in a boson theory one could also have a pair of operators $\tilde{\mu}(x)$ and $\tilde{\sigma}(x)$, where $\tilde{\sigma}(x)$ is now a spinor field [17] satisfying the "abnormal" commutation relations of $\mu(x)$ and $\sigma(x)$. The interesting point is that they can be combined together to form a supersymmetric multiplet of nontrivial relativistic quantum field operators [18].

REFERENCES

1. L. Onsager, Phys.Rev.65, 117 (1944); L. Onsager and B. Kaufman, Phys.Rev. 76, 1244 (1944).
2. T. D. Schultz, D. C. Mattis and E. H. Lieb, Rev.Mod.Phys. 36, 856 (1964).

3. B. McCoy, C. A. Tracy and T. T. Wu, Phys.Rev.Lett. 38, 793 (1977).
4. R. Z. Bariev, Phys.Lett. 64A, 169 (1977), Physica 93A, 354 (1978).
5. D. B. Abraham, Phys.Lett. 61A, 271 (1977), Com.Math.Phys. 59, 17 (1978), 60, 181 (1978), 60, 205 (1978).
6. M. Sato, T. Miwa and M. Jimbo, RIMS Kyoto preprint 207, July 1976, RIMS Kyoto preprint 267, Nov. 1978, to appear in Publication of the Research Institute for Mathematical Sciences, Kyoto University.
7. L. P. Kadanoff and H. Ceva, Phys.Rev. B3, 3918 (1971).
8. B. Schroer and T. T. Truong, Nucl.Phys. B154, 125 (1979).
9. M. Sato, T. Miwa and M. Jimbo, Publ.RIMS 15, 201 (1979).
10. R. A. Ferrell, Jour.Stat.Phys. 8, 265 (1973).
11. S. Coleman, Phys.Rev. D11, 2088 (1975).
12. F. Y. Wu, Phys.Rev. B4, 2312 (1971); L. P. Kadanoff and F. Wegner, Phys.Rev. B4, 3989 (1971).
13. A. S. Wightman in Cargese Lectures in Theoretical Physics 1964, (Gordon and Breach, New York 1966).
H. Lehmann and J. Stehr, preprint Hamburg DESY, 1976.
14. B. Schroer and T. T. Truong, Nucl.Phys. B144, 80 (1978).
15. S. Mandelstam, Phys.Rev. D11, 3026 (1975).
16. B. Berg, M. Karowski and P. Weisz, Phys.Rev. D19, 2477 (1979).
17. M. Jimbo, Proc.Japan Acad. 54A, 263 (1978).
18. B. Schroer, T. T. Truong and P. Weisz, Nucl.Phys. B154, 140 (1979).

STRING VARIABLES IN QUANTUM CHROMODYNAMICS

I. Montvay

University of Bielefeld, ZiF

INTRODUCTION

There has been recently an increasing interest to formulate gauge field theories in terms of the gauge invariant string variables ("path dependent phase factors"). There are several reasons for this. One of them is the close similarity of the classical gauge field dynamics (if formulated in terms of string variables) with the quantum mechanics of relativistic strings [1,2]. In this connection a number of authors [3-6] derived the dynamical equations for string variables in classical gauge field theories. Others [7-8] showed that the gauge fields can be considered as the lowest excitations of a condensate of relativistic strings. Another motivation for introducing string variables is that they are gauge invariant and therefore explicitly solve the otherwise rather complicated constraints resulting from gauge invariance [9-12]. A third important aspect of using string variables is that confinement [12] and duality [13-14] properties of the quantum gauge fields may be simply expressed in terms of them.

In the case of the simplest confined system, namely $q\bar{q}$ bound state meson, the introduction of string operators allows the derivation of gauge invariant approximations to the wave equation. A possible way to do this was suggested by Suura [15] who introduced an infinite set of string operators and derived equations of motion for them. Regularizing the operator products appearing and truncating the infinite system he obtained a wave equation with confinement potential for the $q\bar{q}$ system. In a recent paper Polyakov [16] has suggested to consider the string variables of non-abelian gauge theories as chiral fields defined on closed loops in space time. He proposed that conservation laws analogous to the ones found in twodimensional O(n) Heisenberg spin systems exist and may lead to the solution.

In this lecture I shall briefly review the dynamical equations for the string variables of gauge fields based on two papers [6,17]. In the first part the classical gauge fields will be considered for a general gauge group, whereas the second part will be devoted to the case of quantum chromodynamics on the lattice.

I. EQUATIONS FOR THE PATH DEPENDENT PHASE FACTOR IN CLASSICAL GAUGE FIELD THEORY

The path dependent phase factor in a gauge field theory is defined by

$$U[\sigma] = P \exp \left(-ig \int_{\sigma} A(x)_{\mu}\, dx^{\mu}\right) \tag{1.1}$$

where $\sigma \equiv \sigma(x_1 \leftarrow x_0)$ is some curve (open or closed) starting from x_0 and ending at x_1. $A(x)_{\mu}$ denotes the gauge field in the matrix representation, g is a dimensionless coupling parameter and P stands for matrix ordering (from right to left) along the path. In the classical theory $U[\sigma]$ is an element of the gauge group and it transforms under local gauge transformations U_x like [9-11]

$$U'[\sigma(x_1 \leftarrow x_0)] = U_{x_1}^{-1} U[\sigma(x_1 \leftarrow x_0)]\, U_{x_0} \tag{1.2}$$

In general, the value of $U[\sigma(x_1 \leftarrow x_0)]$ depends both on the end points x_0, x_1 and on the shape of the curve σ. In order to derive dynamical equations for $U[\sigma]$ we have to consider the variation $\delta U[\sigma]$ of $U[\sigma]$ if the curve at the point x is varied by $\delta(x)$. (For simplicity, we shall not consider end point variations: $\delta(x_0) = \delta(x_1) = 0$. That is, we shall concentrate on the more interesting aspect, namely the dependence of $U[\sigma]$ on the shape of σ. The extension to include also end point variations is, however, rather straightforward.)

It can be shown (for details see ref.6) that $\delta U[\sigma]$ can be expressed by the functional derivatives in the form

$$\delta U[\sigma] \equiv U[\sigma+\delta] - U[\sigma] =$$

$$= \int_0^{\bar{s}} ds\, \delta(s)_{\mu} \frac{\delta U[\sigma]}{\delta\sigma(s)_{\mu}} + \int_0^{\bar{s}} ds \int_s^{\bar{s}} ds'\, \delta(s')_{\mu} \delta(s)_{\nu} \frac{\delta^2 U[\sigma]}{\delta\sigma(s')_{\mu} | \delta\sigma(s)_{\nu}}$$

$$+ \frac{1}{2!} \int_0^{\bar{s}} ds \left\{ \delta(s)_{\mu} \delta(s)_{\nu} \frac{\delta^2 U[\sigma]}{\delta\sigma(s)_{\mu} \delta\sigma(s)_{\nu}} + \left[\delta(s)_{\mu} \frac{d\delta(s)_{\nu}}{ds} - \right.\right.$$

$$\left.\left. - \delta(s)_{\nu} \frac{d\delta(s)_{\mu}}{ds} \right] \frac{1}{2} \frac{\delta U[\sigma]}{\delta\sigma(s)_{[\mu\nu]}} \right\} + \ldots \tag{1.3}$$

Here the length of arc parameter s defined by

$$ds = \sqrt{-dx_\mu dx^\mu} \tag{1.4}$$

is used and $\bar{s}$ is the total length of arc of σ.

In eq. (1.3) all the integrals are defined (with $\varepsilon > 0$) like

$$\int_a^b ds \equiv \lim_{\varepsilon \to 0} \int_{a+\varepsilon}^{b-\varepsilon} . \tag{1.5}$$

This means that in $\delta^2 U[\sigma]/\delta\sigma_\mu(s')|\delta\sigma_\nu(s)$ (and in the similar higher order derivatives) s is strictly smaller than s' (the notation is trying to express this). The terms with s=s' are separated and contain antisymmetric tensor pieces like $\delta U[\sigma]/\delta\sigma(s)_{[\mu\nu]}$ and in general like

$$\delta^n U[\sigma]/\delta\sigma(s)_{[\mu\nu]}\delta\sigma(s)_{\mu_2}\dots\delta\sigma(s)_{\mu_n} .$$

(More than one pair of antisymmetric indices is not required.)

The functional derivatives of $U[\sigma]$ can be determined by approximating the curves first with rectangles consisting of finite straight segments and then going to the limit of infinitely short segments. A useful tool in doing this is a simple Taylor-like expansion of $U[\sigma]$ along straight lines. The result depends on the field strength tensor

$$F(x)_{\mu\nu} = \partial_\mu A(x)_\nu - \partial_\nu A(x)_\mu + ig[A(x)_\mu, A(x)_\nu] \tag{1.6}$$

and on its covariant derivatives like e.g.

$$D_\lambda F(x)_{\mu\nu} = \partial_\lambda F(x)_{\mu\nu} + ig[A(x)_\lambda, F(x)_{\mu\nu}] . \tag{1.7}$$

The result of the calculation is [6]:

$$\frac{\delta U[\sigma]}{\delta\sigma(s)_\mu} = -ig\, U[\sigma(x_1 \leftarrow x(s))]\, F(x(s))^{\mu\nu} t(s)_\nu\, U[\sigma(x(s) \leftarrow x_0)]$$

$$\frac{\delta^2 U[\sigma]}{\delta\sigma(s')_\mu|\delta\sigma(s)_\nu} = \frac{\delta U[\sigma(x_1 \leftarrow x(s''))]}{\delta\sigma(s')_\mu}\,\frac{\delta U[\sigma(x(s'') \leftarrow x_0]}{\delta\sigma(s)_\nu}$$

$$\text{for } s < s'' < s' \tag{1.8}$$

$$\frac{\delta^2 U[\sigma]}{\delta\sigma(s)_\mu \delta\sigma(s)_\nu} = -\frac{ig}{2} U[\sigma(x_1 \leftarrow x(s))] \left\{ D^\mu F(x(s))^{\nu\lambda} t(s)_\lambda + \right.$$

$$\left. + D^\nu F(x(s))^{\mu\lambda} t(s)_\lambda \right\} U[\sigma(x(s) \leftarrow x_0)]$$

$$\frac{\delta U[\sigma]}{\delta\sigma(s)_{[\mu\nu]}} = -ig\, U[\sigma(x_1 \leftarrow x(s))]\, F(x(s))_{\mu\nu}\, U[\sigma(x(s) \leftarrow x_0)]$$

Ref. 6 contains the expressions also for the higher order derivatives (in this respect see also ref. 18). In the above equations $t(s)_\mu = dx(s)_\mu/ds$ denotes the tangential unit vector of the curve σ in the point $x(s)$.

The dynamical equation for a pure gauge field is

$$D^\mu F(x)_{\mu\nu} = 0 \; . \tag{1.9}$$

This is supplemented by the identity (following from the Jacobi-identity for the covariant derivative D):

$$D_\lambda F(x)_{\mu\nu} + D_\mu F(x)_{\nu\lambda} + D_\nu F(x)_{\lambda\mu} = 0 \tag{1.10}$$

These equations can be expressed in terms of the functional derivatives of $U[\sigma]$ in the following way:

$$\frac{\delta^2 U[\sigma]}{\delta\sigma(s)_\rho \delta\sigma(s)^\rho} = 0 \tag{1.11}$$

$$\frac{\delta^2 U[\sigma]}{\delta\sigma(s)_{[\mu\nu]}\delta\sigma(s)_\lambda} + \frac{\delta^2 U[\sigma]}{\delta\sigma(s)_{[\nu\lambda]}\delta\sigma(s)_\mu} + \frac{\delta^2 U[\sigma]}{\delta\sigma(s)_{[\lambda\mu]}\delta\sigma(s)_\nu} = 0 \tag{1.11'}$$

In the case of constant

$$C = g^2\, F(x(s))_{\mu\kappa}\, t(s)^\kappa\, F(x(s))^{\mu\lambda}\, t(s)_\lambda \tag{1.12}$$

along the path σ, we obtain from Eq. (1.8)

$$\lim_{s' \to s} \frac{\delta^2 U[\sigma]}{\delta\sigma(s')_\rho | \delta\sigma(s\;)^\rho} + C\, U[\sigma] = 0 \tag{1.13}$$

which is the result of Refs. 1,2 connecting the dynamics of the path dependent phase factor to the one of the quantized relativistic string.

The functional differential equations are, however, somewhat symbolic because one can in reality consider only functions of a finite number of variables. This means that one has to "discretize" the curve somehow (either by "rectification" i.e. by approximating it with a polygon [2,3,6] or truncating the Fourier-series representation of the shape of the curve [4]).

In Ref. 6 the corrections to the dynamical equations were considered in the limit when the segment lengths of the polygon are going to zero. If the curve $\sigma(x_{n+1} \leftarrow x_0)$ is approximated by the polygon $x_0 \rightarrow x_1 \rightarrow \ldots x_n \rightarrow x_{n+1}$ then the functional $U[\sigma]$ is replaced by the function $U(x_0, x_1, \ldots, x_{n+1})$ and the functional derivatives by partial derivatives with respect to x_i. Repeating the steps which lead e.g. to Eq. (1.11) one arrives at the following partial differential equation:

$$\frac{\partial^2 U(x_0,\ldots,x_{n+1})}{\partial x_{j\rho}\,\partial x_j^{\ \rho}} =$$

$$= \frac{\partial U(x_0,\ldots,x_{n+1})}{\partial x_{j\rho}}\, U(x_0,\ldots,x_{n+1})^{-1}\, \frac{\partial\, U(x_0,\ldots,x_{n+1})}{\partial x_j^{\ \rho}} + \ldots \tag{1.14}$$

Here the dots stand for terms vanishing at least cubically with the length of the segments. In fact, it can be easily shown that for Abelian gauge fields these terms vanish identically i.e. Eq. (1.14) is exact as it stands.

II. DYNAMICAL EQUATIONS FOR STRING OPERATORS IN QUANTUM CHROMODYNAMICS

In the quantum theory the path dependent phase factor $U[\sigma]$ in Eq. (1.1) becomes an operator. To avoid ultraviolet divergences in the dynamical equations some regularization procedure is needed. One way of cutting off high frequencies is to make space-time discrete by introducing a lattice instead of the continuum. An obvious way to put field theory on the lattice is to define field operators at the lattice sites. For matter fields (like quarks or leptons) this is indeed the right procedure. For the gauge fields, however, it is much more convenient to put the field variables on links between neighbouring lattice sites [12]. In such a way the gauge field is represented by the phase factor operator $U[\overleftarrow{r}k]$ of the link $\bar{r}k \equiv k(r)$ $\overleftarrow{r}k \equiv k(r)\overleftarrow{r}$ connecting the lattice site k with the neighbouring one $k(r)$ in the direction r. (In what follows only the SU(3) colour gauge group will be considered and the same notations will be used as in Ref. 17: "middle" letters like i,j,k,l,... for lattice sites, "late" letters r,s,t,... for the directions $\pm x, \pm y, \pm z$ and "early" letters b,c,d,... for SU(3) colour octet indices).

In the Hamiltonian formulation [12,19-21] the time is left continuous and the lattice is introduced only in space. The Hamilton-operator in the $A_0=0$ ("temporal") gauge is:

$$H = \frac{1}{2a} \sum_{k,r} \eta(k,r)\; \chi^{+}_{k(r)}\; U[\overleftarrow{r}k]\; \chi_k + \frac{g^2}{4a} \sum_{k,r,b} E^{(k)}_{b,r}\, E^{(k)}_{b,r}$$

$$- \frac{1}{ag^2} \sum_{k[rs]} A(k[rs]) \tag{2.1}$$

Here a is the lattice spacing of the (cubic) lattice and $k[rs]$ denotes the plaquette boundary starting from the site k first in the direction r and then in s : $k[rs] \equiv k\overset{\leftarrow}{-}s\overset{\leftarrow}{-}r\,\overleftarrow{s}\,\overleftarrow{r}\,k$. The path dependent phase factor belonging to this plaquette boundary is denoted by $U[k[rs]] \equiv U[k\overset{\leftarrow}{-}s]\; U[k(s)\overset{\leftarrow}{-}r]\; U[\overleftarrow{s}k(r)]\; U[\overleftarrow{r}k]$ and its trace over SU(3)-indices by $A(k[rs])$. χ_k stands for the quark lattice variable [20] and

$$\eta(k, \pm x) = (-1)^{k_z}, \quad \eta(k, \pm y) = (-1)^{k_x}, \quad \eta(k, \pm z) = (-1)^{k_y} \tag{2.2}$$

if $k_{x,y,z}$ are the (integer) coordinates of the point k . The colour electric flux in the point k is denoted in Eq. (2.1) by $E^{(k)}$.

The above Hamiltonian is supplemented by the commutation relations

$$\{\chi^{+}_{k\alpha}, \chi_{\ell\beta}\} = \delta_{k\ell}\, \delta_{\alpha\beta} \tag{2.3}$$

(if α,β denote the otherwise omitted colour triplet indices) and

$$[U[\overleftarrow{r}k], E^{(\ell)}_{b,s}] = \frac{1}{2}\,(\delta_{rs} - \delta_{-rs})\; \{\delta_{k\ell} U[\overleftarrow{r}k]\, \frac{\lambda_b}{2} +$$

$$+ \delta_{k(r)\ell}\, \frac{\lambda_b}{2}\, U[\overleftarrow{r}k]\} \tag{2.4}$$

(Here λ_b is the usual Gell-Mann matrix for SU(3)). Note that the colour electric flux is associated here to lattice sites, whereas usually it is associated to links. Let us denote the electric flux on the link $\overleftarrow{rk}$ by $E_b[\overleftarrow{rk}] = -E_b[k\overleftarrow{-r}]$, then up to corrections of order a (hence vanishing for $a \to 0$) we have

$$[U[\overleftarrow{rk}],\ E_b[\overleftarrow{s\ell}]] = \frac{1}{2}\,(\delta_{k\ell}\delta_{rs}-\delta_{k\ell(s)}\delta_{-rs})\ \{\ U[\overleftarrow{rk}]\ \frac{\lambda_b}{2} + \frac{\lambda_b}{2}\, U[\overleftarrow{rk}]\}\ . \tag{2.5}$$

One can decompose the flux into the flux "starting from k in the direction r" plus the flux "arriving at k(r) form the direction r" like

$$E_b[\overleftarrow{rk}] = \frac{1}{2}\,\{E_b^{(+)}[\overleftarrow{rk}]-E_b^{(-)}[k\overleftarrow{-r}]\} + \frac{1}{2}\,\{E_b^{(-)}[\overleftarrow{rk}]-E_b^{(+)}[k\overleftarrow{-r}]\} \tag{2.6}$$

where the commutation relations are, by definition

$$[U[\overleftarrow{rk}],\ \ E_b^{(+)}[\overleftarrow{s\ell}]] = \delta_{k\ell}\ \delta_{rs}\ U[\overleftarrow{rk}]\ \frac{\lambda_b}{2} \tag{2.7}$$

$$[U[\overleftarrow{rk}],\ \ E_b^{(-)}[\overleftarrow{s\ell}]] = \delta_{k\ell}\ \delta_{rs}\ \frac{\lambda_b}{2}\ U[\overleftarrow{rk}]$$

Now, let us define the flux associated to the lattice site as the average of the incoming and outgoing flux (in a certain direction), namely

$$E_{b,s}^{(\ell)} \equiv \frac{1}{2}\,\{E_b^{(+)}[\overleftarrow{s\ell}]\ -\ E_b^{(-)}[\ell\overleftarrow{-s}]\} + \frac{1}{2}\,\{E_b^{(-)}[\ell\overleftarrow{s}]-E_b^{(+)}[\overleftarrow{-s\ell}]\}\ . \tag{2.8}$$

This and Eq. (2.7) give the commutator in Eq. (2.4).

The advantage of using the phase operator $U[\overleftarrow{rk}]$ as variable for the gauge field is that with its help it is easy to construct gauge invariant physical states. This is a rather crucial point because in the temporal gauge used here to define the Hamiltonian formalism the (non-abelian analogue of) the Gauss-law has to be implemented as a constraint on physical states. This is, in general, a rather formidable task if the gauge field $A(x)_\mu$ is put at the basis of the theory like it is usually done in continuous space-time (for recent references see e.g. [22-23] The Gauss-law in the quantum theory is equivalent to the invariance of the physical states under time independent local gauge transformations. From the path dependent phase operators, however, it is trivial to construct gauge invariant operators and these operators can be taken as creation (and annihilation) operators producing the physical states from some suitably defined (gauge invariant) vacuum.

In order to define the gauge invariant operators let us consider the "curves" on our lattice. These are sequences of joined links like

$$\overleftarrow{\bar{k}r}_n k_n \ldots k_1 \overleftarrow{r}_o k \equiv \bar{k}\leftarrow k \qquad \text{or}$$

$$i\overleftarrow{r}_n i_n \ldots i_1 \overleftarrow{r}_o i \equiv i\leftarrow i \tag{2.9}$$

depending on whether it is an open or a closed curve. (Sometimes the notations

$$k_o \equiv k,\ \bar{k} \equiv k_{n+1},\ i \equiv i_o \equiv i_{n+1},\ i_{n+j+1} \equiv i_j$$

will also be used). The <u>string operator</u> for an open curve ("$q\bar{q}$-string") is defined putting a quark and an antiquark at the ends of the path dependent phase operator:

$$\chi(\bar{k}\leftarrow k) \equiv \chi^+_{\bar{k}}\, U[\overleftarrow{\bar{k}r}_n k_n \ldots k_1 \overleftarrow{r}_o k]\, \chi_k \equiv \chi^+_{\bar{k}}\, U[kr_n]\ldots$$

$$\ldots U[\overleftarrow{r}_1 k_1] U[\overleftarrow{r}_o k] \chi_k \tag{2.10}$$

In this notation the $q\bar{q}$-string operator appearing in the Hamiltonian (2.1) can be written as

$$\chi(\overleftarrow{r}k) \equiv \chi^+_{k(r)}\, U[\overleftarrow{r}k]\, \chi_k \ . \tag{2.11}$$

For closed curves the "gluon-loop" string operator (or "Wilson-loop operator") is

$$A(i\leftarrow i) \equiv \mathrm{Tr}\, U[i\overleftarrow{r}_n i_n \ldots i_1 \overleftarrow{r}_o i] \ . \tag{2.12}$$

The gauge invariant "mathematical" vacuum state $|0\rangle$ is defined by the requirement that it is annihilated by the electric field operator [21]:

$$E^{(k)}_{b,r}\, |0\rangle = 0 \tag{2.13}$$

The physical states are constructed by acting on $|0\rangle$ any number of times by the string operators A and χ . Among these states the physical vacuum is the state with lowest energy. (It is generally different from $|0\rangle$, except for the strong coupling ($g \to \infty$) case, when the second term in the Hamiltonian (2.1) is dominating).

The dynamical equations for the string operators can be deduced from the Hamiltonian (2.1) and the commutation relations (2.3-4). Let us consider, for simplicity, only non-selfintersecting curves. (Note, however, that this restriction is purely technical and it does not mean that the selfintersecting curves were dynamically less important). In this case the dynamical equations are [17]:

$$\dot{A}(i\leftarrow i) = \frac{-ig^2}{a}\left\{\frac{2}{3}\lambda(i\leftarrow i)A(i\leftarrow i) + \frac{1}{2}\sum_{j=0}^{n}(E^{(i_j)}_{b,r_{j-1}} + E^{(i_j)}_{b,r_j})\cdot\right.$$

$$\left.\cdot\mathrm{Tr}\{\frac{\lambda_b}{2}U[i_j\overleftarrow{r}_{j-1}i_{j-1}\cdots i_{j+1}\overleftarrow{r}_j i_j]\}\right\} \qquad (2.14)$$

and

$$\dot{\chi}(\bar{k}\leftarrow k) = \frac{-i}{2a}\sum_{(r)}\{\eta(k(-r),r)\,\chi(\bar{k}\leftarrow k\overleftarrow{r}) -$$

$$-\eta(\bar{k},r)\,\chi(\overleftarrow{rk}\leftarrow k)\} - \frac{ig^2}{a}\left\{\frac{2}{3}\lambda(\bar{k}\leftarrow k)\,\chi(\bar{k}\leftarrow k) +\right.$$

$$+\frac{1}{2}\sum_{j=0}^{n+1}[(1-\delta_{jo})E^{(k_j)}_{b,r_{j-1}} + (1-\delta_{jn+1})E^{(k_j)}_{b,r_j}].$$

$$\left.\cdot\chi^+_{\bar{k}}\,U[\bar{k}\leftarrow k_j]\,\frac{\lambda_b}{2}\,U[k_j\leftarrow k]\,\chi_k\right\} \qquad (2.15)$$

In these equations λ is defined in the following way (for closed and open curves, respectively):

$$\lambda(i\leftarrow i) = \frac{1}{2}(n+1+\sum_{j=0}^{n}\delta_{r_j r_{j+1}})$$

$$\lambda(\bar{k}\leftarrow k) = \frac{1}{2}(n+1+\sum_{j=0}^{n-1}\delta_{r_j r_{j+1}}) \qquad (2.16)$$

The factor $g^2a^{-1}2\lambda/3$ in Eqs. (2.14-15) is the colour-electric energy of the string. The parameter λ can be considered as a sort of length of arc of the curves on the lattice (measured in lattice units).

The dynamical equations can also be formulated in terms of Schrödinger-like equations for wave functions. If $|0\rangle$ is an arbitrary physical state then there is an infinite set of wave functions characterizing it. These can be defined with the help of the mathematical vacuum $|0\rangle$ in Eq. (2.13) like

$$\psi_{\chi}(\bar{k}\leftarrow k) \equiv \langle 0|\chi(\bar{k}\leftarrow k)|\psi\rangle$$

$$\psi_{A}(i\leftarrow i) \equiv \langle 0|A(i\leftarrow i)|\psi\rangle$$

$$\psi_{A_{\chi}}(i\leftarrow i,\bar{k}\leftarrow k) \equiv \langle 0|A(i\leftarrow i)\chi(\bar{k}\leftarrow k)|\psi\rangle \tag{2.17}$$

and further functions with an arbitrary number of A's and χ's. Taking the matrix elements of Eqs. (2.14-15) and using the commutation relations as well as the expression for $\dot{E}^{(k)}$ it is easy to obtain the Schrödinger-like equations for the wave functions (see Ref. 17).

The dynamical equations for string operators and string wave-functions are still waiting for applications. They can presumably be helpful in understanding the classical limit of QCD as well as the nature and role of conserved quantities in it.

REFERENCES

1. Y. Nambu, Phys. Lett. 80B, 372 (1979)
2. J.L. Gervais, A. Neveu , Phys. Lett. 80B, 255 (1979)
3. E. Corrigan, B. Hasslacher, Phys. Lett. 81B, 181 (1979)
4. J.L. Gervais, A. Neveu , Local harmonicity of the Wilson loop integral in classical Yang-Mills theory, LPTENS 79/1 preprint, (1979)
5. J.L. Gervais, M.T. Jaekel, A. Neveu, Hadronic and supersymmetric string states in chromodynamics, LPTENS 79/4 preprint (1979)
6. I. Montvay, Variation of the path dependent phase factor in gauge field theories, BI-TP 79/05 preprint (1979)
7. F. Gliozzi, T. Regge,M.A. Virasoro, Phys. Lett. 81B, 178 (1979)
8. M.A. Virasoro, Phys. Lett. 82B, 436 (1979)
9. S. Mandelstam, Annals of Phys. 19, 1 (1962); Phys. Rev. 175, 1580 (1968)
10. I. Bialynicki-Birula, Bull. Acad. Polon. Sci. 11, 135 (1963)
11. C.N. Yang, Phys. Rev. Letters, 33, 445 (1974)
12. K.G. Wilson, Phys. Rev. D10, 2445 (1974)
13. G. 't Hofft, Nucl. Phys. B138, 1 (1978)
14. S. Mandelstam, Phys. Rev. D19, 2391 (1979)
15. H. Suura, Equation of motion for string operators in quantum chromodynamics, DESY-79-25 preprint (1979)
16. A.M. Polyakov, Phys. Lett. 82B, 247 (1979)
17. I. Montvay, Dynamical equations for string operators in lattice quantum chromodynamics, BI-TP 79/24 preprint (1979)
18. L. Durand, E. Mendel, Phys. Letters 85B, 241 (1979)

19. J. Kogut, L. Susskind, Phys. Rev. D11, 395 (1975)
20. L. Susskind, Phys. Rev. D16, 3031 (1977)
21. J. Kogut, D.K. Sinclair, L. Susskind, Nucl. Phys. B114, 199 (1976)
22. J. Goldstone, R. Jackiw, Phys. Letters 74B, 81 (1978)
23. S.S. Chang, Phys. Rev. D19, 2958 (1979)

FOURDIMENSIONAL QUATERNIONIC σ-MODELS

Jerzy Lukierski

Institute for Theoretical Physics
University of Wroclaw
Wroclaw, Poland

1. INTRODUCTION

The following two main reasons justify the recent vivid interest in σ-models:

i) In O(N) σ-models (in dimension d = N - 1) and the two-dimensional CP(n) σ-models *) one can introduce the topological quantum numbers and instantons -- as in four-dimensional Yang-Mills theories.
ii) The classical dynamics of the large class of σ-models (including O(N) and CP(n)) is determined by the presence of an infinite number of conserved nonlocal charges [5-8]. One expects that the formalism of σ-models may provide a hint of how to eventually treat the Yang-Mills theory as a completely integrable four-dimensional system.

The main deficiency of the σ-models based on geometries with real and complex coordinates is the lack of instanton configurations for d = 4 **). It appears, however, that if we consider the

---

*) These two series of σ-models are the examples of σ-models with field values in real or complex Grasmanians (see, for example, [1-3]). For complex-valued fields the classification of σ-models with non-trivial field configurations was given by Perelomov[4].

**) An exception is the O(5) σ-model, discussed in [9-11]. It has been shown, however, in [10] that such a O(5) σ-model can be related to the HP(1) quaternionic model, which is described by a part of our Lagrangian.

quaternionic Grassmanians

$$H\,G_{n;k} = \frac{Sp(n)}{Sp(n-k)\times Sp(k)} \tag{1}$$

the σ-fields taking values in (1) have non-trivial topological configurations for d = 4 which are classified by the first non-trivial homotopy group $\Pi_4(HG_{n;k}) = Z$. In such a case, as

$\Pi_i\ (HG_{n;k}) = 0$ for $i < 4$,

the fourth homotopy classes are described due to Hurevic theorem by the Pontriagin index. The topological properties of any quaternionic manifold M can be described by $Sp(1) \simeq SU(2)$ bundle which is attached to every point of M and characterizes uniquely the choice of local quaternionic structure. It will appear useful the following well-known formula for the Pontriagin index classifying topologically inequivalent configurations of SU(2) gauge fields over compactified R_4:

$$Q = -\frac{1}{16\pi^2} \int d^4x \quad Tr\ (F_{\mu\nu}\ \check{F}_{\mu\nu}) \tag{2}$$

where

$$F_{\mu\nu} = \partial_\mu A_\nu - \partial_\nu A_\mu + [A_\mu, A_\nu]$$

$$\check{F}_{\mu\nu} = \frac{1}{2} \epsilon_{\mu\nu\rho\tau} F_{\rho\tau}$$

The first aim of our talk is to explain why the topological properties of quaternionic manifolds make them unique candidates for the description of four-dimensional Yang-Mills fields regarded as composite connection forms in terms of σ-fields. Having established on a purely geometric basis the formula for the topological charge, we shall propose the Lagrangian for subcanonical quaternionic σ-fields, taking values in the quaternionic projective plane HP(n). It appears that the "free" Lagrangian is four-linear, and generalizes the non-linear Skyrme theory [12] ; for a recent discussion see [13]*); one can also break the naive scale invariance and introduce the conventional σ-model Lagrangian density which can be considered as the mass term.

Our construction generalizes the relation between two-dimensional Abelian gauge fields with non-trivial topology and complex Kähler geometries. Recently the properties of a two-dimensional CP(n) σ-model were quoted [16,17] , as providing an insight into four-dimensional Yang-Mills theory (and subsequently QCD) via analogy. We would like to stress here that if we replace complex-

*) There is also a close relation with recently proposed descriptions of the Yang-Mills theory in terms of projection-valued fields [14,15] .

valued σ-fields by the quaternionic ones, we are able to pass from an analogy to the identification of some important solutions in both theories.

2. DIFFERENT RIEMANNIAN GEOMETRIES AND σ-MODELS

There are only three different geometries with affine connection and zero torsion which can be formulated in an arbitrary number of dimensions *) :

i) n-dimensional real manifolds with its holonomy group **) in SO(n). Any oriented Riemann manifold described by the real symmetric metric tensor g_{ij} belongs to this class. The topology of such manifolds is described by real de Rham cohomology classes $H^i(M;R)$ $(i = 1,2,\ldots)$

ii) 2n-dimensional complex manifolds with the holonomy group in $U(n) = SU(n)\times U(1)$.

Such a geometry is described by a complex Hermitian metric $h_{i\bar{j}} = h^*_{\bar{j}\bar{i}}$; the topological properties implied by complex structure are determined by $h_{i\bar{j}}$ via closed non-degenerate fundamental two-form

$$\omega_2 = \frac{i}{2} h_{i\bar{j}} \, dz_i \wedge d\bar{z}_j \qquad (3a)$$

$$d\omega_2 = 0. \qquad \text{(Kähler property)} \qquad (3b)$$

For complex manifolds only even cohomology classes $H^{2i}(M;R) \neq 0$.

Complex Kähler manifolds were used recently (see, for example [4,20]) for the description of two-dimensional composite U(1) gauge fields with non-trivial topology, classified by the two-dimensional analogue of (2) (r,s = 1,2):

$$Q = \frac{1}{4\pi} \int d^2x \, \epsilon_{rs} \, F_{rs} \qquad (4a)$$

---

*) Here we consider only sufficiently general curved geometries, which are not necessarily the symmetric spaces ($\nabla R \neq 0$, where R denotes the curvature tensor field). The classification theorem of geometries is due to Berger [18]; for simpler proofs, see [19] .

**) The holonomy group describes the rotations of a tangent bundle basis obtained by lifting the translation along the closed curves in the base manifold. It "detects" additional constraints obtained by the choice of coordinates (complex or quaternionic in our case).

where the two-dimensional field strength (U(1) curvature form) F_{rs} is expresses in terms of the σ-field $z_i(x)$ as follows

$$F_{rs} = 2\pi i\ h_{i\bar{j}} \left(\frac{\partial z_i}{\partial x_r} \frac{\partial \bar{z}_j}{\partial x_s} - (r \leftrightarrow s) \right) \tag{4b}$$

iii) 4n-dimensional quaternionic manifolds, with the holonomy group in $Sp(n)\times Sp(1)$.

Such a geometry can be described by a quaternionic-Hermitian metric $H_{i\bar{j}} = \bar{H}_{j\bar{i}}$, where the real line element is given by the formula

$$ds^2 = dq_i\ H_{i\bar{j}}\ d\bar{q}_j \tag{5}$$

where $r,s = 1,2,3;\ i,j = 1 \ldots n$ and

$$H_{i\bar{j}} = G_{ij} + e_r\ H_{ij}^{(r)} \qquad e_r\ e_s = -\ \delta_{rs} + \epsilon_{rst}\ e_t \tag{6a}$$

$$G_{ij} = G_{ji} \qquad H_{ij}^{(r)} = -H_{ji}^{(r)} \quad \text{(real)} \tag{6b}$$

In order to introduce the exterior two-form associated with the metric $H_{i\bar{j}}$ we shall use the property that the Riemannian metric and the exterior two-form come respectively from the real and imaginary part of the scalar product

$$H(q,p) = q_i\ H_{i\bar{j}}\ \bar{p}_j \qquad (q_i \neq p_i). \tag{7}$$

One gets

$$\text{Re } H(q,p) = \frac{1}{4}\ \text{Tr } \{H_{i\bar{j}}\ (\bar{p}_j q_i + \bar{q}_j p_i)\} \tag{8a}$$

and

$$\text{Im } H(q,p) = \frac{e_r}{4}\ \text{Tr } \{H_{ij}\ (\bar{p}_j e_r\ q_i - \bar{q}_j e_r p_i)\} \tag{8b}$$

where

$$r\{e_r e_s\} = 4$$

(we use real representation of quaternionic algebra). The formula (8b) defines three two-forms

$$\omega_2^{(r)}$$

depending on the explicit choice of the quaternionic imaginary units, but the quaternionic structure does not depend on the particular choice of quaternionic basis which can be changed by

O(3) rotations

$$e_i \longrightarrow e_i' = a^+ e_i a \qquad a \in Sp(1) \simeq SU(2) \tag{9}$$

where a is a unit quaternion. In order to obtain the Sp(n)xSp(1)-invariant topological characterization of quaternionic manifold where the second factor describes rotations (9), we should introduce the following non-degenerate closed form:

$$\omega_4 = \omega_2^{(1)} \wedge \omega_2^{(1)} + \omega_2^{(2)} \wedge \omega_2^{(2)} + \omega_2^{(3)} \wedge \omega_2^{(3)} \tag{10a}$$

$$d\omega_4 = 0 \qquad \text{(generalized Kähler property)} \tag{10b}$$

The four-form (10a,b) is an invariant element of $H^4(M;R)$, describing the Pontriagin index; for quaternionic manifolds only modulo four cohomology classes

$$H^{4\,i}(M,R) \neq 0.$$

In order to introduce the field theory with topological charge determined by the four-form ω_4, let us assume that $\mathcal{H}$ is a quaternionic Kähler manifold with

$$\Pi_i(\mathcal{H}) = 0 \ (i = 0,1,23) \text{ and } \Pi_4(\mathcal{H}) = Z.$$

The non-trivial topological configurations of quaternionic σ-fields $q_i(x)$ ($x \in S_4$; $q_i(x) : S_4 \to \mathcal{H}$) are classified by the topological charge

$$16\pi^2 Q = -\int_S \omega_4^x = -\int d^4x \, \frac{1}{4} \mathrm{Tr}\, \{H_{i\bar{j}} \frac{\partial q_j}{\partial x_\mu} e_r \frac{\partial q_i}{\partial x_r}\} \cdot \frac{1}{4} \mathrm{Tr}\, \{H_{k\bar{\ell}} \frac{\partial \bar{q}_\ell}{\partial x_\rho} e_r \frac{\partial q_k}{\partial x_\tau}\} \epsilon_{\mu\nu\rho\tau} \tag{11}$$

where ω_4^x describes the four-form (10a) induced on S_4.

Comparing (11) with (2) we see that these formulae can be identified if

$$F_{\mu\nu}^{(r)} = \frac{1}{4} \mathrm{Tr}\, \{H_{i\bar{j}} \frac{\partial \bar{q}_j}{\partial x_\mu} e_r \frac{\partial q_i}{\partial x_\nu} - (\mu \leftrightarrow \nu)\} \tag{12}$$

i.e. we obtain the four-dimensional SU(2) Yang-Mills field strength expressed in terms of quaternionic metric structure.

3. FOURDIMENSIONAL QUATERNIONIC HP(n) MODEL

The quaternionic structure in R_{4n} limits possible choices of Riemann metrics: it has been shown by Alexandrov [22] that any quaternionic manifold for $n > 1$ is an Einstein space.*) Further, we shall consider the simplest quaternionic Grassmanian (1) with k = 1 -- quaternionic projective plane HP(n). One can discuss the HP(n) σ-model in two ways:

a) Consider the atlas of independent quaternionic coordinates and introduce the quaternion-valued Fubini-Study metric (see, for example, [23,24]). In such a case the topological charge is given by formula (11), where

$$H_{i\bar{j}} \quad (q,\bar{q}) = (1 + q_k\bar{q}_k)^{-1} (I + \bar{q}q)^{-1}_{i\bar{j}} \tag{13}$$

b) Parametrize $HP(n) = S^{4n+3}/Sp(1)$ by equivalence classes $\Phi_i \sim a\Phi_i$ where $i = 0, 1, \ldots, n$, $a \in Sp(1)$, $(|a| = 1)$ and

$$\Phi_0 \bar{\Phi}_0 + \Phi_1 \bar{\Phi}_1 + \ldots + \Phi_n \bar{\Phi}_n = |\Phi|^2 = 1. \tag{14}$$

It is easy to check that the quaternion-valued vector field

$$A_\mu = \Phi_i \partial_\mu \bar{\Phi}_i \tag{15}$$

transforms under local Sp(1) transformations as follows

$$A'_\mu = a A_\mu \bar{a} + a\partial_\mu \bar{a} \tag{16}$$

and that the topological charge (11) can be written in the gauge-invariant way as:

$$Q = - \frac{1}{16\pi^2} \int d^4x \; (\nabla_\mu \Phi_i)\overline{(\nabla_\nu \Phi_i)}(\nabla_\rho \Phi_j)\overline{(\nabla_\tau \Phi_j)} \, \epsilon_{\mu\nu\rho\tau} \tag{17}$$

where $\nabla_\mu = \partial_\mu - iA_\mu$. Using (15) it can be checked that

$$F_{\mu\nu} = (\nabla_\mu \Phi_i)\overline{(\nabla_\nu \Phi_i)} - (\mu \leftrightarrow \nu) \tag{18}$$

and so we can identify the expressions (17) and (1).

Let us consider the classical vacua characterized by the choice

---

*) For n = 1 any oriented four-dimensional manifold carries a quaternionic structure, because $SO(4) \simeq Sp(1) \otimes Sp(1)$.

$$\Phi_i(x) \xrightarrow[|x|\to\infty]{} U(x)\cdot C_i + O\left(\frac{1}{|x|}\right) \tag{19}$$

where $|C_i|^2 = 1$ and U is a unit quaternion. One obtains for such fields

$$Q = \lim_{|x|\to\infty} \frac{2}{3} \int_{S^3} d\sigma_\mu \, \mathrm{Tr}\, (U\partial_r \bar{U})(U\partial_s \bar{U})(U\partial_t \bar{U}) \, \epsilon_{rst} \tag{20}$$

in accordance with the relation $\Pi_4(HP(n)) = \Pi_3(SU(2)) = Z$.

Because our quaternionic σ-fields have "naive" canonical dimensionality $d(\Phi_i)$ equal to zero (this follows, for example, from formula (15) and $d(A_\mu) = 1$), the conformal-invariant Lagrangian with first-order derivatives is quartic; one can add also the "conventional" Lagrange density in σ-models multiplied by the mass-like dimensional parameter.*) We obtain

$$\mathcal{L} = -\frac{1}{4g^2}\,\mathrm{Tr}\, F_{\mu\nu} F_{\mu\nu} + \frac{m^2}{2} (\nabla_\mu \Phi_i)(\overline{\nabla_\mu \Phi_i}) + \lambda(|\Phi|^2 - 1) \tag{21}$$

where $F_{\mu\nu}$ is given by the formula (18). By putting n = 0 (i.e. Φ is a unit quaternion) and replacing $\nabla_\mu \to \partial_\mu$, one obtains the Skyrme model [12] which describes the $S^3 = [O(4)/O(3)]$-valued dimensionless σ-fields.**) Here we propose the generalization with fields Φ_i taking values in the quaternionic Stiefel manifold

$$S^{4n+3} \simeq \frac{Sp(n+1)}{Sp(n)} \simeq \frac{O(4n+4)}{O(4n+3)}$$

which is the total space for the Hopf fibre bundle $S^{4n+3} \to HP(n) = HG_{n;1}$. The model is invariant under the Sp(n) global and $Sp(1) \simeq SU(2)$ local gauge transformations.

It should be mentioned that the mass term which occurs in (21) has been proposed (for n = 0) by Slavnov [25] as a way of introducing the gauge-invariant mass term in the Yang-Mills theory; for n = 1 , due to the relations

---

*) Another possibility was considered in [9] (higher order bilinear Langrangians).

**) The sphere S_3 describes the field values for principal O(3) chiral field, and it leads to vanishing curvature form. This property forces the replacement $\nabla_\mu \to \partial_\mu$ in the Skyrme model.

$$\frac{Sp(2)}{Sp(1)\otimes Sp(1)} \simeq \frac{O(5)}{O(3)\otimes O(3)} = \frac{O(5)}{O(4)} = S^4 \tag{22}$$

the second term in (21) was used in [10] for describing the O(5) non-linear σ-model coupled in a particular way with gravity.

We shall now discuss briefly some features of the non-linear theory described by (21) which appear to be interesting:

a) Due to the relation ($S = \int d^4x\mathcal{L}$) valid if m = O

$$\frac{\delta S}{\delta \Phi_j(x)} = \int d^4y \; \frac{\delta S}{\delta A^j_\mu(y)} \; \frac{\delta A^j_\mu(y)}{\delta \Phi_j(x)} \tag{23}$$

the solutions of Yang-Mills equations

$$\frac{\delta S}{\delta A^j_\mu} = D^{ij}_\nu F^j_{\nu\mu} = 0 \tag{24}$$

also solve the equations of the model (21); in particular, the complete set of self-dual instanton solutions of Atiyah et al. [26,27] can be considered as the classical background fields in the discussion of semi-classical approximation, describing one-loop corrections around the classical minima of our action.

b) The Lagrangian (21) in standard perturbation theory, with the term

$$m^2 \partial_\mu \bar{\Phi}_i \partial_\mu \Phi_i$$

used to determine the zero order propagator, is non-renormalizable. It should, however, be possible to introduce a different calculational scheme, in which one can put m = O and consider the propagator in the lowest order consistent for short distances with naive sub-canonical dimensionality of $\Phi_i(x)$: *)

$$U(\lambda)\, \Phi_i(x)\, U^{-1}(\lambda) = \Phi_i(\lambda x) \tag{25}$$

---

*) The "naive" dimensionality (25) is also consistent with the "naive" canonical quantization scheme, where $\Pi_i = \partial\mathcal{L}/\partial\Phi_{i,o}$.

In a conventional (Wigthman) local QFT the fields with the dimensionality $d(\Phi) = 0$ are excluded because they do not satisfy the positivity axiom. Whatever is the nature of Φ_i, owing to the locality of the Lagrangian (21) it is plausible to assume that the Wilson hypothesis about the short-distance behaviour [28] can be applied. One gets from (25)

$$\text{F.T.} \langle o | \, T \, \{\Phi_i(x) \, \Phi_j(o)\} \, | o \rangle \underset{k^2 \to \infty}{\sim} \frac{1}{k^4} \tag{26}$$

Furthermore, one can assume that the exact SD behaviour is obtained if we modify (26) by the anomalous dimensionality ($d\Phi = 0 \rightarrow \Phi = \gamma$ determined by the wave renormalization constant) and logarithmic terms. In such a scheme the mass term in (21) describes the generalized mass term in the sense of Wilson analysis [28] and modifies only the large-distance behaviour of the propagator (26).

c) Let us observe that one can invert the relation (15) for $x \neq x^o$ in the following way: *)

$$\Phi_i^{\text{path}}(x) = \delta_{ii_o} \, T_\xi \, \{ e^{-i \int_o^1 d\xi \, \frac{dx_\mu}{d\xi} A_\mu(x(\xi))} \} \tag{27}$$

where $x_\mu(1) = x_\mu$, $x_\mu(0) = x_\mu^o$ (possibly $|x_\mu^o| \to \infty$) and T_ξ denotes the ordering of quaternion-valued A_μ along the path $x_\mu(\xi)$. Having formula (27) one can connect the relations for the Green functions of Φ_i with the behaviour of the "conventional" Yang-Mills Green functions.

A technique which would, however, involve only the use of the form of the Lagrangian (21) (without "inverse formula" (27)), e.g. for calculating the renormalization corrections to (26), is desirable but at present not known.

---

*) Formula (27) is written in a particular Sp(n) gauge described by the factor δ_{io}. The role of the Sp(n) gauge transformations (including the generalization of our model allowing the local dependence of the Sp(1)x...xSp(1) factor) should be studied in order to clarify the meaning of the parameter n, which describes in the construction of Atiyah et al. [26,27] the Pontriagin index of self-dual solutions.

4. FINAL REMARKS

The HP(n) σ-models, (n = 1,2,...), described by (21) allow us to describe all self-dual SU(2) Yang-Mills solutions in a simple form proposed by Atiyah, Hitchin, Drinfeld and Manin (see e.g. [26]). If we wish to describe all SU(2) connections on S_4, expressed by means of formula (15), due to the theorem of Narasimhan and Ramanan [29] it is sufficient to consider only finite-dimensional (up to some maximal dimension N) complex quaternionic manifolds with coordinates q parametrized by four complex quaternionic components c_o, c_r ($q = c_o + c_r e_r$).

One can also consider generalizations of HP(n) models:

i) The simplest one is to investigate the $HG_{n,k}$ σ-models for $k \neq 1$. Because the quaternionic metric for $HG_{n,k}$ is known (see for example, [23]), such a generalization is in principle straightforward.

ii) It can be shown that all the SU(n) bundles over S_4 are classified by H^4 (M,Z), i.e. by the topological charge (1) with the index i running over the (n^2 - 1) - dimensional adjoint representation of SU(n). If we wish to generalize our geometric scheme, we can consider the geometries with Clifford algebra-valued coordinates and a Clifford algebra-valued metric tensor. In such a way, for example, we can obtain SU(n) gauge potentials as the connection forms on the manifold with the coordinates described by the elements C_N if $n = 2^{N/2}$ (N = 2, 4, 6, ...).

iii) In order to have full "pre-QCD" theory, fermions (quarks) should be added. The usual method of introducing fermions in σ-models is based on consideration of σ-superfields [30,31] . One can, however, keep the dependence of σ-fields on space-time points, and consider fields with values in a coset space of a supergroup [32,33] . In such a formulation, one can introduce composite SU(2) gauge fields constructed out of purely fermionic "pre-gluon fields". In particular, the introduction of quaternion-valued anticommuting pre-gluon fields leads to the appearance of para-fermionic fields as fundamental variables [34]. In such a scheme it is tempting to identify the "pre-gluon" fiels in quaternionic fermionic σ-models with quark degrees of freedom [35].

The author would like to thank Dr. D. Maison and Dr. D. Olive for pointing out an error in first version of this paper (CERN preprint TH-2678) in the definition of the two-form.

REFERENCES

1. W. E. Zahkharov and A. W. Michailov, Zh.Eksp. and Teor.Fiz.74, (1978) (in Russian).
2. A. J. MacFarlane, Phys.Lett. 82B, 239 (1979).
3. E. Brezin, C. Itzykson, J. Zinn-Justin and J. B. Zuber, Phys.Lett. 82B, 442 (1979).
4. A. M. Perelomov, Comm.Math.Phys. 63, 237 (1978).
5. K. Pohlmeyer, Comm.Math.Phys. 46, 207 (1976).
6. M. Lüscher and K. Pohlmeyer, Nucl.Phys. B137, 46 (1978).
7. H. Eichenherr, Nucl.Phys. B146, 215 (1978).
8. H. Eichenherr and M. Forger, Freiburg preprint 79/2 (1979).
9. E. Gava and R. Jengo, Nucl.Phys. B140, 510 (1978).
10. E. Gava, R. Jengo and C. Omero, Phys.Lett. 81B, 187 (1979).
11. V. de Alfaro, S. Fubini and G. Furlan, CERN preprint TH-2584, (1978).
12. T. H. R. Skyrme, Proc. Roy.Soc. 260, 127 (1961).
13. N. K. Pak and H. C. Tze, Ann.Phys. 117, 164 (1979).
14. M. Dubois-Violette and Y. Georgelin, Phys.Lett. 82B, 251 (1979).
15. J. Fröhlich, A new look at generalized, non-linear σ-models and Yang-Mills theory, IHES preprint, 1979.
16. A. d'Adda, P. Di Vecchia and M. Lüscher, Nucl.Phys. B146, 63 (1978).
17. E. Witten, Nucl.Phys. B149, 285 (1979).
18. M. Berger, Bull.Soc.Math.France 83, 279 (1955).
19. H. Wakakuwa, "Differential Geometry", in honour of K. Yano (Kinokuniya Book-Store Co., Ltd., Tokyo, 1972), p.503.
20. M. Forger, Instantons in non-linear σ-models, gauge theories and general relativity, FUB preprint, 1979.
21. V. Y. Krames, Trans.Am.Math.Soc. 122, 357 (1966).
22. D. W. Alexandrov, Funkc.Anal.and Appl. 2, 11 (1968) (in Russian).
23. Y. C. Wong, Proc.Acad.Sci. USA 57, 589 (1967).
24. A. Trautman, Int.J.Theor.Phys. 16, 561 (1977).
25. A. A. Slavnov, Teor.and Mat.Fiz. 10, 305 (1972).
26. M. F. Atiyah, N. J. Hitchin, V. G. Drinfeld and Yu. I. Manin, Phys.Lett. 65A, 185 (1978).
27. E. F. Corrigan, D. B. Fairlie, P. Goddard and T. Templeton, Nucl.Phys. B140, 45 (1978).
28. K. Wilson, Phys.Rev. 179, 1499 (1969).
29. M. S. Narasimhan and S. Ramanan, Ann.J.Math. 83, 563 (1961).
30. P. Di Vecchia and S. Ferrara, Nucl.Phys. B130, 93 (1977).
31. E. Witten, Phys.Rev. 16, 2991 (1977).
32. J. Lukierski, Lett.in Math.Phys. 3, 135 (1979).
33. J. Lukierski, Quarks and fermionic geometry, Proc.4th Workshop on Hadronic Matter, Erice, October 1978, to be published by Plenum Press.
34. J. Lukierski and V. Rittenberg, Phys.Rev. 18, 385 (1978).
35. J. Lukierski, Talk given at the International Seminar on Non-Local QFT, Aluszta, April 1979.

H. Leutwyler

Institute for Theoretical Physics
University of Berne
CH-3012 Berne, Switzerland

ABSTRACT

We discuss the effect of a nonzero gluon expectation value $<o|G^a_{\mu\nu}G^{\mu\nu a}|o>$ on quark bound states. It is argued that this effect is responsible for one of the leading scale breaking power corrections to the potential at short distances.

The first part of these lectures consisted of an introduction to the standard model of strong, weak and electromagnetic interactions. I emphasized the qualitative aspects of asymptotic freedom and discussed the origin and some recent estimates of the quark masses. The interested reader may find an account of this material in the Proceedings of the 1979 GIFT Seminar on QCD [1] .

In the second part I discussed some aspects of the bound state problem in QCD. The binding force is due to gluon exchange; to lowest order (exchange of a single gluon) the force is described by a 1/r potential which is attractive in the colour singlet configuration of a quark-antiquark pair:

$$V = - \frac{4}{3} \frac{g^2}{4\pi} \frac{1}{r} \tag{1}$$

where g is the coupling constant of QCD. As discussed in [1] some of the higher order contributions to V may be taken into account by renormalizing the coupling constant. The relevant strength of g depends on the momentum transferred by the gluon (for large momentum transfer the effective coupling constant is small and vice versa). Repeated exchange of gluons (ladder approximation) leads to the bound state picture familiar from positronium. The quarks are bound within a Bohr radius

$$a = \left(\frac{1}{2}\, m\right)^{-1} \left(\frac{4}{3}\, \frac{g^2}{4\pi}\right)^{-1}$$

and move at a velocity of order

$$v/c = \frac{4}{3} \frac{g^2}{4\pi} \quad .$$

The average momentum transferred in the exchange of a gluon is of order m v. If the quarks are sufficiently heavy, the positronium picture is self-consistent at least for the lowest lying orbits: the average momentum transfer is large and hence the effective coupling strength is small - the lowest order contribution to V should indeed dominate the potential. One therefore expects that the lowest lying states of a heavy $q\bar{q}$ system exhibit the level pattern of positronium.

If the quarks are not heavy, in particular for bound states of u, d and s quarks, the situation is quite different: the Bohr orbits are large, the average momentum transferred by the gluons is small and hence the effective coupling strength will be appreciable - there is no reason why one gluon exchange should provide the dominating contribution. Asymptotic freedom only guarantees that at short distances the potential is given by (1). The distances for which this expression dominates only cover the innermost part of the wave function describing a bound state of two light quarks. Most of the time the quarks are too far from one another for (1) to apply. The behaviour of the potential at large distances is still unknown. In the most popular picture one assumes that if the quarks are sufficiently far from one another then the gluon field takes the shape of a string of finite diameter. The field energy of this string is proportional to the length of the string and the corresponding effective potential between the quarks is therefore proportional to their distance, $V(r) = kr$. The constant k has the dimension of the square of an inverse length. Perturbation theory to any finite order does not give rise to dimensional quantities of this sort - one has to appeal to nonperturbative effects if one wants to squeeze out a potential of this sort and a value for k from QCD. It is unclear at this time whether the classical string picture is indeed appropriate, although some arguments for it can be given on the basis of lattice gauge theories [3] .

A very interesting method to study nonperturbative effects is due to the ITEP group [2] . Shifman, Vainshtein and Zacharov [2] propose to study the short distance expansion of say two currents $j_\mu(x) = \bar{q}(x)\, \gamma_\mu Q\, q(x)$

$$j_\mu(x)\, j_\nu(y) = \sum_n C^n_{\mu\nu}(z)\, O_n(X) \tag{2}$$

where $z = x - y$, $X = \frac{1}{2}(x + y)$. The quantities $C^n_{\mu\nu}$ are c-number functions and $\{O_n\}$ stands for the complete set of gauge invariant local operators in QCD:

$$\{O_n\} = \{\mathbb{1}, \bar{q}(x)\,\Gamma q(x),\ G^a_{\mu\nu} G^a_{\rho\sigma},\ \bar{q}\,\Gamma D_\mu q, \ldots\} \tag{3}$$

and $G^a_{\mu\nu}(x)$ is the gluon strength. One may work out the functions $C^n_{\mu\nu}(z)$ in perturbation theory order by order. Suppose for simplicity that we are dealing exclusively with massless quarks. In this case the quantity $C^0_{\mu\nu}(z)$ e.g. is given by

$$C^0_{\mu\nu}(z) = \frac{\mathrm{tr}Q^2}{8\pi^2}\{g_{\mu\nu}\Box - \partial_{\mu\nu}\}\frac{1}{z^4}\{1 + \frac{g^2}{4\pi^2} + \ldots\} \tag{4}$$

and similarly, the coefficients of the other terms in the expansion are given by some integer power of z^2 multiplying a homogenous polynomial in z and a power series in the coupling constant which in general introduces an additional logarithmic z-dependence.

In the following we will only be interested in contributions from Lorentz scalar operators O_n. The Lorentz scalars of lowest dimension (giving rise to the most singular contribution) are $\bar{q}q = \{\bar{u}u, \bar{d}d, \ldots\}$. For massless quarks the coefficients of these terms vanish however, because neither the currents nor the Lagrangian contain couplings between right- and lefthanded components. The first scalar that does contribute in the expansion (2) is the operator $G^a_{\mu\nu} G^{\mu\nu\, a}$. The corresponding coefficient function is proportional to $g^2\,(g_{\mu\nu}\Box - \partial_{\mu\nu})\ \ln z^2 \sim z^{-2}$.

In the dressed perturbation theory vacuum the expectation values of the operators O_n vanish except for $O_0 = \mathbb{1}$; the quantity $C^0{}_{\mu\nu}(z)$ represents the vacuum expectation value of the product of the two currents. Shifman[2] now observe that two effects must be taken into account before one applies the operator expansion to the real world:

1) Nonperturbative contributions to the coefficients of the operator expansion. Perturbation theory, even if summed to all orders does not necessarily contain all contributions. (Perturbation theory accounts for the fluctuations around the field $G^a_{\mu\nu} = 0$. This field configuration is however not the only one for which the euclidean action has a minimum. Other minima correspond to instanton field configurations; the fluctuations around these give rise to contributions $\sim \exp - (8\pi^2/g^2)$ - they do not show up in perturbation theory). Shifman [2] argue that these non-perturbative effects

only contribute to terms in the operator expansion which involve operators of rather high dimension (i.e. are suppressed by a rather high power of z in comparison to the leading terms).

2) Even if the operator expansion holds up to some given power of z with coefficient functions given by perturbation theory one has to make sure that one is taking matrix elements of this operator relation between the proper states. For QCD to be a realistic theory the dressed perturbation theory vacuum cannot represent the physical ground state of the system. The physical ground state $|0\rangle$ must break chiral symmetry spontaneously, the dressed perturbation theory vacuum $|0\rangle$ does not. Quantitatively, in the limit of massless quarks we should have

$$\langle o|\bar{u}u|o\rangle = \langle o|\bar{d}d|o\rangle = -M_o^3 \tag{5}$$

whereas the corresponding expectation values with respect to $|o\rangle$ vanish. For small, but nonvanishing quark masses the constant M_o determines the pion mass in terms of the quark mass according to

$$m_\pi^2 = \frac{M_o}{F_\pi^2}(m_u+m_d). \tag{6}$$

With $\frac{1}{2}(m_u + m_d) = 5.4$ MeV, $F_\pi = 0.68\ m_\pi$ this gives

$$M_o = 250 \text{ MeV}. \tag{7}$$

Once it is clear that the physical vacuum cannot be identified with $|o\rangle$ the vacuum expectation value of $j_\mu(x)\ j_\nu(y)$ is not any more given by $C^o_{\mu\nu}(z)$. Instead all terms in the operator expansion may contribute. In particular, we get a contribution from the term involving the square of the gluon field strength:

$$\frac{g^2}{4\pi^2}\langle o|G^a_{\mu\nu}G^{\mu\nu a}|o\rangle = M_1^4 \tag{8}$$

The ITEP group has extracted the behaviour of the current correlation function at small values of z from the data on the e^+e^- cross section. They come up with

$$M_1 = 330 \text{ MeV}. \tag{9}$$

In terms of the colour magnetic and electric fields this result states

$$\frac{g^2}{\pi^2}\langle o|\vec{B}^2|o\rangle = -\frac{g^2}{\pi^2}\langle o|\vec{E}^2|o\rangle = M_1^4 \tag{10}$$

[Lorentz invariance requires that the expectation values of B^2 and E^2 are equal in magnitude and opposite in sign. Hence the expectation value of B^2 is positive, the expectation of value E^2 is negative;

magnetic and electric contributions each produce half of the result(8).] One should not be disturbed by a negative value for an apparently positive quantity like E^2; the matrix element $\langle 0|\vec{E}^2|0\rangle$ is only what remains at x = y once one takes away the positive perturbative contribution to $\langle 0|\vec{E}(x)\vec{E}(y)|0\rangle$ that explodes at $x \to y$ like $(x - y)^{-4}$.

Note that the quantity $g^2 G^a_{\mu\nu} G^{\mu\nu a}$ is not renormalization invariant. To make a statement that is independent of the renormalization point, one should instead look at the operator $\beta(g)/g \cdot G^a_{\mu\nu} G^{\mu\nu a}$ that appears in the trace of the energy-momentum tensor. To lowest order in g the β-function is given by

$$\beta(g) = - \frac{g^3}{48\pi^2} N_3 + \ldots \tag{11}$$

where $N_3 = 33 - 2 N_{f1}$. The result of Shifman [2] may be stated in the renormalization group invariant from

$$[- \frac{\beta(g)}{2g}] \langle 0|G^a_{\mu\nu} G^{\mu\nu a}|0\rangle = M_2^4 \tag{12}$$

with M_2 = 340 MeV. It may at first sight seem strange that the quantity which for massless quarks should represent the trace of the energy-momentum-tensor, acquires a nonzero vacuum expectation value. The expectation value of the energy-momentum-tensor of course vanishes. If one insists that the operator G^2 be defined through the operator product expansion discussed above then the relation between $\Theta_\mu{}^\mu$ and G^2 involves an additive c-number which reflects the fact that a change of scale requires renormalization of the vacuum bubbles of the theory.

I hope that these few remarks on the work carried out by the ITEP group invite you to read their papers.

In the following I discuss the effect of a nonvanishing vacuum expectation value for the square of the field strength on a bound $\bar{q}q$ pair. The idea is best understood in analogy with the effect that a magnetic field produces in an atom. Suppose we expose a hydrogen atom to an external magnetic field. The field slightly distorts the electron orbits and induces a magnetic dipole moment opposite to the applied field (this effect is at the origin of diamagnetism)

$$\vec{\mu} = \varkappa \vec{B}, \quad \varkappa = - \frac{e^2}{6m_e} \langle r^2 \rangle \tag{13}$$

where $\langle r^2 \rangle$ is the mean square radius of the electron orbit. The magnetic moment interacts with the magnetic field and gives rise to an interaction energy

$$E = - \frac{\varkappa}{2} \vec{B}^2. \tag{14}$$

This interaction energy amounts to a perturbation of the system by a harmonic potential given by

$$V = \frac{e^2 \vec{B}^2}{12\, m_e} r^2 . \tag{15}$$

In QCD we should find a similar effect: the nonvanishing vacuum expectation value of the square of the colour magnetic field should give rise to a harmonic potential

$$V \sim \frac{g^2}{m} \langle \vec{B}^2 \rangle r^2 .$$

To work this out quantitatively I look at a wave function associated with two scalar quarks and a meson of momentum p:

$$\begin{aligned} \psi(x,y) &= \langle o|\phi^+(y)\ E(y,x)\ \phi(x)|p\rangle \\ E\ (y,x) &= P \exp - ig \int_x^y dz^\mu\ B_\mu \end{aligned} \tag{16}$$

The path ordered gauge factor is introduced to obtain a gauge invariant wave function. The field $\phi(x)$ transforms like a colour triplet and obeys the equation of motion

$$D^\mu D_\mu \phi + m^2\ \phi = 0 \tag{17}$$

where

$$D_\mu \phi = \partial_\mu \phi + ig\ B_\mu\ \phi. \tag{18}$$

The derivative of the wave function may be worked out by means of the formula

$$\partial^x_\mu E(y,x) = ig\ E(y,x) B_\mu(x) + ig\ P\{\int_0^1 d\xi \xi z^\rho\ G_{\mu\rho}(x')E(y,x)\} \tag{19}$$

where $x' = y + \xi z,\ z = x-y$. The first derivative of $\psi(x,y)$ is given by

$$\begin{aligned} \partial^x_\mu\ \psi(x,y) &= \langle o|\phi^+(y)\ E(y,x)\ D_\mu\ \phi(x)\ |p\rangle \\ &+ ig \int_0^1 d\xi\xi \langle o|\phi^+(y)\ P\{z^\rho G_{\mu\rho}\ (x')E(y,x)\}\phi(x)\ |p\rangle \end{aligned}$$

and the second derivative becomes

$$\begin{aligned} \Box^x\ \psi(x,y) &= \langle o|\phi^+ E D^\mu D_\mu \phi\ |p\rangle \\ &- g^2 \int_0^1 d\xi\xi \int_0^1 d\eta\eta\ \ \langle o|\phi^+ P\{z^\rho\ G_{\mu\rho}(x')\ \ z^\sigma G^\mu{}_\sigma\ (x'')E\}\ \phi\ |p\rangle \end{aligned}$$

$$+ ig\int_0^1 d\xi\xi^2 \; <o|\phi^+ P\,\{z^\rho D^\mu G_{\mu\rho}(x')\,E\}\phi\,|p>$$
$$+2ig\int_0^1 d\xi\xi <o|\phi^+ P\{\,z^\rho G_{\mu\rho}(x')\,E\}\,D^\mu\phi\,|p>$$

with x" = y + ηz. Using the equation of motion this result may be rewritten in the form

$$(\Box^x+m^2)\;\psi(x,y) = g^2\int_0^1 d\xi\xi\int_0^1 d\eta\eta <o|\phi^+ P\{z^\rho\, G_{\mu\rho}(x')z^\sigma\, G^\mu{}_\sigma(x'')$$
$$E\}\;\phi\;|p> \qquad (20)$$
$$+\, z^\mu\partial^\nu{}_x\;\psi_{\mu\nu} + z^\mu\psi_\mu$$
$$\psi_{\mu\nu} \sim <o|\phi^+ P\{\,G_{\mu\nu}\,E\}\;\phi\;|p>$$
$$\psi_\mu \sim <o|\phi^+ P\,\{D^\rho\;\;G_{\rho\mu}\,E\,\}\;\phi\;|p>$$

I now make the following approximations:

1) Ignore the correlation between the gluon field and the quarks, i.e. consider only the first term on the r.h.s. of (20)
2) Replace the product of gluon fields by the corresponding vacuum expectation value at x' = x".

The expectation value of $G_{\mu\rho}(x)G^\mu{}_\sigma(x)$ is proportional to the unit matrix and to $g_{\rho\sigma}$ and may therefore be rewritten as

$$<o|\;G_{\mu\rho}(x)G^\mu{}_\sigma(x)\;|\;o>= \frac{1}{3}\mathbb{1}\cdot\frac{1}{4}\,g_{\rho\sigma}\cdot\;<o|tr\;G_{\mu\nu}\;G^{\mu\nu}|o>\,.$$

With

$$G_{\mu\nu} = G^a_{\mu\nu}\;\frac{\lambda^a}{2}\quad;\quad tr\;\lambda^a\,\lambda^b = 2\delta^{ab}$$

this becomes

$$<o|\;G_{\mu\rho}(x)\;G^\mu{}_\sigma\;(x)\;|o> = \frac{1}{24}\,g_{\rho\sigma}\mathbb{1}<o|G^a_{\mu\nu}\;G^{\mu\nu a}\;|o>\,.$$

The ξ and η integrations give rise to a factor of $\frac{1}{4}$ such that (20) takes the form

$$(\Box^x+m^2)\quad\psi(x,y) = \lambda^2\,z^2\;\psi\;(x,y) + \ldots \qquad (21)$$
$$\lambda^2 = \frac{g^2}{96}<o|G^a_{\mu\nu}\;G^{\mu\nu a}\;\;|o> \qquad (22)$$

To check that the approximation used indeed reproduces the correct result in the case of a hydrogen atom exposed to a magnetic field we replace g by e and identify the field $G_{\mu\nu}$ with the electromagnetic field. If there is no electric field, the average

$z^\rho G_{\mu\rho} z^\sigma G^\mu{}_\sigma$ over the directions of $\vec{z}$ is given by $\frac{2}{3}\vec{B}^2\vec{z}^2$. (If the wave function of the system is not spherically symmetric, then the atom has a magnetic moment even in the absence of an external magnetic field and the system will respond by a Larmor precession, not merely by a level shift). The first term on the r.h.s. of (20) therefore becomes $-\frac{1}{6}e^2\vec{B}^2\vec{z}^2$. In the nonrelativistic limit the wave function may be written as

$$\psi(x,y) = e^{-im_e x^0 - im_p y^0}\,\psi_{NR}(x,y)$$

where ψ_{NR} is slowly varying (I identify $\Phi(x)$ with the electron field, $\Phi^+(y)$ with the proton field here; I should of course have used two different symbols to describe the wave function of the hydrogen atom). The nonrelativistic limit of (21) then becomes

$$\frac{1}{i}\frac{\partial\psi_{NR}}{\partial x^0} + \{-\frac{\Delta}{2m_e} + \frac{1}{12}\frac{e^2\vec{B}^2}{m_e}\vec{z}^2\}\,\psi_{NR} + \ldots = 0$$

This verifies that the term we retained in the above approximation procedure indeed reproduces the proper effective potential generated by an external magnetic field.

After this check we now return to chromodiamagnetism. With the value (8) for the expectation value of the field strength (22) becomes

$$\lambda = \frac{\pi M_1^2}{2\sqrt{6}} . \tag{23}$$

The diamagnetic term corresponds to a harmonic potential which, taken by itself, would give rise to a linear level spectrum: $M^2 = M_0^2 + 8\lambda n$, $n = 0,1,2,\ldots$ (compare [1]). The quantity 8λ measures the inverse of the slope. Numerically, using the value $M_1 = 320$ MeV of Shifman [2] we obtain $8\lambda = m_\rho^2$. Experimentally, the slope of the leading trajectory for light quarks corresponds to $8\lambda = 2\,m_\rho^2$. This shows that the chromodiamagnetic effect is sizeable.

To establish the significance of the above crude calculation in a more precise context, consider a bound state of very heavy quarks. As discussed above heavy quarks stay close to one another and remain in the region where the potential is well approximated by the perturbative epxression for the gluon propagator. In this region the correction that we have been evaluating is a small perturbation. It amounts to a power correction to the Coloumb potential of perturbation theory. The operator expansion provides us with a general method to calculate the leading power corrections of the potential in the following manner. One first works out the operator expansion of the product of two heavy quark fields. The matrix element of this expansion between vacuum and a mesonic state provides us with the short distance expansion of the wave function.

From this information it should be possible to read off the behaviour of the potential near the origin. In addition to the power correction due to $\langle G^2 \rangle$ this expansion also contains a power correction due to the vacuum expectation value of the light quark fields $\langle \bar{q}q \rangle$, which yet has to be worked out.

In the remainder of these lectures I discussed a hypothesis put forward in collaboration with Jan Stern: the hypothesis that the collective effects due to soft gluons may be described in terms of a relativistic local potential. For information on this approach the reader is again referred to [1] .

I am indebted to Peter Minkowski for illuminating discussions on the material contained in these notes.

REFERENCES

1. H. Leutwyler, Proceedings of the GIFT Seminar on Quantum Chromodynamics, Jaca (Spain), June 1979.
2. M. Shifman, A. Vainshtein and V. Zacharov, Nucl.Phys. B147, 385, 448 (1979); V. A. Novikov, Phys.Repts. 41C, 1 (1978).
3. J. Fröhlich, Lecture notes included in these Proceedings.

HIGH ENERGY BEHAVIOR OF NONABELIAN GAUGE THEORIES

J. Bartels

II. Institut für Theoretische Physik

Universität Hamburg

ABSTRACT

The high energy behavior (in the Regge limit) of nonabelian gauge theories is reviewed. After a general remark concerning the question to what extent the Regge limit can be approached within perturbation theory, we first review the reggeization of elementary particles within nonabelian gauge theories. Then the derivation of a unitary high energy description of a massive (= spontaneously broken) nonabelian gauge model is described, which results in a complete reggeon calculus. There is strong evidence that the zero mass limit of this reggeon calculus exists, thus giving rise to the hope that the Regge behavior in pure Yang-Mills theories (QCD) can be reached in this way. In the final part of these lectures two possible strategies for solving this reggeon calculus (both for the massive and the massless case) are outlined. One of them leads to a geometrical picture in which the distribution of the wee partons obeys a diffusion law. The other one makes contact with reggeon field theory and predicts that QCD in the high energy limit is decribed by critical reggeon field theory.

I. INTRODUCTION

These lectures intend to give a review of our present understanding of the Regge limit of nonabelian gauge theories, in particular QCD. Since cross sections are large in this kinematic regime, high energy physicists have always been interested in understanding the dynamics behind it (especially the nature of the Pomeron), but a theoretical description which is based on an underlying quantum field theory is still missing. Most previous attempts to understand the Regge limit within a field theory have been based on perturbation theory, and the main difficulty (besides the question which field theory to choose) was that the number of Feynman diagrams that could be handled always turned out to be too small. Now QCD is believed to be the right theory of strong interactions, and we are asked to understand its behavior in the Regge limit. Can we hope that the conventional approach, i.e. the start from perturbation theory, might be successful for this theory? Let me say a few words about this general question, before I come to details. The point I would like to make is that there are good reasons to believe that perturbation theory is a useful starting point, because the Regge limit is not far from that kinematic region in which perturbation theory works [1] (hard scattering processes). But, on the other hand, the Regge limit is also sensitive to certain features which are commonly referred to as nonperturbative.

Let us start with the optimistic part of the argument and consider elastic forward scattering of a very heavy photon off a nucleon. This is the process measured in deep inelastic leptoproduction

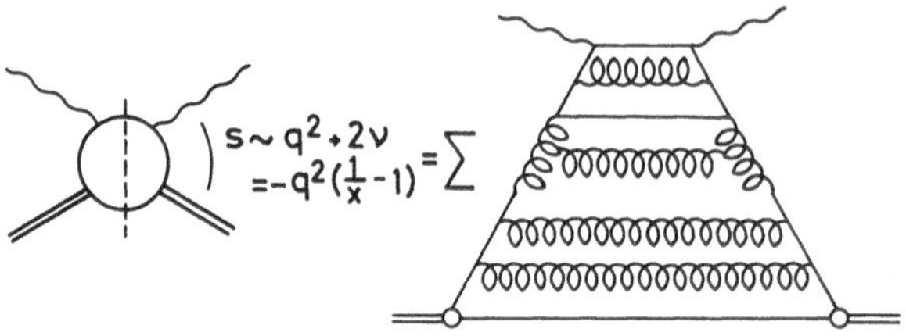

Fig. 1 Hadronic part of the deep inelastic leptoproduction process in QCD

and the standard argument about light cone dominance tells us that in the Bjorken limit

$$(-q^2 \to \infty \;, \quad s \sim -q^2 (\frac{1}{x} - 1) \to \infty \;, \qquad x \text{ fixed})$$

one probes the short distance structure of the nucleon target: if y_1 and y_2 are the two space-time points where incoming and outgoing photons couple to the nucleon, then $(y_1 - y_2)^2 \lesssim 1/q^2$. Within QCD the property of asymptotic freedom then allows to use perturbation theory for this short distance process. Either by means of the operator expansion and renormalization group techniques or, equivalently, by extracting and summing leading logarithmus of Feynman diagrams, one can calculate the q^2 - dependence, i.e. the change of the cross section when we move closer and closer to the light-cone $(y_1 - y_2)^2 = 0$. We now imagine that at some large value of q^2 we take a different limit: keeping now q^2 fixed and taking $x \to 0$, we reach the Regge limit $\cos \Theta_t \sim s/-q^2 \to \infty$. By choosing q^2 large enough, our investigation of the Regge limit can be carried out very close to the light cone, but once q^2 is kept fixed we always stay away from it by some finite distance. In terms of

QCD Feynman diagrams it is not difficult to see that those diagrams (Fig. 1) which govern the leading q^2-behavior of the Bjorken limit cannot be expected to correctly also describe the region of very small x . The tower diagrams of Fig. 1 do not contain "final state interactions" of the produced quarks and gluons and, hence, cannot satisfy unitarity which is known to be important in the Regge limit ($x \to o$ limit). If one wants to investigate the Regge limit within this perturbative approach, it is, therefore, necessary first to find all Feynman diagrams (beyond those of Fig. 1), which are required by unitarity for yielding a sensible $x \to o$ behavior, then to compute their behavior in the limit $x \to o$.

We conclude from this that, since the Regge limit, i.e. the Pomeron, can be investigated very close to the light cone where the (effective) coupling constant is small and perturbation theory works, perturbation theory may be a good starting point also for studying the Pomeron. The problem then consists of two major parts: first one has to decide which terms in the perturbation expansion (Feynman diagrams) have to be taken into account. Because of unitarity which is crucial for the Pomeron physics, these terms will not be the same as those which govern the Bjorken limit. Secondly, one has to find a method for summing them up. As I will make clear later, this part of the problem will require new techniques.

But as I have already indicated before, the Pomeron is also sensitive to certain features of long distance physics ("confinement dynamics"), which implies that at some stage nonperturbative aspects might have to enter the calculations. In the elastic scattering process of a very energetic hadron (say, in the rest frame of the target) the projectile appears as a composite system of partons which are spread out in impact parameter space. The probability of finding a slow parton at distance b is given by the impact parameter transform of the elastic scattering amplitude:

$$\frac{1}{s} T(s,b^2) = \frac{1}{2\pi s} \int d^2k_\perp e^{-i\vec{k}_\perp \vec{b}} T(s,k_\perp^2 = -t). \qquad (1.1)$$

The hadron radius is defined as:

$$\langle b^2 \rangle = \frac{1}{s} \int d^2b \, b^2 T(x,b^2) \qquad (1.2)$$

and, in general, it will depend on the energy s. It might, again, be useful to relate this to the hard scattering process in the Bjorken limit. In the deep inelastic scattering process the photon couples just to those constituents of the hadron which carry the fraction x of the hadron momentum (x is the Bjorken scaling variable). When approaching the Regge limit $x \to o$, these constituents are more and more wee: the Pomeron feels the distribution of the wee partons inside the hadron. When the energy increases, i.e.

the incoming hadron becomes more energetic, more decay processes are necessary before a fast parton slows down and eventually creates wee partons, and this may occupy a larger region in impact parameter space. As a result, the radius $<b^2>$ may grow as a function of s. In order to estimate how fast this growth could be in a realistic model, it may be useful to recall the multiperipheral model where

$$\frac{1}{s} T(s,b^2) = \frac{\text{const}}{\alpha' \ln s} e^{-b^2/4\alpha' \ln s} \tag{1.3}$$

and

$$<b^2> = \text{const} \cdot \alpha' \ln s \tag{1.4}$$

(α' is the Pomeron slope). Lowest order perturbation theory

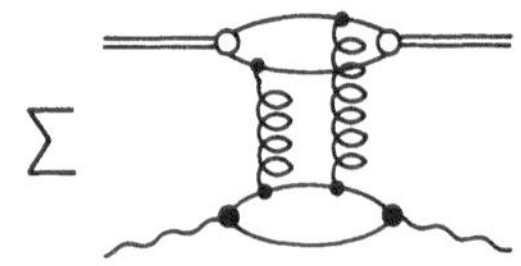

Fig. 2 Simplest model for elastic photon-hadron scattering in QCD: the sum goes over all possibilities of coupling two gluons to the quark lines

in a field theory with massless vector particles [2], on the other hand, leads to

$$\frac{1}{s} T(s,b^2) \underset{|b|\to\infty}{\sim} |b|^{-4} \tag{1.5}$$

$$<b^2> = \infty . \tag{1.6}$$

This indicates that only after summing many more diagrams one may hope to come somewhat close to (1.3), (1.4): the quantity $<b^2>$ can serve as a guide in estimating to what extent the use of perturbation theory alone is sufficient to "confine" the wee partons inside the fast hadron, and the fact that it is infinite in lowest order perturbation theory of QCD indicates how difficult it may be to obtain a correct theory of the Pomeron.

Before I can start to describe how well understood the Pomeron is within nonabelian gauge theories (and this understanding is almost entirely based on perturbation theory), I have to mention the other approach towards a theory of the Pomeron, namely reggeon field theory (RFT). As it is well known [3], physics of the Pomeron can most easily be discussed in terms of singularities in the angular momentum plane, and the interaction of moving pole and cut singularities has been formulated by Gribov in his reggeon calculus (or reggeon field theory). The rules of this formalism follow from certain analyticity properties of the S-matrix (existence of partial wave continuation, and t-channel unitarity equations) and are expected to hold in field theories that contain moving Regge singularities. The values of the parameters of reggeon field theory (intercepts, slopes, and interaction vertices), however, are not very much constrained from these analyticity arguments alone, and as long as RFT has not been considered in the context of a specific underlying field theory, they have been choosen freely. As the most interesting case, the Pomeron with intercept one has been studied extensively, and the best-known result is the critical Pomeron theory with

$$\sigma_{total} \sim (\ln s)^{-\gamma} \quad , \quad \gamma \sim 0.2 \quad . \tag{1.7}$$

Since this solution has also been shown to be consistent with the most restrictive constraints imposed by s-channel unitarity, it is an excellent candidate for a theory of strong interactions at high energies. More recently [4], also the case of the Pomeron intercept being above one has been investigated. Apart from the question how presently available energy ranges fit into these Pomeron field theories, the outstanding theoretical problem remains the derivation of the Pomeron parameters from an underlying field theory, such as QCD. It is not unexpected that these features of angular momentum theory will play an important role in analyzing the high energy behavior of nonabelian gauge theories.

After this introduction I can begin with a brief outline of the program of my talk. The aim is a review of what at present we know about the Regge limit (i.e. the Pomeron) of nonabelian gauge theories, and since the problem has not yet been solved completely, I shall attempt to describe both was has been achieved so far and what seem to be the main strategies for the future. Most of the existing calculations are determined to find the high energy behavior of QCD, the theory of (confined) quarks and gluons, but mainly because of the infrared problems, they start from spontaneously broken gauge theories. The mass of the vector particles then is considered to be an infrared cutoff which at the end of the calculations is taken to zero, hoping that in this limit one reaches QCD. As I have said already, all this will be based on perturbation theory.

In the first two sections of my talk I shall outline our present understanding of what the formal behavior of massive (spontaneously broken) nonabelian gauge theories is in the Regge limit. First I shall discuss the question of reggeization of elementary particles in these theories which divides the gauge theories into two classes: those where (at least) all vector particles reggeize and those where some of them don't. Then I shall describe (for a simple model) how this property of reggeization is seen to lead to a full reggeon calculus: this follows from the requirement of having (asymptotic) unitarity in both s and t channel, and the elements of the reggeon calculus are calculable in the limit of small coupling constant. Although such a reggeon calculus is of interest by itself, I shall consider it mainly as an intermediate step on the way towards finding the high energy behavior of massless Yang-Mills theories. This then requires a study of the zero mass limit of the reggeon calculus, and I shall briefly discuss what we know about this limit. As to the question how the use of perturbation theory may be extended into the Regge limit, this first part of my talk then basically selects all those terms in the perturbation expansion which have to be taken into account for a reliable high energy description: the selection criterion is unitarity, and the Feynman diagrams which are included in the reggeon calculus are just enough to satisfy unitarity.

Section IV deals with the question of how to solve this reggeon calculus, i.e. how to perform the summation of all the terms that we have decided to keep. First I shall briefly sketch an approach which, although it has not been pushed very far yet, has the advantage of asking directly for the distribution of the wee partons in impact parameter space. It allows a rather direct control over $\langle b^2 \rangle$ and, according to what has been said in the introduction, over the validity of the use of perturbation theory. Moreover, this technique seems to be applicable also to QED, where a high energy description which takes full account of unitarity is still missing. Within this approach one sees the possibility that, after summing all terms that have been obtained in the first part, the distribution of wee partons may come close to the multiperipheral picture. Then I shall describe how one might use the full apparatus of reggeon field theory, in particular its phase structure as a function of the bare Pomeron intercept, in order to determine the high energy behavior of massless Yang-Mills theories. Under the assumption that confining QCD can be obtained as the zero mass limit of spontaneously broken gauge theories with a modified $i\varepsilon$-prescription, this approach predicts critical high energy behavior for QCD, i.e. $\sigma_{total} \sim (\ln s)^{-\gamma}$.

II. REGGEIZATION IN YANG-MILLS THEORIES

Let us first consider gauge theories quite in general and ask which of them can be expected to have a "good" high energy behavior. The requirements one would like to impose on a realistic theory are the existence of moving Regge singularities and analyticity of the scattering amplitudes in the complex angular momentum plane. In particular, one would not like to have fixed singularities of the Kronecker delta function type which seem to exist if the theory contains nonreggeizing particles. This leads us to the question of reggeization in Yang-Mills theories.

It might be useful to recall what reggeization of a particle in a given field theory means. Suppose the theory contains a particle with spin J_o and mass μ. The exchange of this particle in lowest order perturbation theory

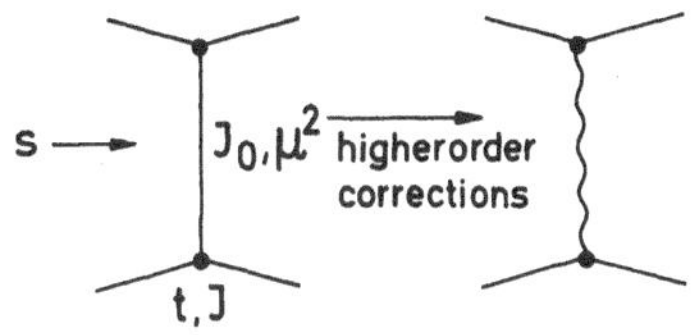

Fig. 3 Reggeization of an elementary particle in field theory: the exchange on the lhs is elementary, on the rhs the particle reggeizes

yields the following contribution to the t-channel partial wave:

$$T(J,t) = \text{const} \cdot \delta_{JJ_o} , \tag{2.1}$$

which is nonanalytic in J. Higher order diagrams for the same amplitude then can have two possible effects: either they leave the lowest order term (2.1) unchanged and simply add some new contributions:

$$T(J,t) = \text{const.}\ \delta_{JJ_o} + \text{terms analytic near } J_o. \tag{2.2}$$

In this case the particle stays elementary and leads to a nonanalytic term in the partial wave amplitude. Alternatively, the higher order contributions remove the δ-function in (2.1), for example:

$$T(J,t) = \frac{\alpha(t)-J_o}{\alpha(t)-J} \cdot \text{const} \tag{2.3}$$

$$\alpha(t) = J_o + (t-\mu^2) \cdot \beta(t) \tag{2.4}$$

($\beta(t)$ is proportional to the coupling constant of the theory and vanishes in lowest order perturbation theory. Eq.(2.3) then reduces to (2.1)). In this case the particle is said to reggeize: it lies on the trajectory (2.4), and the partial wave (2.3) is analytic in J. Experience with strong interaction physics clearly favors this second alternative: there is no evidence that singularities of the type (2.1) should be present. Therefore, one should look for theories in which, if possible, all particles reggeize.

Is there a simple way to decide whether, in a given theory, a particle reggeizes or not? The safest way, of course, is the explicit calculation: one computes the next-to-lowest order term of the partial wave and compares with the power series expansion of (2.3). This is the method by which, in the early sixties, Gell-Mann et.al. [5] found that the fermion in massive QED lies on a

Regge trajectory. Based on these calculations the same authors derived certain criteria which must be satisfied if a particle is to reggeize. One of them implies that the theory must contain particles of spin one (or higher spin). This excludes scalar theories such as ϕ^3 or ϕ^4 (although these theories may still contain moving Regge singularities). Another criterium requires certain factorization properties of the Born amplitudes. Later on, Mandelstam [6] gave counting arguments which say under what conditions a particle must necessarily reggeize. All those methods, when applied to (massive) QED, agree in that the fermion reggeizes but the photon does not. The boson in scalar QED has also been found[7] to reggeize: this result came out only after direct calculation of Feynman diagrams up to eighth order, and it illustrates that factorization and counting arguments [8] have to be applied with great care.

Theories in which also the vector particle reggeizes must be of the nonabelian type. To be more specific let us consider models of the following kind:

$$L = -\frac{1}{4} F^a_{\mu\nu} F^{a\mu}\Big| + \frac{1}{2} (D_\mu \phi)^2 - V(\phi) + \text{spinor part} \tag{2.5}$$

$$F^a_{\mu\nu} = \partial_\mu A^a_\nu - \partial_\nu A^a_\mu + g f^{abc} A^b_\mu A^c_\nu \tag{2.6}$$

$$D_\mu \phi = (\partial_\mu - igA^a_\mu T^a)\,\phi \tag{2.7}$$

$$[T^a, T^b] = if^{abc} T^c . \tag{2.8}$$

The potential $V(\phi)$ is invariant under the gauge group G and has its minimum at some nonzero value $\langle\phi\rangle \neq 0$. The pattern of spontaneous symmetry breaking may be rather complicated (for general symmetry breaking schemes see, for example, Refs.[9 and 10]), and the resulting particle spectrum may mask the original gauge group G. Is there a simple criterium which tells us under which conditions the vector particles reggeize? We first answer the question for two popular models: (i) the Higgs SU(2) model and (ii) the Weinberg-Salam model, and then state the result for the general case.

For the first case the gauge group G is SU(2), and the scalar field comes in two SU(2) doublets. With the scalar potential

$$V(\phi) = -\frac{1}{2} \mu^2 \phi^2 + \frac{\lambda}{4} (\phi^2)^2 \tag{2.9}$$

the Higg's mechanism makes all three vector particles massive:

$$M^2 = g^2 \mu^2 / \lambda \tag{2.10}$$

and leaves one (massive) scalar particle. (Generalization to SU(n) is made in the following way [11]: one starts from the gauge group U(n), adds n complex fundamental representations of scalar fields which make all n^2 vector particles massive, and then restricts oneself to the SU(n) subgroup of U(n)). Using the factorization criterion of the Born amplitudes, Grisaru et al. [11] found that both the fermions and the vector particles of this model reggeize. For the scalars the situation is still somewhat unclear: recently [8] it has been pointed out that it may reggeize, but in a more complicated way, similarly to the boson in scalar QED. This is contrary to the former belief [12] that reggeization of the scalar particle can occur only for special values of the parameters of the theory (masses and coupling constants). Presumably, only calculations similar to those of Ref. 7 will settle this point.

For the Weinberg-Salam model with gauge group G = SU(2)xU(1) one adds one doublet of complex scalar fields:

$$L = -\frac{1}{4} F^a_{\mu\nu} F^{a\mu\nu} - \frac{1}{4} B_{\mu\nu} B^{\mu\nu} + \frac{1}{2} (D_\mu \phi)^+ (D_\mu \phi) + V(\phi)$$

$$+ L(\text{spinors}) \tag{2.11}$$

$$B_{\mu\nu} = \partial_\mu B_\nu - \partial_\nu B_\mu \tag{2.12}$$

$$V(\phi) = -\mu^2 \phi^+ \phi + \lambda (\phi\phi^+)^2 \ . \tag{2.13}$$

As a result of the Higg's mechanism one has the three massive vector bosons:

$$W^{\pm}_\mu = \frac{1}{\sqrt{2}} (A^1_\mu \mp iA^2_\mu) \tag{2.14}$$

$$Z_\mu = \frac{-gA^3_\mu + g'B_\mu}{\sqrt{g^2+g'^2}} \tag{2.15}$$

$$g' = g \cdot \tan \Theta_W \tag{2.16}$$

and the massless photon:

$$A_\mu = \frac{gB_\mu + g'A^3_\mu}{\sqrt{g^2+g'^2}} \ . \tag{2.17}$$

With the same arguments which have been used for the Higg's model one finds [13] that in the Weinberg-Salam model only the W-bosons reggeize whereas the Z and the photon don't. Obviously, it is the U(1) subgroup of G which destroys the reggeization: the W's being purely made out of the nonabelian A-fields still reggeize. The Z and the photon, on the other hand, contain the U(1)-type B-field, which destroys the reggeization.

The general connection between the structure of the gauge group G and the reggeization of the vector particles has recently been investigated by two groups [13, 14]. It turns out that in order to make all vector particles reggeize G must be simple or semisimple. If G does not have this property - in particular if it has an abelian invariant subgroup - , some of the gauge particles lie on Regge trajectories, but others do not reggeize. The Higgs model with G = SU(2) and the Weinberg-Salam model with G = SU(2)xU(1) are examples for these two types of models. It is important to note that this result on the reggeization depends on the gauge group G but not on the way in which the Higgs scalars enter: this may introduce additional (global) symmetries into the theory, which manifest themselves in the mass spectrum of the vector particles but are independent of G. Finally, if after invoking the Higgs mechanism some vector particles are left massless, their trajectory functions (if they reggeize) have to be regularized by some infrared cutoff (cf.(2.4.)). The most important implication of this result concerns the reggeization of the photon within grand unification schemes. In one of the most popular versions [15], weak, electromagnetic, and strong interactions are embedded into a SU(5) gauge theory in which, according to the result stated above, all vector particles reggeize:

$$SU(2)xU(1)xSU(3)_{color} \subset SU(5) \ . \tag{2.18}$$

Such a scheme, for the first time, would allow to get rid of the undesired Kronecker-type partial wave singularities connected with the abelian photon, which according to our present understanding of this problem persist in QED and also the Weinberg-Salam model.

III. CONSTRUCTION OF A (ASYMPTOTICALLY) UNITARY S-MATRIX IN A MASSIVE YANG-MILLS THEORY. THE ZERO MASS LIMIT.

In this section I come to the longest part of my talk. Concentrating on the SU(2) - Higg's model which has been introduced in the previous section, I would like to describe how one can construct a high energy description of this model which satisfies (in an asymptotic sense) unitarity in both the direct and the crossed channel. The mass of the vector particle mainly serves as a convenient way to avoid the problems connected with massless particles and will be kept different from zero until the end of this section where I will mention what is known about the zero mass limit. The logic of this approach for investigating the high-energy behavior of massless Yang-Mills theories is illustrated in

Fig. 4a and b:

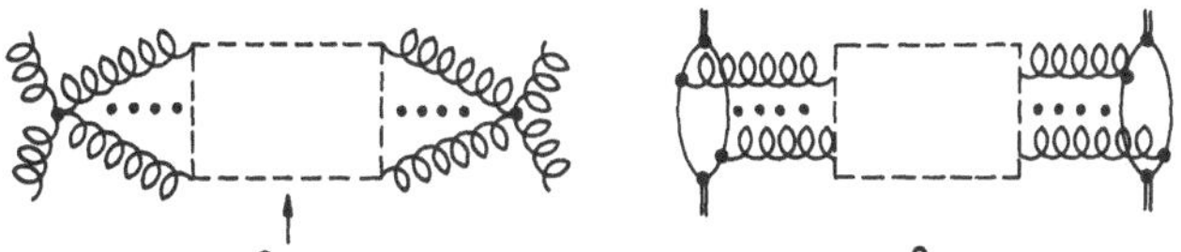

Fig. 4 Model for the Pomeron in QCD: the lhs denotes the unitary high energy expression for vector-vector scattering in the massive Higgs model; on the rhs the external particles are replaced by $q\bar{q}$ bound states, and the gluon mass is taken to zero.

in the massive case one studies the high energy behavior of processes with the (massive) vector bosons as external particles. In order to be able to take the zero mass limit, one has to replace the external particles by appropriate hadron wave function models. The gluon "soup" exchanged between these hadrons (the box of Fig. 4b) is taken to be the zero mass limit of the box of Fig. 4a. Within the line of arguments set up in the introduction the requirement of having for the massive case unitarity serves as a guide in selecting those terms in perturbation theory which one has to sum up for having a reliable description of the Regge limit of massless gauge theories.

After having established that the massive vector particles of the Higgs model reggeize, it seems very natural to expect that the full high energy description of this field theory should come in form of a complete reggeon calculus: the tree reggeons (i.e. the reggeized vector particles) interact with each other through all possible (momentum dependent) interaction vertices which are allowed by signature conservation. This will, in fact, be the result of this section: starting from the requirement of both s and t-channel unitarity, a full reggeon calculus emerges. The method I am going to describe also allows, at least in principle, to compute the elements of this reggeon calculus.

Before I am going into more detail, a few words should be said about the method of calculation. So far the problem of finding a reliable description for the high energy behavior of a field theory has not been solved, and this failure has, at least in part, to do with the calculational technique. For each order of perturbation theory the high energy behavior of a scattering amplitude, say for the $2 \to 2$ process, can be written as:

$$g^{2n}s[(\ln s)^{n-1}f_{n-1}(t) + (\ln s)^{n-2}f_{n-2}(t) + \ldots + \ln s f_1(t) + f_0(t)] + O(s^0) + O(s^{-1}) + \ldots \quad (3.1)$$

Conventionally, the leading term of this expansion, $f_{n-1}(t)$, is found by writing down all Feynman diagrams of this order perturbation theory, and, by means of a clever parametrization (Sudakov variables, infinite momentum variables, α-parameter representa extracting the highest power of lns. Summation over all orders in g then yields the leading -logarithmic approximation (LLA). As it turns out, however, in both QED and nonabelian vector theories this approximation violates the Froissart bound and, hence, is inacceptable. Because of the tremendous technical complications, the nonleading terms in (3.1) $f_{n-2}, \ldots, f_0$ can be computed so far only for very few special cases (for recent progress in this direction see Ref. [16]), but not for the vector theories we are interested in. In order to make further progress it seems, therefore, necessary to look for other methods of computation.

Since the main defect of the LLA was the violation of the Froissart bound, i.e. the lack of unitarity, the first goal must be restauration of unitarity. This suggests to use unitarity for the construction of the amplitudes from the very beginning: the Lagrangian is used only for determining vertices in the tree approximation. Amplitudes and all higher order corrections are then found by means of dispersion relations, i.e. by using our knowledge about the analytic structure of multiparticle amplitudes in the Regge limit. This garantees s-channel unitarity and, thanks to the reggeization of the vector particle, also t-channel unitarity in terms of partial wave unitarity. In terms of the expansion (3.1), presumably only parts of the nonleading coefficient functions $f_{n-2}, \ldots$ can be found in this manner: those which are necessary for achieving unitarity. In other words, what one obtains is likely to be the small- g approximation of the unitary S-matrix (which does not agree with the LLA). Whether this approximation is sufficient to give a reliable high energy theory can, at earliest, be answered after the summation of all these terms has been carried out. This problem will be subject of the next part of my lectures.

The presentation of how the construction of the unitary S-matrix works will be organized in the following way. Since extensive use will be made of the analytic structure of multiparticle amplitudes in the Regge limit, I start (part A) with a short review of those features which will be needed in the following. Then (part B) the construction of $T_{n \to m}$ in the LLA will be described. In part C this approximation will be unitarized, leading to the full reggeon calculus. In the final part D I shall discuss features of the zero mass limit.

A. ANALYTIC STRUCTURE OF MULTIPARTICLE AMPLITUDES AT HIGH ENERGIES.

Throughout this section I shall take a very pragmatic attitude: rather than describing any proofs or derivations, I shall restrict myself to listing those results which will be needed in the following. Those who wish to learn more details I refer to the lectures of Stapp and White [17] in the Les Houches Summer School 1975 and to Refs. 18-20.

Let me first recall a few well known facts about the $2 \to 2$ amplitude at high energies:
(i) $T_{2\to2}$ satisfies a dispersion relation in s with right and left hand cuts. For a theory with vector particles one needs two subtractions.
(ii) Real and imaginary part of the amplitude are connected via the signature factor:

$$T_{2\to2} = \frac{1}{2\pi i} \int dj s^j \left(-i + \frac{\cos\pi j + \tau}{\sin\pi j}\right) F(j,t). \tag{3.2}$$

The partial wave $F(j,t)$ is real in the physical region of the s-channel.
(iii) t-channel unitarity comes in form of partial wave unitarity relations. Starting from the normal t-channel unitarity equations, for example:

$\text{disc}_t T =$  $+ \ldots t \geq 16\mu^2,$ (3.3)

one projects out the partial waves $F(j,t)$, assumes that the two pairs of intermediate state particles couple together to moving Regge poles and continues down to $t < o$:

$\text{disc}_j F(j,t) =$ 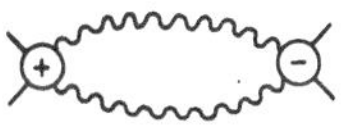 $+ \ldots$ (3.4)

("+" and "-" now refer to the j-variable). Together with

disc_j  $=$ 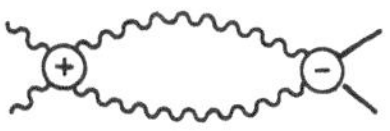 (3.5)

and

disc_j  $=$ 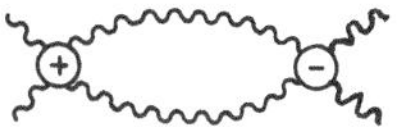 (3.6)

one has a coupled set of "partial wave" unitarity equations: this is the form in which t-channel unitarity enters the region $s \to \infty$, $t \lesssim 0$. In order to obtain a solution to these unitarity equations, it is sufficient that the partial wave function $F(j,t)$ takes the form of a reggeon calculus. It is important to note that at this stage the parameters of the reggeon calculus (trajectory function, interaction vertices) may be rather general: t-channel unitarity alone requires only that the formal rules of the reggeon calculus are satisfied (presence of signature factors etc.).

We now come to the simplest case of an inelastic amplitude: $T_{2\to 3}$ in the double Regge region (Fig.5).

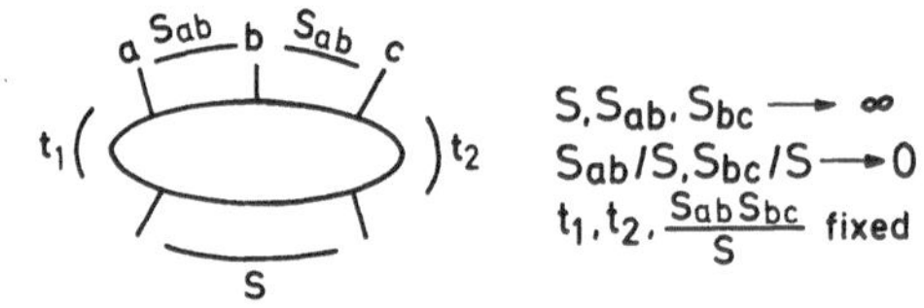

Fig. 5 Kinematics of the 2 → 3 process in the double Regge limit

Before statements analogous to (i) - (iii) of $T_{2\to 2}$ can be made for $T_{2\to 3}$, we have to use one of the key results on the analytic structure of multiparticle amplitudes. It says that the amplitude $T_{2\to 3}$ in the double Regge limit splits into two parts, one having energy discontinuities only in the variables s and s_{ab}, the other one in s and s_{bc} (see Fig.6).

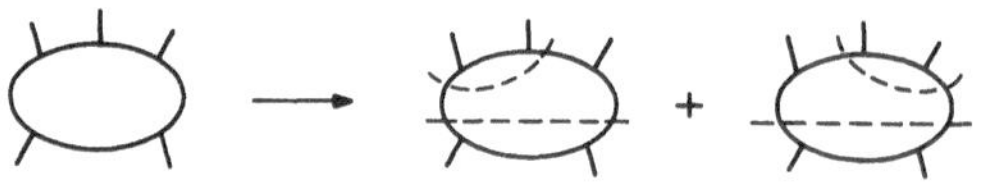

Fig. 6 Analytic decomposition of the 2→3 amplitude in the double Regge limit.

In the partial wave representation one has:

$$T_{2\to 3} = \frac{1}{(2\pi i)^2} \iint dj_1 dj_2 [s^{j_2} s_{ab}^{j_1-j_2} \xi_{j_2} \xi_{j_1 j_2} F_R + s^{j_1} s_{bc}^{j_2-j_1} \xi_{j_1} \xi_{j_1 j_2} F_L] \tag{3.7}$$

$$\xi_j = \frac{e^{-i\pi j} + \tau}{\sin\pi j}$$

$$\xi_{jj'} = \frac{e^{-i\pi(j-j')} + \tau\tau'}{\sin\pi(j-j')} \quad . \tag{3.8}$$

This decomposition is necessary for both s and t-channel unitarity. In the s-channel the Steinmann relations forbid simultaneous discontinuities in energy variables of overlapping channels (in the present case, the (ab) and (bc) channels are mutually overlapping). This problem is avoided when $T_{2\to 3}$ is written as in (3.7). Modern dispersion theory also proves that both pieces in (3.7) (or Fig. 6) have only normal threshold singularities in their respective energy variables: more complicated singularities, such as Landau singularities, are subdominant in the double Regge limit. From the t-channel point of view the decomposition (3.7) is important, because only in this representation the partial waves F_L and F_R are real,i.e. free from internal phase factors. For both F_L and F_R a reggeon calculus exists [20] which satisfies t-channel partial wave unitarity.

We are now in the position to list the properties analogous to (i)-(iii). Once the decomposition (3.7) (or Fig. 6) of $T_{2\to 3}$ has been made, we have:

(i) each of the two terms satisfies a double despersion relation (with both right and left hand cuts and the appropriate number of subtractions).
(ii) Real and imaginary parts are related through the signature factors (3.8).
(iii) For each partial wave a reggeon calculus exists which satisfies t-channel partial wave unitarity.

The generalization to more general multiparticle amplitudes is now rather straightforward. The crucial step in each case is that one first has to find the necessary decomposition of the amplitude, before one is able to write a multiple dispersion relation for the amplitude or a reggeon calculus for the partial wave functions.

In Fig. 7

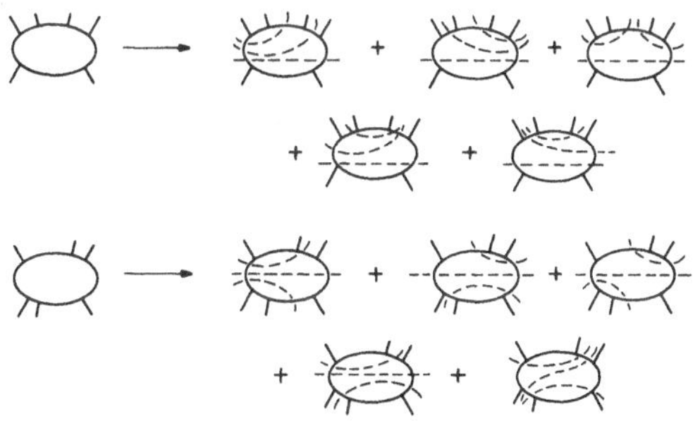

Fig. 7 Analytic decomposition of the 2 → 4 and 3 → 3 amplitudes

this decomposition is illustrated for the two sixpoint amplitudes $T_{2\to 4}$ and $T_{3\to 3}$: it holds in the kinematic region where all energy variables are as large as possible and the momentum transfers and Toller angles kept fixed. A more detailed discussion of these amplitudes (in particular certain subtleties connected with the last two terms in the decomposition of $T_{2\to 4}$ and $T_{3\to 3}$) can be found in Ref. [21].

As it can be seen from these few examples, the number of terms in the decomposition grows rather fast as the number of external particles increases. In practical calculations, however, it seems not necessary to go beyond the sixpoint amplitude: it is believed that these amplitudes already contain all the essential complications of the analytic structure of multiparticle amplitudes (note that some of these complications do not yet show up in $T_{2\to 3}$: the five point amplitude is still "too simple"). Once the correct expression for these amplitudes has been found, it seems possible to generalize to higher order amplitudes.

B. $T_{n\to m}$ IN THE LEADING LOGARITHMIC APPROXIMATION

The construction of the multiparticle amplitudes $T_{n\to m}$ in the LLA which will be described in the following has first been started by the Leningrad group [22] and then been carried through independently in Refs. 23 and 21. A summary of the Leningrad school calculations can be found in Ref. 24 . I will not have the time to present calculations in detail but will concentrate on making the logic as clear as possible. I will use the notations of Ref. 21 , and more details can be looked up there. The starting point is the computation of tree graphs for $T_{n\to m}$ which, in the language of dispersion relations, serve as subtraction constants. For illustration consider the 2 → 2 vector scattering amplitude in lowest order perturbation theory. There are seven Feynman diagrams (Fig.8)

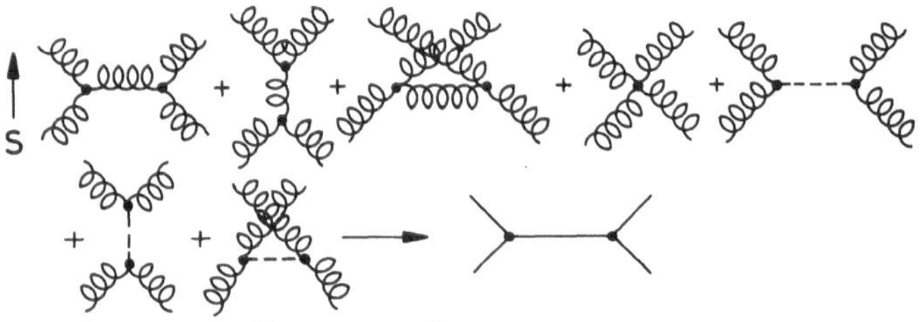

Fig. 8 Seven Feynman diagrams for vector-vector scattering in lowest order perturbation theory and their high energy behavior.

and one finds that some of them individually have a bad high energy behavior (e.g. they grow like s^2). However, when the sum is taken over all diagrams, these unwanted terms cancel and the final result has the appealing form:

$$2g^2 s \frac{1}{t-M^2} \cdot \text{helicity matrices} \cdot \text{group structure} , \quad (3.9)$$

where the helicity matrices are constant (independent of s). As a graphical notation for (3.9) we use the diagram on the rhs of Fig. 8. The fact that all (but one) Feynman diagrams of this order are necessary for obtaining (3.9) illustrates the extensive cancellations between different contributions of perturbation theory which are typical for vector theories. For the next amplitude,

$T_{2\to3}$, the number of Feynman diagrams, which have to be taken into account in order to find the correct high energy behavior of the tree approximation, is already much larger. But the result is again simple (Fig. 9):

Fig. 9 High energy behavior of the 2→3 process in the tree approximation.

$$2g^3 s \frac{1}{t_1-M^2} \vec{\Gamma}(q_1,-q_2) \frac{1}{t_2-M^2} \cdot \text{helicity matrices} \qquad (3.10)$$

$$\cdot \text{group structure},$$

where the three component vector

$$\vec{\Gamma}(q_1,-q_2) = (\Gamma_\sigma e^1_\sigma, \Gamma_\sigma e^2_\sigma, \Gamma_\sigma e^3_\sigma) \qquad (3.11)$$

stands for the production vertex labelling the polarizations of the produced vector particle. It has a nontrivial dependence upon the momenta q_1, q_2 ($q_1{}^2=t_1$, $q^2{}_2= t_2$) and the Toller variable $\eta= s_{ab}\cdot s_{bc}|s$. For $T_{2\to4}$ the result is shown in Fig. 10:

Fig. 10 High energy behavior of the 2 → 4 process in the tree approximation.

it takes the form of a multiperipheral production amplitude, the production vertex being given by (3.11) [21]. $T_{3\to3}$ is obtained from $T_{2\to4}$ by crossing one of the produced particles.

An elegant method for computing these tree approximations for general $T_{n\to m}$ has been suggested by Lipatov [23]. Writing down a t-channel dispersion relation (without subtraction constants), only the particle pole contributes to the tree approximation of $T_{2\to2}$, and for this only the on-shell vertex functions have to be computed. The result agrees with (3.9). For $T_{2\to3}$ a double dispersion relation in t_1 and t_2 is needed which has to be saturated by the pole contributions. The only new element is the production vertex whose off-shell continuation follows from direct computation and the requirement of gauge invariance:

$$\Gamma_\sigma(q_1,-q_2) \cdot (q_1-q_2)^\sigma = 0 . \qquad (3.12)$$

Proceeding in this way it is possible to verify the results of Figs. 9,10 and to show that for general $T_{n\to m}$ the tree approximations always have this multiperipheral structure. Within this approach the Lagrangian is needed only for the calculation of vertices in the tree approximation: tree amplitudes are built up by means of t-channel dispersion relations.

In the next step these tree approximations will be "dressed", and this is done by using s-channel dispersion relations plus unitarity. The amplitudes $T_{n\to m}$ in the LLA are then built up order by order perturbation theory. As a result of this "dressing" procedure, the elementary exchanges of the tree approximation will be reggeized.

Let me illustrate how this happens. To order g^4 one has the one loop contribution to $T_{2\to2}$. For this amplitude the dispersion relation is:

$$T(s,t) = a(t)+b(t)\cdot s+ \frac{s^2}{\pi}\int_{s_0}^{\infty} ds' \frac{\text{disc } T(s',t)}{s'^2(s'-s)} + \text{left hand cut.} \tag{3.13}$$

The discontinuity follows from unitarity which, in this order of perturbation theory, has only the two-particle intermediate state:

disc T =  (3.14)

On the rhs of this equation, only the $2 \to 2$ amplitude to order g^2 - which is the tree approximation - is needed: this we know from the first step of our calculations. Thus eqs. (3.13), (3.14) are sufficient to determine the leading term of $T_{2\to2}$ in fourth order perturbation theory: the integral in (3.13) goes as s.lns and, hence, dominates over the subtraction constants. The result is the term of the order g^4 of the reggeizing vector exchange:

$$T_{2\to2} = -g^2 s^{\alpha(t)} \frac{e^{-i\pi\alpha(t)}-1}{t-M^2} \cdot \text{helicity factors} \cdot \text{group structure,} \tag{3.15}$$

where

$$\alpha(t) = 1 + (t-M^2)g^2\int \frac{d^2k}{(2\pi)^3} \cdot \frac{1}{k_\perp^2+M^2} \cdot \frac{1}{(q-k)_\perp^2+M^2},\quad q^2 = -t. \tag{3.16}$$

Comparison with the tree approximation (3.9) shows that the elementary exchange of (3.9) has been replaced by the reggeizing vector exchange.

In order g^5 we have to calculate the one loop correction to the $2 \to 3$ amplitude. In principle, we could proceed in the same way as we did for $T_{2\to2}$: one uses the decomposition of Fig. 6 and writes down a double dispersion relation for each of the two terms, including the right number of subtraction terms. The various discontinuities and double discontinuities are computed via unitarity, for example:

$\mathrm{disc}_{s_{ab}}\, T_{2\to 3} =$ 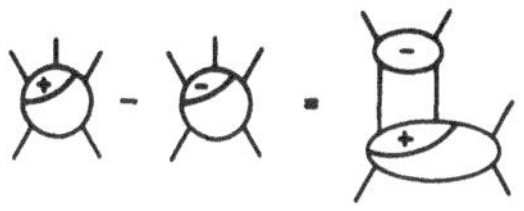 (3.17)

On the rhs, only tree approximations are needed in this order of g. Let me, however, shortcut these calculations a little bit. I directly use the ansatz (3.7) and, anticipating the result that the singularities in j_1 and j_2 will be the poles belonging to the reggeizing vector particle, I simply write:

$$T_{2\to 3} = s^{\alpha_2} s_{ab}^{\alpha_1-\alpha_2} \xi_{\alpha_2} \xi_{\alpha_1-\alpha_2} F_R + s^{\alpha_1} s_{bc}^{\alpha_2-\alpha_1} \xi_{\alpha_1} \xi_{\alpha_2-\alpha_1} F_L \qquad (3.18)$$

(the functions $\alpha_i = \alpha(t_i)$ should, of course, be the same as in (3.16)). The unknown quantities are now the coefficient functions F_L and F_R. F_R, for example, is determined by taking the s_{ab} -discontinuity of eq. (3.18), expanding in powers of g, and comparing the term g^5 with the rhs of eq.(3.17). A consistency check can be made by taking the s-discontinuity of (3.18) and comparing it with the result of evaluating the unitarity equation which yields the s-discontinuity: both F_R and F_L in (3.18) are already fixed by the s_{ab} and s_{bc} discontinuities, resp., and no further freedom is left. What we have found in this way is that, up to this order of perturbation theory, $T_{2\to 3}$ is given by the exchange, in both the t_1 and t_2 channel, of the reggeized vector particle. This has to be compared with the tree approximation (Fig. 9) where the exchanges are the elementary vector particles. In order to make this comparison more explicit (and also for later convenience), we rewrite eq. (3.18). Using the results for F_L and F_R [21], (3.18) can be written:

$$T_{2\to 3} = 2g^3 s \,\frac{s_{ab}^{\alpha_1-1}}{t_1-M^2}\, \vec{\Gamma}(q_1,-q_2)\, \frac{s_{bc}^{\alpha_2-1}}{t_2-M^2} \times \text{helicity matrices} \times$$
$$\times \text{ group structure} \qquad (3.19)$$

(where terms of the order $g^5 \ln \frac{s_{ab}s_{bc}}{s}$ have been neglected). Eq. (3.19) is the form in which the double Regge exchange amplitude (Fig. 11)

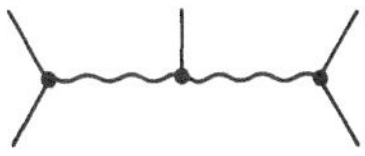

Fig. 11 The leading-logarithm approximation of $T_{2\to 3}$: the wavy lines denote the exchange of a reggeized vector particle.

would conventionally be represented: it is equivalent to (3.18), but it looses the information about the analytic structure in the energy variables.

In order g^6 two contributions have to be calculated: the two-loop correction of $T_{2\to2}$ and the one loop correction to $T_{2\to4}(T_{3\to3})$. For $T_{2\to2}$ we again use the dispersion relation (3.13). The unitarity equation for the discontinuity now has several contributions:

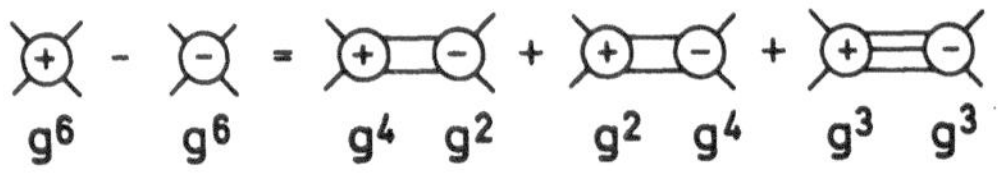

(3.20)

On the rhs of this equation, all amplitudes are known from previous steps: $T_{2\to2}$ in order g^2 and g^4, and $T_{2\to3}$ in the tree approximation. Inserting the result into the dispersion integral, we correctly reproduce the coefficient proportional to g^6 of the amplitude (3.15).

The one-loop contribution to $T_{2\to4}$ (and $T_{3\to3}$) is obtained in the same way as for $T_{2\to3}$: to proceed most generally, one makes the decomposition (Fig. 7) and writes a multiple dispersion relation for each term (with the right number of subtractions). Then unitarity equations are used for computing, in the given order, single and multiple discontinuities. But we again shortcut this procedure and make the ansatz analogous to (3.18). There are now five unknown coefficient functions which can be determined from the (single) discontinuities in the five subenergy variables. The discontinuity across the total energy s again serves as a consistency check. The result for $T_{2\to4}$ can be written (Fig. 12):

Fig. 12 The leading-lns approximation of $T_{2\to4}$.

$$T_{2\to4} = 2g^4 s \frac{s_{ab}^{\alpha_1-1}}{t_1-M^2} \vec{\Gamma}(q_1,-q_2) \frac{s_{bc}^{\alpha_2-1}}{t_2-M^2} \vec{\Gamma}(q_2,-q_3) \frac{s_{cd}^{\alpha_3-1}}{t_3-M^2} \times$$

$$\times \text{ helicity terms } \times \text{ group structure }, \qquad (3.21)$$

and what we have just computed is the term g^6 in the power series expansion of this equation. This result (Fig. 12) is the "dressed" generalization of the tree approximation in Fig. 10.

This procedure of calculating order by order perturbation theory all multiparticle amplitudes $T_{n\to m}$ in the LLA can be continued up to arbitrarily high order (in Ref. 21, this has been done up to the order g^8). Let me, however, stop already here and state the general result. For the four, five, and sixpoint amplitudes we have found that the $T_{n\to m}$ have the simple multiregge form with only pole exchanges, and one should expect that this holds for general $T_{n\to m}$ (Fig. 13).

Fig. 13   The leading-lns approximation of $T_{n \to m}$

This then generalizes the reggeization of the vector particle, as it was found already by Grisaru et al. [11] for $T_{2 \to 2}$ on the level of the Born approximation. A nontrivial feature of this result is the fact that no Regge cuts appear: signature conservation rules would very well allow for two Regge cuts in the central rapidity gap of $T_{2 \to 4}$ (Fig. 12), but as a result of some subtle cancellations [21] these cut contributions drop out for the LLA. As we shall see later, such cut contributions will, however, come in when we go beyond this leading logarithmic approximation, requiring full s-channel unitarity.

One may ask how well justified our extrapolation from $T_{2 \to 2}$, $T_{2 \to 3}$, $T_{2 \to 4}$, $T_{3 \to 3}$ to general $T_{n \to m}$ was. As a "proof" for the correctness of this generalization one can perform a consistency check and test the unitarity content of the $T_{n \to m}$: unitarity puts nonlinear constraints on the elements of the set $T_{n \to m}$ which, on the level of the LLA, must be satisfied if our result is correct. For the simplest case, $T_{2 \to 2}$, it can, in fact, be shown [25] that the elements $T_{2 \to 2}$ and $T_{2 \to n}$ satisfy the "bootstrap" equation:

$$\text{disc}_S \, T_{2 \to 2} = \quad = \sum_n \int d\Omega_n$$

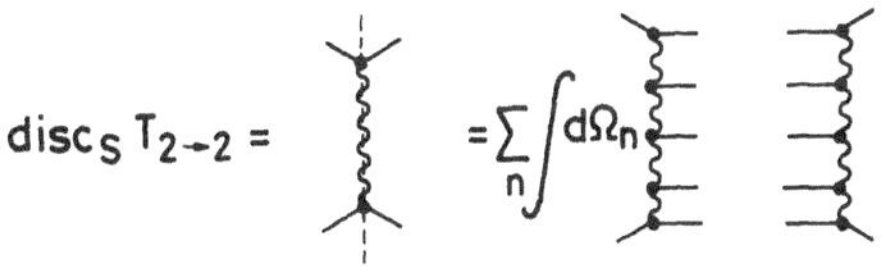

$$(3.22)$$

When "squaring", on the rhs of this equation, the $T_{2 \to n}$ amplitudes, one has to take that quantum number configuration which corresponds to the exchange of the vector particle: from Fig. 13 it is clear, that in the LLA all t-channels carry the quantum number of the vector particle and, in particular, there is no vacuum quantum number exchange yet. (For our SU(2) model we have, after the symmetry breaking due to the Higg's mechanism, a global SU(2) symmetry. If we call this symmetry, for the time being, isospin, than the vector particle carries the quantum number I=1. On the rhs of eq. (3.22) we then have the possibilities I = o, 1, 2, and it is the I = 1 configuration that we must take).

For $T_{2 \to 3}$ we have three constraints given by unitarity. In Ref. 21 it is shown that the following relations hold:

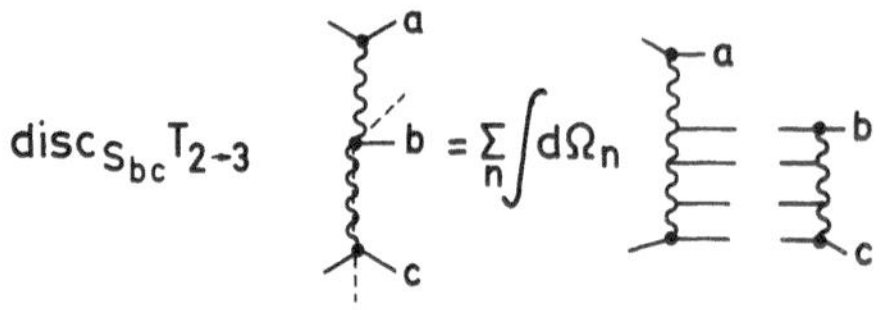

$$\text{disc}_{S_{bc}} T_{2 \to 3} \qquad \qquad = \sum_n \int d\Omega_n \qquad \qquad \qquad \qquad (3.23)$$

$$\text{disc}_S T_{2 \to 3} = \qquad \qquad = \sum_n \int d\Omega_n \qquad \qquad \qquad \qquad (3.24)$$

On the rhs of these equations the t-channel quantum numbers again
have to be that of the vector particle. In the same way it can be
shown that for $T_{2 \to 4}$ and $T_{3 \to 3}$ unitarity holds for all energy variables.

In order to summarize these unitarity properties of the $T_{n \to m}$
in the LLA I use a matrix notation. Let T be the matrix whose elements are the $T_{n \to m}$:

$$T = \begin{pmatrix} T_{2 \to 2} & T_{2 \to 3} & \cdots \\ T_{3 \to 2} & T_{3 \to 3} & \cdots \\ \cdots & \cdots & \cdots \end{pmatrix} \qquad \qquad (3.25)$$

and let the subscript "1" remind us that we are dealing with the
LLA. Then eqs. (3.22) - (3.24) are elements of the following matrix
equation:

$$T^{(1)} - T^{(1)^+} = 2i \cdot T^{(1)} \cdot T^{(1)^+}$$
$$\qquad \qquad \qquad \qquad \text{quantum number restricted}$$
$$\qquad \qquad \qquad \qquad \qquad \qquad \qquad (3.26)$$

On the rhs, all t-channels must have the quantum number of the
vector particle. This restriction signals that $T^{(1)}$ is not yet
completely unitary: to find the missing pieces will be the task
of the following subsection.

C.   UNITARIZATION

In order to find a T-matrix which is fully unitary, i.e.
satisfies eq. (3.26) without any restrictions on the rhs, we make
the ansatz:

$$T = \sum_n T^{(n)} \qquad \qquad (3.27)$$

with all the $T^{(n)}$ being matrices of the form (3.25). The expansion
parameter is the following. $T^{(1)}$ is the LLA which has been obtained
in the previous part; in the sense of the expansion in eq. (3.1)

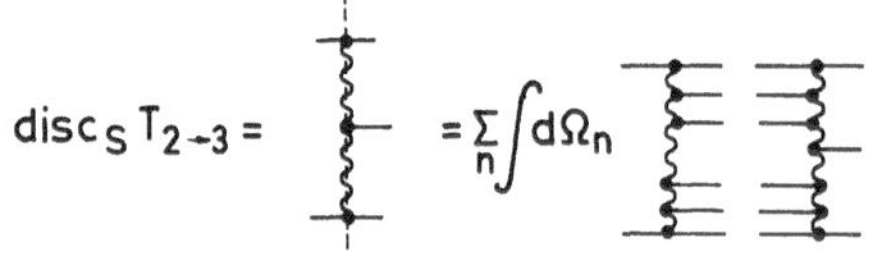

it represents the sum of the leading terms f_{n-1} $(\ln s)^{n-1}$. $T^{(2)}$ is the sum of the next-to-leading terms, but as it was said in the beginning of this section, only those parts of the f_{n-2} will be found which are required by unitarity. Similarly, $T^{(3)}$ corresponds to the f_{n-3}, etc.

To begin with the elements of $T^{(2)}$, we recall that $T^{(1)}$ had nothing but the quantum number of the vector particle in all exchange channels. This was because the leading power of lns in each order of perturbation theory always belongs to odd signature (the expansion of the signature phase factor

$$(e^{-i\pi\alpha(t)} \pm 1)$$

in powers of g^2 starts with the constant -2 for odd signature, but with $O(g^2)$ for even signature). The requirement that the amplitude is odd under s - u crossing projects out the quantum number of the vector particle. $T^{(2)}$ therefore must contain even signature exchanges, in particular the Pomeron. The easiest way to find these amplitudes is via unitarity:

$$T^{(2)} - T^{(2)+} = 2T^{(1)}T^{(1)+}\Big|_{\text{even signature}} \qquad (3.28)$$

This defines $T^{(2)}$: on the rhs, at least one t-channel must have even signature, otherwise we would be back at eq. (3.26) and nothing new would have been found. Eq. (3.28) defines discontinuities; for obtaining the full amplitudes one uses the Sommerfeld-Watson representations, e.g. (3.7). The simplest example is $T_{2\to2}{}^{(2)}$. from (3.28) we have:

$$\text{disc}_s T^{(2)}_{2\to2} = \sum_n \int d\Omega_n \left|T_{2\to n}\right|^2_{\text{even signature}} \qquad (3.29)$$

This determines the partial wave of $T_{2\to2}$ and is illustrated in Fig. 14.

Fig. 14 The leading-lns approximation for even signature amplitudes $T_{2\to2}$, as defined by its discontinuity.

For one of the even signature channels, the Pomeron, the leading singularity in the j-plane comes out as a fixed cut [25] to the right of j=1: it violates the Froissart bound and also dominates the LLA (3.15), both being a clear indication that the expansion (3.27) cannot be truncated after the first or the second term.

In case of the 2 - 3 amplitude the signature degree of freedom allows for three amplitudes contributing to $T^{(2)}$: the configurations (τ_1,τ_2) = (-,+),(+,-),(+,+). For the first case a closer look at the signature factors (i.e. counting powers of g^2 in eq. (3.81)) shows that, out of the two terms in the decomposition of $T_{2\to 3}$ (eq.(3.7) of Fig. 6), the second one which has the s_{bc}-dis continuity dominates over the first one. Hence, the amplitude can be constructed out of the s_{bc}-dicontinuity alone:

$$\mathrm{disc}_{s_{bc}} T^{(2)}_{2\to 3} = \sum_n \int d\Omega_n T_{2\to n+1} T^+_{2\to n} \Big|_{\text{even signature}} \quad . \tag{3.30}$$

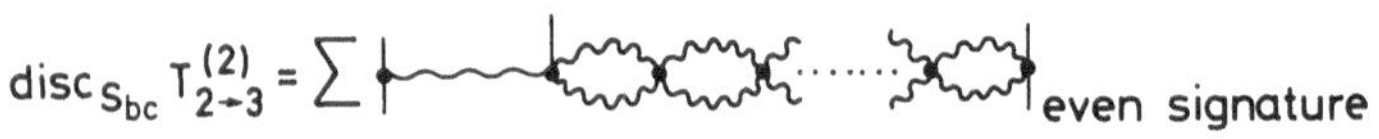

Fig. 15 The leading-lns approximation for the 2→3 amplitude with signatures $(\tau_1, \tau_2) = (-,+)$

In terms of reggeon diagrams, this equation is illustrated in Fig. 15. For the signature configuration (+ , +), the g^2-expansion of the signature factors implies that the amplitude is proportional to its s-discontiunity:

$$T^{(2)}_{2\to 3} \sim \mathrm{disc} T^{(2)}_{2\to 3} = \sum_n \int d\Omega_n T_{2\to n} T^+_{n\to 3} \Big|_{\text{even signature}} \quad . \tag{3.31}$$

The reggeon diagrams for this amplitude are shown in Fig. 16.

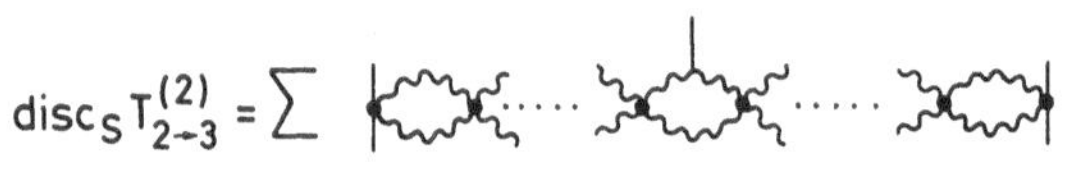

Fig. 16 The leading-lns approximation for the 2→3 amplitude with signatures $(\tau_1, \tau_2) = (+ , +)$

The construction of $T_{2\to 4}$, $T_{3\to 3}$ proceeds in the same way. For the various signature configurations (-, -, +), (-, +, -), (+,-,-), (+,+,-) (-,+,+) (note that (+, -, +) does not belong to $T^{(2)}$ but to $T^{(3)}$) , it is always sufficient to compute single discontinuities, and what one obtains are diagrams similar to Figs.14-16. The following pattern then emerges for the elements of $T^{(2)}$: whereas the elements of $T^{(1)}$ (Fig. 13) have always just one reggeon in the t-channel, those of $T^{(2)}$ (Figs. 14-16) can have either one (for odd signature exchange) or two reggeons (for even signature exchange) in each t-channel. Since the rules, according to which the reggeon diagrams of Figs. 14-16 are constructed, agree with the general reggeon calculus for inelastic production amplitudes [20], the elements of $T^{(2)}$ also satisfy t-channel unitarity. The elements of this reggeon calculus (Fig. 17) are obtained from the defining equations (3.28), (3.29) (analytic expressions will be given in Ref. 26). To complete the construction of $T^{(2)}$, let me

Fig. 17 Elements of the reggeon calculus for $T^{(2)}$

Fig. 18 Reggeon diagrams for $T^{(3)}_{2\to 2}$ with odd signature

Fig. 19 Reggeon diagrams for $T^{(3)}_{2\to 3}$ with signature (- , -).

mention that also certain nonleading terms obtained from expanding the signature factors of the elements of $T^{(1)}$ have to be counted as elements of $T^{(2)}$.

The construction of $T^{(3)}$, $T^{(4)}$... essentially repeats the steps which have led to $T^{(1)}$ and $T^{(2)}$. At the level of $T^{(3)}$ new contributions to the partial waves with only odd signature exchanges appear: this are reggeon diagrams involving the higher order 1-3 reggeon vertex (Fig. 18) or the one reggeon + three reggeons → particle production vertex (Fig. 19). Compared to the diagrams of $T^{(1)}$ (Fig.(3)), these new contributions have two more powers of g^2 (or, in other words, are down by two powers of lns). The lowest order (in powers of g^2) contributions to Figs. 18 and 19 are shown in Fig. 20: these diagrams, having only elementary exchanges, are of the order g^6s and g^7s, respectively, and contain no logarithm of any energy variable. Furthermore, they are real. This implies that they cannot be obtained by just iterating s - channel unitarity (in the language of dispersion relations, they are subtraction constants), but they must be computed by hand: going back to the Lagrangian, one has to use methods which are similar to those which were used for the tree approximations at the level of $T^{(1)}$. Details on this will be found in Ref. 27. To find the elements of $T^{(3)}$, one proceeds very much in the same way as we did for $T^{(1)}$: the lowest order elements (Fig.20) play the same role as the tree approximations, and $T_{2\to 2}$, $T_{2\to 3}$, ... are computed order by order perturbation theory by computing the energy discontinuities from unitarity equations (cf (3.14), (3.17), (3.20)). On the rhs of these equations, a careful counting of powers of g^2 is needed, and contributions from $T^{(1)}$, $T^{(2)}$, and $T^{(3)}$ have to be taken into account. As a result of these calculations

Fig. 20 Lowest order perturbation theory for Figs. 18 and 19.

the elementary exchanges of Fig. 20 are "dressed", i.e. they are reggeized, and they also interact via the quartic reggeon vertex which was found in the previous step. The unitarity content of $T^{(3)}$ is the following:

$$T^{(3)} - T^{(3)^+} = 2i[T^{(1)}T^{(3)^+} + T^{(2)}T^{(2)^+} + T^{(3)}T^{(1)^+}] \quad (3.32)$$

odd signature

This is the analogue of eq. (3.26) for $T^{(1)}$. As in the case of $T^{(2)}$, also nonleading terms of $T^{(1)}$ and $T^{(2)}$ have to be counted as elements of $T^{(3)}$: they are obtained from expanding the signature factors in powers of g^2.

$T^{(4)}$ can be obtained from s-channel unitarity without computing new subtraction constants:

$$T^{(4)} - T^{(4)^+} = 2i[T^{(1)}T^{(3)^+} + T^{(2)}T^{(2)^+} + T^{(3)}T^{(1)^+}] \quad (3.33)$$

even signature

In the matrixelements on the rhs, at least one t-channel must have even signature. Otherwise we would be back at (3.32). Eq. (3.33) is the analogue of (3.28) for $T^{(2)}$. The reggeon diagrams of $T^{(4)}$ contain up to four reggeons in the t-channels.

Repeating these steps, higher and higher $T^{(n)}$ are obtained: at each step the (maximal) number of reggeons in an exchange channel increases by one, and new elements (vertices with a nontrivial momentum dependence) appear. The result for T is a complete reggeon calculus, with the reggeizing vector particle being the reggeon and having (infinitely many) selfinteration vertices. In principle all these vertices are calculable, but so far only a few of them are known, and to find a simple expression for the most general $n \to m$ reggeon vertex remains a subject of future work.

The fact that the result of our unitarization procedure comes in form of a complete reggeon calculus was to be expected as soon as the reggeization of the vector particle had been established. For future investigations it might, however, be useful to mention that, by a slight rearrangement in the expansion of T, a more physical picture of the (elastic) scattering process can be obtained. The idea is simply to reexpand each reggeon diagram in the expansion

$$T_{2\to 2} = \sum_n T^{(n)}_{2\to 2} \quad (3.34)$$

in power of $g^2/(j-1)$ (note that each reggeon line by itself represents a power series in this parameter

$$[j-1-(\alpha(t)-1)]^{-1} = [j-1]^{-1} \sum_m \left(\frac{\alpha-1}{j-1}\right)^m$$

with $\alpha-1 = O(g^2)$):

$$F_{2\to 2}^{(n)} = \sum_m \left(\frac{g^2}{j-1}\right)^m t_m^{(n)} \tag{3.35}$$

($F_{2\to 2}^{(n)}$ is the partial wave of the amplitude $T_{2\to 2}^{(n)}$). Since the variables j-1 and lns are conjugate to each other (cf. eq.(3.2)), (3.35) leads to an expansion of $T_{2\to 2}$ in powers of g^2 lns and has a physical interpretation close to that of the well-known multiperipheral model (for a description of these ideas see Ref. 27). The term proportional to $(g^2 \ln s)^m$ represents the following subprocess of elastic scattering: out of the incoming fast hadron which is a composite system of virtual constituents (partons), some parton has initiated a m-step cascading decay. At the end of this decay slow partons (wee partons) have been produced which can interact with the target at rest. This process is illustrated in Fig. 21. In the introduction I raised the question whether the hadron radius can be made finite: this means that we are interested in the distribution of these wee partons in impact parameter space. As I will explain a little later, the expansion (3.35) may be a better starting point for investigating this question than the reggeon calculus representation of the T-matrix that was obtained in the first instance. This is the reason why I mentioned this second form of representing the matrix T.

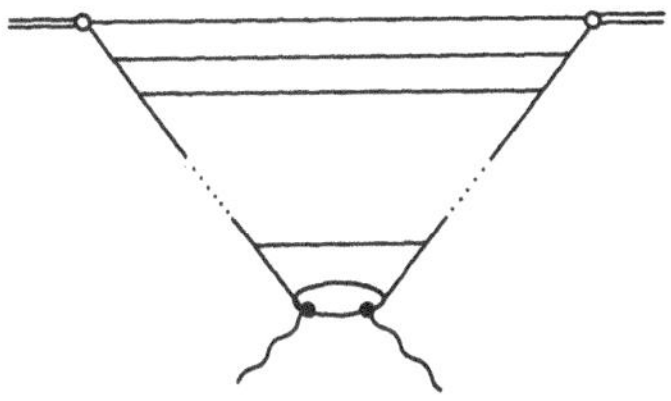

Fig. 21 Space time picture for the elastic photon-hadron scattering process in the Regge limit (rest frame of the photon)

D. THE ZERO MASS LIMIT

In the first three parts of this section I have been dealing with the question of how to select those terms in the perturbation expansion which are needed for obtaining a reliable high energy description. All this was done for the massive SU(2) Higg's model, but our final aim is the pure Yang-Mills case. We therefore have to investigate how our T-matrix behaves under the limit which takes us from the Higg's model to the pure Yang-Mills case.

The T-matrix whose construction I have outlined before depends only on the two parameters g (the gauge coupling) and the mass of the vector particle $M^2 = g^2\mu^2/\lambda$ (cf. eq. (2.10)), but not on the Higg's parameters μ and λ separately. For our purposes it is, therefore, sufficient to demand that $M^2 \to o$, g staying fixed. A brief investigation of low order perturbation theory shows that, in order to decouple the Higgs sector from the gauge particles, on should take λ and μ to infinity such that $\mu^2/\lambda \to 0$. For the time being, I shall concentrate on the question how our T-matrix behaves when M^2 is taken to zero. But it seems to me that a more accurate study of the transition from the Higgs model to the pure Yang-Mills case would be very desirable.

First it is necessary to replace the external states which so far have been taken to be massive vectors and scalars. It is well-known from QED calculus [28] that the simplest case of a high energy scattering amplitude which is finite in the zero mass limit of the photon is that of elastic photon-photon scattering (or elastic photon-electron scattering): the incoming photon dissociates into a electron-positron pair which interacts with the target via photon exchanges (note that elastic electron-electron scattering via multiphoton exchange is not infrared finite). This can easily be generalized to the nonabelian case [24] (Fig. 22a): replace the external photons by hadrons, say vector mesons with some wave functions, and take the fermions to be quarks. It can then be shown [26] for $T^{(2)}_{2\to 2}$, the first term in the expansion (3.27) which contributes to elastic scattering, that in the zero quantum number exchange channel the limit $M^2 \to o$ exists and is finite to all orders of g^2. For higher terms in (3.27), $T^{(4)}_{2\to 2}$ etc., this can be shown [26], so far, only for important subsets of terms (for example those shown in Fig. 22b); but from the results of studying infrared singularities in hard scattering processes [29] it seems likely that in the vaccum quantum number (color zero) channel infrared singularities should always cancel.

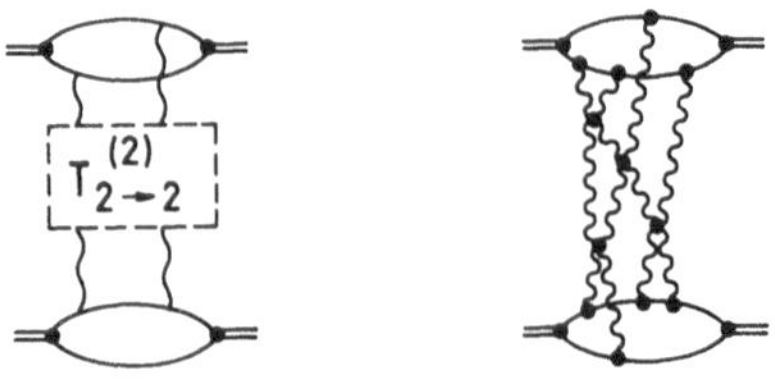

(a) (b)

Fig. 22 Elastic scattering of two $q\bar{q}$-systems in QCD: (a) the zero mass limit of $T^{(2)}_{2\to 2}$; (b) Parts of $T^{(4)}_{2\to 2}$ for which the zero mass limit can be shown to exist.

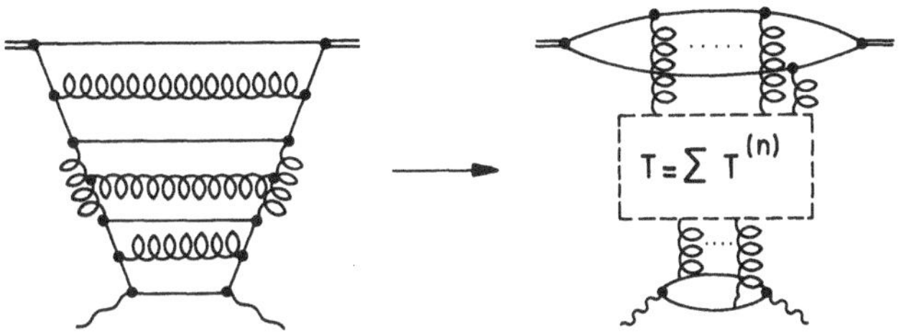

Fig. 23 Elastic photon hadron scattering in QCD: on the lhs in the Bjorken limit, on the rhs in the Regge limit.

Let me assume that this, in fact, is true for all the $T^{(n)}$ in (3.27). Then our present situation can be described as follows (Fig. 23). Starting again from the deep inelastic region where the use of perturbation theory (and in this case even the summation of only leading logarithms) rests on a safe ground, we now have isolated those Feynman diagrams which have to be summed when the Regge limit is taken (q^2 fixed and $x \to o$). They are obtained as the zero mass limit of our T-matrix which is coupled to the quark loop as external source.

I finish this long section with a few comments on other approaches to the same problem. When describing the derivation of the LLA, I have restricted myself to that method which, as I believe, is most suitable for achieving unitarity: the use of the analytic structure of multiparticle amplitudes together with unitarity. Other groups of authors [30,31] have followed the more conventional method of investigating Feynman integrals and extracting the leading term by use of a clever choice of integration variables. This approach has so far been restricted to the $2\to 2$ amplitude (with one exception [32]) in the LLA and one step beyond (in our notation: $T^{(2)}{}_{2\to 2}$). Wherever a comparison can be made, the results of the different approaches agree. As to the next logical step, namely the unitarization of the LLA, Refs. 33 and 34 claim that the fully unitary S-matrix takes a simple eikonal-form, both for QED and the nonabelian case. However, when reexpanding this eikonal representation, it appears that pieces are missing which are necessary for having s and t-channel unitarity. From the s-channel point of view, subchannel unitarity is not satisfied, i.e. rescattering contributions to inelastic production amplitudes are missing. T-channel unitarity (partial wave unitarity) requires that the lowest order g^2-expansion coefficients of the 3-reggeon, 5-reggeon, ... cut diagrams are real. Hence, they cannot be obtained from iterating s-channel unitarity alone, as it is done in the eikonal expression of Ref. 34.

A very different approach has been taken in Ref. 35. For the case of quark-quark scattering, the leading infrared divergent terms are isolated by means of the equations of Cornwall, Tictopoulos and Korthaus-Altes, de Rafael, and then the behavior of these terms in the Regge limit is studied. The result is a fixed cut singularity at j=1. Compared to the procedures which I have been describing so far, this amounts to taking the two limits (Regge limit $s\to\infty$ and zero mass limit $M^2\to 0$) in the reverse order. As it has been shown by Bronzan and Sugar [36], these two limits do not commute: the terms found in Ref.[35](first $M^2\to 0$, then $s\to\infty$) form a subset of those obtained from the other approach (first $s\to\infty$, then $M^2\to 0$) and, hence, do not seem to satisfy unitarity.

IV. SUMMATION OF THE DIAGRAMS

I now come to the final part of my talk: how can one try to sum all the contributions that have been obtained in the previous section? Let me recall the two quantities we wanted to concentrate on: the s - dependence of the total cross section as the most important observable, and the hadronic radius $\langle b^2\rangle$ as a test for the reliability of the calculations. Unfortunately, I will not be able yet to give you final answers. The task of summing all these contributions of T (or, at least, of extracting the relevant information about the leading s-behavior) requires new techniques, and all I can do is to outline the main ideas and mention those results which we already have. For the investigation of the two quantities σ_{total} and $\langle b^2\rangle$ two different approaches seem to emerge: the first one starts from the reggeon calculus representation of the s-matrix and than makes use of the phase structure of reggeon field theory which has been studied within the last few years. For a study of the parton distribution in b - space, on the other hand, the power series (3.35) seems to be a good starting point, and I would like to begin with this approach first.

To be specific, let us consider the model illustrated in Fig. 22 (elastic scattering of two q-$\bar{q}$ bound states via gluon exchanges), assuming that the zero mass limit exists for all $T^{(n)}_{2\to 2}$ in the vacuum exchange channel. As explained before, the expansion in powers of $g^2\ln s$ (c.f. (3.35)) can be related to the parton picture: each power of $g^2\ln s$ stands for a change in rapidity by one unit, and the term $(g^2\ln s)^m\, t_m$ corresponds to a m-step decay of some fast constituent into slower ones such that at the end wee partons emerge. For the rest frame of the target, this situation is illustrated in Fig. 21, for the CM-system in Fig. 24: each horizontal line denotes a point in rapidity (i.e. all points on one line have the same rapidity), but each vertex has its own impact parameter coordinate. The (statistical) distribution of all these points in impact parameter space defines the extension of the hadron. For the lowest approximation to the elastic scattering

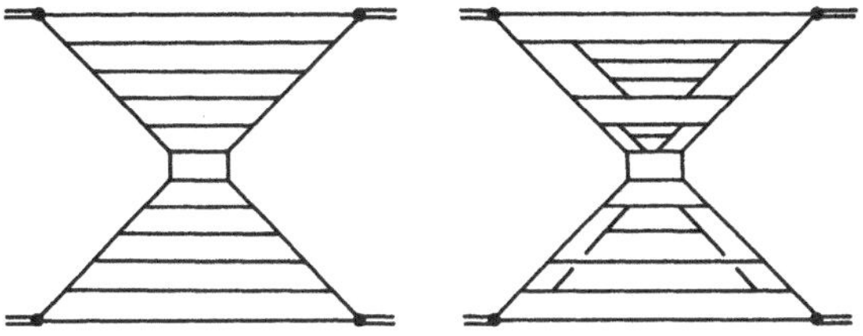

Fig. 24 Space-time picture for the elastic photon-hadron scattering process in the Regge limit (CM-system): (a) $T^{(2)}_{2\to 2}$; (b) $T^{(4)}_{2\to 2}$

processes, $T^{(2)}_{2\to 2}$, there is a direct analogy to Figs. 21 (or 24): two successive steps in the expansion (3.35) are connected by a single two-dimensional transverse momentum integration, i.e. $t^{(2)}_{m+1} = K \cdot t^{(2)}_{m}$ with K being an integral operator (the explicit form of this recursion relation can be found in Ref.[25] $t^{(2)}_{m+1} = K \times K \times \ldots K \times t^{(2)}_{1}$. The lines in Fig. 24a then denote the flow of transverse momentum in $T^{(2)}_{2\to 2}$. At the level of $T^{(4)}_{2\to 2}$, the momentum flow becomes more complicated (Fig. 24b): between two rungs there may be more than two vertical lines, since the Kernel in $t^{(4)}_{m+1} = K \cdot t^{(4)}_{m}$ involves more than one k -integral. With increasing n the number of vertical lines (i.e. the number of k -integrations in $T^{(n)}_{2\to 2}$) increases, giving rise to more and more interaction between partons of different rapidity.

In order to study the b-distribution of the partons in Fig. 24 we shall investigate how the leading j-plane singularity of the partial wave in (3.35) is generated (i.e. we study the behavior of the expansion near the rightmost value of j for which the series diverges). For this we use an observation made by Kuraev et al. [25]) for the case of $T^{(2)}_{2\to 2}$: the divergence of the expansion

$$F^{(2)}_{2\to 2} = \sum_m \left(\frac{g^2}{j-1}\right)^m t^{(2)}_m \tag{4.1}$$

comes from a specific region of phase space of the $k_\perp$-integrations in Fig. 24a. Each $k_\perp$-integration is perfectly finite, but for large m (which is the number of rungs or cells in Fig. 24a) the dominant region of integration in those cells which are far away from the external particles moves more and more towards large $k_\perp$-values:

$$\langle \ln k_\perp^2 \rangle = c_2\sqrt{m} \tag{4.2}$$

where c_2 is a computable number (note that this type of growing transverse momentum is quite different from that found in hard scattering processes). Eq. (4.2) means that the average value of

$\ln k_\perp^2$ obeys a diffusion law as a funtion of the number of steps. Once $k_\perp^2$ is large, the resulting singularity will not depend on finite quantities such as the mass M^2 or the momentum transfer $q^2 = -t$: this explains its nature of being a fixed cut. The j-value j_c for which this singularity arises lies to the right of $j = 1$ and leads to a total cross section which grows like

$$\sigma_{total} \sim s^{j_c - 1}$$

(Fig. 25a). Since with (4.2) also $\langle k_\perp^2 \rangle$ grows, as the number of steps m increases, the variable b^2 conjugate to $k_\perp^2$, which stands for the distance in impact parameter between neighboring vertices in Fig. 24, becomes smaller and smaller, and the parton distribution inside the upper (or lower) hadron is of the form shown in Fig. 25b. As to the zero mass limit $M^2 \to 0$, the most interesting point is that of $q^2 = -t = 0$ [24]: the large b-behavior comes from the small q -region. For $M^2 = 0$, $q^2 = 0$ the diffusion picture of $\ln k_\perp^2$ still holds, but $\langle \ln k_\perp^2 \rangle$ now moves in both positive and negative direction:

$$\langle \ln k_\perp^2 \rangle = \pm c_2 \sqrt{m} \tag{4.3}$$

(c_2 being independent of M^2 is the same as in the massive case (4.21)). In b-space the large negative values of $\langle \ln k_\perp^2 \rangle$, i.e. the small values of $k_\perp^2$, allow for longer and longer steplengths in b-space, and the radius $\langle b^2 \rangle$ grows too fast (as a power of s). This implies that, at the level of the approximation $T^{(2)}$, (a) the radius $\langle b^2 \rangle$ is too large, and (b) the limit $M^2 \to 0$, although it exists order by order perturbation theory, is discontinuous the leading s-behavior. The last point has been made explicit [24] by solving at $t = 0$ the integral equations of $T^{(2)}_{2\to 2}$ for $M^2 \neq 0$ and for $M^2 = 0$: there is a jump in the s-behavior of σ_{total} at the point ($M^2 = 0$, $t = 0$), compared to $M^2 \neq 0$ and / or $t \neq 0$.

As a guideline to what the situation in a realistic high energy theory should be, it might be usefull to recall a few features of the multiperipheral model. Writing the amplitude in the form (4.1), one find that $t_m \sim [\beta(t)]^m$ with $\beta(t)$ being the integral in equation (3.16), and the resulting j-plane singularity is a moving pole. Since the $k_\perp$-integrations in $[\beta(t)]^m$ are always superconvergent, and their mean values do not depend on m at all, the average steplength in b space is constant, and we have the well known random walk picture in b-space with $\langle b^2 \rangle \sim \alpha' \ln s$. This suggests that in our nonabelian gauge theory model we should look for a mechanism which stops the growth of $\langle k_\perp^2 \rangle$ as a function of the number of steps.

Let me briefly outline [37] how the presence of the higher $T^{(n)}$ could lead to a change in the right direction (as long as one does not know the form of general $T^{(n)}$ in full detail I can describe this only qualitatively). A simple dimensional argument for the general $n \to m$ reggeon vertex shows that the diffusion law (4.2) for $\ln k_\perp^2$ will always hold, provided the limit $M^2 \to o$ is finite. The only new feature compared to $T^{(2)}_{2\to 2}$ is that the momentum integration between two steps in Fig. 24b now may consist of two or more $k_\perp$-loops, and the variable which grows is the mean value of these $k_\perp$'s :

$$\ln k_\perp^2 = \frac{1}{n} [\ln k_{\perp 1}^2 + \ldots + \ln k_{\perp n}^2] \sim c_n \sqrt{m}$$

or

$$k_\perp^2 = \sqrt{k_{1\perp}^2 \cdot k_{2\perp}^2 \cdot \ldots \cdot k_{n\perp}^2} \; . \tag{4.4}$$

The numbers c_n in (4.3), belonging to the approximation $T^{(n)}$, will be different from c_2 in (4.2): if the growth of $\ln k_\perp^2$ should come to a stop, we must have $c_n \to o$ as $n \to \infty$. The situation of the n variables $\ln k_{i\perp}^2$, whose "center of mass" coordinate obeys the diffusion law, resembles that of the one-dimensional motion of n atoms, moving in a potential which depends only on the relative distance of the atoms from each other, but not on the center of mass position. In such a case the center of mass coordinate obeys the diffusion law, and depending on whether the relative forces between the atoms are attractive or repulsive the diffusion will be slower or faster than in the absence of those forces. The crucial observation now is that, if the forces are sufficiently attractive, the motion of the center of mass can come to stop when the number of atoms becomes infinite. Applying these ideas to out $T^{(n)}$, we see that if the number of $k_{\perp i}$-variables in (4.4) becomes very large - i.e. in Fig. 24 there is more and more interaction between different horizontal lines, each of which represents a certain rapidity in the "gluon cloud" around the incoming hadron - the growth of $k_\perp$ towards the center of Fig. 24 can come to a stop, and the impact parameter steplength stays finite and constant. About the s-dependence of σ_{total} very little can be said as long as this argument has not been made quantitative yet: if the series of the $T^{(n)}$ converges, the cross section (Fig. 25a) must flatten out at high energies in order to satisfy the Froissart bound.

It is important to mention that the same type of analysis has also to be carried out for the abelian case of QED. At the level of the LLA for the Pomeron channel, the leading singularity of the tower diagrams [38,39] in QED is also a fixed cut to the right of $j = 1$ and has very much the same characteristics as in the nonabelian case. Important differences between the two cases are expected to come in when the effects of the higher approximations $T^{(n)}$ are included (it seems that the sum of all those diagrams which are described in Ref. 38 and Ref. 40 for QED represents the analogue of the T that we have discussed for the nonabelian case. The eikonal graphs of Ref. 39 and even the "operator eikonal"

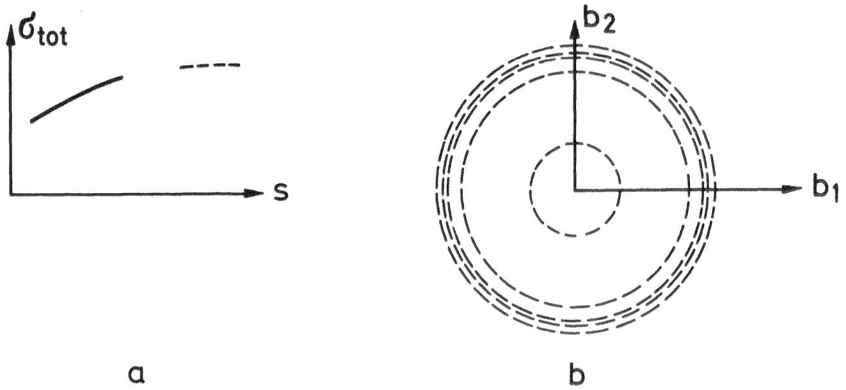

Fig. 25 (a) The total cross section as obtained in $T^{(2)}_{2\to 2}$;
(b) the hadron extension in b-space for $T^{(2)}_{2\to 2}$ (massive case)

expansion of Ref. 33 only form a subset of the more general class of diagrams in Refs. 38 and 40 and do not satisfy full s-channel unitarity).

Let me now describe the other approach towards analyzing the structure of T that I mentioned at the beginning of this section. It starts from the reggeon calculus representation of T and then uses the phase structure of reggeon field theory (FRT) which has been investigated during the last years. As it is well-known [3], RFT lives in two space and one time dimension (impact parameter and rapidity), and it has a nonrelativistic energy-momentum relation: $E = \Delta + \alpha' k_\perp^2$ ($E = 1-j$, j = angular momentum; $\Delta = 1 - \alpha(o)$; $\alpha(o)$ and α' are intercept and slope, respectively, of the trajectory function). Thus it is quite different from relativistic quantum field theory, and since, moreover, the triple interaction vertex (at least for the Pomeron case) is purely imaginary, it is clear that the phase structure of RFT, as a function of the "mass" Δ, is not the same as in usual quantum field theory models. It will, therefore, be useful to first review what we know about the phases of RFT. As I have said in the beginning, RFT is designed to satisfy t-channel unitarity (to be more precise: partial wave unitarity), and, there is no a priori restriction on the parameters such as Δ and α': as long as no connection was made between RFT and a specific underlying theory, it was, therefore, the stategy to vary the RFT parameters and to see for which values a realistic strong interaction theory emerges.

Fig. 26 shows the two phases of RFT: the intercept of the output singularity, i.e. the power of s of the elastic forward scattering amplitude, has been plotted as a function of the nega-

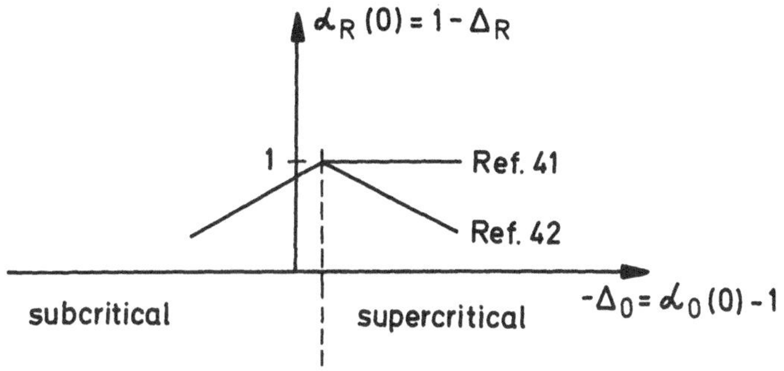

Fig. 26 Phase structure of reggeon field theory: he intercept of the renormalized Pomeron singularity is plotted against the bare negative mass. The two curves to the right of the critical point (dashed line) indicate the two solutions described in Refs. 41 and 42.

tive bare mass: $-\Delta_0 = \alpha(o)-1$. In the subcritical phase to the left of the dotted line the total cross section is falling. There is no problem with s-channel unitarity, but from the physical point of view this phase has little interest, since in nature σ_{total} is far from being falling. When $-\Delta_0$ approaches the critical value slightly to the right of zero (i.e. $\alpha_{oc}(o)$ is slightly above one), the total cross section becomes less and less falling until the power of s reaches zero: at the critical point

$$\sigma_{total} \sim (\ln s)^{-\gamma}, \quad -\gamma \sim 0.2.$$

At this critical point a phase transition occurs: particle production shows long range correlations, and for the elastic scattering amplitude a scaling law with two anomalous dimensions holds. Consistency of this solution with s-channel unitarity is highly nontrivial: it has been checked quite extensively (including the decoupling problems of the Pomeron), and all tests have been passed successfully. Thus this critical RFT is an excellent candidate for strong interaction theory at high energies. Those energies, however, for which asymptopia of strong interactions is expected to set in, lie above presently available energy ranges, and it remains to explain how the finite energy tail of asymptopia connects up with critical RFT. In the supercritical phase to the right of the dotted line in Fig. 26 (the bare mass is now negative) two solutions have been suggested (and there is still disagreement on which of them is the correct one): the first one has been obtained by Amati et al [41] and leads to a total cross section which saturates the Froissart bound $\sigma_{total} \sim (\ln s)^2$. For such a behavior of the total cross section unitarity in the s and t-channel presents certain problems, and a complete check is still missing. The most important physical implication of this solution lies in the fact that the rise of the total cross section as observed at ISR energies, does not require any special value for Δ_0: as long as $\Delta_0 < \Delta_0$ critical, the behavior of σ_{total} has the same s-dependence. The other solution to RFT in the supercritical phase has been presented by A. White [42]. It leads to a falling total cross section, thus making the phase picture in Fig. 26 quite symmetric with respect to the critical point. While this solution has no problems with unitarity, its physical implications are very strong: the only possibility for having a nonfalling total cross section is critical RFT, and this requires a very special reason why the bare Pomeron intercept takes just the critical value. White [43] also gives an explanation for this: he argues that criticality of RFT can be explained within QCD as being equivalent to confinement. I shall now try to explain this argument, which, of course, relies upon the validity of the second solution to supercritical RFT. However, I should emphasize once more that several people consider the first solution to be the correct one.

The basic idea is this: one reformulates the reggeon calculus (which has been derived in the previous section and, as elementary reggeon, only contains the quantum number carrying vector particle but no Pomeron), in terms of a new RFT which now contains, in addition to the vector particle reggeon field, also a Pomeron field (in terms of the vector particle, the Pomeron is a bound state of an even number of vector particles). Then one investigates the structure of this RFT as a function of the parameters of the Yang-Mills theory, in particular the mass M of the vector particle. For the Yang-Mills theory at $M^2 \neq 0$ (one now considers generalization of the SU(2) Higg's model: the gauge group could be SU(3), and the pattern of generating masses for the vector particles may be more complex), it is argued that the normal $i\varepsilon$-prescription should be replaced by a principal value regularization: by assumption, this is the way to reach, in the limit $M \to 0$, the pure Yang-Mills case in the confining phase. The RFT obtained from such a modified Yang-Mills is found to be in the supercritical phase with a falling cross section, as long as $M^2 \neq 0$, and it becomes critical at $M^2 = 0$. As a result, the nonfalling cross section, being a very special feature of strong interaction physics, can be explained only in a massless confining vector theory, where confinement, by assumption, is reached in the zero mass limit of massive Yang-Mills theory with a modified $i\varepsilon$-prescription.

In order to explain this argument in somewhat more detail, it will be necessary to say a few more words about the solution to supercritical RFT, as it has been obtained before any connection to an underlying theory was made. Let us start with a RFT that contains, as the only field, the Pomeron in the subcritical

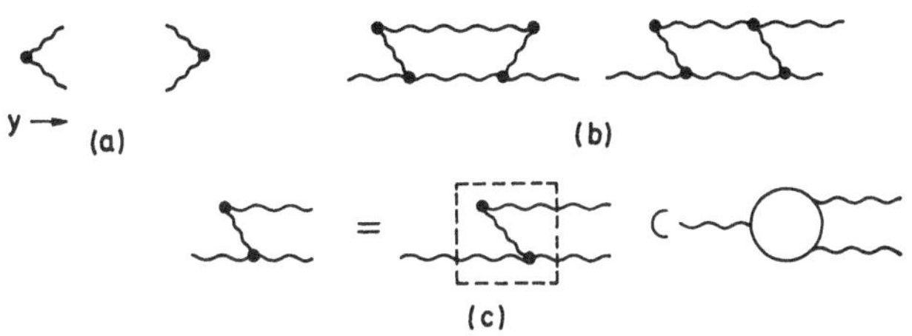

Fig. 27 Elements of supercritical RFT according to Ref. 42: (a) Pomeron creation and annihilation; (b) new diagrams which appear only in this phase of RFT; (c) interpretation of (b): new additions to the triple -Pomeron vertex.

phase, and decrease the mass from positive values to negative ones. Beyond the critical point the effective potential (for simplicity, our RFT contains only a triple interaction) has its stable minimum no longer at the origin, and by redefining the field variables one has to expand around other field configurations (note that in contrast to, say, the simplest Higgs model one cannot simply perform a shift of the field variables by a constant (i.e. time independent) amount: a detailed description of the "generalized shifting" procedure can be found in Ref. 42 . As a result, new interaction terms (Fig. 27a) and new diagrams (Fig. 27b) appear, involving creation and annihilation of Pomeron pairs out of the vacuum, and the mass of the Pomeron propagator is positive again. The "new" elements in Fig. 27a, b lead to additions to the triple Pomeron interaction (Fig. 27c), giving rise to a nontrivial momentum dependence. In fact, this momentum dependence is singular: the reggeon line inside the vertex of Fig. 27c carries a factor

$$[\alpha_o' t_1 - |\Delta_o - \Delta_{oc}|]^{-1}$$

which for positive t_1 produces a pole. This singularity is of the same form as if the upper reggeon in Fig. 27c would be an odd-signature massive vector particle of mass $|\Delta_o-\Delta_{oc}|$, accompanied by its signature factor

$$[\cos \frac{\pi}{2} \alpha(t_1)]^{-1} :$$

this suggests that the supercritical phase has a more complex reggeon content than the subcritical phase we started with. A more detailed investigation (which, via cut reggeon field theory, takes into account the s-channel unitarity content of RFT) shows, in fact, that a consistent interpretation of this solution of supercritical RFT requires the presence of several massive reggeizing vector particles in addition to the Pomeron: in Fig. 27b, for example, the two-reggeon intermediate state receives contributions from both the two-Pomeron cut and the two vector particle cut.

At the critical point $\Delta_o = \Delta_{oc}$ these vector particles become massless (together with the Pomeron), but they completely decouple from the Pomeron because the vertices in Fig. 27a are proportional to $\Delta_o - \Delta_{oc}$.

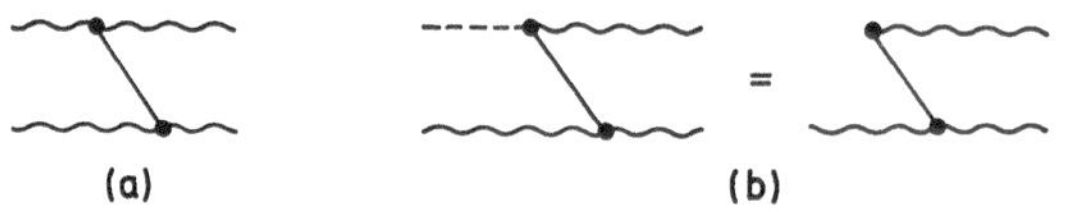

Fig. 28 A part of the 2→2 reggeon vertex (a), as obtained in massive Yang-Mills theory, is identified as the "singular" vertex (b) of RFT (Fig. 27c).

In the next logical step of the argument one wants to identify these massive vector particles with massive gluons that exist in an unconfining phase of Yang-Mills theories (QCD). For this it is necessary to show how the (massive) reggeon calculus of the previous section (the SU(2) Higgs model now being generalized to other gauge groups and Higg's patterns) can be mapped into such a supercritical RFT. Let me show, as an example, that with an appropriate definition of the Pomeron certain elements of the reggeon calculus have, in fact, the same structure as the "singular" RFT vertex of Fig. 27c. One of the simplest elements of the reggeon calculus, the 2→2 reggeon vertex, consists of several contributions one of which is illustrated in Fig. 28. Its momentum dependence comes from the exchange of an elementary gluon between the two reggeized gluons. Each reggeon line in Fig. 28a carries its signature factor which, in the small g approximation is simply a propagator $[t-M^2]^{-1}$. The singularity structure of the two-reggeon state to the left of the interaction vertex is easily analyzed: besides the two-reggeon cut, there is the reggeon particle singularity which for the normal $i\varepsilon$-prescription sits on an unphysical angular momentum sheet, and the two-particle cut. Now it becomes crucial to modify the $i\varepsilon$-prescription such that the reggeon particle singularity appears on the physical sheet simultaneously the two-particle cut disappears on the unphysical sheet): in the limit $M \to o$ it becomes a pole degenerate with the Regge pole of the vector particle, but it still has the quantum numbers of a bound state of two gluons and can be identified as the Pomeron singularity. Taking this singularity on the lhs in Fig. 28a and drawing a single Pomeron line for this bound state of a reggeizing gluon and an elementary gluon, we arrive at Fig. 28b which (always in the limit $M^2 \to o$) is of the same form as Fig. 27c. This shows that a certain part of the reggeon calculus has, after changing the $i\varepsilon$-prescription, the same structure in angular momentum and transverse momentum as supercritical RFT. In the same way more complicated parts of the reggeon calculus can be identified with higher order elements of RFT in the supercritical phase. It is, however, clear that this way of dividing the reggeon calculus of massive Yang-Mills theories into several pieces each of which goes into different elements of the RFT raises counting problems which still remain to be solved: before this can be done it will be necessary to complete the calculation of the most general element of the reggeon calculus which has not been found yet.

Finally, the limit $M^2 \to o$ is taken and by assumption, massive Yang-Mills theory with the modified infrared regularization reaches QCD in the confining phase. At the same time, the masses of the RFT elements, being of the order M^2, approach zero, and the singular elements à la Fig. 28b disappear: from the analysis of the supercritical phase of RFT it then follows that the RFT has become critical with the nonfalling cross section [44]

$$\sigma_{total} \sim [\ln s]^{-\gamma}$$

V. SUMMARY: THE REGGE LIMIT IN QCD

In these lectures I have reviewed the present status of the high energy (Regge) limit of nonabelian gauge theories, distinguishing between what has been achieved already, what sort of strategies and approaches seem to emerge, and what remains to be done in the future. Since (almost) all existing calculations start from perturbation theory of spontaneously broken gauge theories, hoping that at the end the limit, where the Higgs sector decouples, can be taken and reaches pure Yang-Mills theory, I have first tried to illustrate how good perturbation theory can be for this Regge limit: there is hope that perturbation theory is a valid starting point, since the Regge limit can be studied very close to the perturbative regime of QCD. But selection and summation of terms in the perturbation expansion must be much more complicated, because the Regge limit is also sensitive to features that have to do with confinement. It is, therefore, necessary to keep a certain control, throughout all calculations, of how reliable the perturbative approach is, and this can be done by keeping an eye on the hadron radius $\langle b^2 \rangle$.

After dividing gauge models into two classes - those where all vector particles reggeize and those where some of them don't - I have spent some time on describing, for the first type, how unitarity in both the s and t-channel can be used to classify those terms in the perturbation expansion which (at least) have to be summed up for obtaining a valid high energy description. The result (for the massive, i.e. spontaneously broken, Yang-Mills case) comes in form of a complete reggeon calculus, thus generalizing that property of the theory which at a lower level had manifested itself in the reggeization of the vector particles. The elements of this reggeon calculus are computable, but an expression for the general interaction vertex has still to be found. The zero-mass limit seems to exist, provided the external couplings are taken to be a model for hadronic bound states (e.g. $q\bar{q}$).

For the summation of all these contributions two different approaches seem to emerge. The first one, being more geometrical, investigates the distribution in impact parameter space of the wee partons. A diffusion picture then emerges which is quite different from the random walk picture in multiperipheral models: diffusion, as a function of the number of steps, takes place in the variable $\ln k_\perp^2$ rather than b. It is argued that, after summing all contributions required by unitarity, the hadronic radius $\langle b^2 \rangle$ may stay finite when the mass of the vector particles is taken to zero, but a new technique has to be developed in order to put this on a firm ground. Such a technique would also allow to study the abelian case (QED), where the summation of diagrams is still incomplete. The second approach makes use of the phase structure of reggeon field theory, and is based upon one of the two competing solutions that have been advocated for the supercritical phase. By assumming that QCD in the confining phase can be reached from spontaneously broken gauge theories in the zero mass limit, but only after the $i\varepsilon$-prescription of the massive case has been altered, it is argued that such a massive case corresponds to supercritical reggeon field theory with a falling cross section, whereas in the zero mass limit the reggeon field theory becomes critical with the nonfalling cross section

$$\sigma_{total} \sim (\ln s)^{-\gamma}$$

ACKNOWLEDGEMENT:

For very helpful discussions I am indebted to Profs. V.N.Gribov, L. N. Lipatov and A. R. White.

REFERENCES AND FOOTNOTES

1. For reviews of foundation and applications of perturbative QCD see, for example: J. Ellis, Lectures presented at the Les Houches Summer School 1976; H. D. Politzer, Physics Reports 14, 129 (1974).
2. For a discussion of this point I am grateful to Dr. J. Kwieczinsky from Cracov, Poland.
3. A comprehensive review can be found in H.D.I. Abarbanel, J.B. Bronzan, R.L.Sugar, and A.R.White, Physics Reports 21c, 121 (1975).
4. M. Moshe, Physics Reports 37c, 257 (1978) and references therein.

5. M. Gell-Mann and M.L. Goldberger, Phys. Rev. Letters 9, 275 (1962); M. Gell-Mann, M.L. Goldberger, F.E. Low, and F. Zachariasen, Phys. Letters 4, 265 (1963); M. Gell-Mann, M. Goldberger, F.E. Low, E. Marx,and F. Zachariasen, Phys. Rev. 133,B, 145 (1964); M. Gell-Mann, M.L. Goldberger, F.E. Low, V.Singh, and F. Zachariasen, Phys. Rev. 133,B, 949 (1964).
6. S. Mandelstam, Phys. Rev. 137,B, 949 (1965).
7. H. Cheng and C.C. Lo, Phys. Lett. 57B, 177 (1975).
8. M. T. Grisaru, Phys. Rev. D16, 1962 (1977); P.H. Dondi and H.R. Rubinstein, Phys. Rev. D18, 4819 (1978).
9. K. Bardakci and M.B. Halpern, Phys. Rev. D6, 696 (1972).
10. L. F. Li, Phys. Rev. D9, 1723 (1974).
11. M. T. Grisaru, H.J.Schnitzer, and H.-S. Tsao, Phys. Rev. D8, 4498 (1973).
12. M. T. Grisaru, H.J.Schnitzer, and H.-S. Tsao, Phys. Rev. D9, 2864 (1974).
13. M. T. Grisaru and H.J. Schnitzer, Brandeis Preprint 1979.
14. L. Lukaszuk and L. Szymanowski, Preprint of Institute for Nuclear Research, Warsaw (1979).
15. H. Georgi and S.L. Glashow, Phys. Rev. Lett. 32, 438 (1974).
16. M. C. Bergere and C. de Calan, Saclay preprint DPh -T/79-7.
17. H. P. Stapp, in Les Houches Lectures 1975 (North-Holland, Amsterdam) p. 159; A.R. White, ibid. p. 427.
18. V. N. Gribov, JETP 26, 414 (1968).
19. R. C. Brower, C.E. Detar,and J. Weis, Physics Reports 14c, 257 (1974).
20. J. Bartels, Phys. Rev. D11, 2977 and 2989 (1975).
21. J. Bartels, Nucl. Phys. B151, 293 (1979).
22. L. N. Lipatov, Yadernaya Fiz. 23, 642 (1976).
23. E. A. Kuraev, L.N. Lipatov, V.S. Fadin, JETP 71, 840 (1976).
24. Ya.Ya. Balitsky, L.N. Lipatov, and V.S. Fadin in "Materials of the 14th Winter School of Leningrad Institute of Nuclear Research 1979", p.109.
25. E. A. Kuraev, L.N. Lipatov, and V.S. Fadin, JETP 72, 377 (1977).
26. J. Bartels, in preparation.
27. V. N. Gribov in "Materials of the 8th Winter School of Leningrad Institute of Nuclear Research 1973", p.5.
28. S.-J. Chang and S.-K. Ma, Phys. Rev. 188, 2385 (1969).
29 R. K. Ellis, H. Georgi, M. Machacek, H. D. Politzer, and G. G. Ross, CALT 68-684.
30. H. T. Nieh and Y. P. Yao, Phys. Rev. D13, 1082 (1976); B. M. McCoy and T. T. Wu, Phys. Rev. D12, 2357 (1976) and Phys. Rev. D13, 1076 (1976); L. Tyburski, Phys. Rev. D13, 1107 (1976).
31. C. Y. Lo and H. Cheng, Phys. Rev. D13, 1131 (1976) and Phys. Rev. D15, 2959 (1977).
32. J. A. Dickinson , Phys. Rev. D16, 1863 (1977).

33. H. Cheng, J. Dickinson, C. Y. Lo, K. Olausen and P. S. Yeung, Phys. Letters 76B, 129 (1978).
34. H. Cheng, J. A. Dickinson, C. Y. Lo, and K. Olausen, Preprint 1977 and Stony Brook I TP-SB 79-7.
35. P. Carruthers and F. Zachariasen, Physics Letters 62B, 338 (1976).
36. J. B. Bronzan and R. L. Sugar, Phys. Rev. D17, 585 (1978).
37. J. Bartels, unpublished.
38. V. N. Gribov, L. N. Lipatov, and G. V. Frolov, Yad. Fiz 12, 994 (1971); Sov. Journ. of Nucl. Phys. 12, 543 (71).
39. H. Cheng and T.T. Wu, Phys. Rev. D1, 2775 (1970) and Phys. Lett. 24, 1456 (1970).
40. S.-J. Chang and P. M. Fishbane, Phys. Rev. D2, 1104 (1970).
41. For a review of this solution see M. Le Bellac in "19th International conference on High Energy Physics, Tokyo 1978", p. 153 and references therein.
42. A. R. White, Ref. TH 2592-CERN.
43. A. R. White, Ref. TH 2629-CERN.
44. It should be emphasized that this argument is not strictly based on the reggeon calculus which has been derived in the previous section: there it was characterized as the $g \to o$ limit of the unitary S-matrix, and this approximation does not include renormalization of the parameters g, M^2 etc. In order to use the concept of asymptotic freedom of $g^2(k_\perp^2)$ for large values of transverse momentum, as it is done in Ref. 43, it is necessary to go beyond this approximation and include more nonleading terms. Whether this can be done in a consistent way, i.e. without destroying the subtle constraints of unitarity order by order in g^2, remains to be seen. It may also be that some of these new contributions are nonperturbative, i.e. they cannot be expanded in powers of g^2 at all.

VERIFICATION OF PERTURBATIVE QCD IN e^+e^- ANNIHILATION

G. Kramer

II. Institut für Theoretische Physik der

Universität Hamburg

INTRODUCTION

These lectures on the theory of jets in e^+e-annihilation will differ somewhat from the other lectures in this volume. In these lectures theory and experimental results will be mixed. The field of e^+e-annihilation at high energies has so rapidly developed and new discoveries are made all the time that it does not make sense to consider the development of the theory in isolation. We must confront it with recent experimental results. New experimental data will come in the near future and will certainly make some of the reported theoretical predictions obsolete. But perhaps others will be confirmed. Thus these lectures describe the status of high energy e^+e-annihilation as it appears in the fall of 1979.

In the first section I shall give a short introduction into the historical development which led to the notion of jets in e^+e-annihilation and shall describe the gross features of the hadronic final state which are related to jets. In the second section I shall consider the framework of perturbative quantum-chromodynamics (QCD). I shall discuss the connection with QCD predictions for other reactions, in particular deep inelastic lepton-nucleon scattering. In a second part I will explain, on the basis of the work of Sterman and Weinberg [1], how jets can be defined in higher order QCD and give the main definitions for global jet measures. Section 3 contains the theory of jet fragmentation, mostly based on the work of Field and Feynman [2], and the theory of broadening effects caused by the weak decay of charm, bottom and top quarks [3]. The models for these phenomena, fragmentation of the usual quarks u,d,s and the fragmentation of c and b quarks together with their weak interactions are very important for estimating the background which disturbs the signatures of perturbative QCD effects. These perturbative QCD effects which lead in e^+e-annihilation to a third jet, the gluon jet [4], are studied in some detail in section 4. Section 5 contains the generalization to four jets in e^+e-annihilation: $e^+e \to q\bar{q}gg$ and $e^+e^- \to q\bar{q}q\bar{q}$ [5]. In section 6 we describe the evidence for gluon jets in the decay of the $b\bar{b}$ resonance $\Upsilon(9.46)$ into three gluons [6]. The last section is devoted to the analysis of very recent PETRA results in view of effects caused by the gluon emmission [7,8,9].

1. GROSS FEATURES OF THE FINAL STATE AND JET FORMATION

The notion of jets in e^+e-annihilation is closely connected with the discovery of Bjorken scaling in deep inelastic electron-proton scattering in 1968. These experiments showed that the inelastic electron scattering occurs in such a way as if the virtual photon interacts with pointlike constituents of the proton, the partons, now identified with the u and d quarks inside the proton [10]. In the mean-time the parton model has been used to describe numerous so-called hard scattering processes induced by strong, electromagnetic or weak interactions. The easiest example of these is the e^+e^- annihilation process $e^+e^- \to$ hadrons [11]. In this process one first assumes that the e^+e^- system annihilates through a virtual photon into a quark-antiquark pair (see Fig. 1). Thus after the collision of e^+ and e^- one has a free quark and antiquark which begin to move away from each other. This is prevented by the confinement forces, so that at much later times the quark and antiquark transformed into hadrons. This simple picture had its early support from the fact that the total annihilation cross section is given by the squares of the quark charges

$$\sigma(e^+e^- \to \text{hadrons}) = \frac{4\pi\alpha^2}{3q^2} \sum_a 3Q_a^2 \tag{1.1}$$

$$\text{or } R = \frac{\sigma(e^+e^- \to \text{hadrons})}{\sigma(e^+e^- \to \mu^+\mu^-)} = \sum_a 3Q_a^2 = \begin{cases} 2 & \text{for } a = u,d,s \\ \frac{10}{3} & \text{" } \quad a = u,d,s,c \\ \frac{11}{3} & \text{" } \quad a = u,d,s,c,b \\ \frac{16}{3} & \text{" } \quad a = u,d,s,c,b,t \end{cases} \tag{1.2}$$

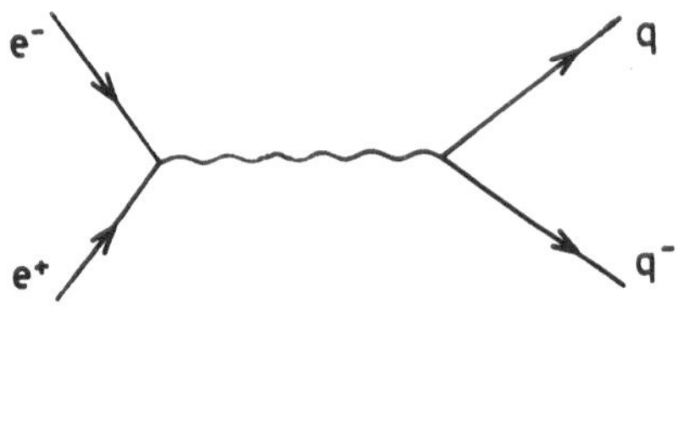

Figure 1 Parton model diagram for $e^+e^- \to$ hadrons.

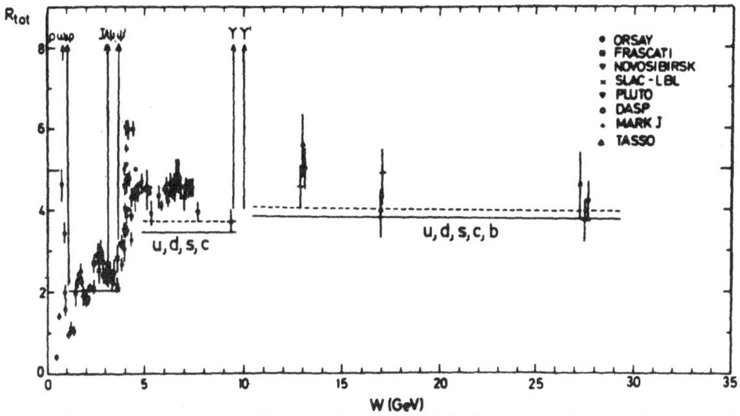

Figure 2 Measurements of R. The solid lines show the prediction from the quark model. The dashed lines show the quark model predictions corrected for gluon emission. $W = E_{cm}$.

A recent compilation of the measured R values and the comparison with (1.2) is shown in Fig. 2 [12]. Other presumed properties of this model are: (i) The produced hadrons have a limited p_T relative to the direction of momentum of the originally produced quarks [13,14]. That means, the produced hadrons come out as jets. (ii) The inclusive cross section as a function of the longitudinal hadron momentum scales according to

$$\frac{d\sigma}{dx} = \frac{8\pi\alpha^2}{q^2} \sum_a Q_a^2 D_a^h(x) \tag{1.3}$$

where (in the e^+e^- center-of-mass frame)

$$x = \frac{p}{p_{quark}} = \frac{2p}{\sqrt{q^2}} \tag{1.4}$$

The $D_a^h(x)$ is the probability of finding a hadron of type h emerging from a quark of type a. For a comparison of single charged hadron spectra up to 27.4 GeV see Fig. 3 [12]. (iii) There should be a central plateau in rapidity if the annihilation energies are large enough. This means, for a given event one defines the longitudinal momentum along the direction of the originally produced quark - antiquark pair and defines the rapidity variable y. Then the rapidity distribution of produced nonleading hadrons should be uniform, just as in ordinary hadron reactions. This is shown in Fig. 4 [12].

The regions near the boundaries of phase space contain the hadrons with largest momentum in the event. In this region we have the scaling behavior (1.3). This is the quark fragmentation region in contrast to the so-called current-fragmentation region near $y \simeq 0$.

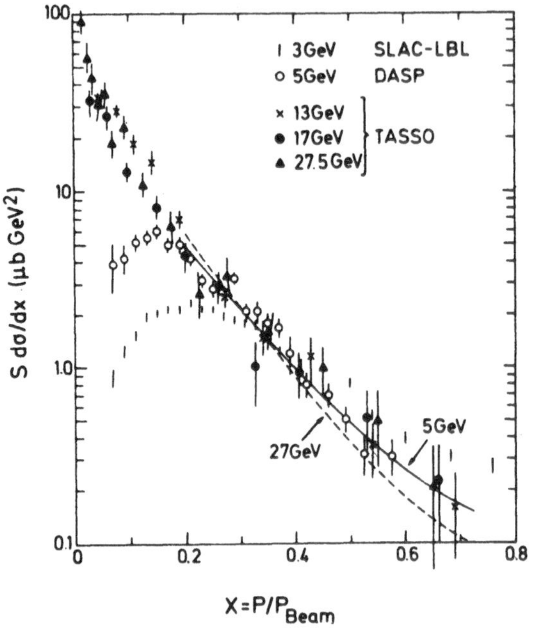

Figure 3 The scaling cross section s dσ/dx, ($x = p/p_{beam}$) for inclusive charged particle production as measured at 3 GeV by SLAC-LBL, at 5 GeV by DASP, and at 13, 17 and 27.4 GeV by TASSO. The curves show the QCD scale breaking effect predicted for going from 5 to 27.4 GeV.

From this, i.e. no increase of $\frac{1}{\sigma} \cdot \frac{d\sigma}{dy}$ near y = o, it would follow, that the multiplicity would grow only logarithmically with beam energy as in ordinary hadron collisions at lower energies. Fig. 4 shows clearly that this is not the case [12]. The plateau near $y \simeq 0$ rises with increasing energy. Therefore the multiplicity must also rise stronger than $\ln q^2$. Recent analysis show that the multiplicity of charged particles behaves like

$$\langle n \rangle_{ch} = 2 + 0.2 \ln q^2 + 0.18 (\ln q^2)^2$$

as shown in Fig. 5 [12].
At low energy it was impossible to see the jets because the two jet cones were too broad. But with transverse momentum being limited and multipliticity growing only logarithmically the jet cone becomes narrower and narrower with increasing q^2. Let

$$\langle n \rangle = a + b \ln q^2 + c (\ln q^2)^2$$

be the average particle multiplicity, $\langle p_T \rangle$ and

$$\langle p_{\parallel} \rangle \simeq p \simeq \frac{\sqrt{q^2}}{\langle n \rangle}$$

the average transverse and longitudinal hadron momenta then the mean half angle δ of the jet cone is (see Fig. 6)

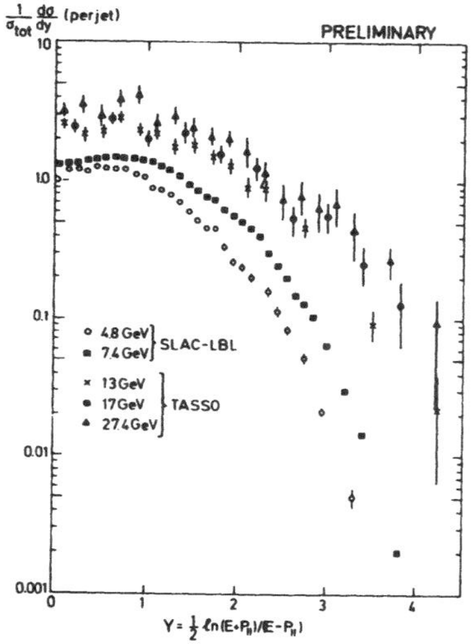

Figure 4 Rapidity distribution for charged particles assuming $m=m_{\pi}$: Yield per jet, normalized to the total cross section. Measurements by SLAC-LBL (4.8 and 7.4 GeV) and TASSO (13,17 and 27.4 GeV).

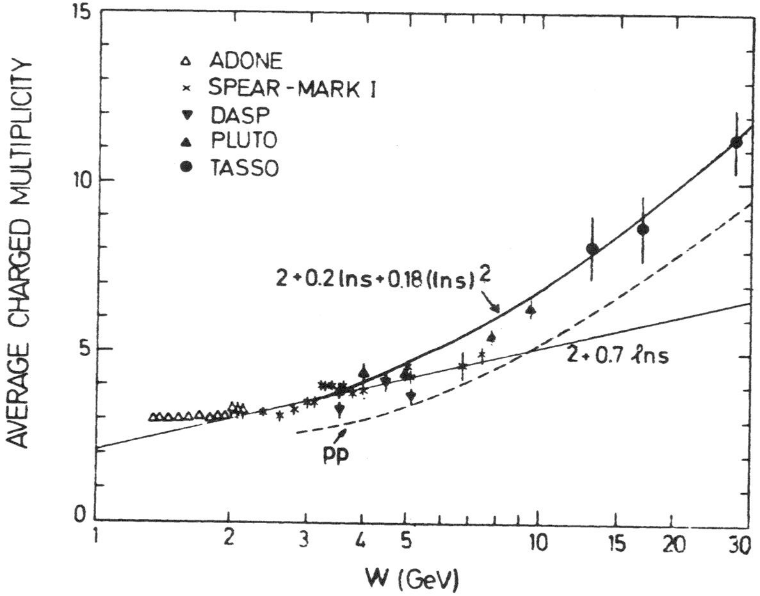

Figure 5 Average charge multiplicity. The dashed line shows the result for pp collisions.

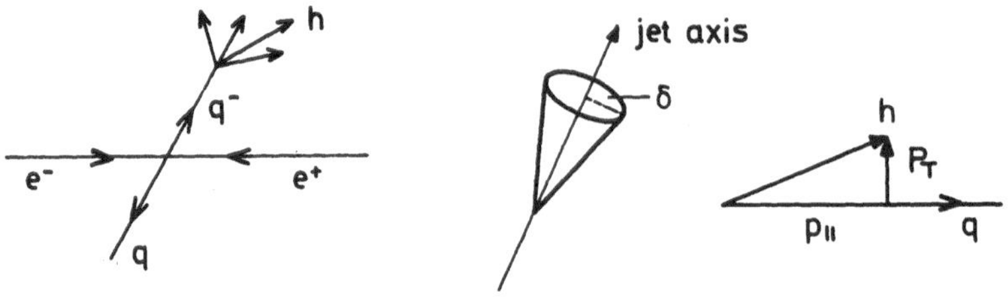

Figure 6 Jet formation and definition of δ

$$\langle\delta\rangle = \frac{\langle p_T\rangle}{\langle p_\parallel\rangle} \simeq \frac{\langle p_T\rangle\langle n\rangle}{\sqrt{q^2}} \sim c\,\langle p_T\rangle \frac{(\ln q^2)^2}{\sqrt{q^2}} \sim \frac{1}{\sqrt{q^2}} \tag{1.5}$$

This means the jet cone opening angle decreases like $^1/_{E_{cm}}$ where $E_{cm} = \sqrt{q^2}$ the total energy of the e^+e^-system.

Thus for E_{cm} = 3GeV the measured multiplicity is 5.3, so that with $\langle p_T\rangle$ = 0.3 GeV we have

$$\langle\delta\rangle \simeq 0.53 = 30^\circ \tag{1.6}$$

Jets in e^+e^- annihilation were first seen in experiments at SPEAR with the MARK J detector [13]. In order to establish the jets it is necessary to prove the limited average transverse momentum with respect to a jet axis. In this early work the jet axis was defined in terms of sphericity [15]

$$\hat{S} = \frac{3}{2}\min \frac{\sum_i |\vec{p}_{i_T}|^2}{\sum_i |\vec{p}_i|^2} \qquad 0 < \hat{S} < 1 \tag{1.7}$$

In (1.7) p_{i_T} are the transverse particle momenta of all particles in an event relative to an axis which is chosen in such a way that

$$\sum_i |\vec{p}_{i_T}|^2$$

is minimal. Comparing (1.7) with (1.5) we see that the sphericity in roughly the average of the square of the jet cone opening angle:

$$\langle\hat{S}\rangle \simeq \frac{3}{2}\langle\delta^2\rangle \simeq \frac{3}{2}\frac{\langle p_T^2\rangle}{q^2}\langle n\rangle^2. \tag{1.8}$$

The first data for the mean sphericity are shown in Fig. 7 of the SLAC-LBL group [13]. These measurements gave the first evidence for jet formation in e^+e^- annihilation. The sphericity is roughly

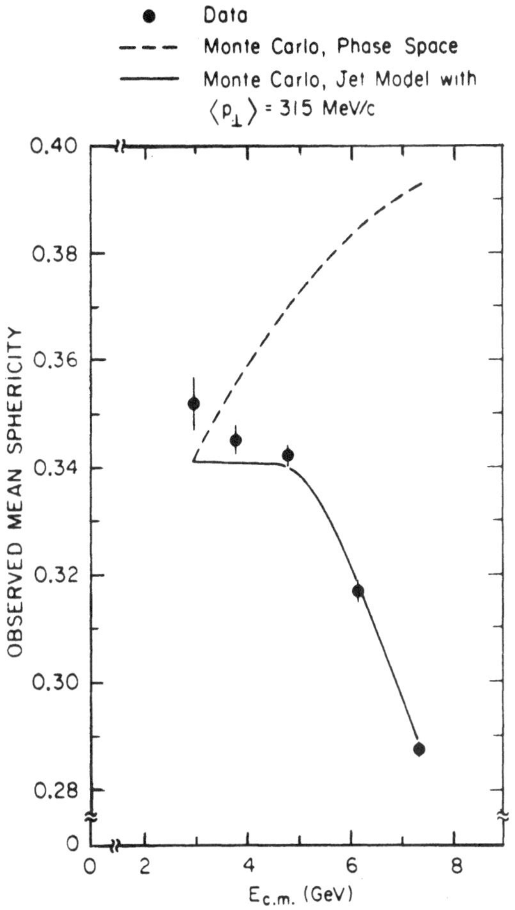

Figure 7 Observed mean sphericity versus total energy. The solid curve is the result of a jet model calculation with $\langle p_T \rangle = 0.315$ GeV. The dashed curve is the invariant phase space prediction.

constant as a function of the total e^+e^- energy up to 4 GeV. The solid and dashed curves show the Monte Carlo results for a jet model with $\langle p_T \rangle = 315$ MeV and for a simple phase space model. At lower energies ($\lesssim$4GeV), where $\langle p_{\parallel} \rangle$ is of the same order as $\langle p_T \rangle$ both models predict the same average sphericity. Above 4 GeV the jet model describes the data very well whereas in the phase space model the sphericity rises with energy in disagreement with the experimental data. In Fig. 8 I show a compilation of average $\langle \hat{S} \rangle$ up to the highest energy E = 31.6 GeV from experiments at PETRA [7,8,9,12]. One finds $\langle \hat{S} \rangle$ to decrease from 0.4 at the J/ψ to 0.14 at 31.6 GeV. The trend to even stronger collimation persists up to the highest energies in agreement with the expectations from the simple quark-parton model. The jet cone opening angle calculated from $\langle \hat{S} \rangle$ is roughly 31° at 4 GeV and decreases to 17.5° at 31.6 GeV.

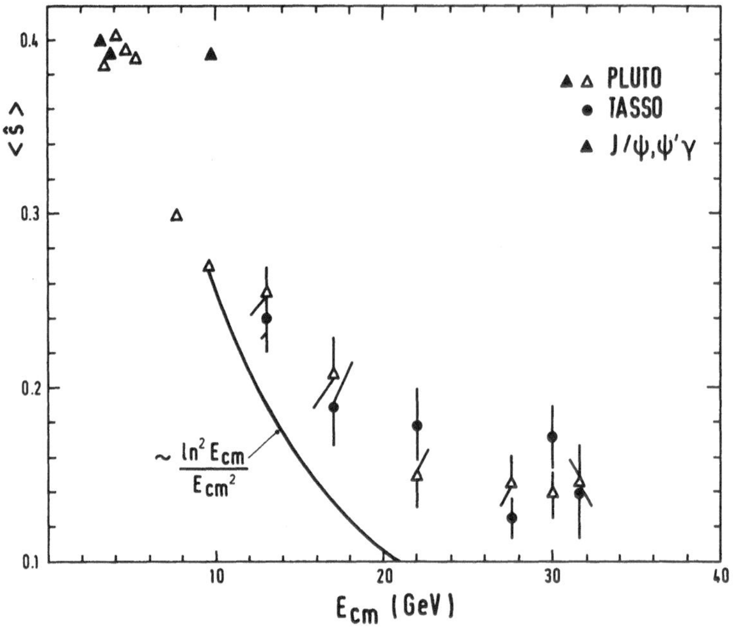

Figure 8 The average sphericity as a function of the total energy $E_{c.m.}$ as observed by PLUTO and TASSO at PETRA.

The shrinkage of the jet cone with increasing beam is slower than expected from the formulas given above $<\delta> \sim 1/q^2$ or $<\hat{S}> \sim 1/q^2$. The reasons for this reduced shrinkage of the jet cones will be in one way or the other the main topic of these lectures.

Another important test of the underlying quark structure of the jets in e^+e^- annihilation is the measurement of the jet axis angular distribution with respect to the beam direction. Neglecting mass effects the polar angular distribution for the production of two spin $1/2$ particles is

$$\frac{d\sigma}{d\cos\Theta} \sim 1 + \cos^2\Theta \tag{1.9}$$

Up to now all experimental data were found to be consistent with (1.9). The best test was made by the SLAC-LBL group at SPEAR with beams polarized transverse to the storage ring plane [13]. In this case the angular distribution is of the form

$$\frac{d\sigma}{d\Omega} \sim 1 + \alpha\cos^2\Theta + \alpha p_+ p_- \sin^2\Theta\cos 2\phi \tag{1.10}$$

where ϕ is the azimuthal angle of the jet axis with respect to the storage ring plane and $p_+(p_-)$ is the degree of polarization of $e^+(e^-)$. Fig. 9 shows the ϕ distribution measured with $p_+p_- = 0.5$.

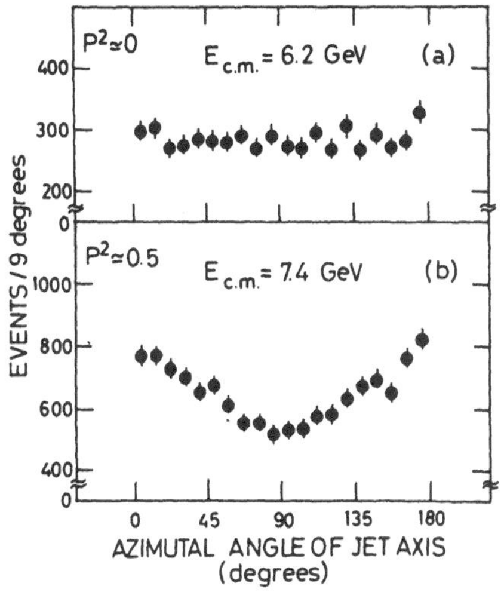

Figure 9
Azimuthal distribution of the reconstructed jet axis; zero degree is in the ring plane = plane of polarization.
(a) for a total energy of 6.2 GeV where the beam polarization is zero;
(b) for a total energy of 7.4 GeV where the product of the e^+ and e^- beam polarization is $p^2 = 0.5$ (SLAC-LBL)

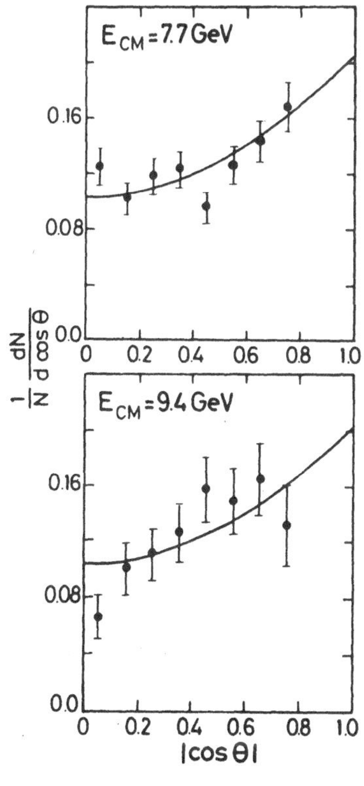

Figure 10
Angular distribution of the jet axis as defined by thrust at 7.7 GeV and 9.4 GeV. The curves are $1+\cos^2\theta$ distributions. (PLUTO data).

A fit to the ϕ distribution yields $\alpha=0.97\pm0.1$ in agreement with the $q\bar{q}$ production mechanism [13]. In Fig. 10 one of the recent measurements of the angular distribution of the Pluto group at the DORIS ring (E = 7.7 and 9.4 GeV) is shown [14]. Since no beam polarization was possible there, the measurement of α has a much less accuracy

$$\left(\begin{array}{l} \alpha = 0.76 \pm 0.3 \ (7.7 \text{ GeV}), \\ \alpha = 1.63 \mp 0.6 \ (9.4 \text{ GeV}) \end{array}\right).$$

In this measurement the jet axis was defined by the thrust axis instead of the sphericity axis. The exact definition of the thrust axis will be given later.

2. THE FRAMEWORK OF PERTURBATIVE QCD.

For quite some time the general opinion was that predictions of QCD for inclusive single particle distributions in e^+e^- annihilations or other hard processes are not possible. It was generally stated that to derive the formulas for the scale breaking of the structure functions in deep inelastic lepton- nucleon scattering two tools: i) the renormalization group equations and the Wilson operator expansion are indispensable. It is clear that there is no Wilson expansion for example for $e^+e^-\to hX$. In the last two years several groups showed that also other methods exist in order to derive the scale breaking pattern in deep inelastic scattering and that these methods can also be applied to other processes as for example: the Drell-Yan process $p\,p \to e^+e^-X$, inclusive distributions in deep inelastic lepton scattering: $ep \to hX$ or single particle distributions in e^+e^- annihilation: $e^+e^- \to hX$. The essence of this work can be stated as follows. The parton model is valid in all hard processes with scaling violations in the quark and gluon distribution functions F and in decay or fragmentation functions D. The distribution functions F and decay functions D are universal, in particular, at least in lowest order in the coupling constant g, F and D have the same q^2 dependence. In all processes the moments of the F's and D's satisfy renormalization group equations with the same anomalous dimensions (in leading order) [16].

It should be emphazised that with these methods one can predict only the q^2 evolution of the fragmentation (or structure) functions and not fragmentation functions themselves. This can be done only when the confinement mechanism of quarks and gluons is understood. How this comes about is closely connected with the infrared singular behavior of QCD. In the following we shall explain this for the process $e^+e^-\to hX$, where h is an arbitrary hadron (see Fig. 11 for the notation of momenta).

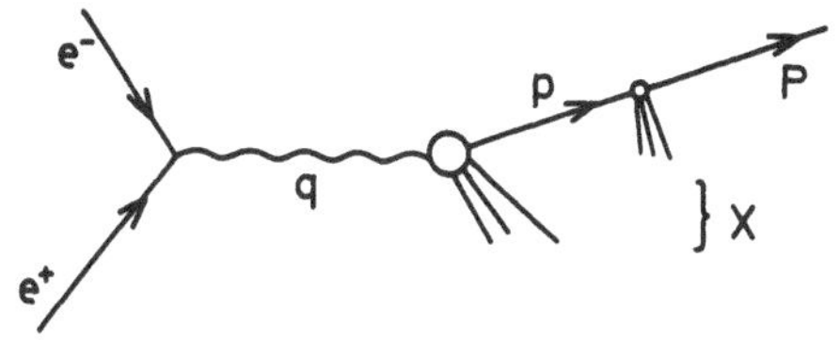

Figure 11 Parton model diagram for single hadron production in e^+e^- annihilation.

We have

$$x = \frac{2Pq}{q^2}$$

as the scaling variable, which in the center-of-mass system:

$$x = \frac{2P_o}{\sqrt{q^2}}$$

and we take $q^2 \to \infty$, $\nu = Pq \to \infty$ with x fixed. The momentum of the produced hadron h receives a fraction ξ of the parton (quark or gluon) momentum p: $P = \xi p$. In the naive parton model the single particle distribution is calculated in the impulse approximation. The result is:

$$\frac{d\sigma(x,q^2)}{dx} = \int dx_p d\xi\delta(x-\xi x_p) \sum_a \frac{d\sigma_a(x_p,q^2)}{dx_p} D_a(\xi)$$

$$= \int_x^1 \frac{d\xi}{\xi} \sum_a \frac{d\sigma_a(\frac{x}{\xi}, q^2)}{d\frac{x}{\xi}} D_a(\xi) \tag{2.1}$$

In zero order the cross section $\frac{d\sigma_a}{dx_p}$ is just the annihilation cross section $e^+e^- \to q\bar{q}$, so that

$$\frac{d\sigma}{d^x/\xi} \sim Q_a^2 \delta(\frac{x}{\xi} - 1)$$

and the sum over a is over quark and antiquark. Now we consider moments of $d\sigma/dx$. They are

$$\int_0^1 dxx^{n-1} \frac{d\sigma(x,q^2)}{dx} = \sum_a \int_0^1 dxx^{n-1} \int_x^1 d\xi\xi^{-1} D_a(\xi) \frac{d\sigma_a(\frac{x}{\xi},q^2)}{d\frac{x}{\xi}}$$

$$= \sum_a \int_0^1 dxx^{n-1} D_a(x) \int_0^1 dyy^{n-1} \frac{d\sigma_a(y,q^2)}{dy} \tag{2.2}$$

$$= \sum_a D_a^n \left(\frac{d\sigma_a^n}{dy} \right)$$

We see that the moments of $^{d\sigma}/_{dx}$ are the product of the moments of D_a and the moments of the parton cross section. Up to order g^2 the moments of the parton cross section $d\sigma_a(y,q^2)/_{dy}$ come from the diagrams in Fig. 12.

Because of a divergence for $y \to 1$ we must take the originally massless quarks off the mass-shell $p^2 \neq 0$. Then the sum of the diagrams in Fig. 11 leads to the following results for the moment of $d\sigma_a/dy$:

$$\int_0^1 dy y^{n-1} \frac{d\sigma_a(y,q^2)}{dy} = Q_a^2 \left(1 + \frac{1}{3} \alpha_s A_n \ln \frac{q^2}{p^2} \right) \tag{2.3}$$

$$= Q_a^2 \left(1 + \frac{1}{3} A_n \alpha_s \ln \frac{q^2}{q_o^2} \right) \left(1 + \frac{1}{3} A_n \alpha_s \ln \frac{q_o^2}{p^2} \right)$$

Eq. (2.3) shows that the mass singularity can be factorized and therefore can be combined with the moments of $D_a(x)$ at $q^2 = q_o^2$. The remaining q^2 dependence leads to

$$D_a^n(q^2) = D_a^n(q_o^2) \left(1 + \frac{1}{3} A_n \alpha_s \ln \frac{q^2}{q_o^2} \right) \tag{2.4}$$

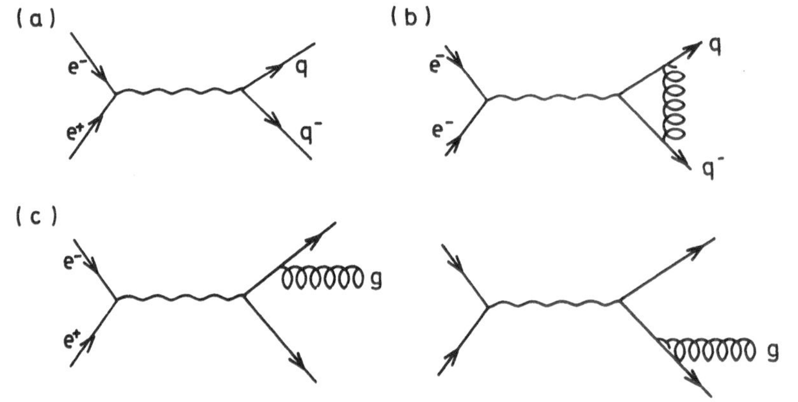

Figure 12 First and second order QCD diagrams for $e^+e^- \to \gamma \to$ hadrons. (a) Born graph, (b) second order diagram (c) diagrams for gluon bremsstrahlung.

Eq. (2.4) predicts the q^2 dependence of the fragmentation function in lowest order. The A_n are the well-known anomalous dimensions which occur also in the structure function of deep inelastic lepton-nucleon scattering. The result (2.4) can be summed with the help of the renormalization group equations. The result is:

$$D_a^n(q^2) = D_a^n(q_o^2)\left(\frac{\alpha_s(q^2)}{\alpha_s(q_o^2)}\right)^{A_n/2b} \tag{2.5}$$

Here b is the constant appearing in the q^2 dependent coupling constant

$$\frac{g^2}{4\pi} \equiv \alpha_s(q^2) = \frac{\alpha_s(\mu^2)}{1 + 4\pi b\alpha_s(\mu^2)\ln\frac{q^2}{\mu^2}} \; ; \; b = \frac{33-2N_F}{48\pi^2} \tag{2.6}$$

We remark that in (2.3) only terms $\sim \ln q^2/p^2$ appear and no quadratic logarithms. The $\ln^2 q^2/p^2$ terms cancel in the sum of the virtual and gluon bremsstrahlung terms. It is clear that our presentation is rather schematic. Because of two kinds of partons, quark and gluons, one obtains a system of coupled equations instead of the single equation (2.4) (or (2.5)). Because of the mass singularities (they are unevadable because the single particle cross section is essentially an exclusive cross section) we have the term

$$\alpha_s \ln q^2/q_o^2$$

in (2.4). This term is O(1) for $q^2 \to \infty$. Therefore one cannot rely on the lowest order perturbation theory and we have summed up the contributions of the higher orders in the leading logarithm approximation with the result (2.5). This result can be tested by measuring $D_a^h(x,q^2)$ for several q^2. As a function of q^2 and x the variation of

$$D_q^{\pi^o}(x,q^2)$$

for q = u and s is shown in Fig. 13 [17]. One notices that q^2 must be very large in order to see a significant effect. For other hadrons, $\pi^\pm$, $K^\pm$ etc. the situation is similar.

In the following we shall consider a more inclusive cross section by summing over all particles contained in a jet. In this case terms of the order

$$\alpha_s \ln q^2/q_o^2$$

are avoided and low order perturbation theory can be applied. Of course, when single jets are observed the singularities coming from infrared (gluon momentum equal zero) or collinear (gluon and

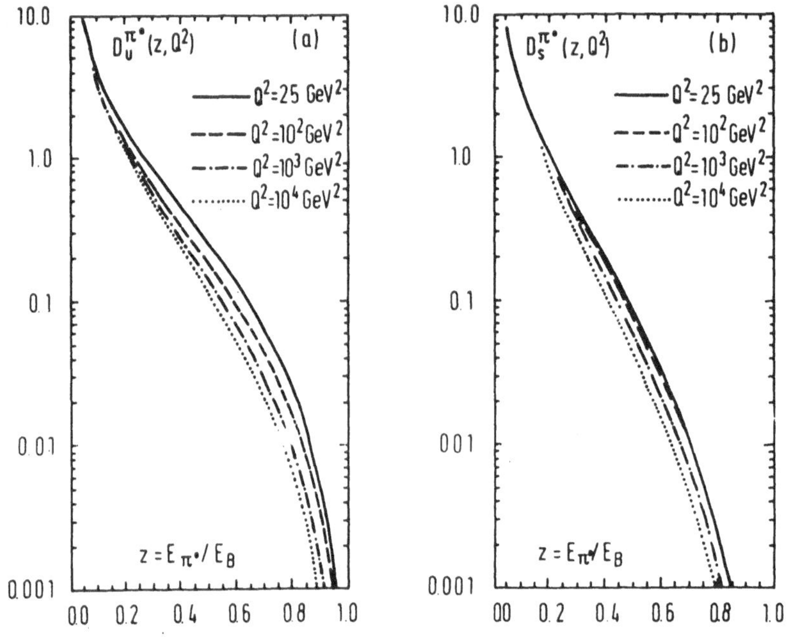

Figure 13 π^0 fragmentation functions at Q^2 = 25, 100, 1000 and 10 000 GeV^2 for (a) u quarks and (b) s quarks.

quark momentum parallel) do not disappear completely. But they can be handled by introducing appropriate parameters which describe the jets. This was shown the first time by Sterman and Weinberg [1]. In the following we shall explain their calculation of the two-jet cross section up to order $g^2 = 4\pi\alpha_s$. The diagrams for $e^+e^- \to q\bar{q}$ and $e^+e^- \to q\bar{q}g$ which occur up to the order g are shown in Fig. 14. We take massless quarks but the gluon has a mass μ. Then the $q\bar{q}$ contribution is as follows (terms of order μ are neglected):

$$\frac{d\sigma}{d\Omega} = \left(\frac{d\sigma}{d\Omega}\right)_0 \left\{ 1 + \frac{2\alpha_s}{3\pi} \left[- \ln^2 \frac{\mu^2}{q^2} - 3\ln \frac{\mu^2}{q^2} - \frac{7}{2} + \frac{\pi^2}{3} \right] \right\} \tag{2.7}$$

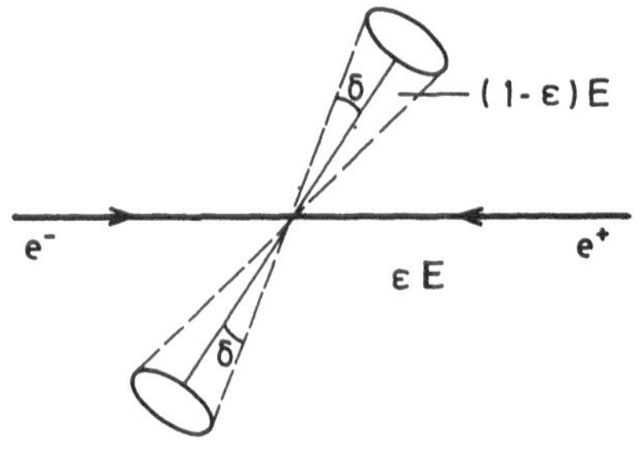

Figure 14 Definition of Sterman-Weinberg parameters ε and δ for e^+e^- annihilation.

where $\left(\frac{d\sigma}{d\Omega}\right)_o$ is the pointlike cross section

$$\left(\frac{d\sigma}{d\Omega}\right)_o = \frac{\alpha^2}{q^2} \sum_a Q_a^2 \frac{3}{4} (1 + \cos^2\Theta) = \sigma_o \frac{3}{16\pi} (1 + \cos^2\Theta) \quad (2.8)$$

Then the lowest order diagram plus the diagrams with the virtual gluon exchange (see Fig.11) contribute to the total cross section (integration over Ω) the following term:

$$\sigma_a = \sigma_o \left\{1 + \frac{2\alpha_s}{3\pi} \left[- \ln^2 \frac{\mu^2}{q^2} - 3\ln \frac{\mu^2}{q^2} - \frac{7}{2} + \frac{\pi^2}{3}\right]\right\} \quad (2.9)$$

We notice that the infrared divergence leads to terms proportional to

$$\ln^2 \frac{\mu^2}{q^2} \text{ and } \ln \frac{\mu^2}{q^2} .$$

The gluon emission part, which is obtained by integration of

$$\frac{d^2\sigma}{dx_1 dx_2} = \sigma_o \frac{2\alpha_s}{3\pi} \frac{x_1^2 + x_2^2}{(1-x_1)(1-x_2)}, \quad x_i = \frac{2p_i}{\sqrt{q^2}} \quad (2.10)$$

over the whole phase space yields the term

$$\sigma_b = \sigma_o \frac{2\alpha_s}{3\pi} \left\{ \ln^2 \frac{\mu^2}{q^2} + 3\ln \frac{\mu^2}{q^2} + 5 - \frac{\pi^2}{3}\right\} \quad (2.11)$$

In the sum of σ_a and σ_b the infrared singular terms which depend on μ cancel and the total annihilation cross section is finite. The result is:

$$\sigma(e^+e^- \rightarrow \text{hadrons}) = \sigma_a + \sigma_b = \sigma_o(1 + \frac{\alpha_s}{\pi}) \quad (2.12)$$

As the next step we calculate the 2-jet cross section [1]. The two-jet cross section is defined as the cross section for all events which lie in two opposite cones with opening angle 2δ around the q or $\bar{q}$ momentum direction respectively and contain all the total energy except a fraction ε which is outside the cones (Fig.14).

This means that from the gluon bremsstrahlung terms all contributions with $p_3 \leq \varepsilon\sqrt{q^2}$ and $\Theta_{13} \leq 2\delta$ ($\Theta_{23} \leq 2\delta$) where Θ_{13} (Θ_{23}) is the angle between the gluon momentum and the quark (antiquark) momentum are included in the 2-jet cross section. Neglecting terms

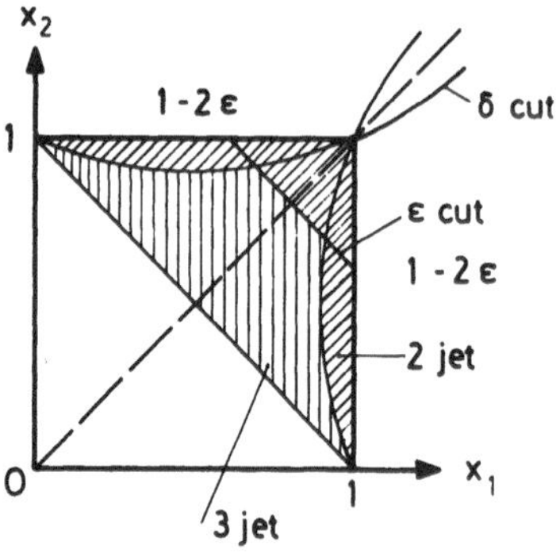

Figure 15 Phase space boundaries of $e^+e^- \to q\bar{q}g$ for two and three-jet cross section.

of order ε (and also δ) the soft gluon emission part to the cross section is (see Fig. 15 for the phase space boundaries)

$$\sigma(\text{soft gluon}) = \sigma_o \frac{2\alpha_s}{3\pi} \left(\ln^2 \frac{\mu^2}{q^2} + 4\ln^2 2\varepsilon - 4\ln \frac{\mu^2}{q^2} \ln 2\varepsilon - \frac{\pi^2}{3} \right) \tag{2.13}$$

Similarly the collinear gluon emission part is:

$$\sigma(\text{collinear gluon}) = \sigma_o \frac{2\alpha_s}{3\pi} \left(4\ln \frac{\mu^2}{q^2} \ln 2\varepsilon - 8\ln\delta \ln 2\varepsilon - 4\ln^2 2\varepsilon + \frac{17}{2} - \frac{2\pi^2}{3} + 3\ln \frac{\mu^2}{q^2} - 6\ln\delta \right) \tag{2.14}$$

The $\ln^2 \frac{\mu^2}{q^2}$ term appears only in the infrared singular part (2.13). Now the sum of (2.13) and (2.14) is:

$$\sigma(\text{soft and collinear gluon}) = \sigma_o \frac{2\alpha_s}{3\pi} \left(\ln^2 \frac{\mu^2}{q^2} + 3\ln \frac{\mu^2}{q^2} + + \frac{7}{2} - \frac{\pi^2}{3} - 8\ln\delta \ln 2\varepsilon - 6\ln\delta - \frac{2\pi^2}{3} + 5 \right) \tag{2.15}$$

By adding the contribution of the virtual gluon diagram (2.9) we see that all terms which depend on the gluon mass cancel and the final result for the 2-jet cross section is:

$$\left(\frac{d\sigma}{d\Omega}\right)_{2jet} = \left(\frac{d\sigma}{d\Omega}\right)_o \left\{ 1 - \frac{4\alpha_s}{3\pi} \left[4\ln\delta \ln 2\varepsilon + 3\ln\delta + \frac{\pi^2}{3} - \frac{5}{2} \right] \right\} \tag{2.16}$$

In this expression terms ~ ε and δ are neglected (see ref.18 for these terms). The fact that the angular distribution is the same as for the $q\bar{q}$ production is due to the omission of terms of order ε or δ. (This means that terms which measure the deviation from the $1 + \cos^2\Theta$ angular distribution are infrared finite).

The result (2.16) shows that a 2-jet cross section, free of infrared and mass singularities, can be defined if jets are defined with an angle and an energy constraint. From (2.16) we can obtain the cross section for 3-jet production by subtracting σ(2-jet) from σ_{tot}:

$$\sigma(\text{3-jet}) = \sigma_{tot} - \sigma(\text{2-jet})$$

$$= \sigma_o\left(1 + \frac{\alpha_s}{\pi}\right) - \sigma_o\left[1 - \frac{4\alpha_s}{3\pi}\left(4\ln\delta\ln2\varepsilon + 3\ln\delta + \frac{\pi^2}{3} - \frac{5}{2}\right)\right]$$

$$= \sigma_o \frac{4\alpha_s}{3\pi}\left[4\ln\delta\ln2\varepsilon + 3\ln\delta + \frac{\pi^2}{3} - \frac{7}{4}\right] \qquad (2.17)$$

Then σ(3-jet) is just the integral over the $q\bar{q}g$ terms given in (2.10) inside the boundary $p_3 \gtrsim \varepsilon\sqrt{q^2}$, $\Theta_{13} \gtrsim 2\delta$, $\Theta_{23} \gtrsim 2\delta$ in Fig.15. We see that the 3-jet cross section depends on two cut off parameters ε and δ chosen in such a way that the singular part of the $\sigma(q\bar{q}g)$ differential cross section is avoided. Of course, in all these considerations ε and δ must be large enough, so that

$$\frac{16\alpha_s}{3\pi}\ln\delta\ln2\varepsilon << 1 ,$$

in order to make perturbation theory applicable.

In the leading logarithm approximation higher order virtual and real gluon emission terms can be summed up. The result is [19]

$$\left(\frac{d\sigma}{d\Omega}\right)_{\text{2-jet}} = \left(\frac{d\sigma}{d\Omega}\right)_o \exp\left\{-\frac{16\alpha_s}{3\pi}\ln\delta\ln\varepsilon\right\} \qquad (2.18)$$

and a similar expression for the multi-jet cross section (number of jets $\gtrsim$ 3) [5]

$$\sigma(\text{multi-jet}) = \sigma_o\left\{\exp\left[\frac{16\alpha_s}{3\pi}\ln\delta\ln\varepsilon\right] - 1\right\} \qquad (2.19)$$

This latter expression diverges for $\varepsilon,\delta \to o$ since the infrared cancellation with the virtual corrections is not considered, the integration over the phase space is from large δ and ε to the boundary near $\delta,\varepsilon \simeq o$. In the 2-jet formula (2.18) the integration is from the exact 2-jet boundary including virtual corrections to finite ε,δ [5]. This cross section vanishes for $\varepsilon,\delta \to o$ as one expects if due to the infrared structure of perturbative QCD. This behavior is analogous to the behavior in usual QED where the cross section for $e^+e^- \to e^+e^-$ vanishes due to the infrared factor if the photon energy cut-off $\in$ is replaced by $\in = 0$.

Instead of using cut-off variables like ε and δ to describe jets it has become customary to use more global measures to characterize the jettiness of an event. Some of these variables have a long history and were used already more than ten years ago in connection with pure hadronic reactions [20]. Most of the variables were invented recently in connection with e^+e^- annihilation into hadrons. The first one is sphericity [15] defined in (1.7). Other ones are thrust [21]

$$T = \max \frac{\sum_i p_L^i}{\sum_i |\vec{p}^{\,i}|} \tag{2.20}$$

and spherocity [22]

$$S = \frac{16}{\pi^2} \min \left(\frac{\sum_i |p_T^i|}{\sum_i |\vec{p}^{\,i}|} \right)^2 \tag{2.21}$$

The sum over i runs always over all particles in an event. All three quantities, $\hat{S}$, T and S define the jet axis and give a measure for the topological structure of the event. The axis is found in a variational method by either minimizing the sum of $|p_T{}^i|^2$ or $|p_T{}^i|$ or maximizing the sum of $|p_L{}^i|$ with respect to a given axis. For ideal two-jet events we have $\hat{S} = S = o$ and $T = 1$ whereas for isotropic events the values are $\hat{S} = S = 2T = 1$. To characterize the planarity of an event one uses acoplanarity [23]

$$A = 4 \min \left(\frac{\sum_i |\vec{p}^{\,i}_{out}|}{\sum_i |\vec{p}_i|} \right)^2 \tag{2.22}$$

The sum runs again over all particles in the final state and p^i_{out} is measured perpendicular to a plane chosen to minimize A. For planarity also other variables can be used [24,25].

Sphericity $\hat{S}$ has the virtue that it can be easily calculated by standard diagonalization methods [24]. Therefore it is used in most of the analysis. However, since the momenta enter quadratically, high momentum particles enter with a stronger weight in the determination of $\hat{S}$. Also it is not invariant against clustering (multiplicity) of particles. It depends stronger on details of the fragmentation. Spherocity, which is linear in the momenta, does not have this drawback. However it suffers from discontinuities and, therefore, cannot be applied easily to data analysis.[25]

3. NONPERTURBATIVE BACKGROUND AND WEAK INTERACTION EFFECTS

§ 1. THE FEYNMAN-FIELD MODEL

The verification of the 3-jet contribution as, for example, given by (2.17) is complicated by the fact that the lower-order term, the 2-jet contribution, overlaps with the contributions coming from 3 jets, 4 jets and so on. For example, if we use the variables thrust T or spherocity S, introduced in the last chapter, the differential 2-jet cross section based on the zeroth-order diagram (see Fig. 1) with massless quarks ($m_u = m_d = m_s = o$) is

$$\frac{d\sigma}{dT} = \sigma_o \delta(1 - T) \quad , \quad \frac{d\sigma}{dS} = \sigma_o \delta(S) \tag{3.1}$$

These thrust and spherocity distributions are not realistic because of several reasons. First at finite total energy the real distributions vanish for $T \to 1$ and $S \to o$ because the observed particles have nonvanishing masses. Second when quarks or antiquarks decay into hadrons nonzero transverse momenta with respect to the quark momentum, which is roughly the jet axis, occur. Since $p^i_{\parallel} < |\vec{p}^i|$, the measured thrust values T are smaller than one (or the spherocity $S \neq 0$). Therefore for finite energies it is necessary to consider also the fragmentation of the quarks (or antiquarks) into hadrons, which actually are the objects observed in experiments. These fragmentation products of a quark (or antiquark), characterized by transverse momenta which are small compared to the longitudinal momenta, are the jets (quark or antiquark) which are experimentally observed. This fragmentation (or decay) of the quarks into hadrons leads to important modifications of the distributions (3.1) [23].

In the following we shall give a more qualitative description of the quark fragmentation or jet formation process. The more quantitative aspects are found in the literature in particular in the work of Field and Feynman. (Feynman-Field model). Quarks and antiquarks in a hadron (meson = $q\bar{q}$, baryon = 3q) are usually regarded as relatively free to move around in a region of the size of the hadron of 1fe $\simeq$ (0.2 GeV)$^{-1}$. In the e^+e^- annihilation process $e^+e^- \to q\bar{q}$, where q stands for u,d,s, the quark and the antiquark aquire a high momentum $\pm \sqrt{q^2}/2 = \pm E_{cm}/2$, sufficient to escape from the production region. This would result in the appearance of fractional charged objects, which are not observed. This is usually explained by saying that the quarks are confined in the hadron, i.e. the forces ($V(r) \sim r$, $F \sim$ const.) prevent the escape of the quarks and are responsible for sorting out quantum numbers. The final result is the transformation (or fragmentation) of the quark into a jet of particles. Thus jets are a consequence of confinement of quarks. The jet formation proceeds through a succession of fundamental breakups of the form

$$\text{quark} \to \text{meson} + \text{quark}. \tag{3.2}$$

The initial quark $q_f(p)$ with flavour f and momentum p breaks up into a meson $\equiv q_f\bar{q}_{f'}$ with momentum xp and a left-over quark $q_{f'}$ with flavour f' and momentum (1-x)p. The meson has flavour content $f\bar{f}'$, where $\bar{f}'$ is the flavour of the antiquark $\bar{q}_{f'}$.

$$q_f(p) \to \text{meson}(xp)_{f\bar{f}'} + q_{f'}\big((1-x)p\big) \tag{3.3}$$

At each further step the quark momentum decreases (see Fig. 16). When it falls below a critical value $\simeq m_\pi$ the process stops and the last quark has no momentum left to escape. Then the jet is complete. The exchange of the quantum number of the last quark occurs with the last antiquark in the equivalent chain of the antiquark jet, so that the complete picture looks like Fig. 17. The separate steps are characterized by the x variable with $0 \leq x \leq 1$ and a nonvanishing transverse momentum p_T of the escaping meson p_T is measured with respect to the jet = quark momentum. On uses

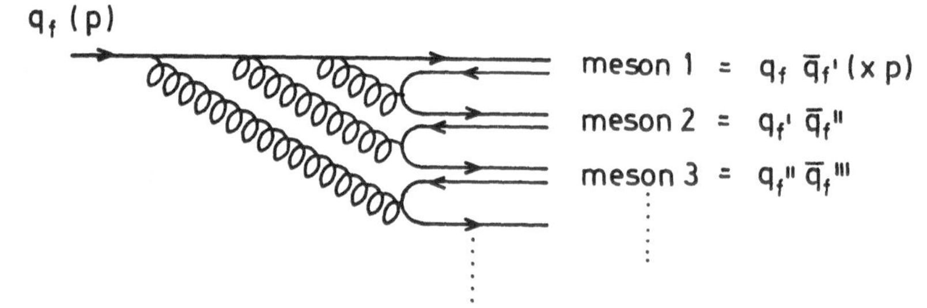

Figure 16 Diagram for quark fragmentation into mesons.

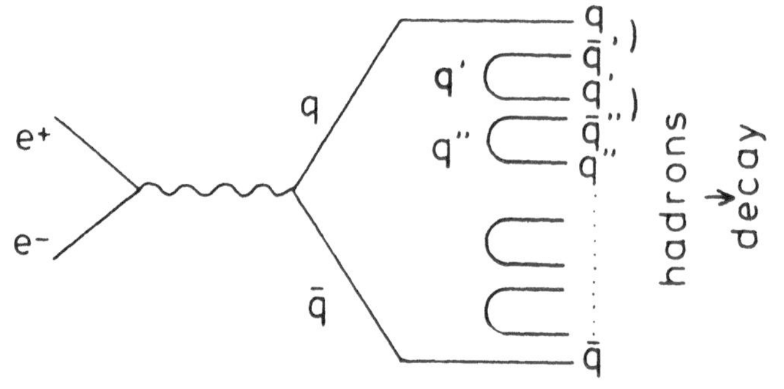

Figure 17 The fragmentation of quarks from $e^+e^- \to q\bar{q}$ in the Feynman-Field model.

as input a gaussian probability distribution

$$f(p_T) = e^{-p_T^2/2\sigma_q^2}$$

with an average transverse momentum of $\langle p_T\rangle = \sigma_q = 300$ MeV. The probability for the occurrence of a particular longitudinal momentum fraction x is usually assumed as $f(x) = 2(1-x)^2$ or some similar function of x. f(x) is tested by calculating the single particle inclusive cross section which is of the form

$$\frac{1}{\sigma}\frac{d^2\sigma}{dx\,dp_T^2} = D^h(x)e^{-bp_T^2} \tag{3.4}$$

This is compared to experimental data. For example, a measured p_T distribution is shown in Fig. 18 for $E_{cm} = 9.4$ GeV [14]. We see clearly the almost gaussian form at smaller p_T^2 with an average transverse momentum as stated above. In this curve the p_T is determined with respect to the sphericity axis.

It is clear that with such model available one can calculate inclusive two-particle, three-particle etc. distributions up to complete events. With the complete event structure at our disposal

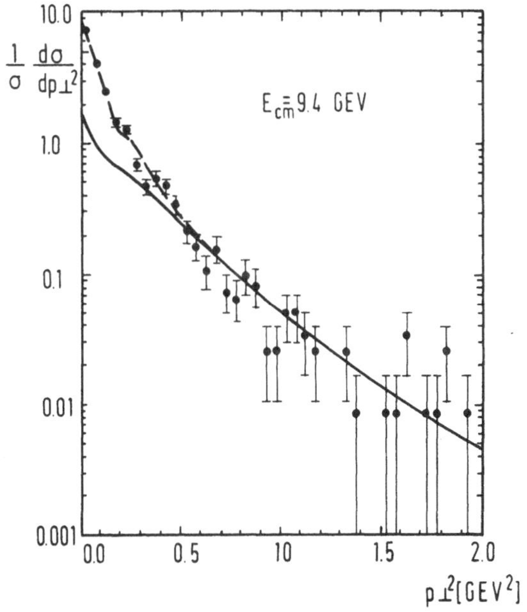

Figure 18 p_T^2 distribution at 9.4 GeV (PLUTO) compared to theoretical curves based on fragmentation of u,d,s quarks ($\langle p_T\rangle = 0.25$ GeV) and c quarks including weak effects. The dashed line is the sum of both contributions. The solid curve is only the charm contribution.

we can also calculate the distributions in the jet measures like thrust T, sphericity $\hat{S}$, spherocity S etc. Such distributions in thrust T for several center-of-mass energies are shown in Fig. 19. We notice that these distributions vanish for $T \to 1$ and become narrower with increasing E_{cm}. This one can understand quite easily if one inspects approximate formulas for the average T. It is

$$\langle 1-T \rangle \simeq \frac{\langle p_T \rangle \langle n \rangle}{2E_{cm}} \tag{3.5}$$

where $\langle n \rangle$ is the average multiplity. If $\langle n \rangle \sim \ln E_{cm}$ (see however Fig. 5) and $\langle p_T \rangle$ = const the average 1-T decreases with E_{cm} like

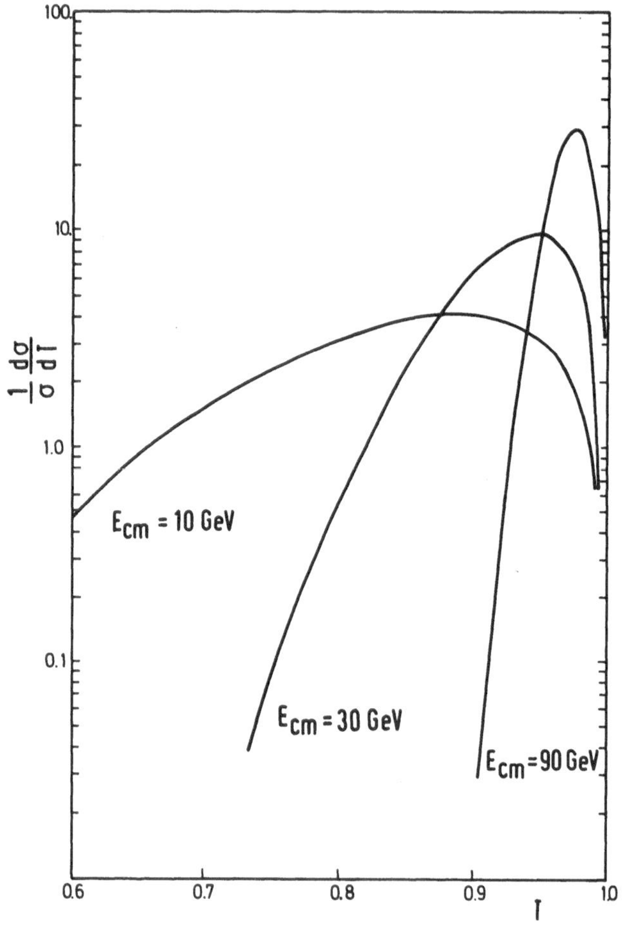

Figure 19 $\frac{1}{\sigma}\frac{d\sigma}{dT}$ for $e^+e^- \to q\bar{q}$ (q= u,d,s) for E_{cm} = 10,30 and 90 GeV from our fragmentation model.

$\ln E_{cm}/E_{cm}$. Similarly

$$\langle S \rangle \simeq \left(\frac{4}{\pi}\right)^2 \frac{\langle p_T \rangle^2 \langle n \rangle^2}{E_{cm}^2} \sim \frac{\ln^2 E_{cm}}{E_{cm}^2} \tag{3.6}$$

In Fig. 20 and 8 we see the recent data for <1-T> and the average sphericity $\langle \hat{S} \rangle$ as a function of E_{cm} between 4 and 31.6 GeV. We compare these data with curves $\ln E_{cm}/E_{cm}$ and $\ln^2 E_{cm}/E_{cm}^2$ respectively fitted to the measured point at E_{cm} = 9.4 GeV. The fall-off with increasing E_{cm} is less than the formulas (3.4) with $\langle n \rangle \sim \ln E_{cm}$ and (3.6) indicate. There exist several reasons for this somewhat softend decrease of <1-T> and $\langle \hat{S} \rangle$. First the average particle multiplicity <n> increases somewhat stronger than $\sim \ln E_{cm}$ as we saw in Fig. 5. Second weak interaction effects in connection with new thresholds, as for example the $b\bar{b}$ production threshold, lead to a broadening of jets and therefore to larger <1-T> and $\langle \hat{S} \rangle$ values. Third the effect of additional hard gluon emission produces also broader jets in average. These latter two effects will be discussed in the following sections.

The deviation of the empirical <n> from the $\ln E_{cm}$ dependence is taken into account in the Feynman-Field model described above

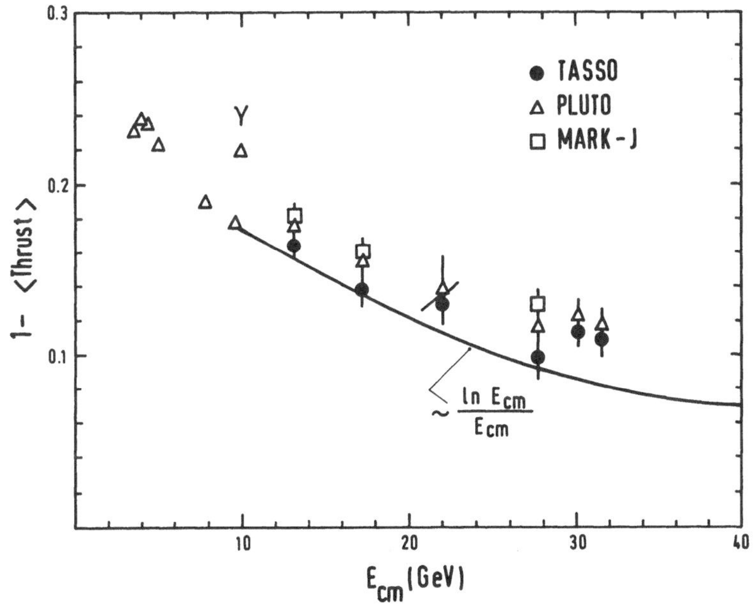

Figure 20 The average (1-T) as a function of E_{cm} as observed by PLUTO, TASSO and MARK J at PETRA.

by choosing appropriate primordial fragmentation functions f(x). Of course a part of the increase in <n> with increasing E_{cm} comes from the weak decays of $b\bar{b}$ quarks and from the hard gluon bremsstrahlung. If we consider only the jets as products of the Feynman-Field fragmentation we conclude that the jets and therefore also the thrust, sphericity etc. distributions become narrower and narrower. We shall see in the next chapter that this narrowing of the simple $q\bar{q}$ jets makes it possible to see eventually the effect of additional hard gluons.

It is clear that the fragmentation description explained above is a model and is not derivable from first principles since the confinement problem has not been solved sufficiently. Of course since the confinement is important for the transformation of quarks and antiquarks into hadrons the fragmentation process cannot be calculated with perturbative methods. Jet formation out of quarks or gluons is a really nonperturbative problem and in the moment can be described only with empirical input like $\langle p_T \rangle$, <n> and $D^h(x)$. After some modification the methods described above are also applicable to gluon fragmentation.

In this section we considered only the production of u,d and s quarks, which go directly into hadrons like π's, K's etc. which are measured. Sometimes the K^o's are not measured and appear as decay products, mostly pions. If necessary this complication can be built into the Feynman-Field model. The situation is more complicated for the production of $c\bar{c}$, $b\bar{b}$ and eventually $t\bar{t}$. This will be studied in the next paragraph.

We remark that so far only the fragmentation of quarks into mesons has been considered. Of course, the decay into baryons is also possible. From experimental data at lower energies one knows that baryon production in e^+e^- annihilation is rather small compared to meson production. Therefore the fragmentation into baryons is usually neglected.

§ 2. WEAK INTERACTION EFFECTS. [3,23,27]

Besides u, d and s quarks also $c\bar{c}$, $b\bar{b}$ and perhaps $t\bar{t}$ will be produced in the PETRA energy range. The decay products of these new quarks will be contained in the final state in $e^+e^- \to$ hadrons + leptons measured at the energies above the threshold for production of $c\bar{c}$, $b\bar{b}$ etc. For very high energies even $m_c \simeq 1.5$ GeV and $m_b \simeq 5$ GeV are neglegible compared to E_{cm} and we have similar conditions as for $u\bar{u}$, $d\bar{d}$ and $s\bar{s}$ production. But near threshold the event pattern is different, because of (i) nonzero mass effects $\left(\sigma \sim (E^2 - 4m_Q^2)^{\frac{1}{2}}\right.$ and the $1 + \cos^2\Theta$ angular distribution is distorted$\left.\right)$ (ii) weak decays $c \to s$ and $b \to c$. In particular weak decays of heavy quarks $Q \equiv (c, b, t, \ldots)$ introduce a new transverse momentum scale of the order

$$\langle p_T \rangle \simeq {}^{m_Q}/_3 \tag{3.7}$$

which leads to a growth of $\langle p_T \rangle$ and jet broadening in direct competition with similar effects coming from hard gluon emission (perturbative QCD effects). In order to be able to distinguish these broadening effects (in $\langle p_T \rangle$, $\langle 1-T \rangle$, $\langle S \rangle$, $\langle \hat{S} \rangle$ or $\langle A \rangle$) from similar ones coming from $q\bar{q}g$ or $q\bar{q}gg$ etc. it is important to have quantitative estimates on these weak effects. Qualitatively at threshold the situation concerning the jet measures $\langle 1-T \rangle$ etc. is as follows [23]:

at threshold (isotropic)	above threshold (boost applied)
$\langle \hat{S} \rangle_{Q\bar{Q}} \simeq 1$	$\langle \hat{S} \rangle_{Q\bar{Q}} \simeq \frac{4m_Q^2}{q^2}$
$\langle S \rangle_{Q\bar{Q}} \simeq 1$	$\langle S \rangle_{Q\bar{Q}} \simeq \frac{4m_Q^2}{q^2}$
$\langle 1-T \rangle_{Q\bar{Q}} \simeq \frac{1}{2}$	$\langle 1-T \rangle_{Q\bar{Q}} \simeq \frac{2m_Q^2}{q^2}$
$\langle A \rangle_{Q\bar{Q}} \simeq 1$	$\langle A \rangle_{Q\bar{Q}} \simeq \frac{4m_Q^2}{q^2}$

($q^2 = E_{cm}^2$). For more quantitative estimates we use a simple model which, we believe, allows us to sensibly calculate topological jet measures as $\hat{S}$, S, T, A etc. We shall explain this model in the following. As an example we take the production of $b\bar{b}$ and the subsequent decay of b and $\bar{b}$. For the weak decay of b and t quarks we assume the Kobayashi-Maskawa (KM) scheme [28]. This is a generalization of the GIM mechanism to six quarks and involves three angles Θ_1, Θ_2 and Θ_3 and one phase δ. In the KM model the three left-handed quark doublets are

$$\begin{pmatrix} u \\ \tilde{d} \end{pmatrix}_L, \quad \begin{pmatrix} c \\ \tilde{s} \end{pmatrix}_L, \quad \begin{pmatrix} t \\ \tilde{b} \end{pmatrix}_L \tag{3.8}$$

In (3.8) the quark states $\tilde{d}$, $\tilde{s}$ and $\tilde{b}$ which appear in the weak interaction Hamiltonian are related to the strong interaction eigenstates by a unitary matrix U

$$\begin{pmatrix} \tilde{d} \\ \tilde{s} \\ \tilde{b} \end{pmatrix} = U \begin{pmatrix} d \\ s \\ b \end{pmatrix}$$

where U has the following form

$$U = \begin{pmatrix} c_1 & -s_1c_3 & -s_1s_3 \\ s_1c_2 & c_1c_2c_3 - e^{i\delta}s_2s_3 & c_1c_2s_3 + e^{i\delta}s_2c_3 \\ s_1s_2 & c_1s_2c_3 + e^{i\delta}c_2c_3 & c_1s_2c_3 - e^{i\delta}c_2c_3 \end{pmatrix} \tag{3.9}$$

This is the most general form of a weak mixing matrix which does not produce flavour changing neutral currents. In addition to the three Euler angles ($c_i = \cos\Theta_i$, $s_i = \sin\Theta_i$, $i = 1,2,3$) it contains one (and only one) phase δ which allows one to describe CP violating effect (such a phase does not occur in the GIM scheme). The matrix (3.9) produces the following weak coupling for the six quarks u, d, s, c, b and t:

$$\begin{array}{c} u \\ c \\ t \\ \\ \end{array} \begin{pmatrix} c_1 & -s_1c_3 & -s_1s_3 \\ s_1c_2 & c_1c_2c_3 - e^{i\delta}s_2s_3 & c_1c_2s_3 + e^{i\delta}s_2c_3 \\ s_1s_2 & c_1s_2c_3 + e^{i\delta}c_2c_3 & c_1s_2c_3 - e^{i\delta}c_2c_3 \end{pmatrix} \tag{3.10}$$

$$\quad d \qquad\qquad s \qquad\qquad\qquad b$$

From (3.10) we see that the $(\bar{b}c)$ coupling is stronger than the $(\bar{b}u)$ coupling, since universality of μ decay and nuclear β-decay ($c_1^2 + s_1^2c_3^2 \simeq 1$) does not allow s_1s_3 to be large. One finds $s_3 < s_1$ and $s_1 \simeq \sin\Theta_C$, where Θ_C is the Cabibbo angle. A limit on the size of s_2 is obtained from the K_0-$\bar{K}_0$ mixing. It follows that also s_2 is small [29]. For simplicity we assume

$$c_1 \simeq 1$$

$$c_1c_2c_3 - e^{i\delta}s_2s_3 \simeq 1$$

and define

$$g_c \equiv c_1c_2s_3 + e^{i\delta}s_2s_3 << 1$$

Then the coupling (3.10) has as dominant contributions the transitions $u \to d$, $c \to s$, $b \to c$ and $t \to b$. The weak Hamiltonian responsible for b decay looks as follows:

(a) nonleptonic decays

$$H_{nl} = \frac{G}{\sqrt{2}} g_c \left[(c\bar{b})_L (\bar{d}u)_L + (c\bar{b})_L (s\bar{c})_L + \text{h.c.} \right] \qquad (3.11)$$

(b) semileptonic decays

$$H_{sl} = \frac{G}{\sqrt{2}} g_c \sum_{\text{leptons}} (c\bar{b})_L (l\bar{\nu}_l)_L \qquad (3.12)$$

Usually the weak Hamiltonian written in (3.11) for noninteracting quarks is modified due to gluon exchange. These contributions which are evaluated in the leading logarithm approximation are not important due to the heaviness of the b quark.

With these basic weak interaction Hamiltonians the decays of the b and c quarks bound in heavy mesons B and D respectively are calculated. Technically one computes the decay in the rest system of the respective heavy mesons and afterwards one transforms to the e^+e^- center-of-mass system with the appropriate Lorentz-boost. The diagram for production and decay is shown in Fig. 21. After production of the $b\bar{b}$ by the virtual photon the b quark (and in the same way the $\bar{b}$, which is not shown) fragments à la Feynman-Field into mesons (mostly pions and kaons) and one heavy meson $B = b\bar{q}$ ($q = u,d,s$). The b quark in the B decays weakly, for example into $c + \bar{u} + d$, via the familiar V-A weak interaction. The resulting c quark undergoes the same steps as the original b quark: fragmentation into mesons, out of which one is a D meson containing the c quark which decays into $s+u+\bar{d}$. The final quarks which appear in connection with the b decay, namly $u+\bar{d}$ and those via c decay, namely $s + u + \bar{d}$ are considered also as jets which fragment into mesons with transverse momenta of the order of $\langle p_T \rangle \simeq 300$ MeV. Actually this latter fragmentation of the weak decay products was neglected and the decay of the B and D mesons was approximated by the free quark decay. Of course the fragmentation of the b just after the production is very important since it leads to a B meson

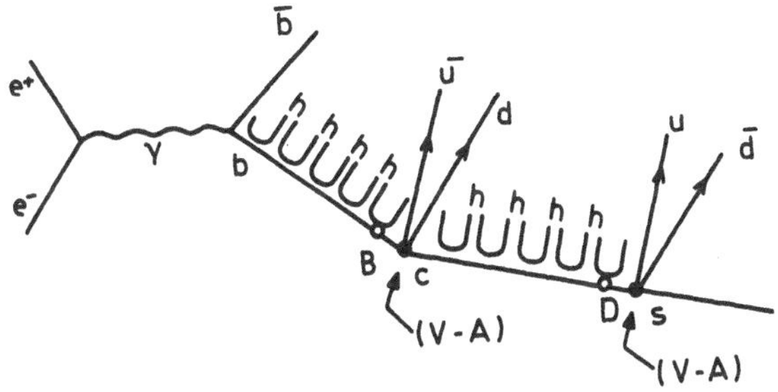

Figure 21 $e^+e^- \to b\bar{b}$ together with decay cascade based on $b \to c$, $c \to s$.

with less momentum than $E_{cm/2}$ so that the broadening through the weak decay becomes more effective. For the fragmentation functions which produce the B meson from the b and the D meson from the c we assumed the following forms

$$D_c^D(x) = 2(1-x)$$

$$D_b^B(x) = 2x \tag{3.13}$$

The explicit form of these fragmentation functions is not really important. Some other functional dependences were also tried. They gave the same results as long as the fragmentation near $x \simeq 1$ was not suppressed. (In a later version of this work the fragmentation of the quarks from B decay was taken into account and the D meson decay was described by experimental data as given in the particle data tables.)

As a first test of the model we computed the sphericity distribution at E_{cm} = 9.4 GeV [30] and compared it with the distribution measured by the PLUTO collaboration at DORIS [31]. This is presented in Fig. 22 where the distribution from u,d and s quarks alone and the distribution coming from c quarks are shown separately. We see that the c-distribution is still broader than the u, d, s-distribution although the measurement is at 5 GeV above charm production threshold. In the next figure, Fig. 23 we show the effect of $b\bar{b}$ production at E_{cm} = 17 GeV on the sphericity distribution. The $b\bar{b}$ production threshold is near 11 GeV. At 17 GeV we see still a broad distribution with maximum around S = 0.6. Since the $b\bar{b}$ channel constributes only with the fraction f (Q_i = quark charge, $Q_b = 1/3$)

$$f = \frac{Q_b^2}{Q_u^2 + Q_d^2 + Q_s^2 + Q_c^2 + Q_b^2} = \frac{1}{11} \tag{3.14}$$

to the total cross section it produces only a shoulder in the sphericity distribution $\sigma^{-1}\, d\sigma/d\hat{S}$ at larger $\hat{S}$ values and not a second maximum. This shoulder is consistent with the experimental result for $\sigma^{-1}\, d\sigma/d\hat{S}$ from the PLUTO [33] and TASSO collaborations at PETRA [34]. With more accurate data one could prove with this method that the threshold for open bottom production had been passed. So far we can say only that the data show an indication for $b\bar{b}$ production. Of course the experimental points disagree with the sphericity distribution for the production of additional $t\bar{t}$ where t has charge $Q_t = 2/3$. In this case $d\sigma/dS$ would have a second maximum near $S \simeq 0.6$ since the additional component coming from $t\bar{t}$ would be a factor 4 larger than the $b\bar{b}$ component shown in Fig. 23. In Fig. 24 we see the mean oberved thrust $\langle 1 - T \rangle$ compared

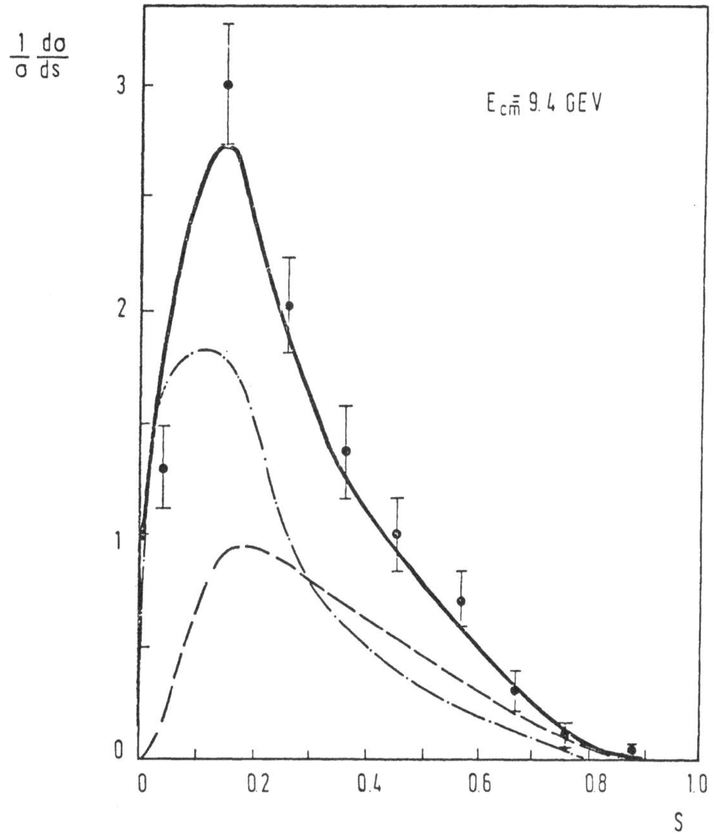

Figure 22 Comparison of PLUTO sphericity distribution at E_{cm} = 9.4 GeV with model calculation. The dash-dot-curve is the acceptance corrected Feynman-Field sphericity distribution for u, d, s quarks with normalization $^6/10$. The dotted curve is the charm distribution with the fragmentation function $D_c{}^D(z) = 2(1-z)$ and normalization $^4/10$. The solid curve is the sum of u, d, s and c contributions.

to a theoretical curve which has also a $b\bar{b}$ component. The agreement is again satisfactory. The experimental points indicate that there is no decrease of <1-T> between E_{cm} = 9.4 GeV and 13 GeV. Without the occurrence of a new threshold one would expect a decrease of <1-T> according to the behaviour $\ln E_{cm}/E_{cm}$. In the last months data where taken also at higher energies between 22 and 31.6 GeV by the three groups Mark J, PLUTO and TASSO. All three groups rule out an open $t\bar{t}$ channel (Q_t = $^2/3$) below 30 GeV. [7,8,9] This conclusion rests on the value of the relative hadronic cross section R and the sphericity and thrust distributions which do not have the second maximum expected from the $t\bar{t}$ production. In Fig. 25 we show the theoretical expectation for the sphericity distribution $^{d\sigma}/d\hat{S}$ and thrust distribution $^{d\sigma}/dT$ for E_{cm} = 30 GeV and an

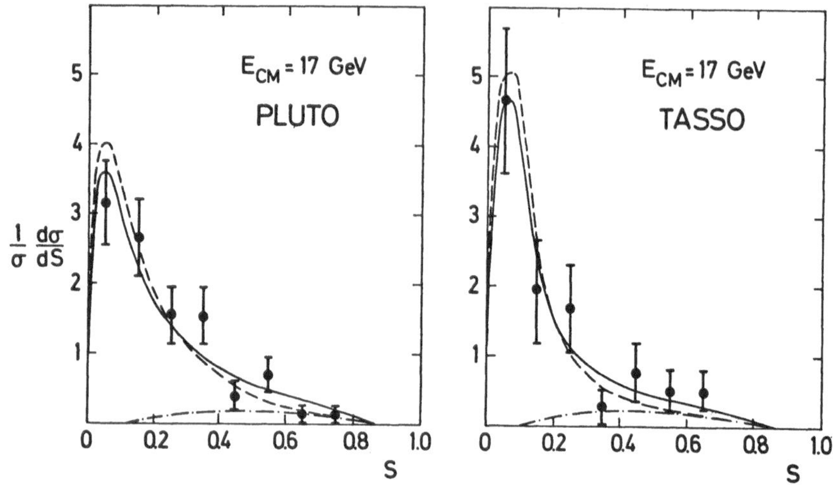

Fig. 23 Comparison of sphericity distributions at 17 GeV with a) PLUTO data and b) TASSO data. The dashed curve is the acceptance corrected Feynman-Field distribution for u,d,s quarks plus charm nomalized to the data points. The dashed-dotted curve is the bottom contribution normalized to $^1/11$. The solid curve is the sum of u,d,s,c and b contributions normalized to the data points.

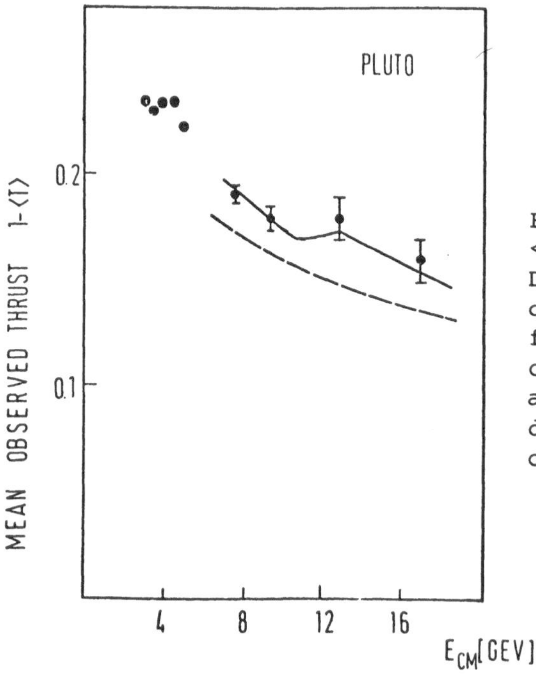

Fig. 24
<1-T> as a function of E_{cm}. Dashed curve is the acceptance corrected Feynman-Field <1-T> for u,d,s quarks. The solid curve is the sum of u,d,s,c and b contributions. The data are from the PLUTO collaboration.

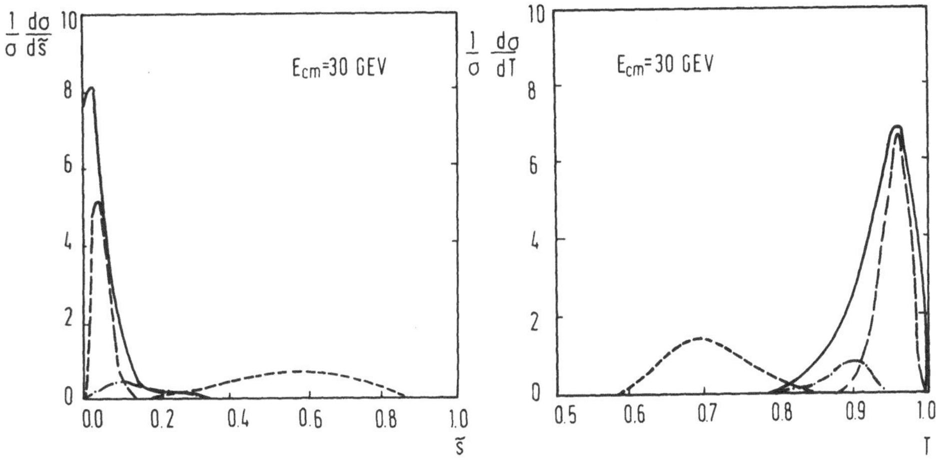

Figure 25 Predicted sphericity (b) and thrust (c) distribution, at the t threshold. The u,d,s contribution (——), the b contribution (-·-) and the sum of u,d,s,c and b contribution (——) is compared to the t part (----).

effective top quark mass m_t = 13 GeV. As an example of measured distributions we show in Fig. 26 the just published PLUTO results at E_{cm} = 22, 27.6, 30 and 31.6 GeV [8]. In all four distributions there is no sign of a second maximum or shoulder in $d\sigma/dT$ near $T \simeq 0.7$. The value of R for the combined data from 27.6, 30 and 31.6 GeV is R = 3.88 ± 0.22 in agreement with the expected value 3.9 obtained from u,d,s,c,b quarks and first order QCD effects according to formula (2.12) with a running coupling constant calculated with Λ = 0.5 GeV and N_f = 5 [8].

The occurrence of $t\bar{t}$ production should be visible already in the mean values of (1-T), $\hat{S}$ or A. The increase in <$\hat{S}$> is roughly 0.1 and in <A> it is ~ 0.06. The energy dependence of these two quantities is seen in Fig. 27 together with the observed values by several groups [3]. Again there is no sign of a top threshold below 32 GeV. In the curves for six quarks we included some QCD effects. The underlying model will be explained in the last section. The jet broadening effects, for example the increase of <1-T>, <S>, <A> etc. by the opening of new thresholds and the subsequent weak decays compete with broadening effects caused by gluon emission which will be considered in the next section. For example, if expressed in <1-T> the two effects, broadening from $b\bar{b}$ threshold and QCD effects, are roughly of the same order of magnitude, so that by detailed analysis one can hope to disentangle them. The broadening effect caused by a Q = $^2/3$ quark threshold are much larger and therefore may hide QCD effects sufficiently. Fortunately the $t\bar{t}$ threshold seems to lie at a higher energy than previously thought so that there is plenty of room to study QCD effects.

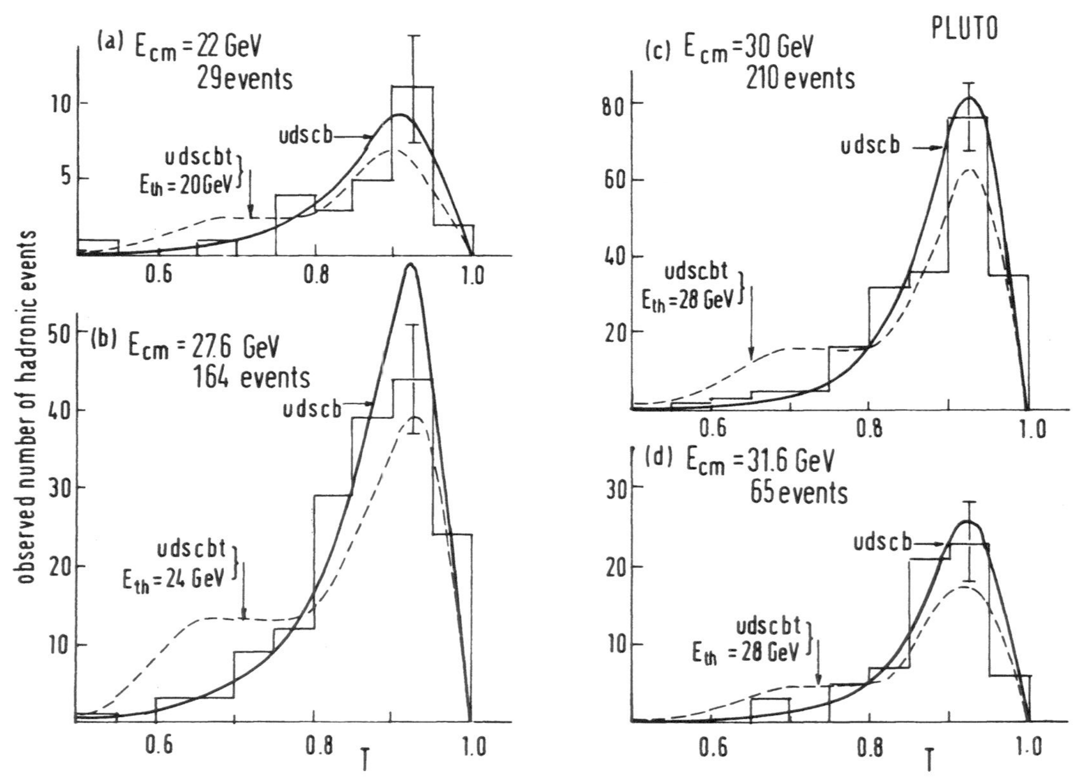

Fig. 26 (a)-(d) Detailed thrust distributions as measured by PLUTO at the specified c.m. energies. Predictions from u,d,s,c,b (——) and u,d,s,c,b,t (----) quarks with specified energy thresholds (E_{th}) for the "top" states are shown.

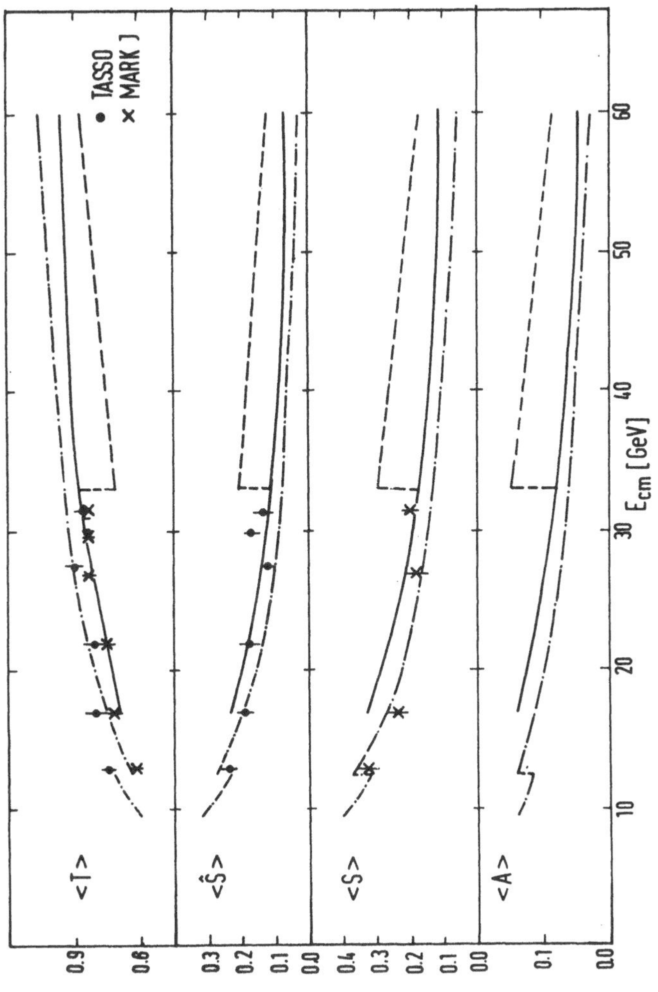

Fig. 27 Average jet measures $\langle T\rangle$, $\langle \hat{S}\rangle$, $\langle S\rangle$ and $\langle A\rangle$ as a function of $E_{c.m.}$ for 5 quarks (dashed-dotted) 5 quarks plus QCD corrections (full) and 6 quarks plus QCD corrections (dashed) compared to experimental data of the TASSO and MARK J group at PETRA.

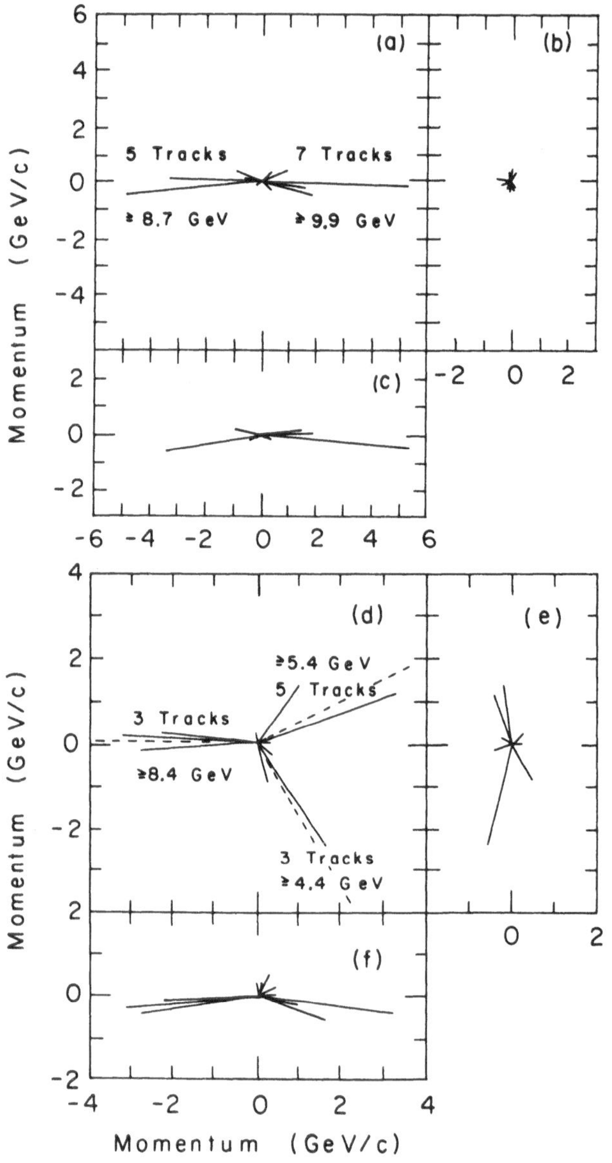

Fig. 28 Momentum space representation of a 2-jet event (a-c) and a 3-jet event (d-f) in each of three projections.
(a,d) = $\hat{n}_2 - \hat{n}_3$ plane
(b,e) = $\hat{n}_1 - \hat{n}_2$ plane
(c,f) = $\hat{n}_1 - \hat{n}_3$ plane.
The dotted lines show the fitted jet axes.

4. SIGNATURES FOR $e^+e^- \to q\bar{q}g$.

In section 2 we interpreted the events outside the two-jet cones with half-angle δ and energy $\varepsilon\sqrt{q^2}$ outside the two cones as three-jet events having a quark, an antiquark and a gluon jet in the final state. Whether these three configurations are visible as three distinct jets is a question of the available cm energy $E_{cm} = \sqrt{q^2}$. This can be seen already from the formula for the 3-jet cross section obtained in (2.16). In this formula δ cannot be smaller than the nonperturbative jet spread. Independent how small δ already is, 2ε must be small also, otherwise the sum of the first two terms in (2.16) becomes negative leading to a very small cross section. $\varepsilon \leq 1/4$ is a reasonable limit to achieve this. So $p_3 \leq \sqrt{q^2}/4$. On the other hand, even in the two-jet case, jets were not really seen below $E_{cm} = 6$ GeV. Therefore $E_{cm} = \sqrt{q^2}$ must be larger or equal 24 GeV before three distinct jets could be seen. Actually this is the energy range where three clearly distinguishable jets have been detected recently at PETRA [7,8,9]. One of such events [7], produced at $E_{cm} = 31.6$ GeV, is shown in Fig. 28 in momentum space projected into a plane containing the largest components of momenta. Projections of this event into two other planes perpendicular to the first plane show quite clearly that the event is planar as one expects it for a three-particle final state. For comparison a typical 2-jet event is also shown. More events of this kind will be measured in the near future. Then one will be able to study in more detail the dynamics of quark, antiquark and gluon jets.

In the following we shall describe the main tests which can be derived from the simple gluon bremsstrahlung diagrams in Fig. 12. The cross section for gluon bremsstrahlung depends on two energies and two Euler angles Θ and χ which describe the production plane with respect to a plane defined by the beam direction and the momentum of one of the final particles. Let the quark, antiquark and gluon have the four momentum p_1, p_2 and p_3 respectively and the normalized momenta be

$$x_i = \frac{2p_i}{\sqrt{q^2}} = \frac{2p_i}{E_{cm}} \qquad i = 1,2,3 \tag{4.1}$$

Then for massless quarks we have the constraints

$$0 \leq x_i \leq 1 \quad , \quad x_1 + x_2 + x_3 = 2 \tag{4.2}$$

Thus in a x_1, x_2-plane the allowed kinematic region is the triangle

$$\begin{aligned} &0 \leq x_1 \leq 1 \\ &1-x_1 \leq x_2 \leq 1 \end{aligned} \tag{4.3}$$

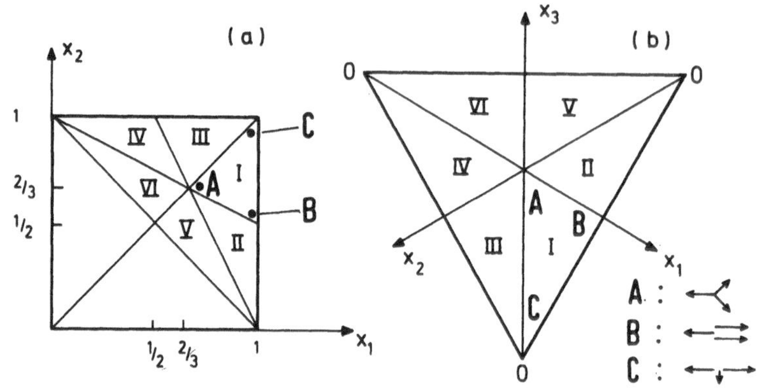

Figure 29 Phase space boundaries for $e^+e^- \to q\bar{q}g$ in terms of the x_i variables. (a) x_1, x_2 plane (b) x_1, x_2, x_3 triangle.

as shown in Fig. 29a or in Fig. 29b in the form of a triangle with equal sides in which $x_1 = 0$, $x_2 = 0$ or $x_3 = 0$ at the three corners respectively. As we know, the quark, antiquark and gluon are not the particles which are measured in the final state. They fragment into hadrons which appear at high enough energy in the form of jets. Therefore we must somehow relate the variables of the $q\bar{q}g$ state, namely the x_i to the variables which describe the final hadron events. Such variables are thrust T, spherocity S and sphericity $\hat{S}$. Their definitions were given in section 1 and 2 ((2.20), (2.21),(1.7)). These global jet measures for the hadron events are identified with the jet measures calculated for the $q\bar{q}g$ state. For example, the thrust for $q\bar{q}g$, using the definition (2.20) is

$$T = \max\,(x_1, x_2, x_3) \tag{4.4}$$

which is just the thrust for a three-particle state with massless particles. Similarly the spherocity is related to the x_i by (see (2.21))

$$S = \frac{64}{\pi^2}\,(1-x_1)(1-x_2)(1-x_3)\Big/x_1^2 \tag{4.5}$$

and the sphericity ((1.7))

$$\hat{S} = 12\,(1-x_1)(1-x_2)(1-x_3)\Big/(x_1^2(x_1^2 + x_2^2 + x_3^2)) \tag{4.6}$$

Of course by identifying T, S, $\hat{S}$ etc. of the $q\bar{q}g$ state with the actual measured jet measures one neglects completely the final fragmentation of q, $\bar{q}$,and g. Therefore the theoretical results for the $q\bar{q}g$ state are only approximate. They are valid for very high energies and in particular kinematic regions of the jet variables T, S, $\hat{S}$ etc. in which the jet spread caused by the fragmentation is small compared to the jet broadening coming from the gluon emission.

In the kinematic plot in Fig. 29 we can distinguish six regions with different orderings of x_1, x_2 and x_3. These regions are:

$$\text{I: } x_1 > x_2 \geq x_3 \qquad \text{III: } x_2 > x_1 > x_3 \qquad \text{V: } x_3 > x_1 > x_2$$

$$\text{II: } x_1 > x_3 > x_2 \qquad \text{IV: } x_2 > x_3 > x_1 \qquad \text{VI: } x_3 > x_2 > x_1 \tag{4.7}$$

According to (4.4) $T = x_1$ in region I and II, $T = x_2$ in III and IV and $T = x_3$ in V and VI. In I the second largest thrust is $T_2 = x_2$ and in II the second largest thrust is $T_2 = x_3$ etc. To obtain the limits for the variables thrust T and spherocity S it is sufficient to consider just one region, for example region I. Then $^2/3 \leq x_1 \leq 1$ and

$$1 - \frac{x_1}{2} \leq x_2 \leq x_1,$$

so that [23]

$$\frac{2}{3} \leq T \leq 1$$

$$\frac{64}{\pi^2} \; \frac{(1-T)^2(2T-1)}{T^2} \leq S \leq \frac{16}{\pi^2}(1-T)$$

$$1 - \frac{T}{2} \leq T_2 \leq T \tag{4.8}$$

In order to describe the kinematic region in which the jet variables change one must choose just two of these variables. For example for T and T_2 the kinematic region is simple. It is just the triangle I with $x_1 = T$ and $x_2 = T_2$. For T and S the boundary is more complicated and is given by (4.8). The most interesting kinematic region is near the middle of the triangle, where $x_1 = x_2 = x_3 = {}^2/3$. In region I it is the left corner, where $T = T_2 = {}^2/3$ and

$$S = \frac{16}{3\pi^2} \quad \text{(A)}.$$

In the other two corners we have configurations equivalent to two-jet events: $x_1 = 1$, $x_2 = x_3 = {}^1/2$ (B) and $x_1 = 1$, $x_2 = 1$, $x_3 = 0$ (C). It is clear that in these kinematic regions the $q\bar{q}g$ events are indistinguishable from $q\bar{q}$ events coming from the zero-order graph. So in order to test details of the gluon bremsstrahlung we must make cuts in T and T_2 or T and S to isolate the events which lie in the vicinity of $T = T_2 = {}^2/3$. Here two of the three momenta have an angle of 120° between them.

In general, the thrust and the spherocity axes are not identical. Of course they coincide for three-particle final state, both aligning with the momentum of the most energetic quantum. The same is true for the sphericity. If nonperturbative effects produced by fragmentation are taken into account the three axes will differ.

The cross section for $e^+e^- \to q\bar{q}g$ integrated over the Euler angles is [4,23]

$$\frac{1}{\sigma_o}\frac{d^2\sigma}{dx_1dx_2} = \frac{2\alpha_s}{3\pi}\frac{x_1^2 + x_2^2}{(1-x_1)(1-x_2)} \tag{4.9}$$

$\alpha_s = g^2/4\pi$ is the running coupling constant given by

$$\alpha_s(q^2) = \frac{12\pi}{(33-2N_f)\ln q^2/\Lambda^2} \tag{4.10}$$

and σ_o is the lowest order cross section 1. The cross section (4.9) is singular for $x_1 \to 1$ (and/ or $x_2 \to 1$). In region I the region $x_1 = 1$ is just the right boundary of this triangle. The possible configurations range from $x_2 = x_3 = 1/2$ (B) over $1 > x_2 > x_3 > 0$ to $x_2 = 1$, $x_3 = 0$ (C). In this last configuration the gluon has zero energy. This is the well-known infrared singularity familiar from QED. In the case $x_3 \neq 0$ and $\frac{1}{2} \leq x_2 < 1$ the antiquark and the gluon have parallel momenta. This is the collinear gluon singularity caused by $m_q = 0$. Such a singularity occurs also in QED with massless electrons. In region III the role of x_1 and x_2 are interchanged whereas in region II we have only the singularity $x_1 = 1$ and $0 \leq x_2 \leq \frac{1}{2}$, i.e. only the mass singularity $m_q = 0$. Since the $q\bar{q}g$ cross section (4.9) diverges for $x_1 \to 1$ or $x_2 \to 1$ the cross section is very large near $x_1 = 1$ or $x_2 = 1$. In this region perturbation theory is not reliable any more. Therefore the formula (4.9) should not be applied near $T = 1$ and/or $S = \hat{S} = 0$. So we have two reasons for using (4.9) only in a limited kinematic region, for example, $2/3 \leq T \leq T_c < 1$. First, near $T \simeq 1$ the formula (4.9), based on first order perturbation theory, becomes unreliable. Second near $T \simeq 1$ the $q\bar{q}g$ contribution cannot be distinguished from the $q\bar{q}$ term, which due to nonperturbative effects covers a finite region of the order $\langle 1-T\rangle_{\text{nonpert.}}$ near $T = 1$ as was discussed in section 3.

Without taking fragmentation of quarks and antiquarks into account the $q\bar{q}$ contribution has the form

$$\frac{1}{\sigma_o}\frac{d^2\sigma}{dTdT_2} = \delta(1-T)\delta(1-T_2)\left[1+O\left(\frac{\alpha_s}{\pi}\right)\right] \tag{4.12}$$

The term proportional to $\frac{\alpha_s}{\pi}$ in (4.12) comes from the diagram with the virtual gluon in Fig. 12. This virtual correction by itself is infrared divergent . In σ_{tot} the divergence cancels against the divergence coming from integrating the bremsstahlung $q\bar{q}g$ contribution. This we had already seen in section 2.

The cross section (4.9) can easily be transformed into a cross section depending on the variables T and T_2 or T and S by translating the variables x_1, x_2 and x_3 into T and T_2 or T and S in the regions I,II ...,VI. The double differential cross section may be too differential to be useful in comparing with low-statistics data. Therefore we integrate over one variable, x_2, in the six regions.

$$\frac{1}{\sigma_0}\frac{d\sigma}{dT} = \frac{2\alpha_s}{3\pi}\left\{2\int_{2(1-T)}^{T} dx_2 \frac{T^2 + x_2^2}{(1-T)(1-x_2)} + 2\int_{2(1-T)}^{T} dx_2 \frac{(2-T-x_2)^2 + x_2^2}{(x_2+T-1)(1-x_2)}\right\} \tag{4.13}$$

The result of the integration is [23]:

$$\frac{1}{\sigma_0}\frac{d\sigma}{dT} = \frac{2\alpha_s}{3\pi}\left\{\frac{2(3T^2-3T+2)}{T(1-T)}\ln\frac{2T-1}{1-T} - \frac{3(3T-2)(2-T)}{1-T}\right\} \tag{4.14}$$

In the limit $T \to 1$ the dominant term, which defines the leading logarithm approximation (LLA), is

$$\left(\frac{1}{\sigma_0}\frac{d\sigma}{dT}\right)_{LLA} = \frac{8\alpha_s}{3\pi}\frac{1}{1-T}\ln\frac{1}{1-T} \tag{4.15}$$

This formula contains the leading singularity coming from collinear and infrared gluon emission. In Fig. 30 we have plotted

$$\frac{1}{\sigma_0}\frac{d\sigma}{dT}$$

as given by (4.14) at E_{cm} = 30 GeV and compare it with the nonperturbative T distribution for $q\bar{q}$ production (q = u,d,s,c and b) [3]. We see that for $T \lesssim 0.75$ the perturbative cross section dominates the nonperturbative contribution. Near T = 0.80 they are of the same order of magnitude. The nonperturbative cross section includes the weak effects from c and b decay described in section 3. In Fig. 30 we also have included the leading logarithm approximation (LLA) (4.15). This approximation agrees roughly with the exact formula (4.14) for $T \gtrsim 0.95$.

From the discussion above it is clear that (4.14) can be used only if it dominates the nonperturbative contribution and in a region of T where perturbation theory makes sense. Without new

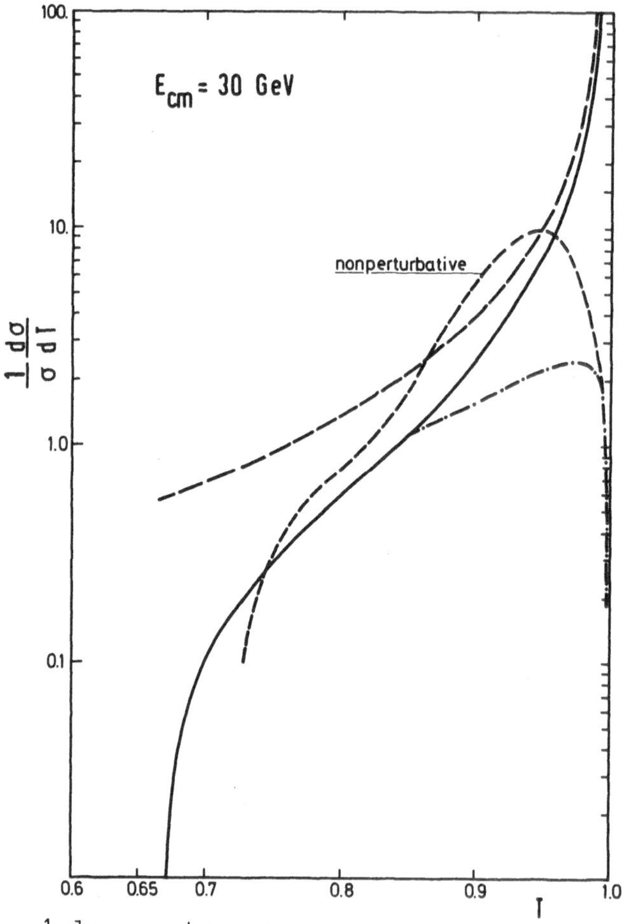

Figure 30 $\frac{1}{\sigma} \frac{d\sigma}{dT}$ for $e^+e^- \to q\bar{q}g$: exact result in order α_s (———),
leading logarithm approximation (– – –),"summed leading
logarithm approximation" (– · – · –) and nonperturbative
2-jet contribution (q = u,d,s,c,b) (– – – – – –).

quark thresholds coming in the width of the nonperturbative T distribution will become narrower and narrower with increasing energy,
so that (4.14) could be used also for T values very near to T = 1.
But in this region the first order perturbation theory ceases to
be valid. In order to have a rough idea, where this will be, we
integrate (4.14) over T up to T_c. The result, called $\sigma(T_c)$, is
given in Table 1 together with the values in LLA coming from

$$\frac{1}{\sigma_o} \int_{\frac{2}{3}}^{T_c} \left(\frac{d\sigma}{dT}\right)_{LLA} dT = \frac{4\alpha_s}{3\pi} \ln^2(1-T_c) \qquad (4.16)$$

T_c	$\sigma(T_c)/\sigma_o$	$\left(\sigma(T_c)/\sigma_o\right)_{11a}$
1.0	∞	∞
0.95	0.36	0.80
0.90	0.14	0.47
0.85	0.06	0.32

Table 1: $q\bar{q}g$ cross section integrated up to T_c at E_{cm} = 30 GeV.

For $T_c = 0.9$ the $q\bar{q}g$ contribution is just 14 % of the zero-order cross section

$$\sigma_o \simeq \sigma_{tot} = \sigma_o \left(1 + \frac{\alpha_s}{\pi}\right)$$

roughly equal to $\frac{2\alpha_s}{\pi}$ with (4.10) $\alpha_s = 0.20$ for $N_f = 5$ and $\Lambda = 0.5$ GeV).

With the formula (4.14) it is a simple task to calculate the average quantities $\langle 1-T\rangle$, $\langle (1-T)^2\rangle$ etc. defined by

$$\langle (1-T)^n\rangle = \frac{1}{\sigma_o} \int_{2/3}^{1} dT\,(1-T)^n \frac{d\sigma}{dT} \tag{4.17}$$

One should notice that $\langle 1-T\rangle$ is not the average of (1-T) of the distribution (4.14). This quantity vanishes since the integral in the denominator is infinite. Therefore we have normalized the integral over

$$(1-T)^n \frac{d\sigma}{dT}$$

with $\sigma_o = \sigma_{tot}$ (up to terms of order α_s). The result of the integration for $\langle 1-T\rangle$ is [23]

$$\langle 1-T\rangle = \frac{2\alpha_s}{3\pi} \left\{- \frac{3}{4} \ln 3 - \frac{1}{18} + 4 \int_{2/3}^{1} \frac{dT}{T} \ln \frac{2T-1}{1-T}\right\} = 1.05 \frac{\alpha_s}{\pi} \tag{4.18}$$

$\langle 1-T\rangle$ is always a finite quantity. It is "infrared save" since the remaining singularity in the integrand $\sim\ln(1-T)$ is integrable.

Similarly one obtains for the average spherocity [23]:

$$<S> = \frac{64\alpha_s}{3\pi^3}\left(-\frac{229}{9} + 64\ln\frac{3}{2}\right) = 1.09\,\frac{\alpha_s}{\pi} \tag{4.19}$$

These averages can be compared with the average quantities of the nonperturbative distributions considered in section 3. The $q\bar{q}g$ averages are always proportional to $\alpha_s(q^2)$ which decreases as $^1/\ln q^2$ with increasing q^2, whereas the nonperturbative averages $<1-T>_{nonpert.} \sim {}^1/\sqrt{q^2}$ and $<S>_{nonpert.} \sim {}^1/q^2$ up to factors $\sim(\ln q^2)^K$ coming from the multiplicity. So, eventually, at very high energies the nonperturbative <1-T> and <S> will be small compared to the $q\bar{q}g$ averages of (1-T) and S. To get an idea for which energy this is going to happen we have plotted <1-T> and <S> for $q\bar{q}g$ and for the nonperturbative background calculated with the model of section 3 (5 quarks: u,d,s,c and b) as a function of $E_{cm} = \sqrt{q^2}$ (Fig.31). We see that $<1-T>_{nonpert} \simeq <1-T>_{q\bar{q}g}$ at 45 GeV. But $<S>_{nonpert.}$ is still larger than $<S>_{q\bar{q}g}$ at this energy. Therefore it will be difficult to unravel the $q\bar{q}g$ effects by looking at <1-T> or <S> alone. This conclusion differs from an earlier analysis [23], where it was thought, that the nonperturbative contributions are already smaller than the $q\bar{q}g$ contribution at 30 GeV.

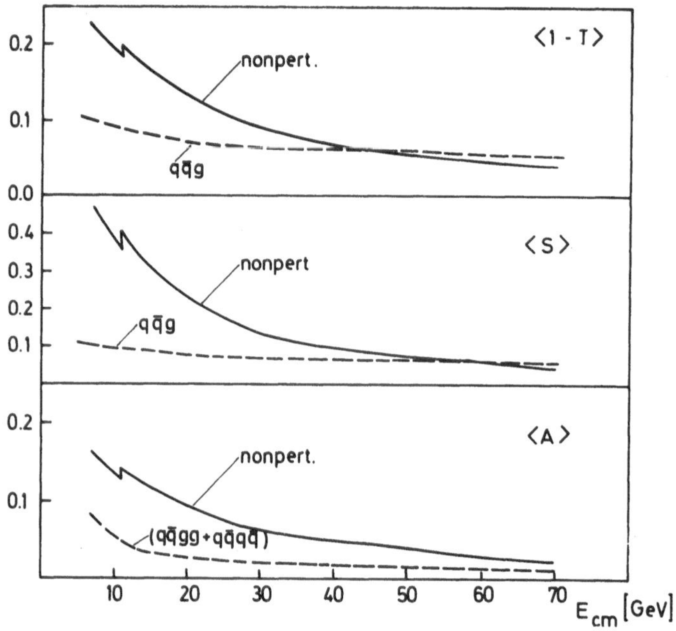

Figure 31 Average jet measures <1-T>, <S> and <A> in lowest order QCD compared to nonperturbative values with 5 quarks and weak effects.

The reason for this change is at least twofold. First, the multiplicity increased much stronger with increasing q^2, $\langle n\rangle_{ch} = 2 + 0.2\ \ln q^2 + 0.18 \ln^2 q^2$ (see Fig. 5), than anticipated from low energy data. This leads to larger values of $\langle 1-T\rangle$ and $\langle S\rangle$ since approximately $\langle 1-T\rangle \sim \langle n\rangle$ and $\langle S\rangle \sim \langle n\rangle^2$ (see (3.5) and (3.6)). Second there is an increase of $\langle 1-T\rangle$ and $\langle S\rangle$ due to the weak decays of c and b quarks which is not included in the estimates (3.5) and (3.6).

From Fig. 31 we can conclude that starting around $E_{cm} = 20$ GeV the measured values of $\langle 1-T\rangle$ and $\langle S\rangle$ should deviate from $\langle 1-T\rangle_{nonpert.}$ and $\langle S\rangle_{nonpert.}$ given in the figure. This deviation should increase with increasing E_{cm}. However it would not be correct to add the nonperturbative and the $q\bar{q}g$ contribution to $\langle 1-T\rangle$ and $\langle S\rangle$. $\langle 1-T\rangle$ as given by (4.18) is certainly an overestimate since $d\sigma/dT$ is used in a region of T where first order perturbation theory is not valid (roughly $T > 0.9 - 0.95$). A better estimate of the $q\bar{q}g$ contribution to $\langle 1-T\rangle$ is given by

$$\langle 1-T\rangle_{T_c} = \frac{1}{\sigma_o} \int_{2/3}^{T_c} dT(1-T)\, \frac{d\sigma}{dT} \tag{4.20}$$

with a cut-off T_c at the upper boundary of the T integration. With $T_c = 0.95$, 0.90 and 0.85 the values of $\langle 1-T\rangle_{T_c}$ are

$$0.59\ \frac{\alpha_s}{\pi}\ ,\ 0.26\ \frac{\alpha_s}{\pi} \text{ and } 0.17\ \frac{\alpha_s}{\pi} \text{ instead of } 1.05\ \frac{\alpha_s}{\pi} \text{ for } T_c = 1.0.$$

Then at $E_{cm} = 30$ GeV and $T_c = 0.95$ the $q\bar{q}g$ contribution to $\langle 1-T\rangle$ is $\langle 1-T\rangle_{0.95} = 0.038$ which is not a very large effect. Similar considerations must be made for $\langle S\rangle$. In order to have a rough idea about the order of magnitude of $\langle 1-T\rangle_{q\bar{q}g}$ and $\langle S\rangle_{q\bar{q}g}$ we have taken the recently measured data of the MARK J[9] and the TASSO [7] collaboration at PETRA and subtracted $\langle 1-T\rangle_{nonpert.}$ and $\langle S\rangle_{nonpert.}$ respectively given in Fig. 31. The result (average of $E_{cm} = 30.0$ and 31.6 GeV data) is for MARK J: $\langle 1-T\rangle_{q\bar{q}g} = 0.032 \pm 0.008$, $\langle S\rangle_{q\bar{q}g} = 0.048 \pm 0.025$ and for TASSO: $\langle 1-T\rangle_{q\bar{q}g} = 0.024 \pm 0.009$. So a cut-off $T \simeq 0.95$ is appropriate to describe the deviations of the measured averages from the nonperturbative values.

Of course a much better method to establish the broadening of thrust or spherocity distributions is to study these distributions in a region where $q\bar{q}g$ effects dominate, i.e. $2/3 \leq T \leq 0.75$ and $0.3 \leq S \leq 0.5$. This has been done recently with $d\sigma/dT$ at $E_{cm} = 31.6$ GeV by the MARK J group [35]. The result can be seen in Fig. 32. The measured points in the lower T region clearly favour the curve which includes gluon contributions.

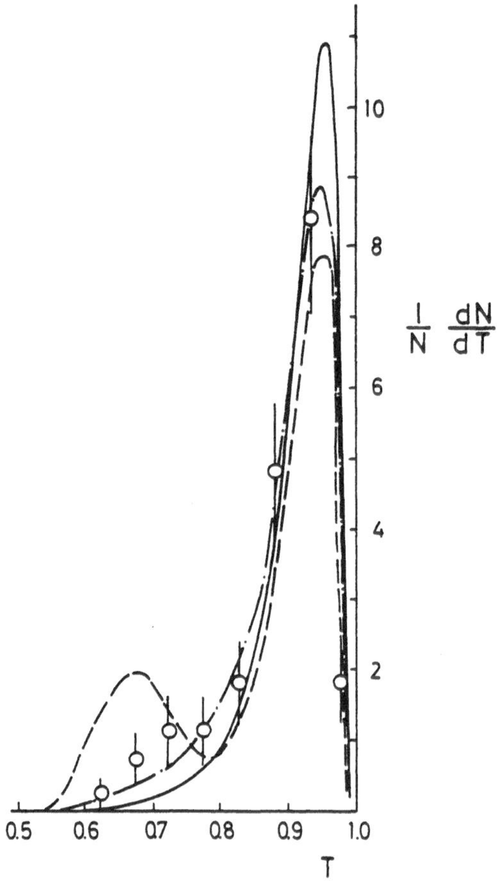

Figure 32 Thrust distribution $\frac{1}{N}\frac{dN}{dT}$. The solid curve is the Monte Carlo prediction based on u,d,s,c and b quarks. The dotted curve includes gluon contributions. The dashed-dot curve has included the t quark contributions. The data and the curves are from MARK J collaboration at PETRA.

So far we looked only at distributions which depend on one jet variable. Of course, in order to separate the three-jet events from the two-jet background it is much better to analyse distributions in two jet variables, as for example T and S or T and T_2. In the T, S - plane genuine three-jet events are characterized by the boundary (4.8). This is a very narrow strip below

$$S = \frac{16}{\pi^2}(1-T)$$

for T between $^2/3$ and 1. The average value of S at fixed T is [23]:

$$\langle S(T)\rangle_{q\bar{q}g} = \frac{32}{\pi^2} \frac{(3T-2)(1-T)(8-20T+18T^2-3T^3)}{T[2(3T^2-3T+2)\ln\frac{2T-1}{1-T} +3(3T-2)T(T-2)]} \qquad (4.21)$$

In the interesting region $^2/3 \leqq T \leqq 0.75$ the spherocity of three-jet events can vary very little and is roughly equal to its average value $\langle S(T)\rangle$.

In order to use the variables T and T_2 one needs a way to determine the axis of the second most energetic jet. For this we can use the triplicity method [25]. The final state hadrons with the momenta $\vec{p}_1, \vec{p}_2, \ldots, \vec{p}_N$ are grouped into 3 nonempty classes C_1, C_2, C_3 with the total momenta

$$\vec{P}(C_\ell) = \sum_{i\in C_\ell} \vec{p}_i \qquad \ell = 1,2,3. \qquad (4.22)$$

Triplicity is defined by:

$$T_3 = \left(1\Big/\sum_{i=1}^{N} |\vec{p}_i|\right) \max_{C_1,C_2,C_3} \{|\vec{P}(C_1)|+|\vec{P}(C_2)|+|\vec{P}(C_3)|\} \qquad (4.23)$$

T_3 varies between $T_3 = 1$ for a perfect 3-jet and $T_3 = {}^{3\sqrt{3}}/8 = 0.65$ for a perfectly spherical event. Those classes C^*_ℓ of particles yielding the maximum T_3 are identified with the hadron jets originating from $q,\bar{q}$ and g. So the jet momenta are $\vec{P}(C_\ell{}^*)$. We denote them by $\vec{P}_1$, $\vec{P}_2$, $\vec{P}_3$ with the convention $P_1 \geqq P_2 \geqq P_3$. Because of momentum conservation these vectors are coplanar and span the triplicity plane. T is the thrust with respect to the direction $\vec{P}_1$ and T_2 with respect to $\vec{P}_2$. The angles between the jets (Fig.33) are interpreted as the angles between the quanta $q,\bar{q}$ and g. Because of $\theta_1 + \theta_2 + \theta_3 = 2\pi$ one can span a triangular Dalitz plot of which only $^1/6$ is populated since $\theta_1 \leqq \theta_2 \leqq \theta_3$ (Fig. 33). The 3 corners A, B, C correspond to the totally symmetric 3-jet case (A: $\theta_1 = \theta_2 = \theta_3 = 120^\circ$) and two 2-jet like configurations (B: $\theta_1 = 0^\circ$, $\theta_2 = \theta_3 = 180^\circ$; C: $\theta_3 = 180^\circ$, $\theta_1 = \theta_2 = 90^\circ$, i.e. $P_3 = 0$). The relation between the normalized jet momenta x_i and the angles θ_i is:

$$x_\ell = \frac{2\sin\Theta_\ell}{\sin\Theta_1 + \sin\Theta_2 + \sin\Theta_3} \qquad \ell = 1,2,3. \qquad (4.24)$$

The range of the θ_i is: $0^\circ \leqq \theta_1 \leqq 120^\circ$, $90^\circ \leqq \theta_2 \leqq 180^\circ$ and $120^\circ \leqq \theta_3 \leqq 180^\circ$. In Fig. 34 we show a scatter diagram θ_3 versus θ_1 for all events measured recently in the high energy region between 27.4 and 31.6 GeV by the PLUTO group at PETRA [8]. In this plot 3-jet events must lie near the corner where $\theta_1 = \theta_2 = 120^\circ$. In Fig. 35 we see a scatter diagram for triplicity T_3 versus thrust T

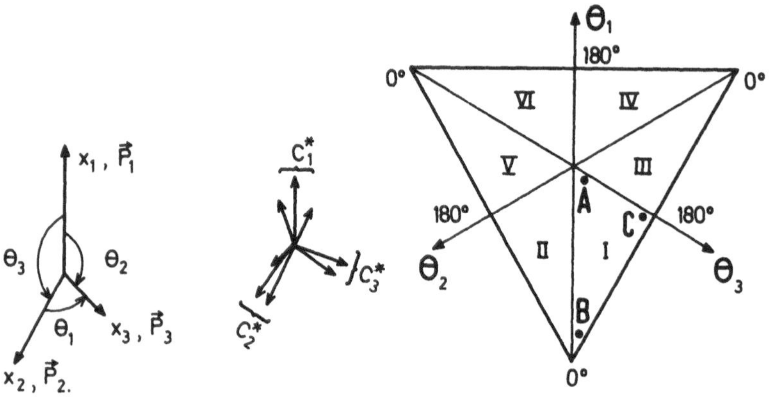

Figure 33 Momentum configurations of hadrons (a) and jets (b) obtained by grouping hadrons into 3 classes according to the triplicity method. Dalitz plot for the angles θ_i between the jet momenta (c). Only area I is populated.

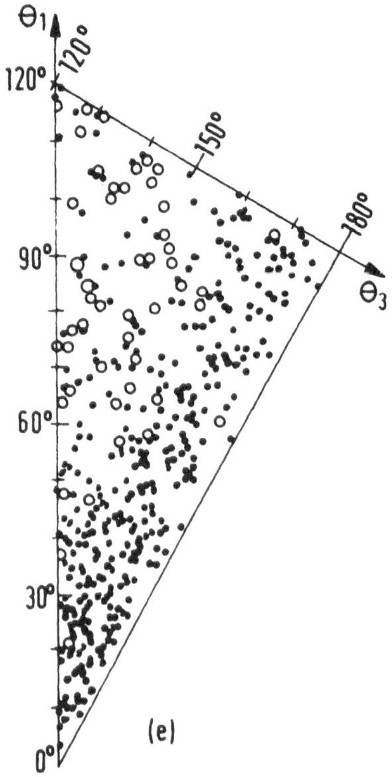

Figure 34 The data of the PLUTO group at $E_{cm.}$= 27.6,30 and 31.6 GeV are shown in the angular Dalitz plot.

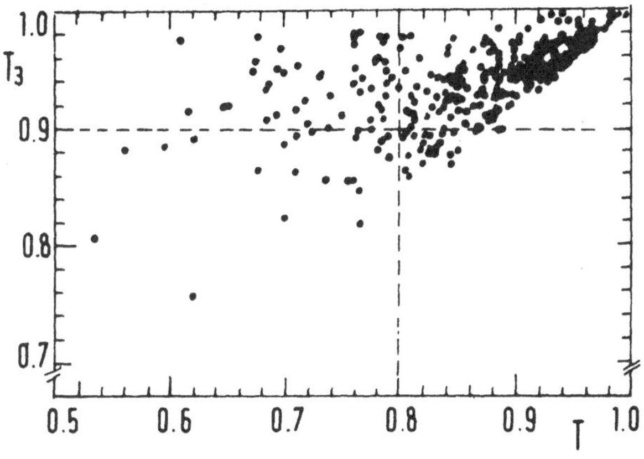

Figure 35 The same data as in Fig. 34 are shown in a scatter diagram of triplicity versus thrust.

for all events in the same energy region [8]. In this plot 3-jet events are concentrated in a band of large triplicity, say $T_3>0.9$. 2-jet events have large thrust and are located at the upper right corner of the diagram. Spherical events have low thrust and triplicity. The 3-jet events defined by $T_3 > 0.9$, $T < 0.8$ appear as large circles, the others as small dots in Fig. 34. Thus in order to isolate and to analyse 3-jet events one needs three variables. First triplicity T_3 or acoplanarity A (see section 2) to select the planar events. The other two variables like θ_3 and θ_1, T and T_2 or T and S are used to separate 3- and 2-jet events.

The most probable shape of the 3-jet events as a function of T can be computed from (4.9). For example, the expectation value of $\cos\theta_3$ is [36]:

$$\langle\cos\theta_3\rangle = 1 - A \qquad (4.25)$$

where

$$A = 2\{2(2-3T+3T^2)\ln\frac{T^2}{(2-T)(1-T)} + 6(2-T)(1-T)\ln\frac{2T}{2-T}$$

$$- 3(2-T)(3T-2)\}\{2(2-3T+3T^2)\ln\frac{2T-1}{1-T} - 3T(2-T)(3T-2)\}^{-1}$$

As a function T the angle θ_3 varies only in a rather narrow strip which is given by

$$-\frac{4T-T^2-2}{T^2} \leq \cos\theta_3 \leq -\frac{T}{2-T} \qquad (4.26)$$

For some characteristic T values, we have

T	θ_3	$\langle\theta_3\rangle$
$\frac{2}{3}$	$\theta_3 = 120^\circ$	120°
0.7	$123^\circ \leqq \theta_3 \leqq 129^\circ$	126°
0.8	$132^\circ \leqq \theta_3 \leqq 151^\circ$	141°
0.9	$145^\circ \leqq \theta_3 \leqq 167^\circ$	157°
1.0	$\theta_3 = 180^\circ$	180°

So in the interesting region $2/3 \leqq T \leqq 0.8$ the range of θ_3 is very limited to the neighbourhood of $\langle\theta_3\rangle$. In Fig. 36 we show for three characteristic thrusts T = 0.7, 0.8 and 0.9 the most likely shape of the 3-jet events if they are coming from quark, antiquark and gluon. We see that with increasing T the most probable shape is

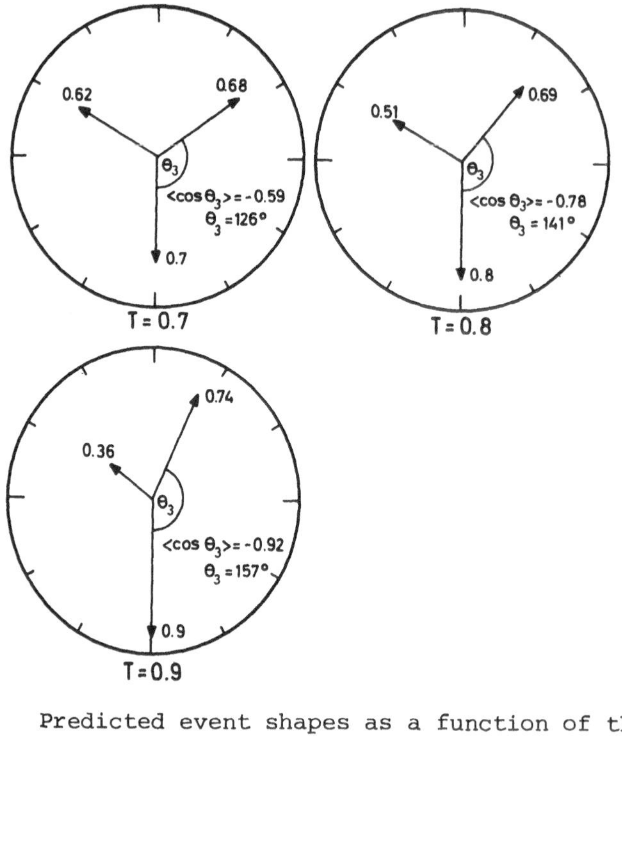

Figure 36 Predicted event shapes as a function of thrust.

the one where $x_3 \to o$, i.e. near the corner C ($\theta_3 = 180^o, \theta_1=\theta_2=90^o$). This configuration is favoured because it is the most singular configuration in (4.9), the infrared singular region x_1, $x_2 \to 1$.

So far we considered only tests of single gluon emission which were based on the cross section formula (4.9). However the matrix element for $e^+e^- \to q\bar{q}g$ has much more structure than is used in (4.9) because the quanta in the initial and final states have spin. The spins of the final state particles are not observed. Therefore the spins of quarks and gluons are summed. But the virtual photon with mass $\sqrt{q^2}$ is always a mixture of components with transverse and longitudinal polarization. As is well known from $e^+e^- \to 3$ particles the cross section for transverse unpolarized (σ_u), longitudinal polarized (σ_L), transverse polarized (σ_T) virtual photons and a term originating from the interference of transverse and longitudinal photons (σ_I) can be disentangled by measuring the angular distribution of the production with respect to a plane defined by the beam direction and the z axis of the final hadron plane (see Fig. 37)[37]. The hadron plane is determined by two vectors. For them one takes the momenta $\vec{p}_1$ and $\vec{p}_2$ of the three final particles either as $\vec{p}_1 \parallel$ z axis, $\vec{p}_1 \times \vec{p}_2 \parallel$ y axis (helicity frame) or as $\vec{p}_1 \times \vec{p}_2 \parallel$ z axis, $\vec{p}_1 \parallel$ x axis (transversity frame). We prefer the helicity frame. But we must remember that the momenta of the final quanta q, $\bar{q}$ and g are not observed, only the hadron jets are measured. Therefore we fix the $O\vec{z}$ axis by the thrust axis. For the $O\vec{x}$ axis we consider two choices:

(A) We distinguish gluon jets from quark and antiquark jets, the latter not being differentiated. This should be possible since we expect the gluon jet to have, for example, a higher multiplicity. We then choose $-O\vec{x}$ to point into the direction of the gluon jet. In other words, $O\vec{x}$ defines the hemisphere in which to find the quark (region I and II in Fig. 29) and antiquark jet (region III and IV in Fig. 29) respectively.

(B) We choose $O\vec{x}$ to point into the hemisphere in which to find the second most energetic jet originating either from a quark, antiquark or gluon. Using the jet momenta $\vec{P}_1$, $\vec{P}_2$, $\vec{P}_3$ found by the triplicity T_3 above we have $\vec{P}_1 \times \vec{P}_2 \parallel O\vec{y}$ and, of course, $\vec{P}_1 \parallel O\vec{z}$.

The cross section for $e^+e^- \to q\bar{q}g$ depends on the Euler angles θ and χ which determine the orientation of the production plane (see Fig. 37) in the following form:

$$2\pi \frac{d^3\sigma}{d\cos\theta d\chi dT} = \frac{3}{8}(1+\cos^2\theta)\frac{d\sigma_u}{dT} + \frac{3}{4}\sin^2\theta\frac{d\sigma_L}{dT}$$

$$+ \frac{3}{4}\sin^2\theta\cos 2\chi\frac{d\sigma_T}{dT} + \frac{3}{2\sqrt{2}}\sin 2\theta\cos\chi\frac{d\sigma_I}{dT} \quad (4.27)$$

Clearly the two choices (A) and (B), for the definition of χ will only affect σ_I. For the partial cross sections the result is [38]:

$$\frac{1}{\sigma_o}\frac{d\sigma_u}{dT} = \frac{2\alpha_s}{3\pi}\left[\frac{2(3T^2-3T+2)}{T(1-T)}\ln\frac{2T-1}{1-T} - \frac{3(3T-2)(2-T)}{1-T} - \frac{2(8T-3T^2-4)}{T^2}\right]$$

$$\frac{1}{\sigma_o}\frac{d\sigma_L}{dT} = \frac{2\alpha_s}{3\pi}\frac{2(8T-3T^2-4)}{T^2}$$

$$\frac{1}{\sigma_o}\frac{d\sigma_T}{dT} = \frac{1}{2}\frac{1}{\sigma_o}\frac{d\sigma_L}{dT} \qquad (4.28)$$

Case (A):

$$\frac{1}{\sigma_o}\frac{d\sigma_I}{dT} = \frac{2\alpha_s}{3\pi}\frac{1}{2\sqrt{2}}\left[-\frac{2(2-T)(3T-2)\sqrt{2T-1}}{T^2} + \frac{2+T}{\sqrt{1-T}}\arcsin\left(\frac{3T-2}{T}\right)\right]$$

Case (B):

$$\frac{1}{\sigma_o}\frac{d\sigma_I}{dT} = \frac{2\alpha_s}{3\pi}\sqrt{2}\,\left(2-2T+T^2\right)\left[-\frac{2}{T^2}\sqrt{2T-1} + \frac{1}{T\sqrt{1-T}}\right]$$

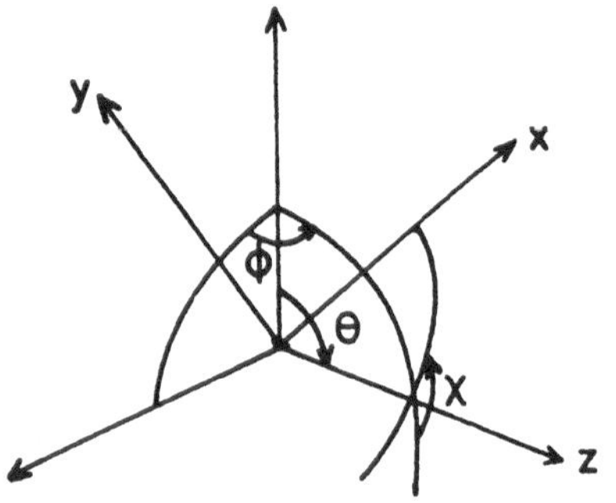

Figure 37 Definition of angles θ, χ and ϕ. The thrust axis is along $\vec{Oz}$ while the q, $\bar{q}$ and g momenta lie in the plane (x, z). The (y, z) plane divides the final state into two hemispheres. $\vec{Ox}$ defines the hemisphere in which to find the antiquark (quark) in case of the thrust axis being given by the quark (antiquark) momentum. If the gluon is most energetic $\vec{Ox}$ defines the hemisphere in which to find the quark. The angles θ, χ and ϕ vary between $0 \leqq \theta \leqq \pi$, $0 \leqq \chi \leqq 2\pi$ and $0 \leqq \phi \leqq 2\pi$. When talking about the thrust distribution $\vec{Ox}$ will be defined according to A and B.

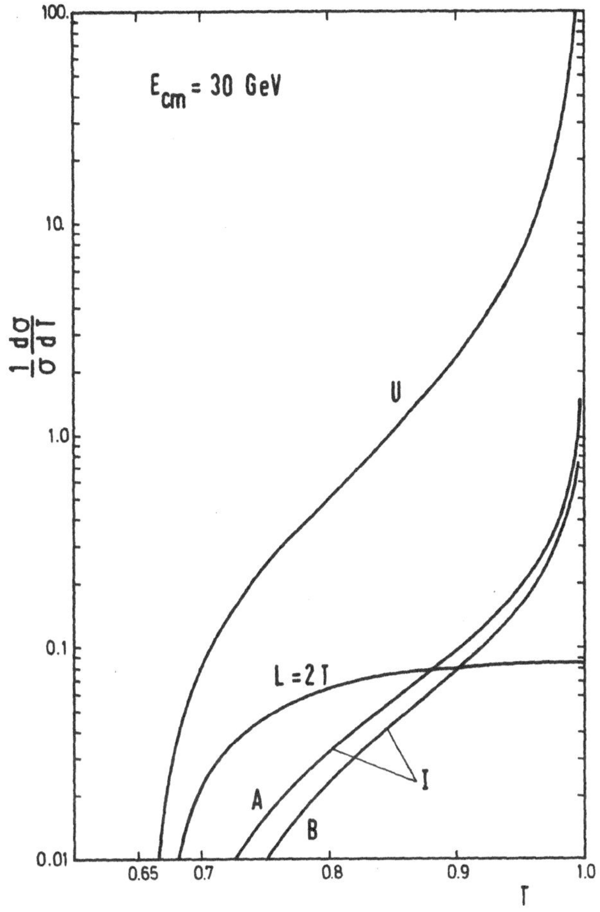

Figure 38 Partial cross sections U: $d\sigma_U/dT$, L: $d\sigma_L/dT$ and I: $d\sigma_I/dT$; $d\sigma_T/dT = 1/2\ d\sigma_L/dT$.

In Fig. 38 we have plotted the various partial cross sections (4.28) for 30 GeV. For higher energies the cross sections decrease with $\alpha_s(q^2)$. $d\sigma_L/dT$ is finite for $T\to 1$ and the singularity for $T\to 1$ in $d\sigma_I/dT$ is integrable. So the infrared singularity is only in $d\sigma_U/dT$ as is to be expected. The other cross sections $d\sigma_L$, $d\sigma_T$ and $d\sigma_I$ are infrared finite.

The measurement of the angular distribution (4.27) is another independent way to verify the $q\bar{q}g$ contribution or if this is established to investigate details of the perturbative matrix element. Of course the effects are largest for small T near the lower boundary. For some characteristic T values the complete angular distribution is:

$$\begin{array}{ll} T & \\ 0.7: f(\theta,\chi) & = 1+0.29\cos^2\theta+0.18\sin^2\theta\cos 2\chi+0.12\,(0.01)\sin\theta\cos\chi \\ 0.8: f(\theta,\chi) & = 1+0.60\cos^2\theta+0.10\sin^2\theta\cos 2\chi+0.14\,(0.10)\sin\theta\cos\chi \\ 0.9: f(\theta,\chi) & = 1+0.87\cos^2\theta+0.03\sin^2\theta\cos 2\chi+0.11\,(0.09)\sin\theta\cos\chi \end{array} \tag{4.29}$$

Near $T = 1$ the angular distribution must approach the characteristic distribution $1 + \cos^2\theta$ of the 2-jet production. The coefficients of $\sin\theta\cos\chi$ are for case A (case B).

The measurement of the angular correlation offers the possibility to test the spin of the gluon. Of course a theory with scalar gluons does not have the appeal which QCD has because of the gauge principle and because of asymptotic freedom. If, however, scalar gluons would exist, they have the following partial cross sections:

$$\frac{1}{\sigma_o}\frac{d\sigma_u}{dT} = \frac{\alpha^*}{3\pi}\left[2\ln\frac{2T-1}{1-T} + \frac{(4-3T)(3T-2)}{1-T} - \frac{2(3T-2)}{T}\right]$$

$$\frac{1}{\sigma_o}\frac{d\sigma_L}{dT} = \frac{\alpha^*}{3\pi}\frac{2(3T-2)}{T}$$

$$\frac{1}{\sigma_o}\frac{d\sigma_T}{dT} = \frac{1}{2}\frac{1}{\sigma_o}\frac{d\sigma_L}{dT} \tag{4.30}$$

Case (A):

$$\frac{1}{\sigma_o}\frac{d\sigma_I}{dT} = \frac{\alpha^*}{3\pi}\frac{2-3T}{2\sqrt{2}}\left[\frac{2(2-T)\sqrt{2T-1}}{T^2} + \frac{1}{\sqrt{1-T}}\,\mathrm{arc\,sin}\,\frac{3T-2}{T}\right]$$

Case (B):

$$\frac{1}{\sigma_o}\frac{d\sigma_I}{dT} = \frac{\alpha^*}{3\pi}\sqrt{2}\left[\sqrt{1-T} - \frac{2(1-T)\sqrt{2T-1}}{T}\right]$$

In (4.30) $g^* = (4\pi\alpha^*)^{\frac{1}{2}}$ is the coupling constant for scalar gluons and quarks. Of course, the application of (4.30) makes sense only if α^* is small enough so that perturbation theory is justified.

For scalar gluons the complete angular distribution has the following form (the numbers in parenthesis are for case (B)):

$$\begin{array}{ll} T & \\ 0.7: f(\theta,\chi) & = 1+0.24\cos^2\theta+0.19\sin^2\theta\cos 2\chi-0.35\,(+0.02)\sin 2\theta\cos\chi \\ 0.8: f(\theta,\chi) & = 1+0.37\cos^2\theta+0.16\sin^2\theta\cos 2\chi-0.32\,(+0.05)\sin 2\theta\cos\chi \\ 0.9: f(\theta,\chi) & = 1+0.58\cos^2\theta+0.11\sin^2\theta\cos 2\chi-0.26\,(\;0.03)\sin 2\theta\cos\chi \end{array} \tag{4.31}$$

Comparing (4.31) with (4.29) we notice that in the scalar gluon case the θ distribution does not approach $\sim 1 + \cos^2\theta$ with increasing T as fast as in the vector gluon case. In case A the scalar and vector gluon theory differ in the sign of the $\sin 2\theta \cos\chi$ term. In the calculation of the angular distribution we noticed that for scalar gluons the regions V and VI in Fig. 29, where the gluon determines the thrust axis, do not contribute to σ_L, σ_T and σ_I.

The results reported so far are all for massless quarks. Away from the threshold our results are also valid for $c\bar{c}g$ and $b\bar{b}g$ production.

From the results in table 1 we have concluded that up to $T = 0.9$ the integrated $q\bar{q}g$ cross section is still small compared to σ_0 so that perturbation theory can be trusted up to this T value. To go to even higher T's at 30 GeV also does not make sense because the nonperturbative contribution is definitely dominant for $T \gtrsim 0.9$ as is seen in Fig. 30. Under the condition that further quark thresholds do not come into play this situation will change for higher energies. For example, for $E_{cm} = 90$ GeV we expect $\langle 1-T \rangle_{nonpert.} \simeq 0.02$ (see $\langle 1-T \rangle$ up to 70 GeV in Fig. 31) so that one would look for a perturbative cross section up to $T \simeq 0.98$. It is clear that for these large T's the $q\bar{q}g$ prediction cannot be used any more and perturbative higher order terms must be included. This is possible only in an approximate way and has been done in the leading logarithm approximation. In this approximation one can sum all virtual and real gluon contributions up to limits ε and δ as defined in chapter 2. The result for the 2-jet cross section is Eq. (2.18). If this method is also applied to the 3-jet cross section and if the variables δ and ε are translated into the thrust variable T we obtain approximately

$$\frac{1}{\sigma_0} \frac{d\sigma}{dT} = \frac{8\alpha_s}{3\pi} \frac{1}{1-T} \ln \frac{1}{1-T} e^{-\frac{8\alpha_s}{3\pi} \ln^2(1-T)} \tag{4.32}$$

Except for the exponential factor it agrees with (4.15). The exponential factor damps the first order cross section so that for $T \to 1$ the cross section vanishes. In Fig. 30 we indicated how the exact first order cross section might go over into the formula (4.32). It is conceivable that (4.32) is a useful approximation in the region $0.9 < T < 0.98$ at the very high energies. But the exact derivation of these higher order terms needs further study.

5. THEORY OF FOUR-JET PRODUCTION. [5]

If we go beyond the order α_s in the gauge coupling constant more than one gluon appear in the final state. These higher order terms are very important since up to order α_s there is no difference between an abelian and a nonabelian gauge theory. All predictions about $e^+e^- \to q\bar{q}g$ are equally valid for massless QED with an abelian gauge coupling (except the colour factor 3 in the formula for $e^+e^- \to q\bar{q}$). QCD shows its full gauge structure only in second or higher order perturbation theory (order $\geq \alpha_s^2$), where the triple gluon coupling comes in. In second order we will be led to four-jet final states: $e^+e^- \to q\bar{q}gg$ and $e^+e^- \to q\bar{q}q\bar{q}$. For these final states one can calculate various differential cross sections in terms of these jet measures as thrust, sphericity, acoplanarity etc. as was done for the $q\bar{q}g$ final state. For the four-jet calculation acoplanarity is a very useful variable since four-jet events stand out against two- and three-jet final states by having a nonvanishing acoplanarity. Thus $d\sigma/dA$ is the canonical quantity to analyse as it allows one to cut off the dominant 2- and 3-jet events experimentally. Of course to consider events with nonvanishing A does not eliminate the 2- and 3-jet background completely since these events have some finite A through the nonperturbative jet spread near $A \simeq 0$. With an appropriate cut on A one can reduce this nonperturbative background contribution and thereby enhance the true QCD perturbative contribution. This nonperturbative background includes also the broadening effect in A originating from the weak decay of heavy quarks considered in section 3.

To order α_s^2 the four-jet cross section is given by the two sets of diagrams shown in Fig. 39 which correspond to the final states

$$e^+e^- \to q(p_1) + \bar{q}(p_2) + g(p_3) + g(p_4) \tag{5.1}$$

and

$$e^+e^- \to q(p_1) + \bar{q}(p_2) + q(p_3)+\bar{q}(p_4) \tag{5.2}$$

The differential cross section

$$d\sigma = \frac{e^4}{(2\pi)^8 2q^6 N_s} \{p_+, p_-\}^{\mu\nu} \prod_{i=1}^{4} \frac{d^3p_i}{2p_{io}} \delta^{(4)}(p_+ + p_- - \sum_{k=1}^{4} p_k) H_{\mu\nu} \tag{5.3}$$

where $q = p_+ + p_-$ and

$$\{p_+, p_-\}^{\mu\nu} = p_+^{\mu} p_-^{\nu} + p_-^{\mu} p_+^{\nu} - g^{\mu\nu} \frac{q^2}{2} \tag{5.4}$$

is the lepton tensor for unpolarized beams. The hadron tensor $H_{\mu\nu}$ contains summations over the final spin, colour and flavour states including the appropriate quark charge factors $^1/_9$ or $^4/_9$. N_s is a statistical factor due to the identity of final state particles. The formulas, for $H_{\mu\nu}$, for example, for $e^+e^- \to q\bar{q}gg$ have been writ-

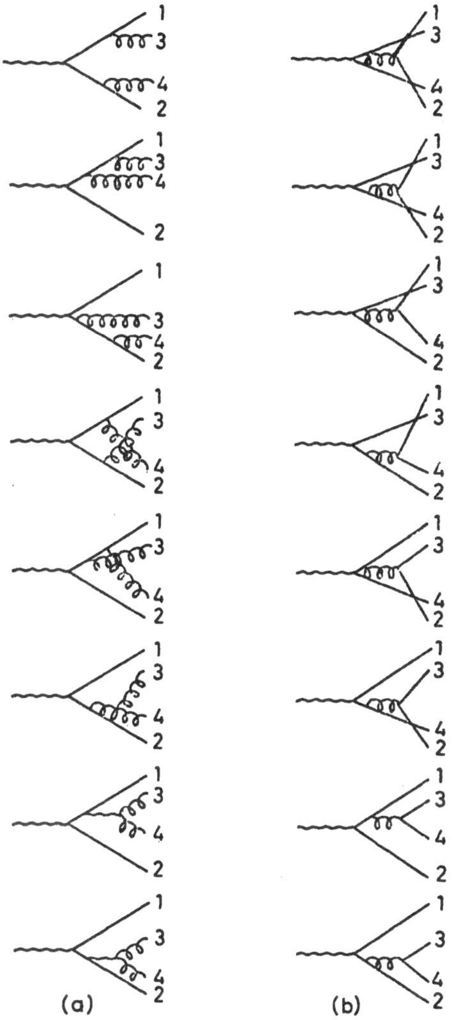

Figure 39 Tree diagrams for four-jet production (a) $q\bar{q}gg$ and (b) $q\bar{q}q\bar{q}$.

ten down in ref. 5. They have been derived, as well as those for $e^+e^- \rightarrow q\bar{q}q\bar{q}$ with the help of the computer program REDUCE.

The differential cross section (5.3) depends on five independent "hadronic" variables describing the 4-jet final state which one could choose as x_1, x_2, x_3, x_{12} and x_{13} where

$$x_i = 2|\vec{p}_i| / \sqrt{q^2} \quad , \quad x_{ij} = 2|\vec{p}_i + \vec{p}_j| / \sqrt{q^2} \tag{5.5}$$

and two angles θ and χ which describe the orientation of the 4-jet event relative to the leptonic beam direction. The integration over the angle variables and 4 of the 5 hadronic variables is done with a Monte Carlo routine. For the remaining variable the acoplanarity, see (2.22).

Of course, instead of A, other variables can be chosen, like thrust T or sphericity $\hat{S}$. However, differential T and $\hat{S}$ distributions are not singularity free at the tree graph level since the remaining phase space include singular configurations of quarks and gluons. Therefore, from this point of view, it is natural to study the dependence of the cross section on A first. We show the A distribution $d\sigma/dA$ normalized to the zeroth order cross section

$$\sigma_0 = 4\pi\alpha^2 \sum_a e_a^2/q^2$$

for $E_{cm} = \sqrt{q^2} = 40$ GeV in Fig. 40 for $e^+e^- \to q\bar{q}gg$ and $e^+e^- \to q\bar{q}q\bar{q}$ separately.

The differential cross section $d\sigma/dA$ diverges for $A \to 0$. For $e^+e^- \to q\bar{q}gg$ the leading log behavior is

$$\frac{1}{\sigma_0}\frac{d\sigma}{dA} = \frac{8}{9}\left(\frac{\alpha_s}{\pi}\right)^2 \frac{1}{A} \left|\ln A\right|^3 \tag{5.6}$$

as $A \to o$. The leading log formula can be seen to give a good description of the differential A-distribution up to rather large A values. For four massless final state particles A is bounded bet-

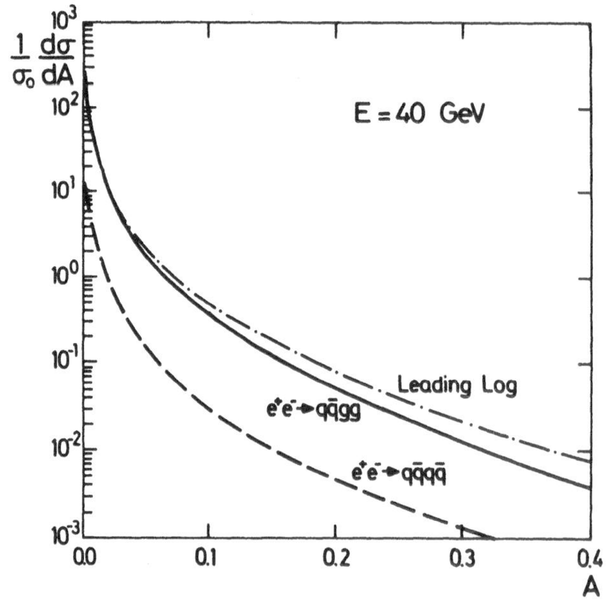

Figure 40 $d\sigma/dA$ for $e^+e^- \to q\bar{q}gg$ (full curve), $e^+e^- \to q\bar{q}q\bar{q}$ (dashed curve) and $e^+e^- \to q\bar{q}gg$ in the leading log approximation (dashed-dotted curve).

ween O and $^2/3$. The maximal A value occurs for the configuration where the four momenta point from the center of a tretrahedron (with side length $\sqrt{q^2}/\sqrt{6}$) to its four corners which gives $A = {}^2/3$ according to Eq. (2.22). For $e^+e^- \rightarrow q\bar{q}q\bar{q}$ the leading behavior is $A^{-1}(\ln A)^2$. Latter cross section can be seen to be ten times smaller than $e^+e^- \rightarrow q\bar{q}gg$ over most of the A region considered.

Because of the singularity for $A \rightarrow 0$ (see (5.6)) the differential cross section is not integrable over A. This is to be expected because of the infrared singularities associated with collinear and soft emission of quarks, antiquarks and gluons in (5.1) and (5.2). Of course, when calculating the total cross section to order α_s^2, these singularities cancel against the corresponding singularities of the one- and two-loop virtual contributions. Due to the singular behavior of $^{d\sigma}/dA$ as $A \rightarrow 0$ the differential distribution should be considered reliable only for values of A above some cut-off A_C. Integrating $^{d\sigma}/dA$ from $^2/3$ to A_C one obtains a cut-off dependent 4-jet cross section $\sigma(A_C)$ which is presented in Fig. 41. A cut-off value for which

$$\sigma(A_c)/_{\sigma_0} \simeq \alpha_s^2 \simeq 0.04$$

should be considered a reasonable choice above which a perturbatively calculated $^{d\sigma}/dA$ can be trusted. According to Fig. 41 this corresponds to $A_C = 0.07$. We remark that this reasoning which led us to the cut-off value $A_C = 0.07$ is completely analogous to our considerations in the $q\bar{q}g$ case which gave us $T_C \simeq 0.9$.

One may ask how large the influence of the three-gluon coupling is. In some sense this is an ill-defined question since on the one hand the relative contribution of the three-gluon coupling depends on the gauge choice and on the other hand theories with only global SU(3) symmetry are not renormalizable (although for the tree diagrams this is not relevant). Therefore we compare to an abelian theory (i.e. QED with massless and colourless quarks) which is gauge invariant in itself. For such a theory we get for the $q\bar{q}gg$ cross section

$$\sigma(q\bar{q}gg,QED) \simeq 0.15\ \sigma(q\bar{q}gg,QCD) \tag{5.7}$$

From (5.7) we would estimate the effect of coloured gluon in the four-jet cross section to be 85 %. Of course, if we take into account that $\sigma_O(QED) = {}^1/3\sigma_O(QCD)$ the effect of the three-gluon coupling in the relative four-jet rate is only 55 %. From this point of view the cross section for four jets constitutes an important check on the nonabelian nature of QCD.

Finally we present the dependence of the 4-jet cross section on some other variables. As we mentioned already the four-jet cross section depends on five variables so that multidifferential

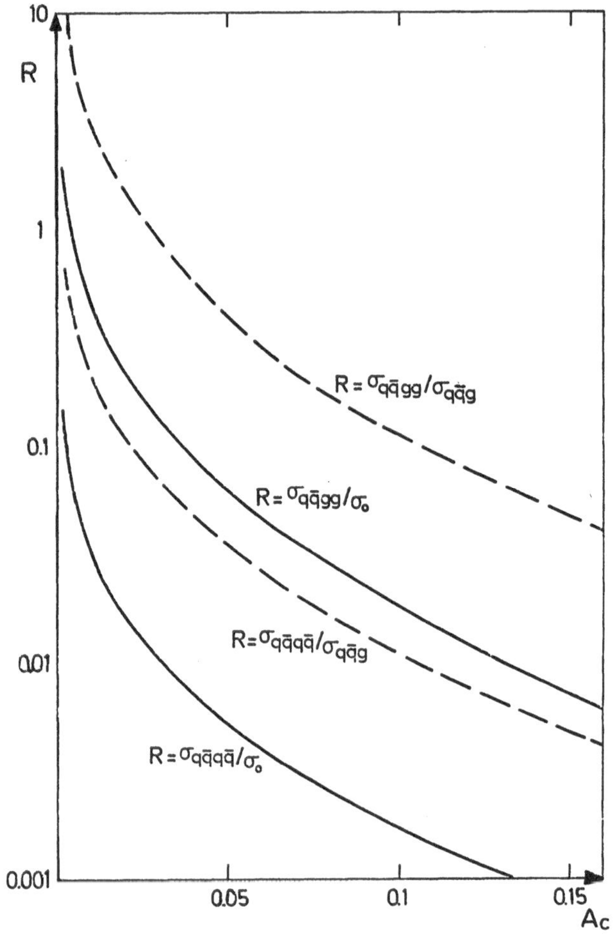

Figure 41 Integrated four-jet cross section as a function of the acoplanarity cut-off A_c for $e^+e^- \to q\bar{q}gg$ and $e^+e^- \to q\bar{q}q\bar{q}$ separately. Ratio of σ(four-jet) to σ(three-jet) as a function of A_c. The $q\bar{q}g$ cross section is cut off at $T_c = 0.9$.

cross sections as for example $d^2\sigma/dAdT$ or $d^2\sigma/dAd\hat{S}$ could be measured in principle. As this requires high statistics data which will not be available for some time to come we have explicitly calculated these two dimensional distributions but have integrated them over A. As discussed earlier, the singular region $A \to o$ has to be excluded in this integration. In Figs 42 and 43 we present differential thrust and sphericity distribution for an acoplanarity cut $A_c = 0.05$. The curve are normalized in such a way, that the area under the curve gives the integrated cross sections in Fig. 41. Comparing with $\frac{1}{\sigma}\frac{d\sigma}{dT}$ for $e^+e^- \to q\bar{q}g$ as shown in Fig. 30

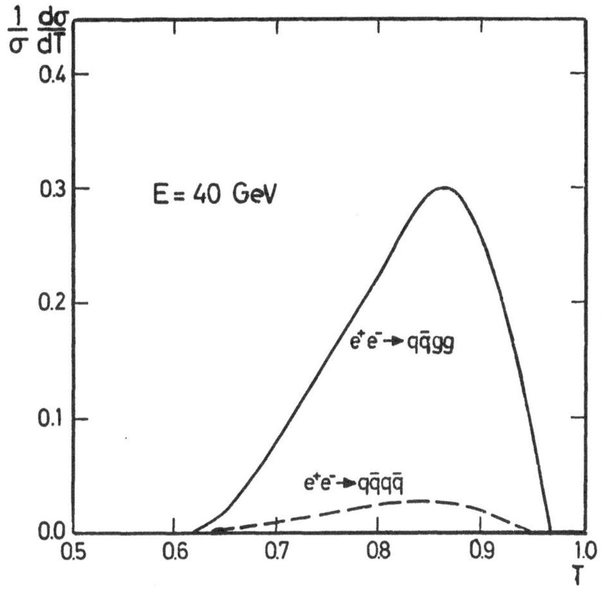

Figure 42 Differential thrust-distribution with acoplanarity cut A_C = 0.05. Normalization described in text.

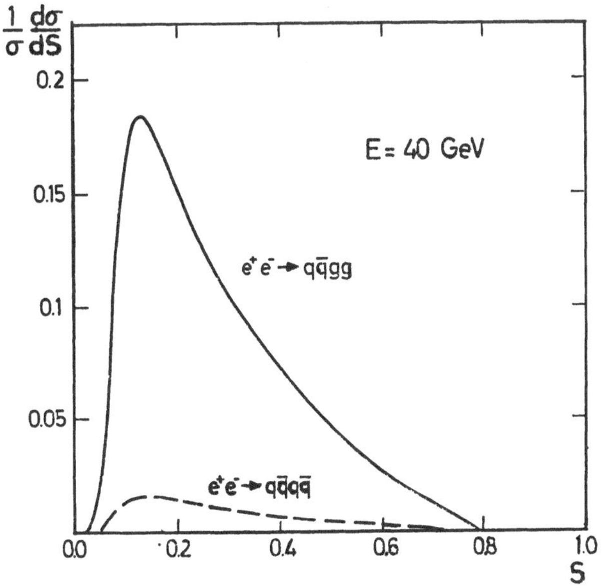

Figure 43 Differential sphericity-distribution with acoplanarity cut A_C = 0.05. Normalization described in text.

we see that $e^+e^- \to$ 4 jets produces a tail towards smaller thrust values. Of course a similar tail appears in the sphericity distribution.

It is clear that the tests of four-jet behaviour in QCD discussed in this report are rather limited. They presumably test very little of the vector character of the gluons and finer details of the three-gluon coupling. For this purpose angular correlations and asymmetries of the final jets would be much more suitable. Such tests have been discussed in ref 5.

As the last point we discuss the magnitude of $d\sigma/dA$ for four jets in relation to the background coming from nonperturbative 2-jet production and $e^+e^- \to q\bar{q}g$ with fragmentation of quarks and gluon included. In Fig. 44 we show $d\sigma/dA$ for nonperturbative 2-jet production as calculated in a Feynman-Field model including weak interaction effects from c and b decay as described in section 3. This distribution is still rather broad at $\sqrt{q^2}$ = 40 GeV. The average A for this distribution is $\langle A \rangle_{nonpert.} = 0.05$. The decrease of the nonperturbative $\langle A \rangle$ with increasing energy can be looked at in Fig. 31. It decreases rather slowly with E_{cm}. For comparison the perturbative $\langle A \rangle$ for $e^+e^- \to (q\bar{q}gg + q\bar{q}q\bar{q})$ is also shown there. Near E_{cm} = 80 GeV they are of comparable magnitude. Therefore at E_{cm} = 40 GeV we can expect that $d\sigma/dA$ for four jets is larger than $d\sigma/dA$ for two jets only for rather large A. As seen from Fig. 44 this is the case for $A > 0.2$. But then $d\sigma/dA$ has decreased already by a factor 100 compared to its maximum value. The contribution of the three-jet state to $d\sigma/dA$ is also shown in Fig. 44. This distribution is even broader than the 2-jet contribution. But it is reduced in magnitude. It is calculated with a thrust cut-off $T_c = 0.9$, as explained in section 4, by demanding $\sigma_{q\bar{q}g}(T_c)/\sigma_0 \simeq \alpha_s$. In order to see how large the influence of the fragmentation of quarks and gluons into hadrons is on the four-jet distribution we have included it in Fig. 44 also. The fragmentation procedure is applied only to the four-jet terms with $A \geq 0.05$, the small A region is cut off. Fig. 44 shows the change in $d\sigma/dA$ if we compare with $d\sigma/dA$ in Fig.40. The fragmentation has the effect that the small A region is reduced and the larger A region is enhanced. Near $A \simeq 0.2$ the distribution $d\sigma/dA$ increases by a factor four compared to $d\sigma/dA$ from 2 jets when we add the contributions of 2 jets, 3 jets and 4 jets. The normalization of $d\sigma/dA$ is such that the integral is one if the three components are added up. The relative contributions of 2, 3 and 4 jets are 83 %, 13 % and 4 %.

For $A > 0.25$ the perturbative 3- and 4-jet cross section dominates over the 2-jet cross section. Therefore this would be the region where the perturbative QCD predictions can be tested. Of course in order to isolate the four-jet contribution one has to go to even higher A, of the order $A \simeq 0.35$. Unfortunately for this background free region $d\sigma/dA$ for four jets is already rather small.

Presumably one has to go to much higher energies in order to see genuine 4 jet effects.

If it should turn out that one is above $t\bar{t}$ threshold at 40 GeV the interpretation of large A events becomes more difficult due to the large-acoplanarity t-background. One then has to go to higher energies where the t-originated acoplanarity distribution shrinks or try to remove events with t-signature from the data sample.

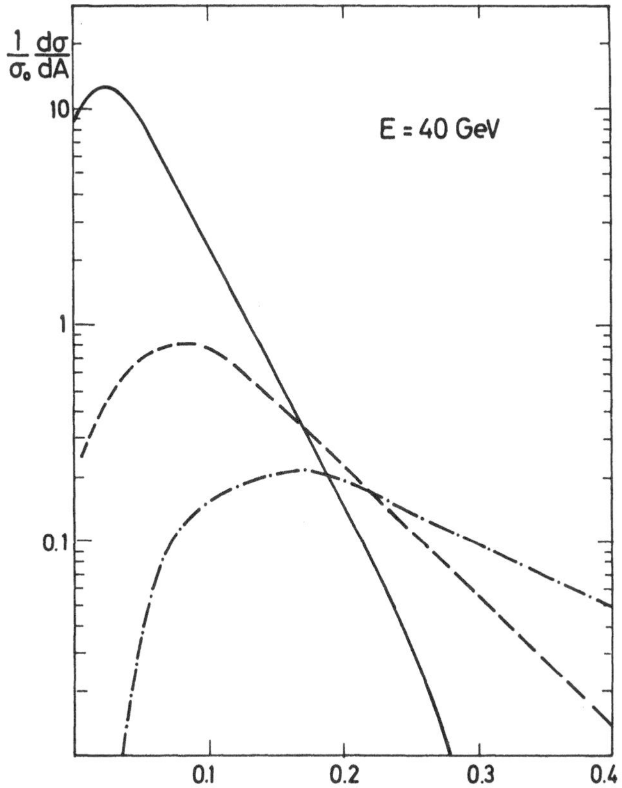

Figure 44 $d\sigma/dA$ for $e^+e^- \rightarrow q\bar{q}$ (full curve), for $e^+e^- \rightarrow q\bar{q}g$ (dashed curve) and for $e^+e^- \rightarrow q\bar{q}gg + q\bar{q}q\bar{q}$ (dashed-dotted curve) supplemented with Feynman-Field fragmentation of quarks and gluons.

6. ANALYSIS OF $\Upsilon(9.46) \to$ 3 GLUONS. [6,23,39]

It is well known that gluon jets can occur in the decay of resonances which are bound states of new quark - antiquark pairs, as for example $c\bar{c}$, $b\bar{b}$ or $t\bar{t}$. For $c\bar{c}$ the resonance mass is only of the order of 3 GeV so that the description of the final state in terms of jets would be out of place. In the $b\bar{b}$ case resonance masses are of the order of 10 GeV so that even for two gluon jets (in the case of C = 1 resonances $^1S_0$, $^3P_0$, $^3P_1$, $^3P_2$ etc.) and even more so for 3 gluon jets ($^3S_1 = \Upsilon, \Upsilon'$ etc.) it is difficult to see clearly separated jets. Of course for $t\bar{t}$ resonances if they exist, whose mass must be above 30 GeV three or two clear gluon jets, depending whether it is a C = 1 or C = -1 state which is studied, will be observed. Up to now only the decay of the $\Upsilon(9.46)$, the lowest lying $^3S_1$ $b\bar{b}$ resonance has been measured and analysed in great detail by the PLUTO collaboration at DESY. The measurement had been done at the DORIS storage ring which was pushed up to beam energies of about 5 GeV.

In this section we shall describe shortly the underlying theory for 3 gluon jets in the decay of $^3S_1$ states of bound heavy quark - antiquark pairs. After this we shall present a comparison of this theory with the PLUTO data based on the report given by Brandt at the Geneva conference [6].

The kinematics of the process $\Upsilon(9.46) \to 3g$ is the same as $e^+e^- \to q\bar{q}g$ explained in section 4. One uses

$$x_i = \frac{2E_i}{M_\Upsilon}, \quad i = 1,2,3$$

where E_i (i=1,2,3) are the energies of the three gluons. $x_1 + x_2 + x_3 = 2$. The x_i vary inside the boundary as shown in Fig. 29 again with 6 equivalent regions with the ordering $x_1 > x_2 > x_3$ etc. These sectors are completely identical, also dynamically, since we have three identical quanta in the final state. Thus it is sufficient to consider only sector I in Fig. 29. As for $q\bar{q}g$ one can define the angles θ_i between the jet momenta. The corresponding Dalitz plot is as in Fig. 33. In the sector I we have $\theta_1 > \theta_2 > \theta_3$.

The density $W(x_1,x_2,x_3)$ in the Dalitz triangle which gives the probability for 3g decay with normalized momenta x_1, x_2, x_3 is identical to the well-known formula for positronium decay in 3 photons (Ore-Powell formula) and has the following form [23,39]

$$W(x_1,x_2,x_3) = \frac{1}{x_1^2 x_2^2 x_3^2}\{x_1^2(1-x_1)^2 + x_2^2(1-x_2)^2 + x_3^2(1-x_3)^2\} \qquad (6.1)$$

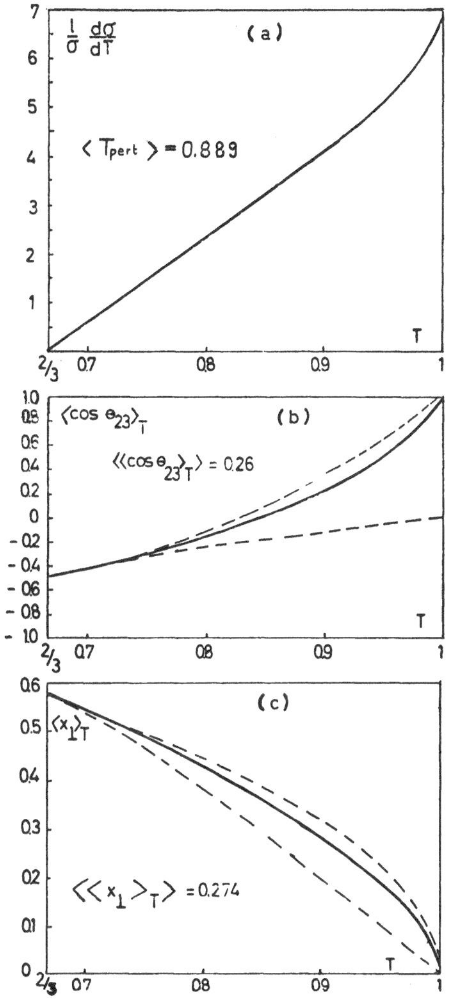

Figure 45 (a) The normalized distribution of 3g events versus thrust. (b) The mean $<\cos\theta_{23}>_T$ of the two less energetic gluons as a function of thrust. The dashed lines show the kinematic limits. (c) The average $<x_\perp>_T$ as a function of thrust. The dashed lines show the kinematic limit.

It is easy to calculate the thrust distribution from (6.1) by integrating over x_2 and $x_1 = T$ in sectors I and II. The result is [23,39]

$$\frac{1}{\sigma}\frac{d\sigma}{dT} = \frac{1}{\Gamma_{3g}}\frac{d\Gamma_{3g}}{dT} = \frac{3}{\pi^2-9}\int_{2(1-T)}^{T} dx_2\, W(T,x_2)$$

$$= \frac{6}{\pi^2-9}\left[\frac{(3T-2)(2-T^2)}{T^3(2-T)} + \frac{2(1-T)}{T^2(2-T)^3}(5T^2-12T+8)\frac{\ln 2(1-T)}{T}\right] \tag{6.2}$$

This distribution which is plotted in Fig. 45a is maximal at $T = 1$. It is not singular there [39]. Infrared singularities do not occur in $\Upsilon \to 3g$. The average of the T-distribution is $\langle T\rangle = 0.889$. As to be expected the "two-jet" configuration with $T = 1$ has the highest probability. The more interesting configuration with $T = 2/3$ with three jets of equal momenta and angles $\theta_i = 120^\circ$ occurs with vanishing probability.

Instead of T or spherocity one can use also the angles between the jets as kinematical variables. As an example we consider $\theta_{23} \equiv \theta_1$ the angle between the jets with momenta $\vec{P}_2$ and $\vec{P}_3$. The kinematically allowed range of $\cos\theta_{23}$ is shown in Fig. 45b together with the average calculated as a function of T from the formula [39]

$$\langle\cos\theta_{23}\rangle = 1 - 2\,\frac{\int_{2(1-T)}^{T} dx_2\, \frac{2(1-T)}{x_2(2-T-x_2)}\, W(T,x_2)}{\int_{2(1-T)}^{T} dx_2\, W(T,x_2)} \tag{6.3}$$

The average of $\langle\cos\theta_{23}\rangle$ in the interval $2/3 \leqq T \leqq 1$ is equal to 0.26 which corresponds to $\theta_{23} \equiv \theta_1 = 75^\circ$, roughly in the middle of the allowed θ_1 range: $0 \leqq \theta_1 \leqq 120^\circ$. A similar plot for the variable

$$x_T = x_2\sin\theta_j = \frac{2}{x_1}\left[(1-x_1)(1-x_2)(1-x_3)\right]^{\frac{1}{2}}$$

which is the normalized transverse momentum of jet $\vec{P}_2$ with respect to jet $\vec{P}_1$ $(2P_1/M = x_1 = T)$ is seen in Fig. 45c [39]. The average of this curve is $\langle\langle x_T\rangle\rangle = 0.274$ which corresponds to an average transverse momentum of $\langle p_T\rangle_{jet} = 1.30$ GeV. The interesting events are those with $T \geqq 0.85$. For them $\theta_1 \geqq 90^o$, $x_T \geqq 0.35$ and (p_T) $(p_T)_{jet} \geqq 1.75$ GeV. These are 30 % of all events.

Two jets nearby in angle like jet 2 and 3, if $x_1 > x_2 > x_3$, are not resolved any more if half the energy of one jet lies inside a cone of halfangle θ_{jet}. Therefore resolved jet require $\theta_1 >> 2\theta_{jet}$. θ_{jet} is the opening angle of the jet caused by the nonperturbative smearing in p_T. For the resonance it has the following value

$$\theta_{jet} \simeq \frac{\langle p_T \rangle}{\langle p \rangle} \simeq \frac{0.3}{0.9} = 0.3 = 17^o \qquad (6.4)$$

which is not small compared to $\frac{1}{2}\langle\theta_1\rangle = 37.5^o$. Therefore the average event of the decay will not show a resolved 3 jet structure. For larger resonance masses, like $(t\bar{t})_{^3S_1}$, which are above 30 GeV, this will be less of a problem since then

$$\theta_{jet} = \frac{\langle n \rangle \langle p_T \rangle}{M_{tt}} \leqq 0.01 \ \langle n \rangle \qquad (6.5)$$

where $\langle n \rangle$ is the average multiplicity.

The decay of the $\Upsilon(9.46)$ can also be used to test the spin of the gluon. The angular distribution of the thrust axis with respect to the beam direction is of the form

$$g(\theta) = 1 + \alpha(T) \cos^2\theta \qquad (6.6)$$

$\alpha(T)$ has a characteristic T dependence which is shown in Fig. 46 [39]. We have $\alpha(T=1) = 1$ as for $q\bar{q}$ 2jet events. The average is $\langle\alpha(T)\rangle = 0.39$. For scalar gluons one obtains $\langle\alpha(T)\rangle = -0.995$ [40] whereas the measured value is 0.83 ± 0.23 [31]. This is such a drastic difference to $\langle\alpha\rangle \simeq -1$ for scalar gluons that we are inclined to conclude that the measured angular distribution is incompatible with spin zero gluons. The deviation of $\alpha = 0.83 \pm 0.23$ from the theoretical value $\alpha = 0.39$ can be explained presumably by fragmentation corrections. We remark that the scalar gluon model leads to a divergent thrust distribution in the limit $T \to 1$ and a cut-off is needed for the total decay width.

We already mentioned that the decay products of the resonance of $\Upsilon(9.46)$ which according to our understanding does not decay into two jets (the resonance is below threshold for $b\bar{b}$) but at most in three jets, cannot show the distinct three-jet pattern. The angular width of the jets originating from the fragmentation of the gluons into hadrons is much too large. In this case we must expect large corrections due to the fragmentation process to the ideal theory which we have described so far. The decay of the $\Upsilon(9.46)$ into 3 gluons and the subsequent fragmentation of the gluons into actual hadrons has been calculated by the PLUTO group and compared with the measured data.

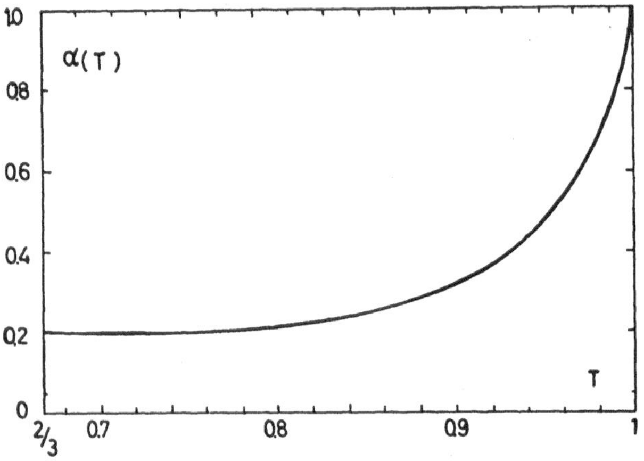

Figure 46 The angular distribution parameter α(T) as a function of thrust.

The input is the Ore-Powell formula (6.1) and conventional assumptions about the gluon fragmentation. The gluon fragmentation is calculated according to a Feynman-Field type model. The data are also compard to a pure phase space model (with no extra transverse momentum dependence). For these two models, the three-gluon model and the phase space model, the thrust T, the triplicity T_c distribution and the distributions of the reconstructed gluon energies x_1 and x_3 and reconstructed gluon angles θ_1 and θ_3 have been calculated. The results are shown in Fig. 47 [6]. The corresponding average values are collected in table 2 [6].

As was explained in section 4 the triplicity T_3 tests the planarity of the events. $T_3 = 1$ selects the real planar events. Because of the finite p_T smearing of the events coming from the fragmentation we do not expect real planar events although the original 3 gluon decay is certainly planar. This is seen in Fig. 47. The average T_3 of the data is $\langle T_3 \rangle = 0.86$. The phase space model has a triplicity distribution which is shifted more to smaller T_3. But the average T_3 is still large $\langle T_3 \rangle = 0.84$ and larger than $T_3 = 0.65$ expected for spherical events. The reason is the still rather small multiplicity at 9.46 GeV which prevents that real spherical events are the dominant ones. Furthermore the T_3 distribution of the 3 gluon model agrees much better with the experimental points than the T_3 distribution of the phase space model. This is also the case for the five other distributions in T, x_1, x_3, θ_1 and θ_3. In the thrust distribution we can study the influence of the fragmentation process. The T distribution for the pure 3 gluon decay shown in Fig. 45a has a maximum at $T = 1$. After fragmentation corrections (see Fig. 47) the maximum is at $T = 0.7$ and the region near $T = 1$ is completely depopulated. For this low jet energy, around 3 GeV, the fragmentation process is very essential and modifies the distributions of the original

Table 2

Observed mean values and corresponding predictions of 3-gluon and phase space models.

	T direct data	3-gluon MC	phase space MC
$\langle T \rangle$	0.715 ± 0.004	0.712 ± 0.003	0.671 ± 0.003
$\langle T_3 \rangle$	0.858 ± 0.002	0.850 ± 0.002	0.838 ± 0.002
$\langle x_1^J \rangle$	0.855 ± 0.004	0.853 ± 0.003	0.819 ± 0.002
$\langle x_2^J \rangle$	0.722 ± 0.004	0.724 ± 0.003	0.700 ± 0.002
$\langle x_3^J \rangle$	0.423 ± 0.006	0.422 ± 0.005	0.481 ± 0.004
$\langle \theta_1^J \rangle$	$84.1^\circ \pm 1.0^\circ$	$85.5^\circ \pm 0.8^\circ$	$93.2^\circ \pm 0.6^\circ$
$\langle \theta_2^J \rangle$	$125.6^\circ \pm 0.7^\circ$	$124.3^\circ \pm 0.5^\circ$	$122.9^\circ \pm 0.4^\circ$
$\langle \theta_3^J \rangle$	$150.3^\circ \pm 0.6^\circ$	$150.2^\circ \pm 0.5^\circ$	$144.0^\circ \pm 0.4^\circ$

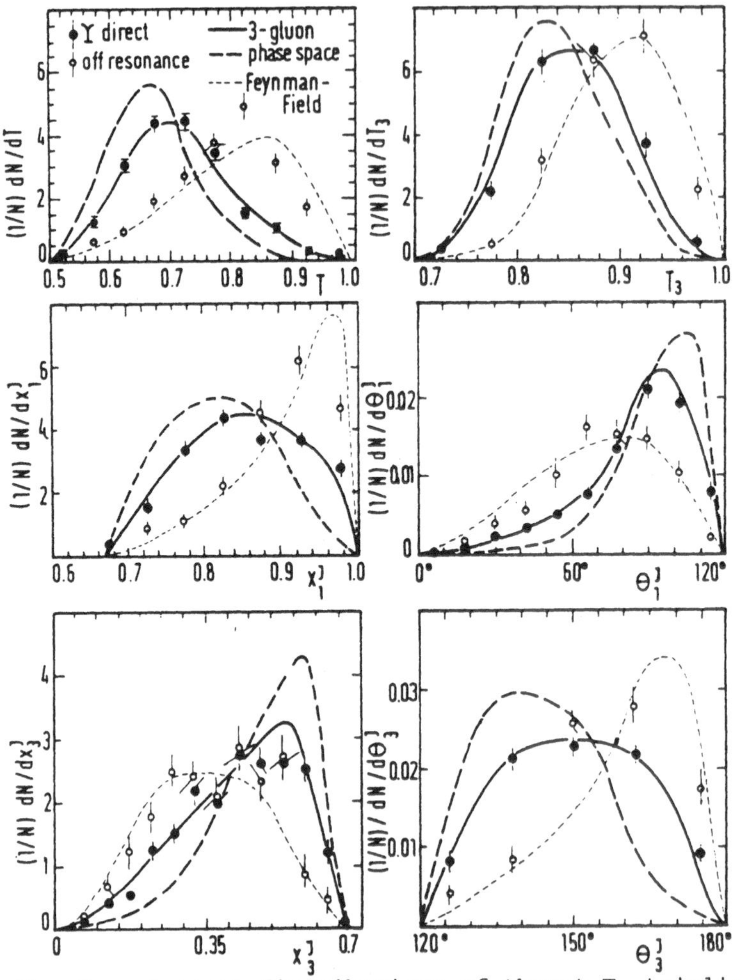

Figure 47 Experimental distributions of thrust T, triplicity T_3, reconstructed gluon energies x_1^J, x_3^J and reconstructed angles θ_1^J , θ_3^J between gluons compared to Monte-Carlo calculations based on various models.

model drastically. In the same figure also the corresponding distribution off resonance at 9.4 GeV are plotted. Here we expect mostly two jet distributions for $u\bar{u}$, $d\bar{d}$, $s\bar{s}$ and $c\bar{c}$ production which should be peaked near $T = T_3 = x_1 \simeq 1$, $x_3 \simeq 0$, and $\theta_3 \simeq 180^0$. Although this is the case the shift of the distributions if compared to the ideal two jet distributions is seen quite clearly. Of course these data at resonance cannot be explained with the 2 jet Feynman-Field model. But this is to be expected.

In conclusion we can say that the 3 gluon model with vector gluon explains the decay characteristics of the ϒ(9.46) quite nicely and a simple phase space model is in disagreement with the data.

7. SEARCH FOR QCD EFFECTS IN RECENT EXPERIMENTAL DATA.

We discussed already in section 4 some evidence for extra gluon emission effects in $\frac{1}{\sigma}\cdot\frac{d\sigma}{dT}$ coming from the MARK-J experiment [35]. Other tests were reported recently in several conference talks and papers by the JADE [41], the MARK-J [9], the PLUTO [8] and the TASSO [7,12,42] Collaborations at PETRA. It is beyond the scope of this review to discuss all these very interesting results, which give evidence for the existence of gluons. So we shall restrict ourselves to a few remarks and then shall consider the so called "sea-gull effect" in some detail.

Before we come to this I must make some remarks about the theoretical model which we shall use to verify the gluon bremsstrahlung contribution. In the previous sections we treated the $q\bar{q}g$ contribution in the tree-graph approximation in which the subsequent decay of quark, antiquark and gluon is not explicitly taken into account. Since in reality hadrons are observed and not gluons or quarks this transformation of quarks and gluons into hadrons must have some effect on the distributions which one observes in the experiment.

Therefore the formulas in section 4 which are derived from the perturbative bremsstrahlung matrix element must be supplemented with the fragmentation procedure. This has been done in the meantime. More details can be found in ref. 43. In this work the fragmentation of u,d,s quarks, the fragmentation of c and b quarks including subsequent weak decays (as described in section 3) and the fragmentation of the gluon in the final states $q\bar{q}g$ and $q\bar{q}gg$ (see section 5) has been put in. One unknown input in this model is the primordial gluon fragmentation for which different assumptions were tried out. This dependence on the gluon fragmentation $D_g(x)$ was studied in connection with the sea-gull effect. The curves shown in Fig. 44 for the $q\bar{q}g$ and the $q\bar{q}gg$ cross section were already calculated with this model.

The most direct evidence for gluons seem to be the real three jet events in the data sample at the highest PETRA energies (27.4-31.6) GeV. One example of such events as measured by the TASSO group is displayed in Fig. 28. Similar events have been reported also by the PLUTO group [8]. It is clear that at high energies some events of this kind are always present in any data sample and would occur also in a pure phase space model. So more stringent tests of the hypothesis that the observed three-jet events are quark - antiquark - gluon jets are really needed. One of the tests would be the fact that they occur with a probability expected for $q\bar{q}g$ final states as determined by the gluon coupling constant $\alpha_s(q^2)$. Such tests are reported in the publication of the PLUTO group [8]. They made cuts in the T_3 - T - scatter plot (see Fig. 35) $T_3 > 0.9$, $T < 0.8$, to isolate three-jet events, and compared the number of observed events in this strip in the $T_3 - T$ plane with the number expected for pure $q\bar{q}$ and for $(q\bar{q} + q\bar{q}g)$ final states. At the higher energies, (27.4-31.6) GeV, the number of observed events agree with the number predicted by the $(q\bar{q} + q\bar{q}g)$ model but disagreed with the pure $q\bar{q}$ model. The predictions from both models agree, however, with the lower-energy data at 13-17 GeV. Similar checks have been made with cuts in the $\theta_1 - \theta_3$ Dalitz plot (see Fig. 34). Very similar to these tests are studies of the distributions of $\langle p_T^2\rangle_{out}$ and $\langle p_T^2\rangle_{in}$ where $p_{T\,out}$ is computed with respect to the normal to the triplicity plane and $p_{T\,in}$ with respect to a unit vector in the triplicity plane perpendicular to the fastest jet axis, respectively. Instead of the triplicity plane one can choose also the acoplanarity plane or the plane defined with the help of the conventional sphericity tensor [15]

$$T^{\alpha\beta} = \sum_{i=1}^{N} (p_i^2\delta_{\alpha\beta} - p_i^\alpha p_i^\beta) \tag{7.1}$$

where the p_i are the momentum vectors of all hadrons and α,β are the coordinate indices. The eigenvalues λ_k of $T^{\alpha\beta}$ are ordered so that $\lambda_1 \geqq \lambda_2 \geqq \lambda_3$ and the corresponding eigenvectors are denoted by $\hat{n}_1, \hat{n}_2, \hat{n}_3$. The sphericity axis is $\hat{n}_3$ [24,31]. Then p_{Tout} and p_{Tin} are defined as $p_T{}^2{}_{out}=(\vec{p}\hat{n}_1)^2$ and $p_T{}^2{}_{in}=(\vec{p}\hat{n}_2)^2$ over all charged particles of an event as a measure of the momentum out of the plane and in the plane in a direction perpendicular to the sphericity axis. As can be seen from Fig. 48 [7], the measured $\langle p_T^2\rangle_{out}$ distribution is narrow and **roughly** independent of energy whereas the $\langle p_T^2\rangle_{in}$ distribution is much broader and broadens if one goes from low energy (13-17 GeV) to the higher energies (27.4-31.6) GeV. This behavior is to be expected since the extra gluon emission which causes the increase of $\langle p_T^2\rangle \sim \alpha_s\cdot q^2$, is most operative in the plane and not out of the plane. Out of the event plane we see only the effect of the nonperturbative p_T effects. At even higher energies we expect here to see the two-gluon bremsstrahlung contribution. The broadening of the $\langle p_T^2\rangle_{in}$ distribution with increasing energy is observed quite clearly in the data and agrees with the curve calculated with the $q\bar{q}g$ model referred to above [43]. The same conclusion can be drawn from the PLUTO data [8].

Analysis of a similar nature have been performed also by the MARK-J group. Since they cannot measure the momenta of individual hadrons they used more global variables like oblateness O instead of $\langle p_T^2\rangle_{out}$ and $\langle p_T^2\rangle_{in}$. The oblateness distribution $\frac{1}{\sigma}\cdot\frac{d\sigma}{dO}$ becomes broader with increasing energy if the events become more planar. This is the case. The oblateness distribution for the high energy runs (27.6, 30 and 31.6 GeV) [9] is shown in Fig. 49 and compared with our QCD model. The agreement is again satisfactory.

The effects of single gluon emission are even more dramatic if one studies the transverse momentum with respect to the jet

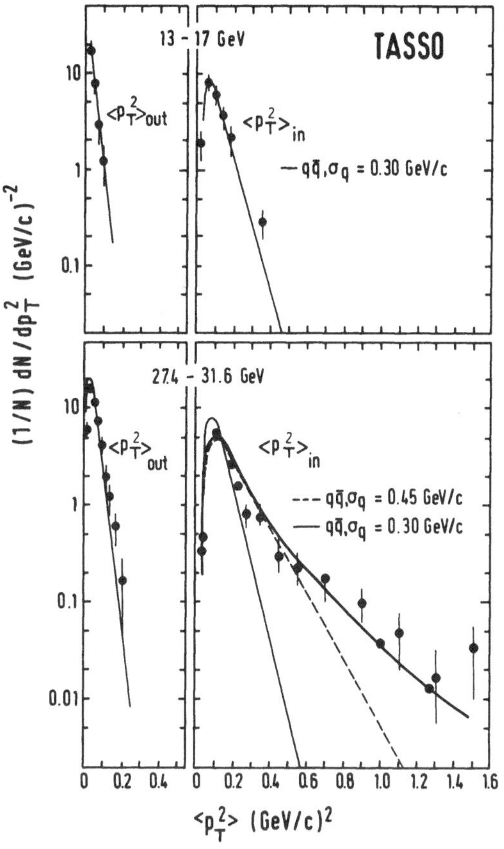

Figure 48 The mean transverse momentum squared normal to the event plane $<p_T{}^2>_{out}$ and in the event plane $<p_T{}^2>_{in}$ per event for the low energy and the high energy data. The predictions from the $q\bar{q}$ model are shown assuming σ_q = 0.3 GeV (solid curves) and σ_q = 0.45 GeV (dotted curve) q = u,d,s,c,b. The curve through the data points at high energy is the model including $q\bar{q}g$ [43].

axis (thrust or sphericity) of single hadrons as a function of the total momentum or $x = {}^2p/E_{cm}$, the so called "sea-gull" plot. This sea-gull effect was emphasized as a test for gluon contributions in e^+e^- annihilation in ref. 44.

Let us neglect for a moment the finite p_T effect coming from the fragmentation of quarks and gluons and let the quark momentum in the $q\bar{q}g$ final state determine the jet axis. Then the p_T of the hadron originating from the quark vanishes whereas the p_T of the hadron coming from the antiquark (gluon) are proportional to the transverse momentum of the antiquark (gluon) with respect to the quark momentum (see Fig. 50). They are:

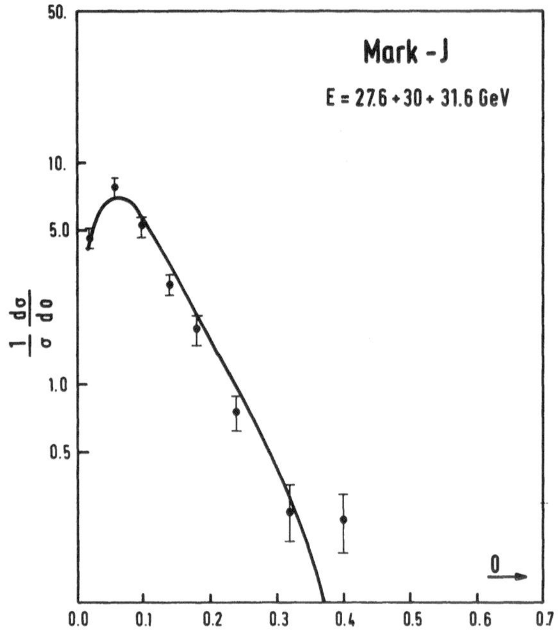

Figure 49 The distribution $\frac{1}{\sigma}\frac{d\sigma}{dO}$ as a function of oblateness O at $E_{c.m.}$ = 27.4 - 31.6 GeV. The data are from MARK-J group [9] and compared to the $q\bar{q}g$ model [43].

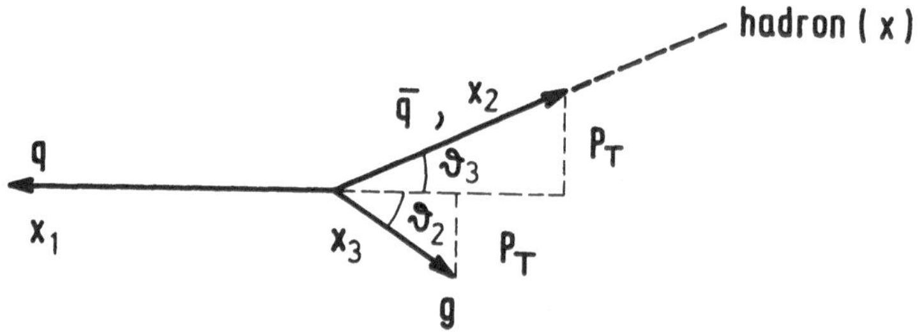

Figure 50 Kinematic plot for hadron p_T in $e^+e^- \rightarrow q\bar{q}g$.

$$\text{antiquark:} \quad p_T = p \sin\theta_3 = \frac{\sqrt{q^2}}{2} \, x \sin\theta_3$$

$$\text{gluon:} \quad p_T = p \sin\theta_2 = \frac{\sqrt{q^2}}{2} \, x \sin\theta_2 \tag{7.2}$$

where

$$\frac{1}{2} x_i x_j \sin\theta_k = [(1-x_1)(1-x_2)(1-x_3)]^{\frac{1}{2}} \tag{7.3}$$

with i,j,k = 1,2,3 cyclic. With (7.2),(7.3) and the probability for gluon emission as given in section 4 the average p_T^2 of a hadron of momentum fraction x is calculated from:

$$\langle p_T^2(x)\rangle = \frac{\alpha_s}{\pi} q^2 F(x) \tag{7.4}$$

where

$$F(x) = \frac{2}{3} \frac{x^2}{D_q(x)} \int_{2/3}^{1} \frac{dT}{T^2} \int_{\max(x,2(1-T)}^{T} \frac{dy}{y^3} \, g(x,T,y) \tag{7.5}$$

and

$$g(x,T,y) = (\rho(T,y) + \rho(y,2-T-y)) \, D_q(\frac{x}{y}) + \rho(2-T-y,T) D_g(\frac{x}{y}) \tag{7.6}$$

In (7.6) the function $\rho(x_1,x_2)$ is:

$$\rho(x_1,x_2) = \frac{x_1^2 x_2^2}{4} \sin^2\theta_3 \frac{d^2\sigma}{dx_1 dx_2} = (x_1^2 + x_2^2)\cdot(x_1 + x_2 - 1) \tag{7.7}$$

and $D_q(x)$ and $D_g(x)$ describe the fragmentation of the quark and gluon respectively into the hadron h (see section 1 and 2). $\rho(x_1,x_2)$ is not singular for $x_1, x_2 \to 1$. Therefore $\langle p_T^2(x)\rangle$ is infrared finite. We see that $\langle p_T^2(x)\rangle$ vanishes for $x \to o$, because of kinematic reasons. It depends on the fragmentation functions $D_q(x)$ and $D_g(x)$. From lower energies one has some information about the quark fragmentation function. A simple and realistic representation is $xD_q(x) = 3(1-x)^2$. Concerning the gluon fragmentation function little information exists. The simplest hypothesis is $D_g(x) = D_q(x)$. With $D_q(x)$ and $D_g(x)$ given the $\langle p_T^2(x)\rangle$ can be calculated from the Eq. (7.4)-(7.7). The result for $xD_g(x) = xD_q(x)$ and two energies E_{cm} = 17 GeV and 27.4 GeV [45] is shown in Fig. 51 and compared to preliminary experimental data from the TASSO group [42]. We see that the increase with q^2 which is proportional to q^2 (see Eq. (7.4)) is roughly in agreement with the experimental data. $p_T^2(x)$, as a function of x, shows (half of) a sea-gull structure familar from hadronic processes and from electroproduction.

More detailed studies showed that the influence of the explicit form of the gluon fragmentation function on the shape of the sea-gull plot is not very great [45,46].

The results for $\langle p_T^2(x)\rangle$ based on Eqs.(7.4)-(7.7) should not be directly compared to experimental data for several reasons. First in our formulas above the p_T effects in the fragmentation function $D_q(x)$ and $D_g(x)$ are not taken into account. Second the contribution from $e^+e^- \to q\bar{q}$ with subsequent fragmentation produces a background term of roughly $\langle p_T^2\rangle \simeq 0.3\ \text{GeV}^2$ (near $x = 0.4$), which should be added to (7.4). In this case the gluon emission contribution to $\langle p_T^2(x)\rangle$ must be reduced by introducing a thrust cut-off $T_c < 1$ into Eq. (7.5). Third it is more realistic to calculate the fragmentation of the hadron whose $p_T^2(x)$ is looked at in a Feynman-

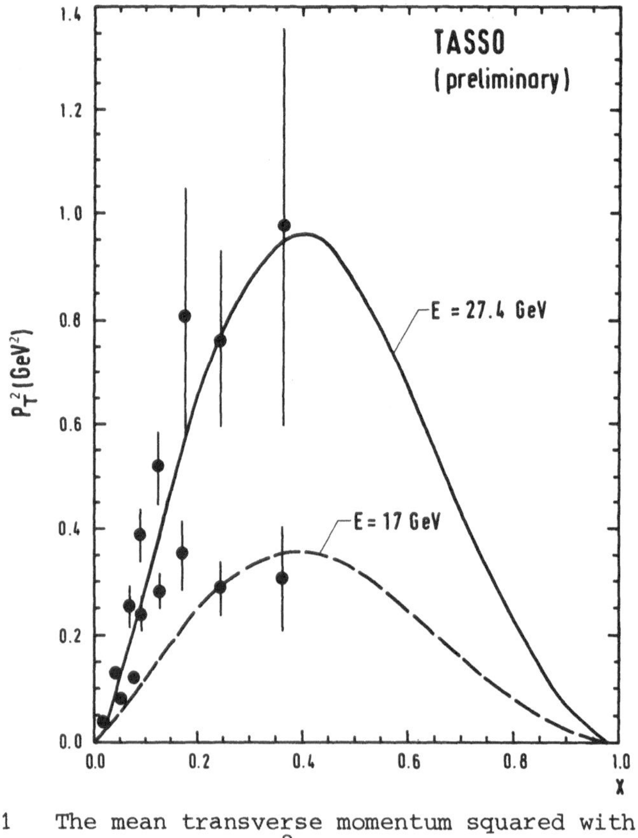

Figure 51 The mean transverse momentum squared with respect to the jet axis $\langle p_T^2(x)\rangle$ as a function of the scaled hadron momentum x for E_{cm} = 17 and 27.4 GeV compared to $q\bar{q}g$ theory Eq. (7.4)-(7.7).

Field fragmentation model including weak interaction effect for c and b quarks and in the same way the fragmentation of quarks, antiquarks and gluons of the $q\bar{q}g$ component. All these modifications have been built into the model [43] described in the second paragraph of this section. The $\langle p_T^2(x)\rangle$ computed with this model for the two energy ranges (13-17 GeV and 27.4 - 31.6 GeV) can be seen in Fig. 52 a,b (here $z \equiv x$). The agreement with the TASSO experimental data [7] is satisfactory. One notices that the increase of $\langle p_T^2(x)\rangle$ with q^2 is well accounted for. In Fig. 52b we test the influence of the gluon fragmentation function by considering (i) $D_g(x) \sim [x^2 + (1 - x)^2]$ and (ii) $D_g(x) \sim x(1 - x)$. The agreement with the data is better for (i) which is the fragmentation function based on $g \to q\bar{q}$. In Fig. 53 we show the dependence on the coupling constant $\alpha_s(q^2)$ parametrized by Λ (see section 2 and 4). The agreement with the experimental data is better with a somewhat smaller coupling constant (Λ = 0.2 GeV is favoured compared to Λ = 0.8 GeV). A better determination of Λ seems to be possible by looking at

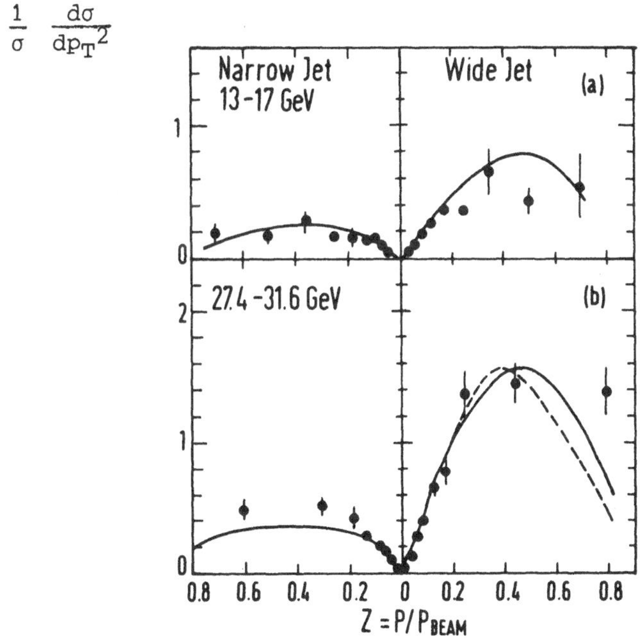

Figure 52a,b $\langle p_T^2(z)\rangle$ as a function of the scaled hadron momentum $z = 2p/E_{cm}$. for the wide and narrow jet separately, for the low energy (a) and the high energy (b). The curves show the predictions from the $q\bar{q}g$ model [43]. In (b) the solid curve is for primordial gluon fragmentation function $D_g(x) \sim x^2 + (1-x)^2$, the dashed curve is for $D_g(x) \sim x(1-x)$.

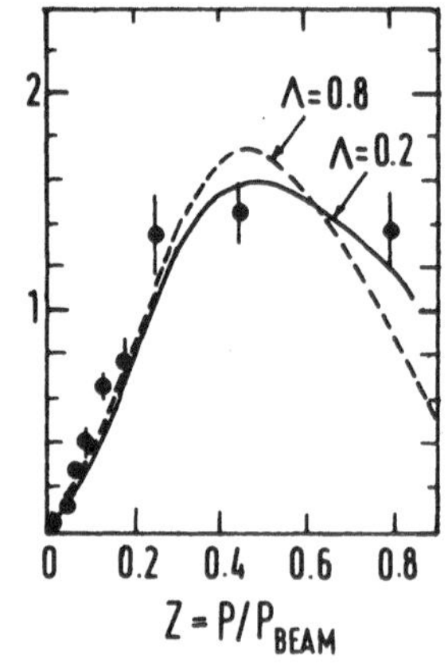

Figure 53 $<p_T^2(z)>$ as a function of z for the wide jet and $E_{c.m.} = 27.4 - 31.6$ GeV. solid curve is for a coupling constant $\alpha_s(q^2) = \frac{12\pi}{23} \ln^{-1} q^2/\Lambda^2$ with $\Lambda = 0.2$ GeV. The dashed curve is with $\Lambda = 0.8$ GeV. Both curves are for $D_g \sim (x^2+(1-x)^2)$ [43].

which is plotted in Fig. 54. In order to account for the measured points at large p_T^2 at the higher $E_{c.m.}$ energies a larger Λ is better. S_o $\Lambda = 0.35$ GeV seems to be a good compromise. Of course experimental data with better accuracy are needed before $\alpha_s(q^2)$ can be determined with good accuracy.

In Fig. 52a,b the sea-gull plot is drawn for the "narrow" and the "wide" side separately. If hard noncollinear gluon emission is a rare process, then there should usually be only one such gluon in the observed events. Dividing each event by a plane perpendicular to the jet axis and determining $<p_T^2(x)>$ separately for the two sides, the "narrow" side should rarely have a non-collinear hard gluon [46]. Therefore the $<p_T^2(x)>$ at the narrow side will increase with energy less strongly than $<p_T^2(x)>$ on the wide side. This is also observed in the experimental data shown in Fig. 52a,b. Of course part of this narrow-wide asymmetry is due to statistical fluctuations as was shown by explicit calculations based on the $q\bar{q}$ model (see refs. 7 and 8). This statistical effect is roughly of the order of magnitude as the asymmetry in the low energy data. But one cannot describe the data at high energy with such a model.

We conclude that a change in the p_T^2 distribution with increasing energy and a strong increase of $<p_T^2>$ with increasing energy is observed. This increase occurs predominantly in only one of the two jets.

In the mean time many more tests for QCD effects in e^+e^- annihilation have been proposed. They are always particular con-

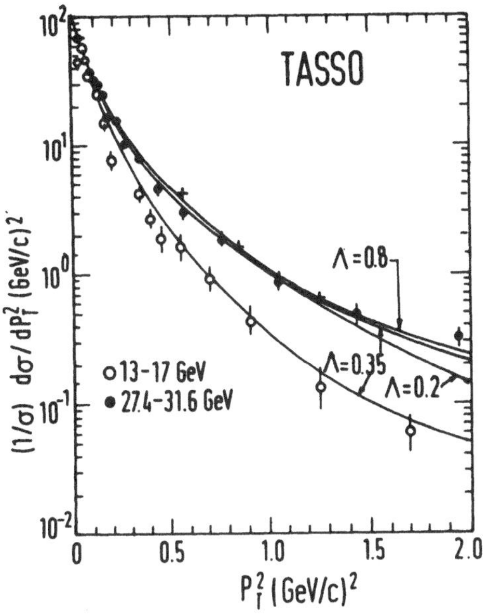

Figure 54 $\frac{1}{\sigma}\frac{d\sigma}{dp_T^2}$ at 13 and 17 GeV combined (o) and at 27.4, 27.7, 30.0 and 31.6 GeV combined (o) as a function of p_T^2 compared to $q\bar{q}g$ model of ref. 43 for various Λ values Λ = 0.2, 0.35 and 0.8 GeV.

sequences of the gluon bremsstrahlung formula introduced in section 4. We mention two-particle correlations in jets [47], the analysis of event shapes [48], energy correlations [49] and further spin tests of the gluon [50].

SUMMARY AND CONCLUSIONS.

In this report we reviewed the theory of jet phenomena in e^+e^- annihilation and confronted it with experimental data coming from the SPEAR, DORIS and PETRA rings. The underlying theory is perturbative QCD which makes definite predictions concerning the existence and the properties of more than two jets, i.e. three, four, and multi jets with increasing order of the perturbative expansion. We discussed in detail the limitations of perturbative QCD connected with the perturbative nature of the theory and with the confinement character of quarks and gluons. Because of the latter it is necessary to apply models to describe the fragmentation of quarks and gluons into hadrons in order to interpret the experimental data. Such models were discussed and also an important background to all higher order QCD effects, the broadening of jets due to the weak decay of heavy quarks.

Just before and after the summer school the discovery of three-jets in e^+e^- annihilation at 30 GeV was reported at several conferences by the JADE, MARK-J, PLUTO and TASSO collaborations at PETRA. We reviewed some of their evidence for quark - antiquark - gluon jets and discussed QCD tests, in particular the asymmetric sea-gull plot which shows very convincingly the increase of the transverse momentum of hadrons with increasing beam energy as predicted by QCD.

ACKNOWLEDGEMENTS

I am grateful to my colleagues at DESY and the University of Hamburg for numerous interesting discussions regarding the topics of this review. In particular I thank A. Ali, I. G Körner, Z. Kunszt, E. Pietarinen, G. Schierholz and I. Willrodt for their constant help and encouragement.

REFERENCES

1. G. Sterman and S. Weinberg, Phys. Rev. Lett. 39, 1436 (1977)
2. R. D. Field and R. P. Feynman, Nucl. Phys. B136, 1 (1978)
3. A. Ali, J. G. Körner, G. Kramer and J. Willrodt, DESY 79/63.
4. J. Ellis, M. K. Gaillard and G. Ross, Nucl. Phys. B111, 253, (1976), Erratum B130, 516 (1977)
5. A. Ali, J. G. Körner, Z. Kunszt, J. Willrodt, G. Kramer, G. Schierholz and E. Pietarinen, Phys. Lett. 82B, 285 (1979); A. Ali, J. G. Körner, Z. Kunszt, E. Pietarinen, G. Kramer, G. Schierholz and J. Willrodt, DESY 79/54.
6. PLUTO Collaboration,Ch. Berger et al. (Paper presented by S. Brandt at the International Conference on High Energy Physics Geneva 27 June - 4 July 1979) DESY 79/43
7. TASSO Collaboration, Brandelik et al., Phys. Lett. 86B, 243 (1979); TASSO Collaboration (Invited Talk given at the 1979 International Symposium on Lepton and Photon Interactions at High Energies, August 23-29, FNAL, presented by G. Wolf) DESY 79/61
8. PLUTO Collaboration, Ch. Berger et al., Phys. Lett. 86B, 413, 418 (1979)
9. D. P. Barber et al., Phys. Rev. Lett. 43, 830 (1979); D. P. Barber et al. (Invited Talk given at the 1979 International Symposium on Lepton and Photon Interactions at High Energies, August 23-29, FNAL, presented by H. Newman)
10. R. P. Feynman, Photon-Hadron Interactions, W. A. Benjamin, New York, 1972
11. S. Drell, D. Levy and T. Yan, Phys. Rev. D1, 1617 (1970); N. Cabibbo, G. Parisi and M. Testa, Lett.Nuovo Cimento 4, 35 (1970); S. Berman, J. Bjorken and J. Kogut, Phys. Rev. D4, 3388 (1971)

12. G. Wolf (Rapporteur talk given at the 1979 EPS International Conference on High Energy Physics, Geneva, 27 June - 4 July 1979) DESY 79/41
13. G. Hanson et al., Phys. Rev. Lett. 35, 1609 (1975)
14. Ch.Berger et al., Phys. Lett. 78B, 176 (1978)
15. J. D. Bjorken and S. J. Brodsky, Phys. Rev. D1,1416 (1970)
16. H. D. Politzer, Nucl. Phys. B129, 301 (1977); H. Georgi and H. D. Politzer, Nucl. Phys. B136, 445 (1978); R. K. Ellis, H. Georgi, M. Machacek, H. D. Politzer and G. C. Ross, Phys. Lett. 78B, 281 (1978) and CALT 68-684 (1978); D. Amati, R. Petronzio and G. Veneziano, Nucl. Phys. B140, 54 (1978), B146, 29 (1978); C. H. Llewellyn-Smith, Oxford preprint 67/78 (1978); J. Gunion and W. Frazer, UC Davis preprint 10 P 10 - 94 (1978); Yu. L. Dokshitser, D. I. D'yakanov and S. I. Troyan, lectures at the 13th Leningrad Winter School, SLAC-TRANS 183; A. H. Mueller, Phys. Rev. D18, 3705 (1978)
17. K. Koller, T. F. Walsh and P. Zerwas, Z. Physik C, Particles and Fields 2, 197 (1979)
18. P. M. Stevenson, Phys. Lett. 78B,451 (1978); B. G. Weeks, Phys. Lett. 81B, 377 (1979); B. Binêtruy and G. Girardi, Phys. Lett. 83B, 382 (1979)
19. A. V. Smilga, preprint ITEF-78 (1978); E. Curci and M. Greco, Phys. Lett. 79B, 406 (1978)
20. S. Brandt, Ch. Peyron, R. Sosnovski and A. Wroblewski, Phys. Lett. 12, 57 (1964)
21. E. Farhi, Phys. Rev. Lett. 39, 1587 (1977)
22. H. Georgi and M. Machacek, Phys. Rev. Lett. 39, 1237 (1977)
23. A. De Rujula, J. Ellis, E. G. Floratos and M. K. Gaillard, Nucl. Phys. B138, 387 (1978)
24. S. L. Wu and G. Zobernig, Z. Physik C, Particles and Fields 2, 107 (1979)
25. S. Brandt and H. D. Dahmen, Z. Physik C, Particles and Fields 1, 61 (1979)
26. B. Anderson, G. Gustafson and C. Peterson, Nucl. Phys. B135, 273 (1978); Z. Physik C, Particles and Fields 1, 105 (1979). See also: J. Engels,J. Dabkowski and K. Schilling, Bielefeld preprint BI-TP 79/21, WUB 79-17 (1979)
27. A. Ali, J. G. Körner, G. Kramer and J. Willrodt, Z. Physik C, Particles and Fields 1, 203 (1979)
28. M. Kobayashi and M. Maskawa,Progr. Theor. Phys. 49, 652 (1973)
29. J. Ellis, M. K. Gaillard, D. V. Nanopoulos and S. Rudaz, Nucl. Phys. B131, 285 (1977)
30. A. Ali, J. G. Körner, J. Willrodt and G. Kramer, DESY 79/16 (1979)
31. Ch.Berger et al., Phys. Lett. 82B, 449 (1979)
32. A. Ali, J. G. Körner, J. Willrodt and G. Kramer, Phys. Lett. 83B, 375 (1979)
33. PLUTO Collaboration, Ch. Berger et al., Phys. Lett. 81B, 410 (1979)

34. TASSO Collaboration, R. Brandelik et al., Phys. Lett. 83B, 261 (1979)
35. D. P. Barber et al., Phys. Lett. 84B, 463 (1979)
36. G. Schierholz, private communication.
37. A. C. Hirschfeld and G. Kramer, Nucl. Phys. B74, 211 (1974); N. M. Avram and D. Schiller, Nucl. Phys. B70, 272 (1974)
38. G. Kramer, G. Schierholz and J. Willrodt, Phys. Lett. 78B, 249 (1978); Erratum, Phys. Lett. 80B, 433 (1979)
39. K. Koller, H. Krasemann and T. F. Walsh, Z. Physik C, Particles and Fields 1, 71 (1979). See this paper for earlier references on this subject.
40. K. Koller and H. Krasemann, DESY 79/52 (1979)
41. JADE Collaboration, Talks presented by S. Orito at the International Symposium on Lepton and Photon Interactions at High Energies (FNAL, 23-29 Aug. 1979) and by R. Felst at the Workshop about Quarks, Leptons and QCD (DESY, 8-10 Oct. 1979)
42. TASSO Collaboration, Talks presented by R. Cashmore and P. Söding at the Int.Conf. on High Energy Physics (Geneva 27 June - 4 July 1979), DESY 79/50
43. A. Ali, E.Pietarinen, G. Kramer and J. Willrodt, DESY 79/X (in preparation)
44. G. Kramer and G. Schierholz, Phys. Lett. 82B, 108 (1979)
45. G. Schierholz and J. Willrodt, private communication.
46. P. Hoyer, P. Osland, H. G. Sander, T. F. Walsh and P. Zerwas, DESY 79/21
47. G. Schierholz and J. Willrodt, DESY 79/32 (1979)
48. G. C. Fox and S. Wolfram, preprint CALT-68-723
49. C. L. Basham et al., Phys. Rev. D17, 2298 (1978) Phys. Rev. 41, 1585 (1978), Phys. Rev. D19, 2018 (1979)
50. J. Ellis and I. Karliner, Nucl. Phys. B148, 141 (1979)

ARBITRARY PARTON CROSS-SECTIONS - UNIQUE ASYMPTOTIC FREEDOM PREDICTIONS

B. Humpert and W. L. van Neerven

CERN

Geneva, Switzerland

ABSTRACT

In the framework of asymptotic free (AF) field theories (mainly ϕ_6^3) we expose by simple examples the arbitrariness of perturbative parton cross sections due to the ultraviolet (UV), mass (M) and the infrared (IR) divergences and their regularisation. Establishing a one-to-one correspondence between the renormalization group (RG) approach and the Feynman diagram calculations we gain insight into the importance of M-factorization to arrive at unique AF-predictions. We give details of the parton cross sections for deep-inelastic (DI) scattering and Drell-Yan (DY) massive lepton-pair production.

I. INTRODUCTION

The theoretical justification of perturbative calculations in the framework of Quantum Chromodynamics [1] triggered in the recent past a phletora of theoretical investigations and exploratory analyses which determined the integrated/differential distribution of the quark-gluon sub-processes in the framework of the parton model [2]. In a consistent calculation M-factorization [3] would have to be carried out with all M-singularities [4], originating from the initial partons, being absorbed in the momentum distribution. Once we go beyond a leading-log approximation this last step becomes essential since the quark cross sections as they are determined from the Feynman graphs involve a high degree of arbitrariness which in these phenomenological calculations is often ignored. Although the quark-gluon cross sections are <u>arbitrary</u> the AF predictions [5] resulting from a correct application of the M-factorization program are unique.

With this short write-up we aim to demonstrate the dependence of the parton cross sections on the regularisation procedures to prevent the M-,UV-and the IR-singularities. In its second part we focus on the AF-program and expose a general and elegant method to extract from the perturbative parton cross sections the M-factorized functions of the RG-approach. In order to keep all unnecessary technicalities apart from the discussion we present our reasoning in the framework of ϕ^3-theories (mainly ϕ^3_6 since it is asymptotically free and has no IR-divergences)[6].

2. THE PERTURBATIVE PARTON CROSS SECTIONS

In this first part our strategy is as follows: we present the QCD Feynman diagrams, analyze their singularity structure and explain how the appearing divergences are correctly removed by pointing to the crucial assumptions.

Considering DI-scattering with the lowest order QCD-diagrams exposed in Fig.1 we distinguish the virtual-gluon corrections due to the self-energy and the vertex graphs, the radiative-gluon graphs and the initial-gluon graphs. The analogous Feynman diagrams contribution to the DY-process are shown in Fig.2.

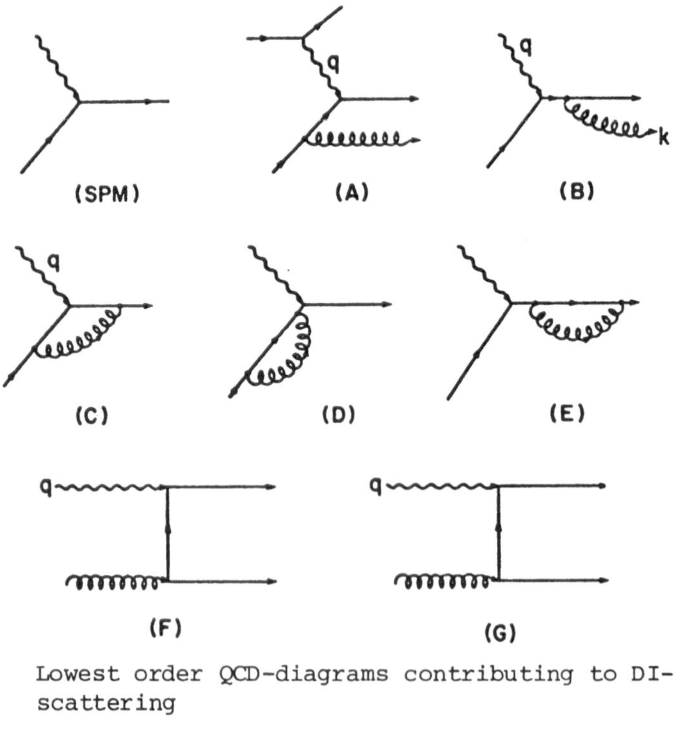

Fig.1 Lowest order QCD-diagrams contributing to DI-scattering

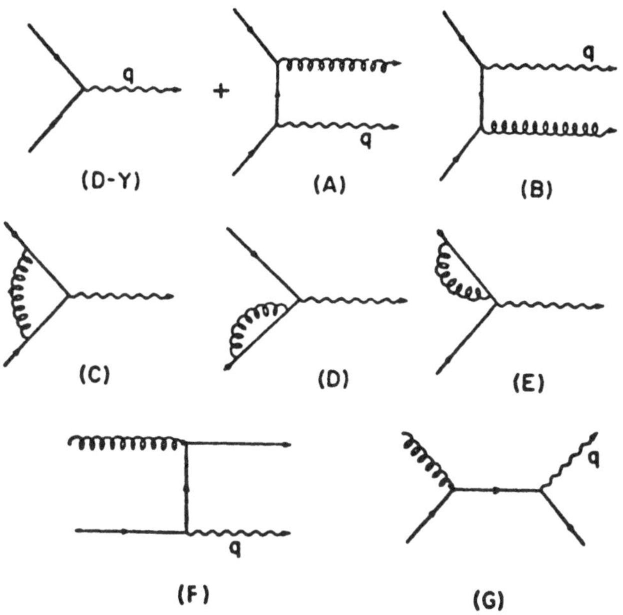

Fig.2 Lowest order QCD-diagrams contributing to the DY-process

Feynman diagrams are well-known to give rise to UV-,M- and IR-singularities. We analyze their origin by introducing simple methods for their analysis which permit immediate recognition of the singularity structure of each graph. Whilst attempting to carry out the loop integration in the virtual-gluon terms we notice a logarithmic singularity as $k^{\mu} \to \pm \infty$ - the UV-divergence. It is removed in two steps: UV-regularisation and renormalization. UV-regularisation can be achieved by several different techniques where one of them, dimensional regularisation [7], is particularly attractive for gauge theories. It has the advantage of maintaining the Ward-identities, respects unitarity and causality, and it is simple in its practical applications.

We apply it on the self-energy contribution and find

$$1 - Z_2^{-1} \equiv \frac{\partial \Sigma}{\partial p^2}\bigg|_{p^2=m^2} \overset{n\to 6}{=} \frac{\alpha_s}{6\pi} \left[\frac{2}{n-6} + \ell n \frac{m^2}{M_{reg}^2} + \text{constants} \right]. \quad (1)$$

This result was obtained for ϕ_6^3. There is an (n-6)-pole which reflects the UV-divergence and it exhibits a logarithmic dependence on the quark mass m. The regulator mass M_{reg} was introduced to maintain a dimensionless action Lagrangian. Once we have technically prevented the singularities we can proceed to the renormalization step. Denoting $\mathcal{L}_0$ as the bare, unrenormalized Lagrangian we introduce the subtraction term $\Delta\mathcal{L}$ leading to additional terms which subtract out the UV-divergences. There is a certain freedom in its definition which reflects itself in the fact that in eq.1 the pole-term alone or additional constant pieces can be removed. The minimal subtraction scheme [7] choses the 1/(n-6) pole term alone to disappear whereas renormalization at the propagator mass imposes that for $p^2 \rightarrow m^2$ the propagator is given by

$$\frac{Z_2}{p^2 - m^2} \quad \text{imposing} \quad \Sigma^R(p^2=m^2) = 0. \tag{2}$$

Of course, any other choice would have been equally valid although the latter two alternatives are particularly convenient. We could go through the analogous analysis of the vertex diagram with precisely the same insight.

We therefore come to our <u>first conclusion</u>:

The perturbatively calculated parton cross sections depend on the particularly chosen subtraction scheme whilst performing renormalization.

We focus on the <u>M-divergences</u>[4] and analyse for this purpose the Feynman diagram describing gluon-radiation (Fig.1, A+B).

The partons are given the following masses: the intermediate propagator mass is denoted by m_x^2, p^2 stands for the initial quark mass and k^2 is the mass of the final gluon which for demonstration purposes was chosen non-vanishing. q^2 is the off-shell photon mass. Straightforward integration reveals (in ϕ_6^3) for the parton structure function the form

$$\hat{\mathcal{F}}_2^R = \frac{\alpha_s}{\pi}\left[P(z)\ln\frac{q^2}{M^2(z)\cdot z^2} + h(z) + \left(\frac{1/6}{1-z}\right)\right] \tag{3}$$

where

$$P(z) = z(1-z),\quad h(z)=\frac{11}{3}(1-z)^2 - 3(1-z) - \frac{1}{2}$$
$$M^2(z) = \frac{z(1-z)p^2 - zk^2 - (1-z)m_x^2}{z(1-z)} \tag{4}$$

For the definitions and conventions we refer to eqs.15-19 later in the text. This result exposes part of the mass dependence of the parton cross section. If all masses vanish the cross section diverges logarithmically due to its first term. The third term exhibits a pole at z=1 which also gives rise to a mass divergence manifesting itself however in a slightly different form. Whilst doing QCD calculations in the parton model one assumes that if the infinity of the contributing Feynman diagrams is summed up the parton masses may be ignored as compared to q^2 and the only remaining mass is the renormalization point q_0^2. Using the operator product expansion and renormalization group techniques this picture is indeed analytically verifiable. In our Feynman diagram calculations we see that the parton cross sections diverge if all masses are assumed to vanish. We thus begin to realize that there are M-singularities which technically are prevented by giving the partons on- or off-shell masses in order to arrive at finite mathematical expressions; this is called M-regularisation. As an alternative one also can use dimensional regularisation assuming that <u>all</u> parton masses vanish giving

$$\hat{\mathcal{F}}_2^R = \frac{\alpha_s}{\pi}\left[P(z)\left\{\frac{2}{n-6} + \ln\frac{-q^2}{M^2_{reg}} \cdot \frac{1-z}{z} + \text{constants}\right\} + h(z) + \right.$$
$$\left. + \left(\frac{1/6}{1-z}\right)\right]. \qquad (5)$$

The M-divergence explicitly appears as an (n-6)-pole which however can not be distinguished from the (n-6)-pole of UV-origin. The above regularization procedures introduce an arbitrariness in the cross section since there is a priori no preferred mass assignment apart from reasons of simplicity and elegance. The first expression in eq.4 demonstrates clearly this point.

What about the (1-z)-singularity? It is spurious and only appears since z-integration and subsequent Bj-limit with $q^2 \to -\infty$ were interchanged. In order to prevent all difficulties we split the z-integration in a "soft" gluon part with $z \approx 1$ (which corresponds to the threshold region) and a "hard" gluon part. In the latter contribution the Bj-limit may be determined first, giving a simple form of the parton cross section without corrective masses. In the "soft"-part all masses are kept finite in order to prevent the M-singularities, and z-integration is carried out first; subsequently the scaling limit is imposed. One immediately might ask how we then can dispose of these M-singularities and arrive at unique predictions? The answer to this question comes from M-factorization which will be the subject of the second part of this write-up.

We come to our <u>second conclusion</u>:

Perturbative parton cross sections are M-singular with the singularities being prevented by giving the partons on/off-shell masses or by using dimensional regularisation in the totally massless case. These regularisation procedures lead to results which suffer from a high degree of arbitrariness.

Whilst analyzing the perturbative-QCD diagrams we come across a third type of divergence - the IR-singularities [4] . They are absent in ϕ^3_6 but reappear as we lower the dimension of space-time to n=4. ϕ^3_6 is asymptotically free but not ϕ^3_4 which is IR-singular instead. We therefore find it here suitable to keep for demonstration purposes the space-time dimension n open. One might criticize these simplified analyses as academic since QCD involves spin with n=4. ϕ^3theories with variable n however are convenient to demonstrate all the difficulties of QCD in their most simple form. In order to exhibit the origin of the IR-singularities we use time-ordered perturbation theory [8] and analyze all graphs with this calculation technique. The negative energy denominators are ignored since they cannot give rise to M- and IR-divergences. All contributions can be cast in the general form

$$d\sigma = \int d^n PS \frac{N(\theta_i \ldots)}{[k(1-\cos\theta_1)]^\alpha [k(1-\cos\theta_2)]^\beta} \cdot D(q_0,\ldots) \qquad (6)$$

k stands for the loop-momentum and θ_i is the angle between the loop-momentum and the inital or final state partons. All factors which contain a dependence on the photon energy q_0 are included in $D(q_0,..)$ which in specific cases involves a δ-function due to energy-conservation. The remaining two denominators result from the propagator

$$\frac{1}{E_{ext} - E_{int} + i\epsilon} \Rightarrow \frac{1}{p-k-|\vec{p-k}|} \overset{k\to o}{\Rightarrow} \frac{1}{k(1-\cos\theta)} \quad . \qquad (7)$$

$N(\theta_i,..)$ in the numerator is due to spin and the phase space factor, generalized to n dimensions, reads

$$d^n PS \equiv |\vec{k}|^{n-3} dk \, (1-\cos^2\theta_i)^{\frac{n-4}{2}} d\cos\theta_i \quad . \qquad (8)$$

A simple calculation permits us to establish for each diagram the powers α and β, in table 1 we give an example, and to decide about the (1/k)-divergence which gives rise to the IR-singularity. In completely analogous manner we then also determine the angular singularity which is at the origin of the M-singular behaviour of the Feynman diagrams. Since we now know where and how IR-singularities occur we wonder how they are removed. We first are concerned with their regularisation.

DEEP INELASTIC SCATTERING

$q\gamma^* \to qg$ Graph	Fig.	(α,β)	ϕ_6^3 (n=6) M-singul.	Q.C.D. (n=4) M-singul.	IR-singul.
$\hat{t}$ - graph	3 a_2	(2,0)	yes	yes!	yes
$\hat{s}t$-,Vx - graph	3 b_2,b_3	(1,1)	no	yes	yes
$\hat{s}t$-,Vx - graph	3 c_2,c_3	(1,1)	no	yes	yes
$\hat{s}$ -,S_2 - graph	3d_2,d_3,d_4	(2,0)	yes	yes!	yes
S_1 - graph	3 e_2	(2,0	yes	yes!	yes
S_1 - graph	3 f_2	(2,0)	yes	yes!	yes

Table 1

Analysis of the DI-graphs in Fig. 3 on their M- and IR-singular behaviour

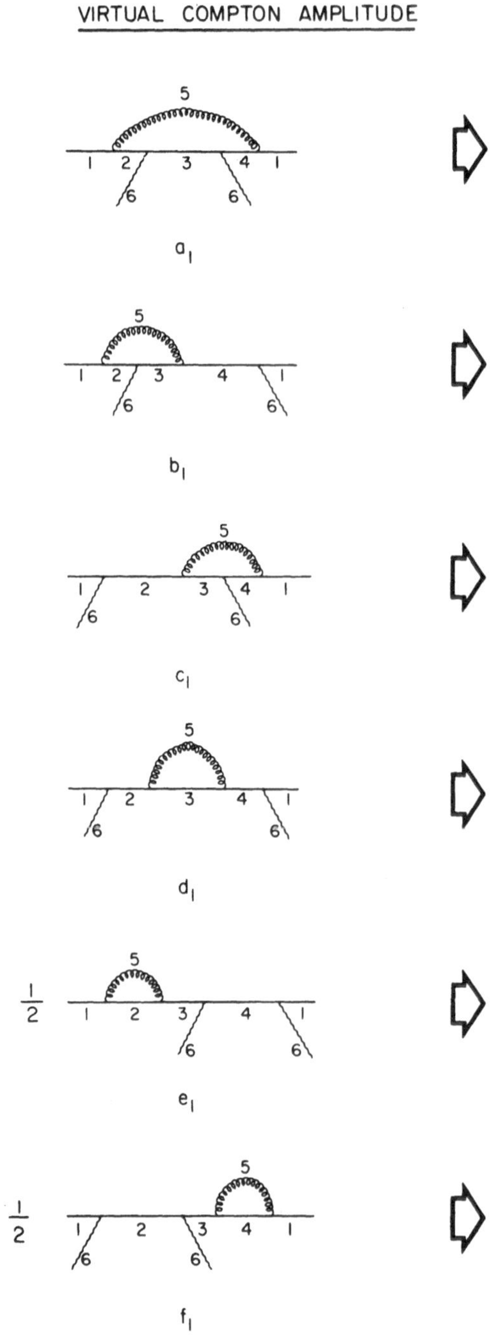

Fig.3 Lowest order QCD-diagrams contributing to the virtual Compton amplitude (left). Their discontinuities (right) provide the graphs contributing to the DI-cross sections.

DEEP INELASTIC CROSS SECTION

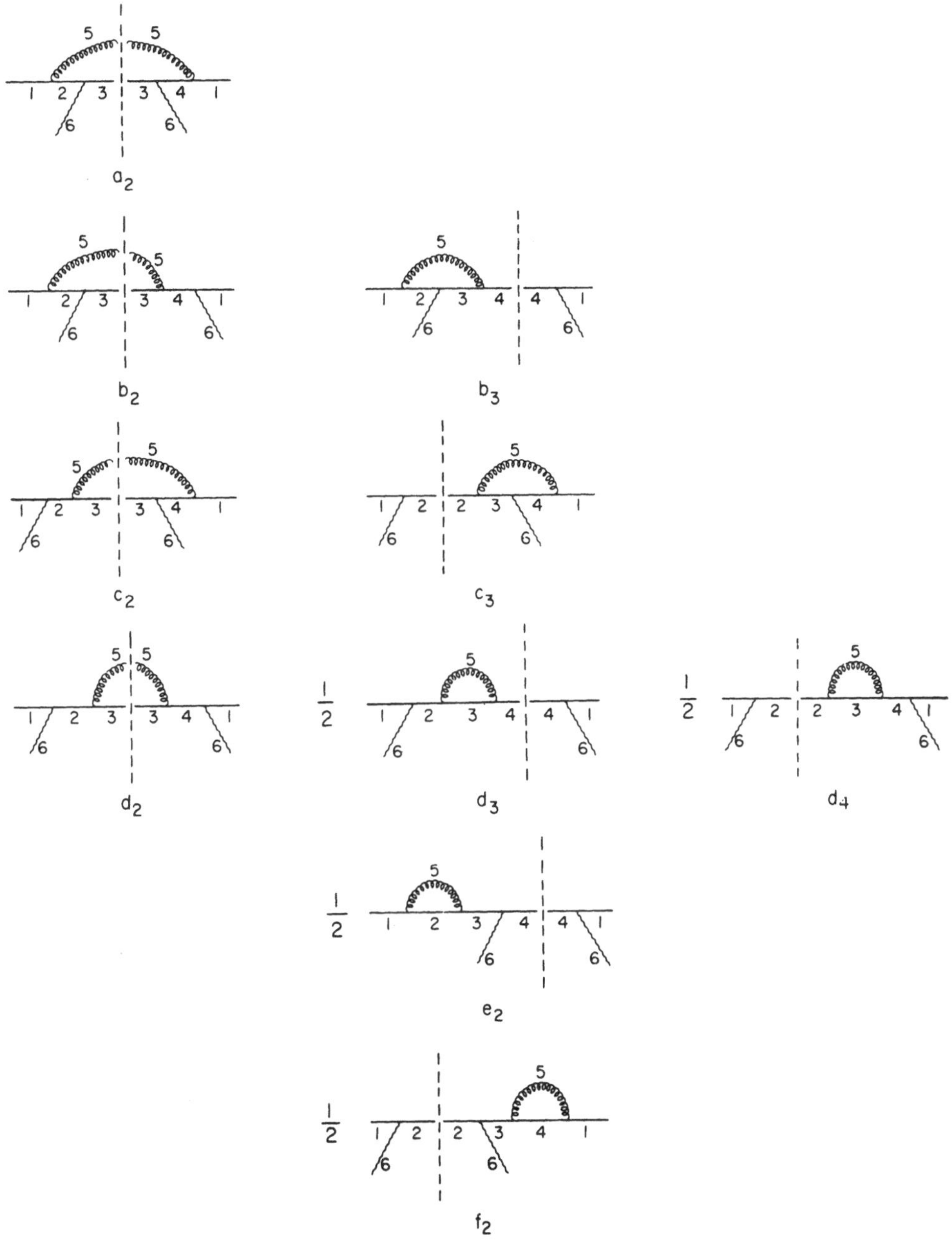

The IR-singularities can be regularized by different techniques. Either we introduce an infinitesimal gluon regulator mass λ leaving the external quark masses on-shell [17,18] or we assume that all fermion masses slightly differ. This latter possibility was earlier referred to as "off-shell" regularisation leading to (intuitively motivated) confusion and incorrect results [9] . We also could have used dimensional regularisation to prevent this divergence type [10] . IR-singularities cannot be argued away but may only be removed by their proper cancellation between different contributions (of the same order) to the cross section [4] . In many phenomenological calculations these cancellations were not properly worked out and this divergence was prevented by "natural" cut-offs. Predictions based on such manipulations bear a high degree of arbitrariness; it is to our mind doubtful whether they are even of qualitative value as their proponents claim. Going back to the cancellations we give in the following an example how they take place.

We consider the Feynman graphs for the DI-cross section as shown shown in Fig. 3 and specialize for demonstration purposes to Figs. $3b_2, 3b_3$. The contributions of the two graphs read

$$d\sigma_1 = (2\pi i)^2 \int d^n p_5 \frac{1}{p_2^2 - m^2 + i\epsilon} \delta^+(p_5^2)\,\delta^+(p_3^2 - m^2) \times \frac{1}{(p_3+p_5)^2 - m^2 + i\epsilon} \tag{9}$$

$$d\sigma_2 = (-2\pi i) \int d^n p_5 \frac{1}{p_2^2 - m^2 + i\epsilon} \frac{1}{p_5^2 + i\epsilon} \frac{1}{p_3^2 - m^2 + i\epsilon} \delta^+((p_3+p_5)^2 - m^2). \tag{10}$$

Since the singular behaviour of the integrands in eqs.9,10 originates only from the δ-function part of the propagators we use

$$\begin{aligned} \frac{1}{p^2 - m^2 + i\epsilon} &= -i\pi\,\delta(p^2 - m^2) + P\frac{1}{p^2 - m^2} \\ &= -\,2\pi i\;\delta^+(p^2 - m^2) + P\frac{1}{p^2 - m^2} \end{aligned} \tag{11}$$

The second equality in eq.11 is only possible since the positive and negative frequency parts contribute equally in the $d^n p_5$-integration. Eq.10 is then changed to

$$d\sigma_2 = (2\pi i)^2 \int d^n p_5 \frac{1}{p_2^2-m^2+i\epsilon} \left\{ \frac{\delta^+(p_5^2)}{p_3^2-m^2} + \frac{\delta^+(p_3^2-m^2)}{p_5^2} \right\}$$

$$\times \; \delta^+((p_3+p_5)^2-m^2) \qquad (12)$$

The omitted pieces do not contribute to the singular behaviour since they emerge from the principle value parts which account for the off-shell contribution. We drop all masses and take the δ-constraint into account which permits us to write

$$d\sigma_1+d\sigma_2 = (2\pi i)^2 \int d^n p_5 \frac{1}{8p_3p_5} \left[\frac{\delta(p_{30}-|\vec{p}_3|)\,\delta(p_{50}-|\vec{p}_5|)}{|\vec{p}_3|\;|\vec{p}_5|} - \right.$$

$$- \frac{\delta^+(p_{30}+p_{50}-|\overrightarrow{p_3+p_5}|)}{|\overrightarrow{p_3+p_5}|\,|\vec{p}_5|}$$

$$\left. - \frac{\delta^+(p_{30}+p_{50}-|\overrightarrow{p_3+p_5}|)\,\delta^+(p_{30}-|\vec{p}_3|)}{|\overrightarrow{p_3+p_5}|\,|\vec{p}_3|} \right] \frac{1}{p_2^2-m^2+i\epsilon} \qquad (13)$$

If the angle θ between $\vec{p}_3$ and $\vec{p}_5$ goes to zero eq.13 can be simplified to

$$d\sigma_1 + d\sigma_2 = (2\pi i)^2 \int d^n p_5 \frac{1}{8p_3p_5} \left[\frac{1}{|\vec{p}_3|\,|\vec{p}_5|} - \frac{1}{|\vec{p}_5|\,(|\vec{p}_3|+|\vec{p}_5|)} - \right.$$

$$\left. - \frac{1}{|\vec{p}_3|\;(|\vec{p}_3|+|p_5|)} \right] \frac{1}{p_2^2-m^2+i\epsilon} \qquad (14)$$

We observe that the IR-singularity is cancelled between the first and second term ($|\vec{p}_5| \to 0$) whereas we need all three terms to cancel the M-singularity connected with the final quark lines 3,4.

We come to our <u>third conclusion</u>:

All IR-divergences of the Feynman diagrams contributing to a given order in perturbative QCD cancel. The form of the remaining finite cross section depends on how the IR-singularities were regularized. If the cancellation is not properly worked out the resulting parton cross sections strongly depend on the regularization procedure.

We so far have not mentioned that any mass assignment to prevent the M-singularities is subject to consistency constraints. This insight emerges from considering the lowest order perturbative diagrams contributing to the virtual Compton amplitude $\gamma^* q \to \gamma^* q$. Its discontinuity gives the parton cross sections contributing to the lowest order QCD-corrections in DI-scattering such as $\gamma^* q \to qg$. We consider a set of illustrative graphs shown in fig. 3.
The virtual Compton graph is not M-divergent in the intermediate quark lines 234 whereas the corresponding DI-graphs are M-divergent. All M-singularities related to the lines 234 therefore must cancel in their sum. This can only be achieved if the M-singularities are regularized in a consistent way. Unitarity imposes that the regulator masses m_2, m_3, m_4, which slightly may differ, must be chosen as indicated in Figs. 3. It clearly implies that the final state fermion mass is not necessarily the same in all diagrams! The mass assignment was fixed in the M-finite graphs and then is consistently carried over to the cut-graphs. Explicit evaluation of all DI-diagrams shows that in their entire sum all final state M-singularities indeed cancel whereas those of the initial parton lines persist.

In this example of DI-scattering we immediately understand and accept the need of this constraint, but its extension to other processes was not so obvious.

The DY-subprocesses as $q\bar{q} \to \gamma^* g$ have of course no analogue to the virtual Compton amplitude in DI-scattering and still all final state M-singularities are absent or cancel. The remaining initial state M-singularities of this process are however absorbed in the scale-dependent momentum distributions; they therefore are replaced in the M-factorization program by the analogous ones of the DI-structure function. It therefore is of utmost importance to realize that the M-singularities in both type of processes must be regularized in precisely the same way. The essential point is to start from one and the same M-finite Feynman graph - the vacuum bubble - and to derive the corresponding DI- and DY-diagrams through appropriate cuttings (in the sense of Cutkosky) [11,4]. We give a simple example in fig. 4. The mass assignment is chosen in the vacuum-bubble and then consistently carried over to the two parton processes.

We come to our fourth conclusion:

The (auxiliary) mass assignment must be chosen on an M-finite amplitude which via cuttings leads to the parton processes under consideration.

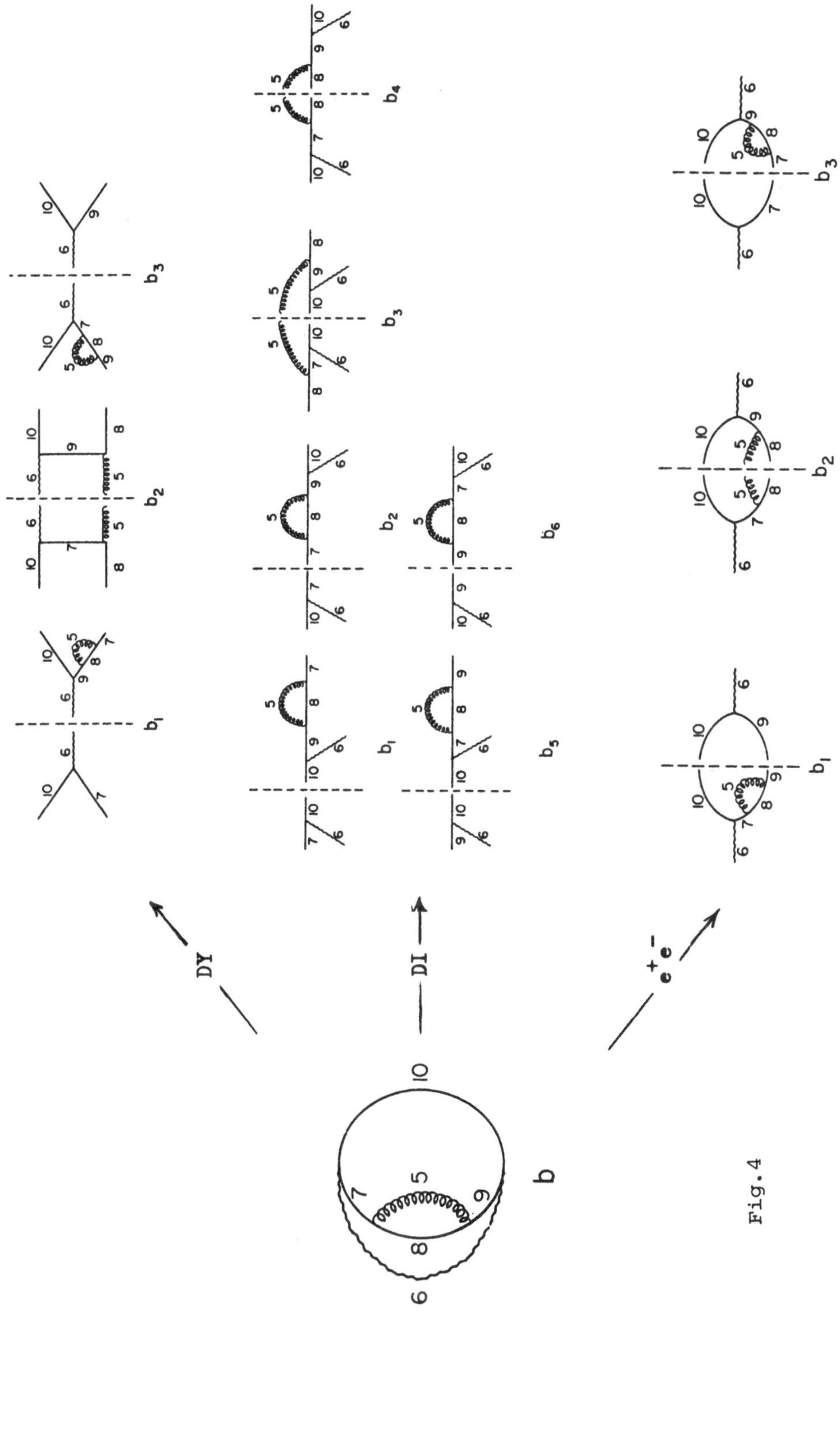

Fig.4

We have determined the DI-structure function and the DY-cross section on the parton level and give in the following some of the intermediate and final results for ϕ_6^3 - theory [12]. We have selected the on-shell and the completely massless cases which satisfy the above constraints and used for the latter mass assignment n-dimensional regularisation. The DI-structure function is

$$\frac{\nu W_2}{x} \equiv \mathcal{F}(x,q^2) = \int_x^1 \frac{dz}{z} f\left(\frac{x}{z}\right) \hat{\mathcal{F}}_2(z,q^2) \tag{15}$$

with the scaling variables defined as

$$x = \frac{-q^2}{2Pq}, \quad z = \frac{-q^2}{2pq} \quad \text{where} \quad p = y\,P. \tag{16}$$

The quark-part of the parton structure function is parametrized as

$$\hat{\mathcal{F}}_2(z,q^2) = \delta(1-z) + \alpha_s \{P \cdot \ell n \frac{-q^2}{p^2} + f\} \tag{17}$$

with P(z) and f(z) given in table 2. The DY-cross section reads

$$sQ^2 \frac{d\sigma}{dQ^2} = \frac{4\pi\alpha^2}{g} \int \frac{dx_1}{x_1}\frac{dx_2}{x_2} [f_1\bar{f}_2 + \bar{f}_1 f_2]\, \hat{W}(z,q^2) \tag{18}$$

where $z = {}^{\tau}/x_1x_2$, $\tau = Q^2/s$ with the analogous parton cross section parametrized as

$$\hat{W}(z,q^2) = \delta(1-z) + \alpha_s \{\sum_1^2 P \cdot \ell n \frac{Q^2}{p_j^2} + f\} + \ldots \tag{19}$$

and P(z) and f(z) listed in table 2.

3. THE ASYMPTOTIC FREEDOM APPROACH

We come to the second part of our presentation which concerns the connection between the RG-approach and the lowest order perturbative calculations [13]. In order to expose clearly the essential reasoning we ignore all mixing complications. Our arguments can of course be extended to the singlet case where mixing may not be ignored anymore. The AF-approach [5] to the DI-structure functions is based on

- the operator product expansion [14]
- the solution to the RG-equation for Wilson coefficients [15]
- and analyticity establishing the connection [16]

$$M_i^{(n)} = f^{(n)} A^{(n)} \mathbb{C}_i^{(n)} \tag{20}$$

$M_i^{(n)}$, $f^{(n)}$, $A^{(n)}$, $\mathbb{C}_i^{(n)}$ stand for the moments of the structure function i, the parton momentum distribution, the reduced operator matrix element and the Wilson coefficient function. All dependence on the parton momentum p^2 is contained in $A^{(n)}$, whereas $\mathbb{C}_i^{(n)}$ only depends on q^2. The renormalization point q_0^2 appears in both.

Instead we could have carried out our analysis in the parton model using QCD perturbation theory with

$$\begin{aligned} M_i^{(n)} &= f^{(n)} d\sigma_i^{(n)} \\ &= f^{(n)} \Gamma^{(n)} d\bar{\sigma}_i^{(n)} \end{aligned} \tag{21}$$

The eqs. 20 and 21 are distinct in that all parton momentum dependence on one hand and the dependence on the off-shell photon mass are separated by the operator product expansion whereas this is not the case in a perturbative calculation. $d\sigma_i^{(n)}$ is a function of p^2, q^2 (and the renormalization point q_0^2 in higher orders). We can achieve their separation through M-factorization as is indicated by the second equality in eq. 21. All parton mass dependence is absorbed in $\Gamma^{(n)}(p^2, q_0^2)$ whereas $d\bar{\sigma}_i^{(n)}(q^2, q_0^2)$ depends only on the dynamical variable q^2; the renormalization point q_0^2 appears in both even in a lowest order calculation. In phenomenological applications the product $f^{(n)} \Gamma^{(n)} \equiv u_0^{(n)}$ is treated as one function and considered as the input momentum distribution which is fixed by the experiment at a chosen value q_0^2.

Since we calculate perturbatively

$$d\sigma_i = \text{const} \left[\delta(1-z) + g^2 \cdot \Lambda + \,..\, \right]_i \tag{22}$$

the explicit form of $\Lambda(p^2, q^2)$ emerges from evaluation of all QCD Feynman diagrams to order g^2 as indicated above. It thus is possible to specify $d\sigma_i^{(n)}$ and to carry out M-factorization.

The arbitrariness mentioned earlier due to the correction term $\Delta\mathcal{L}$ in the Lagrangian in order to achieve renormalization permits the formulation of the Callan-Symanzik (differential) equation [15] for $A^{(n)}$ (in the parton model) and for $\mathbb{C}_i^{(n)}$, with the solution for the latter quantity given by

$$\mathbb{C}_i^{(n)}(q^2) = c_i^{(n)}(\bar{g}) \cdot \exp\left[\int_g^{\bar{g}} \frac{\gamma^{(n)}(x)}{\beta(x)} dx \right] . \tag{23}$$

	$2F^{Vx}(q^2)$	$\Sigma_1(p_1^2)$
on shell	$2\frac{\alpha_s}{6\pi}\left[\frac{2}{6-n} + \ln\frac{M_{reg}^2}{-q^2} - \gamma + \ln 4\pi + \frac{11}{3}\right]$	$-\frac{\alpha_s}{6\pi}\left[\frac{2}{6-n} + \ln\frac{M_{reg}^2}{m^2} - \gamma + \ln 4\pi + \frac{5}{3}\right]$
massless	↓	o

	$\Sigma_2(p_2^2)$	$2\cdot F^{V}(q^2)$
on shell	$-\frac{\alpha_s}{6\pi}\left[\frac{2}{6-n} + \ln\frac{M_{reg}^2}{m^2} - \gamma + \ln 4\pi + \frac{5}{3}\right]$	$\frac{2\alpha_s}{6\pi}\left[\ln\frac{m^2}{-q^2} + 2\right]$
massless	o	$\frac{2\alpha_s}{6\pi}\left[\left\{\frac{2}{6-n} + \ln\frac{M_{reg}^2}{-q^2} - \gamma + \ln 4\pi\right\} + \frac{11}{3}\right]$

	$\hat{\mathfrak{F}}_2^{R\text{-soft}}$
on-shell	$\delta(1-z)\,\frac{\alpha_s}{6\pi}\left[\ln\frac{-q^2\epsilon}{m^2} - 1\right]$
massless	$\delta(1-z)\frac{\alpha_s}{6\pi}\left[-\frac{2}{6-n} + \ln\frac{-q^2\epsilon}{M_{reg}^2} + \gamma - \ln 4\pi - \frac{8}{3}\right]$

Table 2

Formulas for the parton structure function in DI-scattering and for the DY-parton cross section calculated in ϕ_6^3 -field theory.

	$\alpha_s P_q$	$\alpha_s f_q$ (on-shell)
$\mathcal{F}_2$	$\frac{\alpha_s}{\pi}\left[z(1-z) - \frac{1}{6}\,\delta(1-z)\right]$	$\frac{\alpha_s}{\pi}\left[- z(1-z)\,\ell n\, z(1-z) + \frac{11}{3}(1-z)^2 - 3(1-z) - \frac{1}{2} + \frac{\frac{1}{6}}{1-z} + \delta(1-z)\{\frac{1}{6}\,\ell n\, \epsilon + \frac{1}{2}\}\right]$
$\hat{W}_i$	$\frac{\alpha_s}{\pi}\{z(1-z) - \frac{1}{6}\,\delta(1-z)\}$	$\frac{\alpha_s}{\pi}\left[- z(1-z)\,\ell nz + (1-z)(2-3z) + \frac{1}{3}\,\delta(1-z)\right]$

	$\alpha_s f_q$ (massless)	$\alpha_s \cdot \Delta_q$
$\hat{\mathcal{F}}_2$	$\frac{\alpha_s}{\pi}\Big[\{z(1-z)-\frac{1}{6}\,\delta(1-z)\}\{\frac{2}{n-6}+\gamma-\ell n\,4\pi\}$ $+ z(1-z)\,\ell n\,\frac{1-z}{z}+\frac{11}{3}(1-z)^2-3(1-z)-\frac{1}{2}$ $+\frac{\frac{1}{6}}{1-z}+\delta(1-z)\{\frac{1}{6}\,\ell n\,\epsilon+\frac{7}{9}\}\Big]$	o
$\hat{W}_i$	$\frac{\alpha_s}{\pi}\Big[\{z(1-z)-\frac{1}{6}\,\ell n(1-z)\}\{\frac{2}{n-6}+\gamma-\ell n\,4\pi\}$ $+z(1-z)\,\ell n\,\frac{(1-z)^2}{z}+(1-z)(2-3z)$ $+\frac{11}{18}\,\delta(1-z)\Big]$	$2\cdot\frac{\alpha_s}{\pi}\Big[z(1-z)\,\ell n(1-z)$ $-\frac{2}{3}(1-z)^2+2(1-z)+\frac{1}{2}$ $-\delta(1-z)\{\frac{1}{6}\,\ell n\,\epsilon+\frac{1}{6}\}\Big]$

Table 2 (contd.)

Formulas for the parton structure function in DI-scattering and for the DY-parton cross section calculated in ϕ_6^3 -field theory.

$\bar{g} \equiv \bar{g}(Q^2)$ is the running coupling constant and $\gamma^{(n)}(g)$, $\beta(g)$ stands for the anomalous dimensions and the β-function characterizing the field theory under consideration. By this calculation technique one is able to determine the influence of an infinity of Feynman diagrams as is indicated by eq.20. We seek connection between these two calculation methods in lowest order of the strong coupling constant g using the general Ansätze

$$\begin{aligned}
\beta(g) &= [\, g^3 b_1 + g^5 b_2 + \ldots] \\
\gamma^{(n)}(g) &= [\, g^2 d_1 + g^4 d_2 + \ldots]^{(n)} \\
A^{(n)}(p) &= [\, 1+g^2 \{a_{11} \, \ln \frac{p^2}{\mu^2} + a_{10}\} + \ldots \,]^{(n)} \qquad (24) \\
C_i^{(n)}(\bar{g}) &= [\, 1+c_1 \bar{g}^2 + \ldots \,]^{(n)}
\end{aligned}$$

The running coupling constant is given by

$$\bar{g}^2(t) = \frac{g^2}{1-2b_1 g^2 t} \quad , \; t = \frac{1}{2} \ln \, Q^2/Q_o^2 \qquad (25)$$

where $b_1 < 0$ is the necessary condition for asymptotically free field theories. We insert these expansions in eq. 23 and 20 and find

$$M_i^{(n)} = f^{(n)} \; [1 + g^2 \cdot \Lambda \; + \ldots \,]^{(n)} \qquad (26)$$

with

$$\Lambda^{(n)} = [\, \frac{1}{2} d_1 \; \ln \frac{Q^2}{p^2} + c_1 + a_{10} \,]^{(n)} \qquad (26)$$

By this simple expansion in the fixed coupling constant g we have established the connection between the renormalization group result and a lowest order perturbative calculation. One-to-one identification therefore allows us to determine the expansion coefficients in eq. 24 and in this way to specify the contribution of an infinity of Feynman diagrams via the RG-equation in eq.20. We realize that all perturbative cross sections have been absorbed in $A^{(n)}$ (or $\Gamma^{(n)}$ since we ignore all mixing complications) and finally were shifted in the phenomenological input momentum distributions.

But what about the arbitrariness of the perturbative parton cross section mentioned earlier due to UV-renormalization and the

M- and IR-divergences? The RG-equations are just a consequence of the subtraction ambiguity whilst performing renormalization. If we therefore specify $\mathfrak{C}_i(q^2)$ as sketched above via a perturbative calculation we have determined a renormalization independent function; whatever subtraction procedure we select $\mathfrak{C}_i(q^2)$ is invariant. The same applies for $A^{(n)}$. Mass factorization has a double function: it splits off all M-singularities from the perturbative cross section leaving a term which is not dependent on the subtraction procedure of renormalization with q^2 being its only dynamical variable. The above procedure applies for the structure function $\mathfrak{F}_2(x,q^2) \equiv u(x,q^2)$ being conventionally defined as the q^2-dependent momentum distribution with $C_2^{(n)}(\bar{g}) \equiv 1$ in eq.23 which imposes that there are no constant terms but only logs. Of course any other could have been chosen but this one is experimentally best known.

Using the fact that the moments of all other structure functions i can be written as

$$M_i^{(n)} = u^{(n)}(q^2) \quad [1 + \bar{g}^2 \cdot \Delta_i \ldots]^{(n)} \quad (i \neq 2), \tag{27}$$

which is easily obtained with eq.20 using $C_i(\bar{g})$ in eq.24, we realize that all other structure functions are parametrized by $u(x,q^2)$ plus a correction term which is responsible for all next-to-leading corrections. In this way all M-singularities of $d\sigma_i$ are subtracted (or divided out) by those of $d\sigma_2$ since the factorization matrix $\Gamma^{(n)}$ is universal and in particular is <u>not</u> i-dependent. The above form is particularly suitable for phenomenological applications [17] . $u(x,q^2)$ is determined via Mellin- inversion from $M_2^{(n)}$ with simple parametrizations being proposed for its practical use [18] . $\Delta_i \equiv f_i - f_2$ is available from the lowest order QCD-calculations. This latter expression is independent of any regularisation procedure!

We briefly indicate the completely analogous procedure for the DY-process. The moments of its cross section read

$$d\sigma^{(n)} = \text{const } f^{(n)} \; \bar{f}^{(n)} \; \hat{W}^{(n)} \tag{28}$$

with the lowest order QCD-result given in eq. 19. Its mass singularities and part of its Q^2-dependence is absorbed in the scale-dependent momentum distributions

$$d\sigma^{(n)} = \text{const. } u^{(n)}(Q^2) \; \bar{u}^{(n)}(Q^2) \quad [1 + \bar{g}^2 \cdot \Delta_{DY} + \ldots]^{(n)} \tag{29}$$

in complete analogy to eq.27 in DI-scattering. The general form of the parton cross section then is

$$\hat{W}^{(n)} = [\,1 + g^2\ \{\Lambda + \bar{\Lambda} + \Delta_{DY}\} + \ldots\,]^{(n)} \tag{30}$$

The Λ's stand for the order-g^2 corrections in DI-scattering (see eq.22). Comparison between the eqs.30 and 19 determines

$$\Delta_{DY} = f_{DY} - 2\,f_2 \quad \text{where } a_{10} \equiv f_2$$

was identified earlier. Δ_{DY} is expected to be independent of any regularisation procedure; this could indeed be verified for on/off-shell mass assignment and using n-dimensional regularisation in the case of all vanishing parton masses (see Table 2).

4. CONCLUSIONS

In this short note we exposed the main assumptions in calculating parton cross sections in the framework of perturbative QCD. We pointed to the divergence problems and the arbitrariness emerging from technically preventing the UV-, M- and IR-divergences. Subsequently the DI-structure functions and the DY-cross sections on the parton level are determined in order g^2. We use on-shell mass assignment and n-dimensional regularisation with all partons massless. Detailed results are given for ϕ_6^3-theory.

In the second part of this lecture we exposed the connection between the RG-approach and perturbative evaluation. Establishing a one-to-one correspondence between these two calculation methods we demonstrate that regularisation independent predictions still can be obtained. We outline the arguments for DI-scattering and briefly indicate their extension to the DY-process.

REFERENCES

1. H. Fritzsch, M. Gell-Mann and H. Leutwyler, Phys.Lett. 47B, 365 (1973); H. Fritzsch, CALT-68-524 (1975), Nordita Lectures 1975; H. D. Politzer, Phys.Rep. 14C, 129 (1974); W. Marciano and H. Pagels, Phys.Rep. 36, 137 (1978); S. Weinberg, Rev.Mod.Phys. 46, 255 (1974).
2. For reviews see:

   Y. L. Dokshitzer, D. I. D'Yakonov and S. I. Troyan, Lectures at the Leningrad Winterschool 1978, SLAC-TRANS-183 (1978); Proc. of EPS Internat.Conf. on High Energy Physics, Geneva, 1979; Proc. of the 19th Internat. Conf. on High Energy Physics, Tokyo, 1978; H. Fritzsch, 18th Internat. Universitätswochen für Kernphysik, Schladming, 1978; C. H. Llewellyn-Smith, 18.Internat.Universitätswochen für Kernphysik, Schladming 1978, Cargèse Summer

Institute 1977; J. Ellis, SLAC-PUB-2121 (1978); B. Humpert, Helv.Phys. Acta 51, 542 (1978); and ref. 18 below.

3. H. D.Politzer, Nucl.Phys. B129, 301 (1977); D. Amati, R. Petronzie and G. Veneziano, Nucl.Phys. B140, 54 (1978) and Nucl.Phys. B146, 29 (1978); R. K. Ellis, H. Georgi, M. Machacek, H. D. Politzer and G. G. Ross, Phys.Lett. 78B, 281 (1978) and Nucl.Phys. B152, 285 (1979); S. Libby and G. Sterman, Phys.Lett. 78B, 618 (1978); C. T. Sachrajda, Phys.Lett. 73B, 185 (1978).

4. T. Kinoshita, J.Math.Phys. 3, 650 (1962); T. D. Lee and M. Nauenberg, Phys.Rev. 133, B1549 (1964); N. Nakanishi, Progr.Theor.Phys. 19, 159 (1958).

5. M. Glück and E. Reya, Phys.Lett. 69B, 77 (1977); W. A. Bardeen, A. J. Buras, D. W. Duke and T. Muta, Phys. Rev. D18, 3998 (1978); E. G. Floratos, D. A. Ross and C. T. Sachrajda, Nucl.Phys. B129, 66 (1977), E: Nucl. Phys. 139, 545 (1978), and Nucl.Phys. B152, 493 (1979); J. Ellis, Lectures at the Les Houches Summerschool 1976; A. Peterman, CERN-Ref.Th. 2581 (1978), to appear in Physics Report; M. K. Gaillard, XII Rencontre de Moriond 1977, p. 485; and conference reports cited in ref. 2.

6. G. C. Marquez, Phys.Rev. D9, 386 (1974); P. M. Fishbane, C. S. Lam and T. M. Yan, Cornell University 1979 (and references cited therein).

7. G. 't Hooft and M. Veltman, Nucl.Phys. B44, 189 (1972); C. G. Bollini and J. J. Giambiagi, NC. 12B, 20 (1972); G. 't Hooft and M. Veltman, CERN-Report 73-9 (1977) "Diagrammar".

8. H. Heitler, "The Quantum Theory of Radiation", Oxford Clarendon Press 1959, International Series of Monographs on Physics; J. J. Sakurai, "Advanced Quantum Mechanics", Reading, Mass., Addison Wesley 1967 (Addison Wesley Series in Advanced Physics); S. J. Brodsky, R. Roskies and R. Suaya, Phys. Rev. D8, 4574 (1973).

9.a B. Humpert and W. L. van Neerven, Phys.Lett. 84B, 327 (1979), E: Phys.Lett. 85B, 471 (1979).

9.b G. Altarelli, R. K. Ellis and G. Martinelli, Nucl.Phys. B143, 521 (1978).

9.c K. Harada, T. Kaneko and N. Sakai, Nucl. Phys. B155, 169 (1979) and details in CERN-Preprint TH. 2619 Erratum (1979).

9.d J. Kripfganz, "Comments on Regularisation Prescription Dependence of QCD Corrections to the Drell-Yan Formula", McGill University 1979.

10. W. J. Marciano, Phys.Rev. D12, 3861 (1975) and references therein; G. Altarelli, R. K. Ellis and G. Martinelli, Nucl.Phys. B157, 461 (1979; and ref. 9 a.

11. B. Humpert and W. L. van Neerven, Ref.Th. 2738-CERN,Sept.1979, to appear in Phys.Lett.

12. B. Humpert and W. L. van Neerven, "IR-and M-Regularisation in AF-Field Theories I:ϕ_6^3", Ref.Th. 2785-CERN, Nov. 1979.
13. We assume that the reader is familiar with the basics of these approaches.
14. Y. Frishman, Phys.Rep. 13C, 1 (1974); J. Ellis and R. L. Jaffe, U.C. Santa Cruz Summerschool Lectures, SLAC-PUB-1253 (1973).
15. C. G. Callan, Phys.Rev. D2, 1541 (1970); K. Symanzik, Comm. Math.Phys. 18, 227 (1970); G. 't Hooft, Nucl.Phys. B61, 455 (1973); M. J. Holwerda, W. L. van Neerven and R. P. van Royen, Nucl.Phys. B75, 303 (1974); R. G. Crewther, Cargèse Summer Institute Lectures, CERN-Preprint TH.2119 (1976); S. Coleman, "Dilatations", Lectures given at the 1971 International Summerschool of Physics "Ettore Majorana" (Ed.Compositori, Bolognia 1973).
16. N. Christ, B. Hasslacher and A. H. Mueller, Phys.Rev. D6, 3543 (1972).
17. J. Abad, B. Humpert and W. L. van Neerven, Phys.Lett. 83B, 371 (1979); B. Humpert and W. L. van Neerven, Phys.Lett. 85B, 293 (1979); B. Humpert, Zeitschr.f.Physik 2C, 73 (1979); A. N. Schellekens and W. L. van Neerven, Univ. of Nijmegen, THEF-NYM-79.8; J. Kubar-André and F. E. Paige, Phys.Rev. D19, 221 (1978).
18. B. Humpert, Workshop on Lepton-Pair Production in Hadron-Collisions, Bielefeld, Germany, 1978 (and references therein), Ref.Th. 2639-CERN, April 1979.

CHIRAL PROPERTIES OF QUANTUM CHROMODYNAMICS

R. J. Crewther

CERN

Geneva, Switzerland

1. INTRODUCTION AND SUMMARY

The success of PCAC and current algebra more than a decade ago was an important constraint on field-theoretic attempts to construct a realistic model of the strong interactions. Our best candidate, quantum chromodynamics (QCD), satisfies part of this requirement: Gell-Mann's current commutators are valid for renormalized operators of the fully interacting theory. What is not clear is how to realize the approximate chiral SU(n)x SU(n) symmetry of the QCD Hamiltonian in the Nambu-Goldstone mode. The hadronic ground state (vacuum) should yield expectation values of quark mass operators which do not vanish in the chiral limit; for example, we should find

$$\langle vac\, |\bar{u}\, u|\, vac\rangle \neq 0 \tag{1.1}$$

for the up quark u. It is difficult to find a systematic expansion of the QCD generating functional in which the leading approximation exhibits the property (1.1). However, some progress can be made by assuming (1.1) to be true; then generally valid equations such as anomalous Ward identities give conditions which must be satisfied by any self-consistent expansion of QCD.

Long before the invention of QCD, Glashow [1] recognized that the spontaneous breaking of chiral symmetry implied by (1.1) involves U(n)x U(n), not just SU(n)x SU(n). This difficulty became known as the U(1) problem; (see Ref.2 for references). Instantons were claimed to solve the problem, but, as I noted two years ago [3], these claims involved two important misunderstandings:

(a) The U(1) problem was not recognized to be associated with the property (1.1). Instead, it was assumed that it was sufficient to observe an example of a non-zero change of U(1) chirality without the generation of a corresponding Nambu-Goldstone boson. That is a fake problem.

(b) The spontaneous breaking of axial U(1) symmetry was confused with explicit symmetry breaking. Then this breaking was assumed to be represented by an effective Lagrangian [4] which is valid for a particular Green's function computed for an SU(n)x SU(n) invariant vacuum. The spontaneous breaking of SU(n) x SU(n) was supposed to arise from a Hartree-Fock treatment of this effective Lagrangian [5] . This procedure is not self-consistent because it does not respect the anomalous Ward identities of QCD [3] : It amounts to the replacement of the topological charge density *)

$$(g^2/32\pi^2)F.F^* ,$$

which has zero U(1) chirality in anomalous Ward identities, by a sum of operators with non-zero chiralities.

The effect of the present discussion will be to reinforce the case against instantons being a cure for the U(1) problem. As in all of my work [2,3,6,7] on the problem, the θ parameter is introduced as a coupling constant in the QCD Lagrangian

$$L = -\frac{1}{4}F^2 + \sum_{i=1}^{N} \bar{q}_i(i\not{D} - m_i)q_i - \theta(g^2/32\pi^2)F.F^* \tag{1.2}$$

(where q_i are the quark fields u,d,s,... and D_μ is the gauge covariant derivative). It is not introduced in terms of linear combinations of pure-gauge states - indeed, I believe that sort of assumption to be hopelessly unrealistic. According to (1.2), the parameter θ acts as an x_μ independent source of the topological charge operator (connected insertions):

$$i\frac{\partial}{\partial\theta} \leftrightarrow \frac{g^2}{32\pi^2}\int d^4x\, F.F^*(x) \tag{1.3}$$

The basic observation of Ref. 3 was: if one assumes the absence of massless U(1) bosons in the SU(n) x SU(n) limit for all values of θ, the condition (1.1) can be realized only if the topological charge operator takes fractional values

*) Here g is the gluon coupling constant, $F^a_{\mu\nu}$ is the field strength tensor, and $F^*_{\mu\nu}$ is the dual tensor $\frac{1}{2}\varepsilon_{\mu\nu\alpha\beta}F^{\alpha\beta}$ with $\varepsilon_{0123} = +1$.

$$\nu = \pm {}^{1}/n \tag{1.4}$$

and integer multiples thereof. This remark will be confirmed and made more precise below.

The condition (1.4) posed a logical difficulty which I was unable to resolve in Ref.3 except by relaxing the assumption about U(1) bosons. Eq.(1.3), which is generally valid, seems to imply that Green's functions must have periodicity $2\pi n$ in θ as the SU(n) x SU(n) limit is approached. However, apart from the constraint

$$n \leqq N \tag{1.5}$$

the integer n can be chosen at will, so the period seems to depend on which SU(n) x SU(n) limit we choose to consider. This conclusion is impossible, because the period cannot be 4π and 6π in the overlapping vicinities of the SU(2) x SU(2) and SU(3) x SU(3) limits.

Finding no satisfactory alternative, I proposed [3] that massless U(1) bosons should exist for all values of θ except for the neighbourhood of the P and T conserving value $\theta = 0$. As far as I know, there are no theoretical inconsistencies in this proposal. For example, the θ dependence of the S-matrix in the SU(n) x SU(n) limit is not obviously wrong: the conventional argument for θ independence assumes the absence of massless U(1) bosons.[See Refs. [2,8] and Sect. 2 below.] However, the proposal is not really satisfactory because:

(a) There is no reason to suppose that the case $\theta = 0$ is special.

(b) Even if massless U(1) bosons are absent for $\theta = 0$ in pure QCD, a renormalization of θ by weak interactions is likely to cause their reappearance.

[At no stage have I argued that the problem is merely that ν is not an integer. In general, there is no reason to require finite classical action, or more accurately, compactification, so ν is not restricted by any classical theorem [2, 3]. Of course, people who insist that gluonic functional space is dominated by mixtures of instantons and anti-instantons cannot live with Eq.(1.4).]

These lectures concern a more attactive alternative [6,7]*) in which the fractional topological charge of Eq.(1.4) is still required but the θ period is 2π, whatever the value of n. The loophole which permits this possibility is contained in some 1971

---

*) I thank S. Coleman for his generous assistance. See Ref. [9] for the large-N_c point of view.

remarks of Dashen [10] on spontaneous CP violation in SU(3)×SU(3) current algebra.

Normally, one supposes that if none of the quark mass parameters vanishes, all Nambu-Goldstone bosons aquire masses and a unique vacuum state is picked out by the mass term which breaks chiral symmetry in the Hamiltonian. In the vicinity of the SU(n) ×SU(n) limit where the first n mass parameters $m_1,\ldots,m_n$ are small but not zero, one expects the vacuum state to be uniquely defined: hence the operation $i\partial/\partial\theta$ should yield an unambiguous answer. If that is true for all θ, there is no escape from the periodicity problem apart from the proposal of Ref. [3].

However, we will now permit part of the vacuum degeneracy associated with a chiral limit to remain for isolated values of θ even when all m_i are positive. It is then possible to require the absence of massless U(1) bosons for all values of θ in any SU(n) ×SU(n) limit. This requirement results in a two-fold vacuum degeneracy at $\theta = \pi \pmod{2\pi}$ if the condition

$$m_1^{-1} \leq \sum_{i=2}^{n} m_i^{-1} \tag{1.6}$$

is obeyed; here the small mass parameters $m_1,\ldots,m_n$ are ordered in the following way:

$$0 < m_1 \leq m_2 \leq \ldots. \leq m_n \tag{1.7}$$

The operator $\hat{U}$ relating these two vacua is a non-trivial SU(n)×SU(n) transformation which depends on the mass ratios m_i/m_j and exists only if (1.6) is satisfied. Eq. (1.6) is the general condition for spontaneous CP violation of the type considered by Dashen [10] . For the completely degenerate case $m_i = m_j$, Eq.(1.4) is valid and $\hat{U}$ becomes an element of the chiral centre group $Z_n \times Z_n$. For the special case

$$n = 2,\ m_u = m_d\ ,\ \theta = \pi \pmod{2\pi} \tag{1.8}$$

there is an additional continuous degeneracy - the SU(2)×SU(2) Nambu-Goldstone bosons remain massless in the first order of chiral symmetry breaking.

Despite the validity of Eqs.(1.3) and (1.4), the θ period satisfied the requirement sought in Ref.3: it does not depend on n. In the vicinity of any SU(n)× SU(n) limit, the period is 2π irrespective of whether ν is fractional or not. If Eq.(1.6) is satisfied, there is a two-fold ambiguity in $i\partial/\partial\theta$ at $\theta = \pi$ associated with the two vacua related by $\hat{U}$. The sign of the perturbation

$$\Delta L = (\pi-\theta)\ (g^2/32\pi^2)\ F.F^* \quad , \quad (\theta \simeq \pi) \tag{1.9}$$

determines which of these vacua is chosen. Thus the system jumps from one vacuum to the other as θ passes through π, and so the simple Fourier relation between θ dependence and the spectrum of ν (which normally fixes the period) ceases to be valid just at the isolated points $\theta = \pi \pmod{2\pi}$. This observation allows us to reconcile fractional values of ν with period 2π.

This scheme contains its own assumptions, so there is no guarantee that it can be realized in an explicit calculation. For example it is assumed that tunnelling occurs between the two vacuum candidates at $\theta = \pi$. It is only for the special case (1.8) that the existence of a transition is obvious: then the two vacua are connected by a continuously degenerate set of states, so there is no potential barrier and hence no need to postulate tunnelling.

I hasten to add that the periodicity 2π has nothing to do with finite action or instantons. The existence of contributions to the generating functional with fractional topological charge (1.4) is essential for the success of this scheme. It is true that the physical values of the quark masses probably do not satisfy (1.6),

$$\sum_{i=2}^{n} (m_u/m_i)_{phys.} \simeq 0.6 < 1 \tag{1.10}$$

but conclusions like (1.10) depend on chiral perturbation theory being true for arbitrary values of the mass ratios. In other words, the chiral perturbation expansion should be valid in the neighbourhood of the origin $m_i=0$, and that neighbourhood contains points satisfying (1.6).

These conclusions are explained in Section 4. Previous sections concern some aspects of anomaly theory (Section 2) and the U(1) problem (Section 3) which I believe to be poorly understood in recent literature. In particular, the subject of anomalous commutators [11,12] seems to have been generally ignored, despite my attempts [2,8] to emphasize its importance:

(a) Gauge-invariant chiralities appearing in anomalous Ward identities are generated by a chiral current which is necessarily gauge-dependent [13,14].

(b) The original operator derivation [4] of the connection between chirality and topological charge does not work [8]: the operator explanation implies that left-handed fermions couple to an instanton, whereas the explicit calculation [4,15] correctly shows that they must be right-handed.

(c) It is necessary to assume the absence of massless U(1) bosons in order to obtain a connection between the θ parameter and chiral U(1) rotations or prove the θ independence of the S-matrix when some quark mass parameters vanish [8] .

Section 3 also concerns the problem of specifying boundary conditions for QCD functional integrals with realistic chiral properties. This leads to a consideration (Section 5) of calculational methods which may exhibit these properties explicitly. The importance of allowing the functional integral to choose its own spectrum of topological charge (in order to satisfy (1.4) and (1.9)) is emphasized: infrared boundary conditions in functional space have to be determined by self-consistency, not by conventional but irrelevant constraints such as finiteness of the classical action.

Finally, the phenomenology [16,17] of CP violating processes induced by a small θ parameter is considered in Section 6. These effects may be observable in strangeness-conserving amplitudes, of which the most important is the electric dipole moment of the neutron.

2. ANOMALIES, ABELIAN CHIRALITY, AND THE Θ PARAMETER

Anomalies are "anomalous" because they have no canonical explanation. Nevertheless, there is a tendency in present-day accounts of topological charge or instantons to rely on canonical arguments which are at best misleading. For example, one continually sees the statement that "the anomaly breaks chiral U(1) invariance". As a statement about operators, this is wrong: the fact is [3] that the U(1) chirality of $F.F^*$ is zero. It is not correct to suppose the anomaly to be the U(1) analogue of the soft divergence of a partially conserved SU(n) x SU(n) current - there is no anomalous analogue of the σ-term [3] . The crucial point [13,14] is that U(1) chiralities are gauge invariant, but the chiral U(1) current which generates them is necessarily gauge dependent (because of the anomaly). Changes in chirality observed in explicit calculations [4,15] are caused by the presence of a continuously degenerate set of vacua associated with a spontaneous realization of chiral U(1) symmetry.

These remarks are essential for the derivation of the oft-quoted connection between chiral U(1) rotations and the θ parameter when some of the quark mass parameters m_i vanish. It is not true that the derivation is just a matter of inspecting the anomalous divergence equation. The correct statement [2,8] is that θ is equivalent to a chiral U(1) angle only if massless U(1) bosons are assumed to be absent for all θ. The derivation involves anomalous Ward identities.

The explanation of these points depends on some subleties of anomaly theory which will now be reviewed. To simplify symmetry properties, I shall consider the case

$$\begin{aligned} &m_i = 0 \qquad , \ (i = 1, \ldots, n) \\ &m_{n+1}, \ldots, m_N \neq 0 \end{aligned} \tag{2.1}$$

Everyone agrees that SU(n) x SU(n) is then a symmetry of the QCD Lagrangian, but there is controversy about the candidate symmetry

$$\{U(1)_L \times U(1)_R\}_n$$

given by

$$\begin{aligned} \frac{1}{2}(1 \pm \gamma_5)\, q_i &\rightarrow e^{i\alpha_\pm}\, \frac{1}{2}(1 \pm \gamma_5) q_i \ , \ (i=1,\ldots,n) \\ q_i &\rightarrow q_i \qquad , \ (i=n+1,\ldots,N) \\ A^a_\mu &\rightarrow A^a_\mu \end{aligned} \tag{2.2}$$

where A^a_μ is the gluon field and $q_L = \frac{1}{2}(1-\gamma_5)q$ is a left-handed quark.

Let us denote the axial part of the Abelian group of transformations (2.2) by

$$U_n(1) = U(1)_L \times U(1)_R / U(1)_{L+R} \tag{2.3}$$

In general, renormalization of the corresponding $\gamma_\mu\gamma_5$ vertex produces ambiguities of the form

$$\varepsilon_{\mu\nu\alpha\beta}\, A^\alpha \partial^\nu A^\beta \quad \text{and} \quad \varepsilon_{\mu\alpha\beta\gamma}\, A^\alpha\, A^\beta \times A^\gamma$$

generated by triangle and box diagrams respectively. The anomaly reflects the absence of a renormalization procedure which simultaneously preserves gauge and chiral U(1) covariance of Green's functions. However, $\gamma_\mu\gamma_5$ is an external operator, so nothing prevents us from using inequivalent renormalization procedures (one gauge-invariant, the other chiral U(1) invariant) to construct two renormalized $\gamma_\mu\gamma_5$ vertices, or "normal products": *)

$$J^n_{\mu 5} = N_{g.inv.}\left[\sum_{i=1}^{n} \bar{q}_i \gamma_\mu \gamma_5 q_i\right] \tag{2.4}$$

*) As in Ref. [8] , I am deliberately avoiding the standard notation $\sum_i \bar{q}_i \gamma_\mu \gamma_5 q_i$ for the gauge-invariant operator because of the confusion that this formula arouses in the canonically minded.

$$J_{\mu 5}{}^{n}{}_{sym} = N_{ch.inv.}\left[\sum_{i=1}^{n} q_i \gamma_\mu \gamma_5 q_i\right] \qquad (2.5)$$

The difference between these operators

$$J_{\mu 5}{}^{n} = J_{\mu 5}{}^{n}{}_{sym.} + 2n\, K_\mu \qquad (2.6)$$

involves an operator K_μ which (in unrenormalized form) is a definite linear combination of the triangle and box diagram ambiguities:

$$K^\mu{}_{Bare} = \frac{g^2}{32\pi^2}\, \varepsilon^{\mu\alpha\beta\gamma} A^a_\alpha (F^a_{\beta\gamma} - \frac{1}{3} g\, c^{abc} A^b_\beta A^c_\gamma) \qquad (2.7)$$

A gauge-invariant way of expressing the inequivalence of $J_{\mu 5}{}^{n}$ and $J_{\mu 5}{}^{n}{}_{sym}$ is to consider the divergence

$$\partial^\mu K_\mu = (g^2/32\pi^2)\; F.F^* \qquad (2.8)$$

The chiral invariance of the renormalization procedure for the gauge-dependent current $J_{\mu 5}{}^{n}{}_{sym}$ preserves conservation of the $\gamma_\mu\gamma_5$ vertex,

$$\partial^\mu J_{\mu 5}{}^{n}{}_{sym.} = 0 \qquad (2.9)$$

so the familiar equation [13,14]

$$\partial^\mu J_{\mu 5}{}^{n} = 2n\; (g^2/32\pi^2)\; F.F^* \qquad (2.10)$$

follows from Eqs.(2.6),(2.8) and (2.9).

Anomaly theory is not just a matter of memorizing Eq.(2.10): a precise understanding of operator chirality and anomalous Ward identities is essential if inconsistencies are to be avoided. Consider a local gauge (such as a Lorentz invariant gauge) in which commutativity at space-like separations is preserved. When $\partial/\partial x^\mu$ is applied to Green's function of $J_{\mu 5}{}^{n}(x)_{sym.}$, the only non-zero contribution is a sum of δ^4 functions associated with short-distance singularities of the T-product:

$$\partial^\mu_x T < vac \mid J_{\mu 5}{}^{n}(x)_{sym} \prod_k O_k(x_k) \mid vac >$$

$$= - \sum_\ell \delta^4(x-x_\ell)\; \vec{\chi}_\ell(n).T < vac \mid \vec{O}_\ell(x_\ell) \prod_{k\neq\ell} O_k(x_k) \mid vac > \qquad (2.11)$$

A constant χ_ℓ produced in this way by any conserved or partially conserved current cannot depend on the gluon coupling constant g because [2,18]:

(a) It is m_i-independent (being derived from a leading short-distance singularity) and dimensionless, so it cannot depend on the renormalization subtraction point μ .

(b) The conservation properties of the current ensure that is has no wave-function renormalization, so χ_ℓ must be renormalization-group invariant:

$$\beta(g) \frac{\partial}{\partial g} \vec{\chi}_\ell = 0 \tag{2.12}$$

Thus χ_ℓ equals its value in free-field theory.

A non-renormalization theorem such as (2.12) is an essential ingredient in the derivation of Ward identities. For Abelian chiralities, it implies the obvious corollary that χ_ℓ is gauge invariant, despite being generated by the gauge-dependent current $J^n_{\mu 5}(x)_{sym}$.

Thus it makes sense to talk about operators O_k of definite $U_n(1)$ chirality $\chi_k(n)$,

$$\partial^\mu_x \, T \langle vac \mid J^n_{\mu 5}(x)_{sym} \prod_k O_k(x_k) \mid vac\rangle$$

$$= - \sum_\ell \chi_\ell(n)\, \delta^4(x-x_\ell)\, T \langle vac \mid \prod_k O_k(x_k) \mid vac\rangle \tag{2.13}$$

and (in local gauges) use the notation

$$\left[Q^n_5, O_k\right] = - \chi_k(n)\, O_k \tag{2.14}$$

$$Q^n_5 = \int d^3x \; N_{ch.inv.} \left[\sum_{i=1}^n q_i^+ \gamma_5 q_i\right]$$

$$= \int d^3x \; J_{05}{}^n(x)_{sym} \tag{2.15}$$

The minus sign in Eqs. (2.13) and (2.14) corresponds to the right-handed up quark $\frac{1}{2}(1+\gamma_5)u$ having positive $U_n(1)$ chirality $\chi = +1$.

The procedure outlined above for $J_{\mu 5}(x)_{sym}$ does not work for the gauge-invariant operator $J_{\mu 5}$: it has a hard F.F.* divergence whose T-product ambiguities compete with commutator terms proportional to $\delta^4(x-x_k)$ which we want to isolate. The correct way to obtain anomalous Ward identities in this general fashion [2,8] is to combine Eqs. (2.6) and (2.13):

$$\partial_x^\mu \langle vac| J_{\mu 5}^{\ n}(x) \prod_k O_k(x_k) |vac\rangle$$

$$= \quad 2n\, \partial_x^\mu\, T \langle vac|\ K_\mu(x) \prod_k O_k(x_k)\ |vac\rangle \tag{2.16}$$

$$- \sum_\ell \chi_\ell(n)\ \partial^4(x-x_\ell)\ T \langle vac| \prod_k O_k(x_k) |vac\rangle$$

For gauge-invariant operator O_k, the anomalous term involving K_μ is gauge invariant because the remaining terms in the identity are gauge invariant.

The most interesting form [3] of the anomalous Ward identity (2.16) is found by considering the limit in which the operator $J_{\mu 5}^{\ n}(x)$ carries zero momentum. Let us apply $\int d^4x$ to Eq. (2.16). Obviously, this tests whether massless U(1) bosons are physically present or not. It also yields a connection with the θ parameter, because the substitution (1.3) can be applied to the anomalous term. Thus Eq. (2.16) becomes a differential equation in θ for a soft-meson amplitude:

$$\int d^4x\ \partial_x^\mu\ T \langle vac| J_{\mu 5}^{\ n}(x) \prod_k O_k(x_k)\ |vac\rangle \tag{2.17}$$

$$= \left[\ 2ni \frac{\partial}{\partial\theta} - \sum_\ell \chi_\ell(n)\ \right]\ T \langle vac| \prod_k O_k(x_k)\ |vac\rangle$$

Note that the anomalous term in (2.16) is a total divergence with ∂_x^μ <u>outside</u> the time-ordering operation. Therefore the anomaly in (2.17) can be written as a surface integral with the out and in winding-number operators

$$K_{\pm} = \int d^3x\ K_0(\vec{x},\ t = \pm\, \infty) \tag{2.18}$$

acting directly on $\langle vac|$ and $|vac\rangle$. That reflects the dual role of θ as a coupling constant [Eqs. (1.2) and (1.3)] and a vacuum parameter, and leads to the general result [2] that vacuum states are mixtures

$$|\theta'\rangle = \sum\!\!\!\!\int_m e^{im\theta'} |m\rangle \tag{2.19}$$

of eigenstates $|m\rangle$ of $K_\pm$ with eigenvalues m.

Remarks:

(a) in general, $|m\rangle$ has nothing to do with pure-gauge configurations $G^{-1}\partial G$ in functional integrals [2]. Certainly,

$G^{-1}\partial G$ is an eigenstate, and it is relevant for ordinary instanton calculations about SU(n) x SU(n) invariant vacua; but there is no theorem with realistic hypotheses which requires all eigenstates to be pure-gauge states. Indeed, it is likely that $G^{-1}\partial G$ is not contained in the vector space of states generated by a realistic approximation to QCD, i.e., one which obeys Eq. (1.1).

(b) The formula (2.19) holds for a range of θ values with unique vacua. As indicated in Section 1, a degeneracy at a particular value of θ requires a separate treatment. See Section 4.

It is naive to say that the Lagrangian is "naively" chiral $U_n(1)$ invariant in the SU(n) x SU(n) limit. There is nothing naive about its $U_n(1)$ invariance, as is evident from the corresponding Ward identities (2.13). The point is that gauge and chiral $U_n(1)$ transformations are <u>both</u> symmetries of L but because of the anomaly, they <u>do not commute</u> -- as is evident from the impossibility of <u>simultaneously</u> preserving gauge and chiral U(1) covariance in Green's functions.

Thus, in the presence of anomalies, topological charge, instantons or whatever, we have (in the SU(n) x SU(n) limit):

$$e^{i\alpha Q}\, L(\theta)\, e^{-i\alpha Q} = L(\theta) \quad . \tag{2.20}$$

A chiral transformation does <u>not</u> transform the operator $L(\theta)$ into $L(\theta-2n\alpha)$. A change of θ to $(\theta-2n\alpha)$ induced by the chiral transformation $\exp\,(i\alpha Q_5)$ concerns <u>states</u> $|\theta\rangle$ and <u>matrix elements</u> $\langle\ldots\rangle_\theta$, not operators, and involves <u>assuming</u> the absence of massless U(1) bosons [2,8].

The muddle in the literature on this point arises from a canonical misunderstanding of the conventional but unfortunate notation

$$\sum_i \bar{q}_i \gamma_\mu \gamma_5\, q_i \tag{2.21}$$

for the gauge-invariant operator $J_{\mu 5}$. The subject of anomalous commutators [11,12] cannot be ignored: U(1) chiralities have nothing to do with $J_{\mu 5}$. Chiral anomalies and the failure of canonical commutation relations are both consequences of renormalization.

For example, consider commutators of the correct generator Q_5 of Eq. (2.15) and the wrong generator

$$X(t) = \int d^3x\, J_{05}(\vec{x},t) \tag{2.22}$$

with the time derivative of spatial components of the gluon field (Feynman gauge):

$$\left[Q_5, \partial_o A_i \right] = 0 \tag{2.23}$$

$$\lim_{t'\to t} [X(t'), \partial_o A_i(\vec{x},t)] = -(ing^2/4\pi^2)F^*_{oi} + O(g^4 \ell n|t'-t|) \tag{2.24}$$

[This is the non-Abelian generalization of an example of Adler and Boulware [12]]. Eq. (2.23) is a consequence of the non-renormalization result (2.12). In Eq. (2.24), the higher-order $\ell n|t'-t|$ terms are caused by wave-function renormalization [2,8,13] of $J_{\mu 5}$. This also means that the normalization of X(t) in higher orders is arbitrary [8]. It should be obvious that X(t) does not generate chirality.

Another illustration of this point is the inconsistency in the original analysis [4] of instantons and chirality which was noted and resolved in Section 8 of Ref. [8].

Explicit calculation [4,15] shows that an instanton with topological charge $\nu = +1$ gives rise to a right-handed quark amplitude

$$T < \prod_{i=1}^{N} \bar{q}_L q_R(x_i) >_{\text{inst.}} \neq 0 \tag{2.25}$$

even if all mass parameters m_i vanish. The simplest and best way to explain this phenomenon is to consider the anomalous Ward identity for the vanishing *) soft-meson amplitude

$$\int d^4x\, \partial^\mu_x\, T < J_{\mu 5}(x) \prod_{i=1}^{N} \bar{q}_L q_R(x_i) >_{\text{inst.}} = 0 \tag{2.26}$$

However, the original explanation depended on a confusion of X(t) with Q_5 and on an attempt to use the anomalous divergence equation (2.10) (for n=N) without considering matrix elements and time-ordering. The total U(1) chirality in (2.25) was supposed to be given by

$$\int d^4x\, \partial^\mu J_{\mu 5}$$

and hence by $+2N\nu$. This interpretation requires instantons to couple

*) It vanishes because [2] the calculation is based on compactifying Euclidean space to the hypersphere in five dimensions.

to left-handed quarks; i.e., it gives the wrong answer.

A quick way [8] of getting the right answer is to write

$$X(t) = Q_5 + 2N \int d^3x\, K_o(x,t) \quad ,(\text{for } n=N) \tag{2.27}$$

and hence

$$\Delta X = \Delta Q_5 + 2N\Delta K = \Delta Q_5 + 2N\nu \tag{2.28}$$

where ΔQ means the out eigenvalue of an operator $Q(t)$ at $t = +\infty$ minus the in eigenvalue at $t = -\infty$; (so in general, a physical amplitude contains a mixture of transitions with various values of $\Delta X, \Delta Q_5, \Delta K$). Eq. (2.26) written as a surface integral implies that all transitions satisfy [8]

$$\Delta X = 0, \quad (\text{no } U(1) \text{ bosons}) \tag{2.29}$$

so Eq. (2.28) becomes:

$$\Delta Q_5 = -2N\nu \tag{2.30}$$

This is the desired result because, according to Eq. (2.14), ΔQ_5 and $\Sigma\chi$ have opposite signs:

$$\sum_k \chi_k = 2N\nu \quad . \tag{2.31}$$

Note, that the value of ΔQ_5 or ΔX has nothing to do with conservation (or lack thereof) of the corresponding $\gamma_\mu\gamma_5$ operator. In fact, the formulas

$$\begin{aligned} \dot{Q}_5 &= 0 \\ \Delta Q_5 &\neq 0 \end{aligned} \tag{2.32}$$

precisely state [19] what is meant by <u>spontaneously</u> realizing the symmetry (2.20).

Because of its SU(n) x SU(n) invariant vacua, the above example is not directly relevant for the U(1) problem. However, it does demonstrate the necessity of setting a soft-meson amplitude to zero in order to derive a result -- a necessity not evident to those who confuse $X(t)$ with Q_5. The same logical distinctions have to be understood when one tackles the U(1) problem.

Let us now consider the circumstances under which the S-matrix becomes θ independent.

In general, the anomalous Ward identity in the SU(n) x SU(n) limit is given by Eq.(2.17). This can be rewritten in terms of a $U_n(1)$ chiral angle ϕ_n which parametrizes a vacuum-state degeneracy generated by transformations

$$U(\phi_n) = \exp\,(-iQ_5^n\,\phi_n) \tag{2.33}$$

From the identity

$$i\,\frac{\partial}{\partial\phi_n}\{U^{-1}\prod_k O_k\,U\} = \sum_\ell \chi_\ell U^{-1}\prod_k O_k U \tag{2.34}$$

we find:

$$\int d^4x\,\partial_x^\mu\,T\,\langle vac|\,J_{\mu5}^n(x)\prod_k O_k(x_k)\,|vac\rangle$$

$$= (2ni\,\partial/\partial\theta - i\,\partial/\partial\phi_n)\,T\,\langle vac|\prod_k O_k(x_k)\,|vac\rangle \tag{2.35}$$

The same equation applies to the $U(1)_{PQ}$ - invariant unified theories proposed by Peccei and Quinn [20] , with n replaced by the number of fermion flavour chirally transformed by $U(1)_{PQ}$.

Now the crucial point is that θ and ϕ_n are related in the usual way only if soft-meson amplitudes on the left-hand side of Eq.(2.35) are assumed to vanish. In other words, if zero-mass $U_n(1)$ bosons are supposed to be absent in the SU(n) x SU(n) limit, the operator

$$2n\,\frac{\partial}{\partial\theta} - \frac{\partial}{\partial\phi_n} \tag{2.36}$$

annihilates all Green's functions, so any change $\Delta\theta$ in θ is the same (for matrix elements) as performing an equivalence transformation $U(\Delta\theta/2n)$ on L. In the SU(n) x SU(n) limit, U does not change L, so S-matrices with different values of θ are described by the same theory; i.e., the S-matrix is θ independent. Similarly, it is necessary to assume the absence of massless $U(1)_{PQ}$ bosons in order to derive P and T conservation of the S-matrix of a Peccei-Quinn theory.

For many people, this emphasis on having to assume the absence of massless U(1) bosons to derive something seems to be misplaced. They assume (or claim [21,22]) the existence of a theorem that zero-mass bosons cannot couple to the gauge-invariant operator $J_{\mu5}$. Typical arguments are that this follows from the lack of conservation of $J_{\mu 5}$, or from the gauge dependence of the Kogut-Susskind pole [23] coupled to $J_{\mu 5_{sym}}$, or from the fact that the generator Q_5 rotates state vectors out of the physical Hilbert space.

In fact, there is a theorem which demonstrates that these arguments are not valid: it is impossible to prove the absence of massless U(1) bosons using Ward identities alone.

The proof is trivial. Let L_A be a Lagrangian with gauge-invariant operator $J^A_{\mu 5}$ which is not conserved because of an anomaly, and has <u>no</u> massless boson coupled to it. Let L_B be another Lagrangian which exhibits the same symmetries as L_A but without anomalies, which does not depend on the fields in L_A, and which has a conserved current $J^B_{\mu 5}$ with a massless Nambu-Goldstone boson coupled to it. Then the combined system (A+B) described by the Lagrangian

$$L_{A+B} = L_A + L_B \tag{2.37}$$

and non-conserved operator

$$J^{A+B}_{\mu 5} = J^{A}_{\mu 5} + J^{B}_{\mu 5} \tag{2.38}$$

satisfies the same Ward identities as system A (or as QCD), but there is a zero-mass particle coupled to $J^{A+B}_{\mu 5}$.

For example, Q^{A+B}_5 rotates state vectors out of the physical space for system (A+B), but zero-mass particles are present.

Claims [21,22] to prove a theorem to the contrary involve two common errors:

(a) The wrong generator X(t) is confused [21] with Q_5. Then $\int d^4x$ is applied to the anomalous divergence equation (2.10), with the left-hand side incorrectly identified with $i\,\partial/\partial\phi_n$ and the right-hand side with $2n\,i\,\partial/\partial\theta$. This argument is then supposed to apply to Eq.(2.35), so the claimed result is that the left-hand side of (2.35) must vanish.

(b) It is incorrectly assumed [21,22] that the equation

$$e^{i\alpha Q_5}\,|\theta\rangle \overset{?}{=} |\theta - 2n\alpha\rangle \tag{2.39}$$

is generally true. Eq.(2.39) has nothing to do with anomalous divergences.

The fact is that the unexplained assumption [5]

$$Q_5\,|m = o\rangle \overset{?}{=} 0 \tag{2.40}$$

is essential for the derivation of (2.39): indeed, it is a necessary and sufficient condition. An equivalent formula is

$$X|\theta> \stackrel{?}{=} 0 \qquad (2.41)$$

However, we see from Eq.(2.29) that the assumption (2.40) or (2.41) is a special case of the assumption that massless U(1) bosons are absent. Hence arguments based on Eq.(2.39) are circular/ they assume what they should prove.*)

Let me summarize the conclusions of this Section. Operator chiralities in anomalous Ward identities are coupling-constant and hence gauge independent, but they are generated by a (partially) conserved, gauge-dependent charge Q_5. Despite the effects of topological charge, the QCD Lagrangian is U(1) symmetric for $m_i = 0$. Changes of chirality observed in instanton calculations [4,15] are due to a spontaneous breaking of chiral U(1) symmetry. Derivations necessarily include an assumption or assertion about amplitudes of the gauge-invariant operator $J_{\mu 5}$ in the soft-meson limit.

3. U(1) PROBLEM

The U(1) problem dates back to 1967, when Glashow [1] found that the spontaneous realization of SU(3)×SU(3) symmetry envisaged by Glashow and Weinberg [26] and Gell-Mann, Oakes and Renner [27] is difficult to reconcile with the light meson spectrum if additional Abelian chiral commutators are abstracted from the quark model. The problem became more serious in 1969, when it was noticed by Glashow, Jackiw and Sehi [28] and Gell-Mann [29] that, irrespective of the size of SU(3) symmetry breaking, the SU(2)×SU(2) limit generates four massless particles -- three pions, and an unwanted 0^- isoscalar meson. Of course, gluonic anomalies were not part of the subject at that time, and they certainly changed the analysis. However, I would like to postpone their inclusion for a few paragraphs in order to review some important features of the original work.

At that time, it had already been agreed [26,27] that the spontaneous realization of chiral symmetry should be signalled by quark mass operators acquiring vacuum expectation values, as in Eq.(1.1). In the absence of anomalies, it was concluded that the axial-vector operator

$$\text{two-quark current} = \bar{u}\gamma_\mu\gamma_5 u + \bar{d}\gamma_\mu\gamma_5 d \qquad (3.1)$$

*) These points were made in Sect.8 of Ref. 8 with the aim of forestalling incorrect analyses. See also Ref. 24. The fact that (2.39) is not generally true was also noticed by Ida [25],but he incorrectly claims that this point was overlooked in Ref.2 and 3.

is conserved in the SU(2)×SU(2) limit

$$m_u = m_d = 0 \; ; \; m_s, m_c, \ldots, m_N \neq 0 \tag{3.2}$$

The resulting axial U(1) symmetry, which I call $U_2(1)$ to indicate that only two quarks (u and d) are rotated, is spontaneously realized because of Eq. (1.1) and the fact that the left- and right-handed components of $\bar{u}u$ have definite non-zero $U_2(1)$ chiralities. Thus the crucial point of the analysis was that pions and the unwanted fourth boson were being generated by the same mechanism.

Referring to Fig. 1, we see that according to the 1969 analysis there had to be four massless bosons at every point on the horizontal axis except the origin. As long as the origin was avoided, the η meson (whose mass m_η is 549 MeV in the real world)remained massive,

$$m_\eta^2 \propto m_s \, , \quad (\text{horizontal axis } m_u = m_d = 0) \tag{3.3}$$

so it could not influence the $U_2(1)$ Goldstone theorem associated with Eqs.(1,1),(3.1) and (3.2):

$$\begin{aligned} &\text{Amplitude for zero-mass bosons coupled to current (3.1)} \\ &= \; U_2(1) \text{ commutator amplitude (1.1)} \end{aligned} \tag{3.4}$$

So, in addition to the massive particles η and $\eta'(958)$, there had to be a massless isoscalar 0^- boson everywhere on the horizontal axis except at the origin.

I prefer to distinguish the "$U_2(1)$ problem" discussed above from the "$U_3(1)$ problem", which concerns what happens at the origin of Fig.1, i.e. in the limit of SU(3)×SU(3) symmetry:

$$m_u = m_d = m_s = 0 \; ; \; m_c, \ldots, m_N \neq 0 \tag{3.5}$$

Here the η meson is a member of the massless Nambu-Goldstone octet (π, K, η) associated with the spontaneous realization of SU(3)×(SU(3) invariance. Since η couples to the two-quark current (3.1) and m_η is now zero, it contributes to the left-hand side of Eq.(3.4) and so upsets the $U_2(1)$ analysis. However, we can circumvent this complication by considering the Abelian symmetry $U_3(1)$ generated by the conserved three-quark current

$$\bar{u}\gamma_\mu\gamma_5 u + \bar{d}\gamma_\mu\gamma_5 d + \bar{s}\gamma_\mu\gamma_5 s \tag{3.6}$$

The left- and right-handed components of $\bar{u}u$ have non-zero $U_3(1)$ chiralities, so Eq.(1.1) forces the spontaneous realization of $U_3(1)$ symmetry. Therefore, in the absence of anomalies, the $U_3(1)$

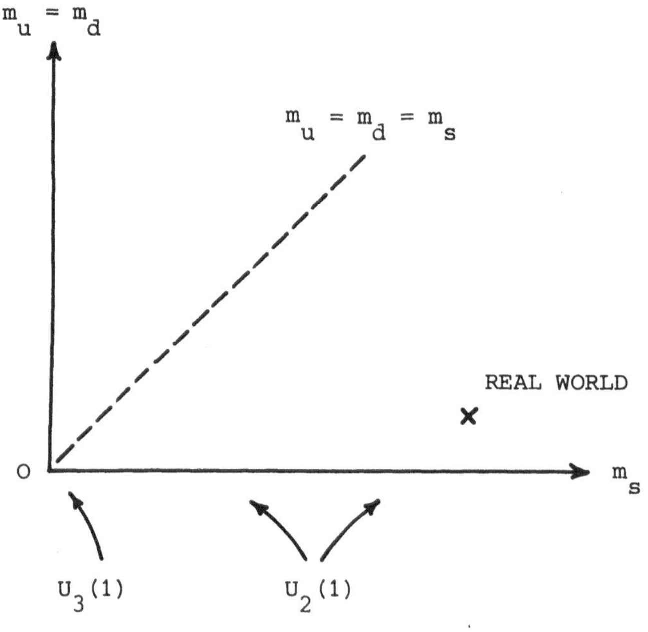

Fig. 1 Mass-parameter space for the U(1) problem. The u - d mass difference is neglected here

analogue of Eq.(3.4) would require the presence of an SU(3)-singlet massless 0^- boson in addition to the massless octet (π, K, η). Because of the success [27] of approximate SU(3)×SU(3) symmetry in providing a proper theoretical foundation for the octet mass formula

$$3m_\eta^{\ 2} \simeq 4m_K^{\ 2} - m_\pi^{\ 2} \tag{3.7}$$

and explaining why η-η' mixing is negligible, it made no sense to attempt to identify this extra boson with $\eta'(958)$. [Even if that course had been taken, it would have left the $U_2(1)$ problem unresolved.]

In 1969, the solution to these problems was to argue that there might be an extra SU(3)×SU(3) invariant $U_3(1)$ breaking term in the Hamiltonian, or that one should not abstract too many commutators from the naive quark model. Neither of these suggestions is now acceptable. We cannot tamper with the Hamiltonian because we believe that QCD is the correct theory for strong interactions. Similarly we cannot ignore the existence of conserved currents and their commutators in the renormalized theory, because that would amount to a rejection of the renormalization-group assumption which underlies all predictions of asymptotic freedom.

On the other hand, the modern theory contains gluonic anomalies which must be systematically included from the beginning of the analysis. When there is an anomaly, Goldstone's theorem (i.e. Eq.(3.4) or its $U_3(1)$ analogue) has to be replaced by the soft-meson theorem (2.17), which is a formula for the coupling of zero-mass $U_n(1)$ bosons to the gauge invariant operator $J_{\mu 5}^{n}$ in the SU(n)×SU(n) limit (2.1). We are interested in the consequences of Eq. (1.1), so let us replace the operator product $\prod_K O_K$ in (2.17) by the right-handed mass operator for the up quark:

$$\int d^4x\, \partial_x^{\mu}\ T <vac|\ J_{\mu 5}{}^{n}(x)\ \bar{u}_L u_R\ |vac> = 2(i\frac{\partial}{\partial\theta} - 1)<vac|\bar{u}_L u_R|vac> \tag{3.8}$$

One might object that Eq.(3.8) is not well formulated: there is an SU(n)× SU(n) vacuum degeneracy which implies a corresponding ambiguity in the meaning of $<\bar{u}_L u_R>$ and $i\partial/\partial\theta$. This objection can be met by perturbing the Lagrangian such that the SU(n)×SU(n) vacuum direction is appropriately fixed.*) The form of Eq.(3.8) is preserved if:

(a) The perturbation conserves U(1)×U(1) and breaks SU(n)×SU(n) down to the centre group $Z_n \times Z_n$.

(b) The perturbation does not depend on θ. This condition is necessary because Eq.(3.8) depends on the substitution (1.3) of $i\,\partial/\partial\theta$ for the topological charge operator. That substitution is correct only if all other sources (such as the coupling constant g) are held constant while θ is varied.

For example, a suitably perturbed Lagrangian is

$$L\,[\vec{J},\theta] = L_o\ [\theta] + \vec{J}^{\mu} \cdot \vec{I}_{\mu} \tag{3.9}$$

where $L_o\ [\theta]$ is the QCD Lagrangian (1.2) in the SU(n)×SU(n) limit (2.1), and $\vec{J}_{\mu}(x)$ are θ independent sources for the SU(n)×SU(n) currents

$$\vec{I}_{\mu} = \bar{q}\gamma_{\mu}\ (1 \pm \gamma_5)(\vec{\lambda}/2)q \tag{3.10}$$

The vacuum direction in SU(n)×SU(n) space (modulo the centre group and the vacuum-annihilating SU(n) subgroup) is fixed by a sufficiently general choice of $\vec{J}_{\mu}(x)$.

---

*) Consideration of perturbations by SU(n)× SU(n) violating mass terms is postponed to Section 4.

Let us now assume the absence of massless U(1) bosons in the SU(n)×SU(n) limit. There are (n^2-1) massless Nambu-Goldstone bosons coupled to $\vec{I}_\mu$, but they belong to the adjoint representation of the SU(n) subgroup which leaves vacua invariant, so they do not couple to the singlet operator $I_{\mu 5}$ in the limit $\vec{J}_\mu \to 0$. Therefore the left-hand side of (3.8) vanishes. The resulting homogeneous differential equation has general solution

$$\langle vac|\, \bar{u}_L u_R |vac\rangle = e^{-i\theta/n}\, C\, e^{i\phi} \tag{3.11}$$

where C and ϕ are real constants which do not depend on θ.

The angle ϕ is arbitrary. It reflects the existence of an axial U(1) degeneracy of vacua associated with rotations $U(\phi|2)$ defined by Eq. (2.34). Thus Eq. (3.11) exhibits the familiar feature that a change in θ can be compensated by an axial U(1) rotation. As emphasized in Section 2, this connection exists only if the absence of massless U(1) bosons is assumed.

The essential but widely misunderstood point is that the breaking of axial U(1) symmetry due to topological charge has to be spontaneous, not explicit. In other words, Glashow's observation that Eq. (1.1) implies a spontaneous breaking of U(n)× U(n) instead of just SU(n)× SU(n) remains correct in the presence of of anomalies. Claims to the contrary in the literature reflect a lack of familiarity with anomalous commutators.

Thus the full chiral degeneracy of vacua associated with the SU(n)× SU(n) invariant Lagrangian is given by [2]

$$\text{vacua} \in [U(n)/Z_n] \times [U(n)/Z_n] \;/\; [U(n)/Z_n]_{ann} \tag{3.12}$$

where generators of $U(n)_{ann}$ annihilate vacua. The factors $1/Z_n$ in Eq. (3.12) are needed in order to avoid double counting: elements of the centre group $Z_n \times Z_n$ in the quark representation belong both to SU(n)×SU(n) and to U(1)×U(1). The effect of the term $\vec{J}\cdot\vec{I}$ in (3.9) is to fix the direction

$$[SU(n)/Z_n] \times [SU(n)/Z_n] \;/\; [SU(n)/Z_n]_{ann} \tag{3.13}$$

leaving an axial U(1) degeneracy described by

$$\phi \in U(1) \times U(1)/U(1)_{ann} \tag{3.14}$$

[Compare this with instanton calculations of the 't Hooft [15] type. These involve SU(n)× SU(n) invariant vacua constructed from linear combinations of pure-gauge states. The vacuum degeneracy parameter is simply

$$\theta \in [U(1)/Z_n] \times [U(1)/Z_n] \;/\; [U(1)/Z_n]_{ann} = [U(1)/Z_n]_{axial} \qquad (3.15)$$

Non-zero changes in chirality are a consequence of this degeneracy. The Lagrangian remains $U(1) \times U(1)$ symmetric - the symmetry is spontaneously broken.*) The same point was made (for massless two-dimensional quantum electrodynamics) as long ago as 1971 by Lowenstein and Swieca [30]].

It is completely wrong to say that the effect of topological charge is to fix the axial U(1) direction of $U(n)\times U(n)$ vacua and hence reduce the number of massless bosons from n^2 to (n^2-1). That misunderstanding is based on self-consistent calculations [5,31] with 't Hooft's effective determinantal interaction [4] . Such calculations are not self-consistent for QCD because [3] they do not respect anomalous Ward identities; (see Sections 1 and 5). The fact that the wrong vacuum degeneracy $SU(n)\times SU(n)/SU(n)$ is obtained illustrates the point. Once again, it is the confusion between $X(t)$ and Q_5 which has caused these inconsistencies.

The effect of topological charge ν is to expand the number of available states from which vacuum candidates may be constructed (as appropriate linear combinations of winding-number eigenstates). In other words, the vacuum degeneracy is increased if the density of the topological charge spectrum is increased. If a sufficiently large number of states is produced in this way, there is no need to have a massless U(1) particle in order to create the degeneracy (3.12). The condition for this situation to be realized is obvious from the θ dependence of Eq.(3.11) for fixed U(1) phase ϕ : topological charge must be quantized in units of [3] $1/n$, as in Eq.(1.4).

There is no obvious theoretical reason for supposing that this result should be restricted to the phenomenologically interesting cases n=2,3. Even the bound

$$n \leqq N \leqq 16 \qquad (3.16)$$

required for asymptotic freedom seems to play no role in the analysis. So I expect that a practical calculation exhibiting the property (1.1) without U(1) bosons and obeying all requirements of self-consistency will involve gluonic boundary conditions which permit all rational values of ν . Of course, the fermionic boundary conditions will also be complicated and have nothing to do with the compactifiable conditions assumed in present-day calculat-

*) Please avoid semantic confusion in the use of the terms "explicit" and "spontaneous". The calculation of 't Hooft is an explicit example of spontaneous symmetry breaking. The example is explicit, not the breaking.

ions. ["Compactifiable" means that the relevant configurations in Euclidean space can be smoothly mapped onto a compact surface without boundary. A typical surface is S_4, the hypersphere in five dimensions. The rationale for this procedure is that determinants are easier to calculate because eigenvalues become discrete. However, the value of a determinant depends on the choice of boundary conditions. The compactifiable choice involves a physical assumption about infrared behaviour which is not acceptable for the present problem. See Sect. 6 of Ref. 2 and Sect. 2 of Ref.8.]

These conclusions disturb semi-classical calculators, who suppose finiteness of the Euclidean classical action S to be a fundamental requirement. The argument for this is that functional integrals are weighted by e^{-S}, so non-zero amplitudes cannot be generated by expanding about an infinite-action configuration. This idea is false, especially for systems which exhibit the Nambu-Goldstone mechanism. For example, the σ-field in the linear σ-model develops a vacuum expectation value and hence satisfies the boundary condition

$$\sigma(x) \longrightarrow \text{constant} \neq 0 \quad , \; (x_\mu \longrightarrow \infty) \tag{3.17}$$

The classical action diverges quartically, but physical amplitudes are certainly not zero.

In this case, the explanation is trivial. Physical Green's functions are computed with the vacuum normalized to unity. Therefore the functional representation of a Euclidean amplitude is a ratio of two functional integrals:

$$\langle \prod_k O_k \rangle = \int \prod_k O_k \, e^{-S} \Big/ \int e^{-S} \tag{3.18}$$

The classical factor $e^{-\infty}$ simply cancels between numerator and denominator.

There is no reason to suppose that a similar mechanism is forbidden in QCD. Spontaneous breaking of SU(n)xSU(n) (if it occurs) must involve an unusual choice of function space over which functional integrals are to be performed. The minimum value of the Euclidean action S in this space is likely to be infinite, but one can imagine $e^{-\infty}$ being cancelled off as in the σ-model. In other words, the absolute value of S is less important than its relative magnitudes for different values of ν. The question of whether the difference between $S(\nu\neq 0)$ and $S(\nu=0)$ should be finite or not depends on calculational details; for example, if there are Gaussian integrations, the relative magnitudes of fluctuation metrics for different ν become important.

The privileged role of finite-action configurations in the literature arises from an initial assumption of conventional calculations [15] that the $\nu \neq 0$ effects to be computed are merely small corrections to ordinary perturbation theory. In particular, the denominator

$$\int e^{-S}$$

is supposed to be dominated by the perturbative contribution (fluctuations about S=0). Contributions to the numerator involve classical minima (or approximate minima) with finite *) values of S relative to the perturbative value S=0. However, if we are not interested in the perturbative phase of QCD, we have no reason to require finite S or integer ν.

Accordingly, let us assume (as in Ref.3) that there is a "practical analytic method" for dealing with the QCD functional integral such that rational values of ν contribute. Almost certainly, such a calculation would involve boundary conditions inconsistent with finite action. I tried [3, 8] to find configurations with $S < \infty$ and $\nu \neq$ integer, but without success. Also, it is reported [34] that all smooth finite-action solutions of the equations of motion are necessarily compactifiable and hence must have integer ν. Therefore it is very unlikely that $\nu \neq$ integer effects can contribute to the perturbative phase studied in conventional instanton calculations. So, although some spontaneous breaking of axial U(1) symmetry can be exhibited as a correction to the perturbative phase [Eq.(3.15)], this does not seem to be possible for SU(n)×SU(n). [Again, I emphasize that Ref. 3 does not overlook self-consistent methods — it gives conditions on ν which must be satisfied in such calculations, and explains why self-consistent calculations with the determinantal interaction are not self-consistent for QCD.]

We now come to the problem which has bothered me from the beginning: if the $U_n(1)$ problem is solved by having topological charge quantized in units $1/n$ in the SU(n)×SU(n) limit, what is the θ-period of the theory in the neighbourhood of this limit? It is not possible to conclude that the period is $2\pi n$, because then the solutions to the $U_2(1)$ and $U_3(1)$ problems clash -- the period cannot be 4π and 6π simultaneously. To examine this problem, it is necessary to determine which member of the degenerate U(n)×U(n) set (3.12) is picked out by the mass perturbation

---

*)Fluctuation metrics about S=0 and instantons turn out to be relatively finite after renormalization [15] (if the size integral is truncated - but see Ref. 32 for progress on this question). Note that compactification is assumed for the ratio (3.18). The overall metric normalization involves $S=\infty$ configurations (App.C of Ref. 33), but this factor cancels in (3.18).

$$H' = \sum_{i=1}^{n} m_i \bar{q}_i q_i \tag{3.19}$$

which breaks U(n)×U(n) in the QCD Hamiltonian density.

4. PERTURBATION BY MASS TERMS

In the SU(n)×SU(n) limit (2.1), the U(n)×U(n) vacuum degeneracy (3.12) corresponds to the appearance of an arbitrary U(n) matrix V in in the amplitude

$$\langle vac|(\bar{q}_L)_j(q_R)_i|vac\rangle = CV_{ij}, \text{ (SU(n)×SU(n) limit)} \tag{4.1}$$

Different choices of vacua yield different n×n matrices V. In other words, the matrix V depends on the manner in which U(n)×U(n) symmetry is broken in the Lagrangian before the symmetry limit is taken.

The reason for my emphasis on spontaneously breaking axial U(1) symmetry with $\nu \neq 0$ effects should now be apparent. Anomalies do not break U(1) symmetry explicitly, so the phase of V in (4.1) is not fixed. The effect of topological charge is to create the U(1) ambiguity in V, not remove it. The assumption is that the required ambiguity (4.1) can be achieved with only n^2-1 massless bosons. The same degenerate set {V} is obtained for all values of θ .

In Section 3, the perturbed Lagrangian (3.9) was used to fix the SU(n)×SU(n) vacuum direction. This procedure established the θ independence of the real constant C:

$$\partial C/\partial\theta = 0, \quad C < 0 \tag{4.2}$$

It is convenient to choose a negative sign for C because the amplitude (1.1) is negative for the P and T conserving value θ = 0 of the QCD Lagrangian (1.2). In the following, the perturbation J.I in (3.9) will be replaced by the mass term εH' of Eq.(3.19).

In general, the perturbation εH' fixes both the SU(n)×SU(n) and U(1)×U(1) vacuum directions because it simultaneously breaks both of these symmetries. Let $\bar{V}$ be the U(n) matrix picked out by this perturbation. The mass matrix has been chosen to be diagonal and γ_5 free, so εH' is invariant under transformations

$$q \longrightarrow \exp(i\vec{\xi}\cdot\vec{\lambda}_{diag})\, q \tag{4.3}$$

where $\vec{\lambda}_{diag}$ denotes the traceless diagonal n×n matrices $\lambda_3, \lambda_8, \lambda_{15}, \ldots \lambda_{(n^2-1)}$. Therefore the same U(n) matrix $\bar{V}$ should result after performing the substitution (4.3) in $\bar{q}_L q_R$:

$$\exp(i\vec{\xi}\cdot\vec{\lambda}_{diag})\ \bar{V}\ \exp(-i\vec{\xi}\cdot\vec{\lambda}_{diag}) = \bar{V} \qquad (4.4)$$

This constraint must be true for all values of the parameters $\vec{\xi}$, so $\bar{V}$ must be diagonal:

$$<vac\ |(\bar{q}_L)_j (q_R)_i\ |\ vac> \ = C\delta_{ij}\ e^{-i\phi_i} + O(\varepsilon) \qquad (4.5)$$

Since $\bar{V}$ is a U(n) matrix, ϕ_i is real.

The phase differences $(\phi_i - \phi_j)$ characterize the portion of SU(n)×SU(n) space occupied by the vacuum-annihilating subgroup $SU(n)_{ann}$. Let $\vec{F}_L$ and $\vec{F}_R$ be the generators of left- and right-handed SU(n),

$$\vec{F}_R = \int d^3x\ q^+ \frac{1}{2}(1+\gamma_5)(\vec{\lambda}/2)q \qquad (4.6)$$

and let $SU(n)_{L+R}$ be the subgroup generated by the parity-conserving combination

$$(\vec{F}_L + \vec{F}_R) .$$

Then the SU(n)×SU(n) transformation

$$W(\phi) = \exp\{\frac{i}{2}\sum_{i=1}^{n}(\phi_i - n^{-1}\sum_j \phi_j)\int d^3x\ q_i^+\ \gamma_5 q_i\} \qquad (4.7)$$

relates the two subgroups:

$$SU(n)_{ann.} = W(\phi)^{-1} SU(n)_{L+R}\ W(\phi) \qquad (4.8)$$

[Eq. (8.61) of Ref. 2 is in error: it supposes continuity of $<\bar{q}_i q_j>_\theta$ at the origin in m_i space for arbitrary θ . The unintentional effect of this *) was to prevent $SU(n)_{ann}$ from being chirally rotated with respect to $SU(n)_{L+R}$. However, (8.61) was not used in the original discussion [3], and its removal does not affect the remainder of Ref. 2. The main issue, the relation between the θ period and fractional topological charge, remains to be explained; see below.]

In the leading term of Eq. (4.5), the dependence on θ and the mass ratios m_i/m_j is entirely contained in the phases

---

*) I thank S. Coleman for this remark and for pointing out the advantages of formulating the problem in terms of the phases ϕ_i.

$$\phi_i = \phi_i(\theta, m_j/m_k) \tag{4.9}$$

This dependence can be found by applying chiral U(n)×U(n) Ward identities and assuming the absence of U(1) bosons with zero or $O(\sqrt{\varepsilon})$ mass.

The SU(n)× SU(n) direction is best handled by applying a theorem of Dashen [10] . According to this theorem, the correct vacuum state has the property that it minimizes the expectation value of $\varepsilon H'$ with respect to small SU(n)×SU(n) rotations

$$U(\vec{\omega}) = \exp i\vec{\omega}.\vec{F} \quad , \quad (|\vec{\omega}| << 1) \tag{4.10}$$

This statement is verified by examining the expansion

$$\begin{aligned} &\langle vac|\ U(\vec{\omega})\ \varepsilon H'\ U(\vec{\omega})^{-1}\ |vac\rangle \\ &= \langle vac|\varepsilon H'|\ vac\rangle + i\vec{\omega}\ .\ \langle vac|[\vec{F},\varepsilon H']\ |vac\rangle \\ &\quad - \langle vac|\ [\vec{\omega}.\vec{F},\ [\vec{\omega}.F,\ \varepsilon H']\ ]|vac\rangle \quad + O(\omega^3) \end{aligned} \tag{4.11}$$

There is a stationary point at $\vec{\omega} = 0$ because the first-order term is proportional to

$$\langle vac|\ [\vec{F},\varepsilon H']\ |vac\rangle \quad = i\ \langle vac|\ \partial^\mu I_\mu\ |vac\rangle \ = 0 \tag{4.12}$$

The $O(\omega^2)$ contribution to Eq.(4.11) is a σ term for which there is a standard Ward identity:

$$\begin{aligned} &- \langle vac|\ [\vec{\omega}.\vec{F},[\vec{\omega}.\vec{F}\ ,\ \varepsilon H']]\ |vac\rangle \\ &= i\int d^4x\ T\ \langle vac|\ \vec{\omega}.\partial^\mu\ \vec{I}_\mu(x)\ \vec{\omega}.\partial^\nu\vec{I}_\nu(o)\ |vac\rangle \\ &\geq 0 \end{aligned} \tag{4.13}$$

Equality is permitted only if the operator $\vec{\omega}.\ \partial^\mu\vec{I}_\mu$ vanishes. Therefore the stationary point is a local minimum.

It is tempting to speculate [10] that $\vec{\omega} = 0$ is a global minimum. This depends on whether states differing from $|vac\rangle$ by a finite rotation $U(\vec{\omega})$ can tunnel to $|vac\rangle$ in the presence of the perturbation $\varepsilon H'$. If tunnelling occurs, the usual variational principle is applicable and $\vec{\omega} = 0$ is a global minimum.

Consider the consequences of the theorem for U(n)×U(n) vacua perturbed by the mass term $\varepsilon H'$. Because of the Kronecker delta δ_{ij} in (4.5), it is sufficient to consider right-handed SU(n) rotations involving the diagonal flavour matrices $\vec{\lambda}_{diag}$. These rotations vary the phases

$$\phi_i \longrightarrow \phi_i + \omega_i \tag{4.14}$$

subject to the SU(n) constraint

$$\sum_{i=1}^{n} \omega_i = 0 \tag{4.15}$$

Therefore the expectation value for vacua rotated through $\vec{\omega}$

$$< \varepsilon H'>_{\vec{\omega}} = 2C\sum_{i=1}^{n} m_i \cos(\phi_i + \omega_i) + O(\varepsilon^2) \tag{4.16}$$

must have a local minimum at $\vec{\omega} = 0$.

The general method for solving this problem is to use Lagrange multipliers [35] , but the following procedure is simple and sufficient. All first-order variations $\vec{\omega}$ subject to the constraint (4.15) are linear combinations of variations

$$\omega_i = - \omega_j = \omega; \ \omega_k = 0 \ , \ (k \neq i,j) \tag{4.17}$$

for arbitrary pairs of indices (i,j). Substituting (4.17) into (4.16) and requiring the derivative d/dω to vanish, we find the result

$$m_i \sin \phi_i = m_j \sin \phi_j + O(\varepsilon^2) \tag{4.18}$$

Therefore, to lowest order in the SU(n)×SU(n) symmetry breaking parameter ε, the combination $m_i \sin\phi_i$ does not depend on the flavour index i.

Dashen's theorem is not applicable in the axial U(1) direction. If $U(\vec{\omega})$ is replaced by an infinitesimal version of the Abelian chiral transformation (2.33), the first-order analogue of (4.12) is the vacuum expectation value of the divergence of the gauge-dependent <u>symmetry</u> current $J^n_{\mu 5 \ sym}$:

$$< vac \ [Q^n_5, \varepsilon H'] \ |vac> = i<vac|\partial^\mu J^n_{\mu 5 \ sym}|vac> \tag{4.19}$$

There is no reason to suppose that this amplitude vanishes. The symmetry current operates on the large state-vector space which contains the complete set of U(n)×U(n) vacua and other states associated with the full range of values of θ . The projection [2] of this space onto the physical Hilbert space with unique ground state |vac > and unique value of the parameter θ is possible for gauge-invariant operators such as $J^n_{\mu 5}$, but not for the symmetry current. So it is certainly correct to write

$$\langle vac|\partial^\mu J_{\mu 5}^{\ n}\ |vac\rangle = 0 \tag{4.20}$$

but the same argument cannot be carried through for the symmetry current. Indeed, Eqs. (2.6) and (4.20) imply

$$\langle vac|\partial^\mu J_{\mu 5\ sym}^{\ n}\ |vac\rangle = -\ 2n\langle vac|(\frac{g^2}{32\pi^2})\ F.F^*|vac\rangle \tag{4.21}$$

and even the simplest instanton calculations yield a non-zero answer for this amplitude.

[This example shows the danger of an unrestricted use of variational methods. One should vary over states connected by physical transitions. In general, variations in the large vector space are not acceptable because this space describes a set of independent physical systems labelled by the continuous parameter θ.]

A proper treatment of the U(1)×U(1) direction involves deriving the anomalous Ward identity for the matrix element (4.5) in the presence of the mass perturbation εH'. As in Section 2, the derivation proceeds from the axial U(1) Ward identity for the symmetry current. The difference is that now this current is only partially conserved because of the presence of εH':

$$\begin{aligned}\partial^\mu J_{\mu 5\ sym}^{\ n} &= i\ [\varepsilon H', Q_5^n\ ] \\ &= 2i \sum_{i=1}^{n} m_i\ \bar{q}_i\ \gamma_5\ q_i\ = D_n\end{aligned} \tag{4.22}$$

The soft mass term D_n contributes an extra term to the axial Ward identity (2.11) and hence to the anomalous Ward identities (2.16) and (2.17). Remembering that i $\partial/\partial\theta$ induces a connected insertion of the topological charge operator (1.3)

$$\begin{aligned}i\frac{\partial}{\partial\theta}\ T\ \langle vac|\ \prod_k O_k|vac\rangle &= T\ \langle vac|\ (\frac{g^2}{32\pi^2})\int d^4x\ F.F^*(x)\ \prod_k O_k|vac\rangle \\ &\quad -\ \langle vac|(\frac{g^2}{32\pi^2})\int d^4x\ F.F^*(x)|vac\rangle \\ &\quad \times\ T\ \langle vac|\prod_k O_k\ |vac\rangle\end{aligned} \tag{4.23}$$

and using (4.21), we see that the generalization of (2.17) is

$$\int d^4x\ \ \partial_x^\mu\ T\ \langle vac|J_{\mu 5}^{\ n}(x)\ \prod_k O_k\ |vac\rangle$$

$$= \{ 2ni \frac{\partial}{\partial\theta} - \sum_{\ell} \chi_{\ell} (n) - \langle vac| \int d^4x\, D_n(x)|vac\rangle\} T \langle vac| \prod_k O_k |vac\rangle$$

$$+ T \langle vac| \int d^4x\, D_n(x) \prod_k O_k |vac\rangle \qquad (4.24)$$

The anomalous Ward identity for the amplitude (4.5) is found by replacing the operator product $\prod_k O_k$ by the right-handed mass operators

$$(\bar{q}_L)_j (q_R)_i .$$

We assume that zero-mass bosons do not couple to the gauge-invariant operator $J^n_{\mu 5}$

$$\int d^4x \quad \partial^\mu_x T \langle vac| J^n_{\mu 5} (x)(\bar{q}_L)_j (q_R)_i |vac\rangle = 0 \qquad (4.25)$$

so the relevant Ward identity becomes

$$0 = 2(in \frac{\partial}{\partial\theta} - 1) \langle vac| (\bar{q}_L)_j (q_R)_i |vac\rangle$$
$$+ T \langle vac| \int d^4x\, D_n(x) (\bar{q}_L)_j (q_R)_i |vac\rangle_{CONN} \qquad (4.26)$$

where the label "CONN" indicates that the connected part is to be taken.

In general, the two-point function

$$T \langle \int D_n \bar{q}_L q_R \rangle$$

does not vanish in the limit $\varepsilon \to 0$, despite the fact that D_n is proportional to ε. If there are poles due to bosons b with mass

$$M_b^2 = O(\varepsilon) \qquad (4.27)$$

they contribute

$$T \langle vac| \int d^4x\, D_n(x) (\bar{q}_L)_j (q_R)_i |vac\rangle_{pole}$$

$$= - i \sum_b \langle vac| D_n |b\rangle \langle b| (\bar{q}_L)_j (q_R)_i |vac\rangle / M_b^2 \qquad (4.28)$$

at zero momentum transfer; (see Fig.2). As ε tends to zero, the ratio D_n/M_b^2 remains finite.

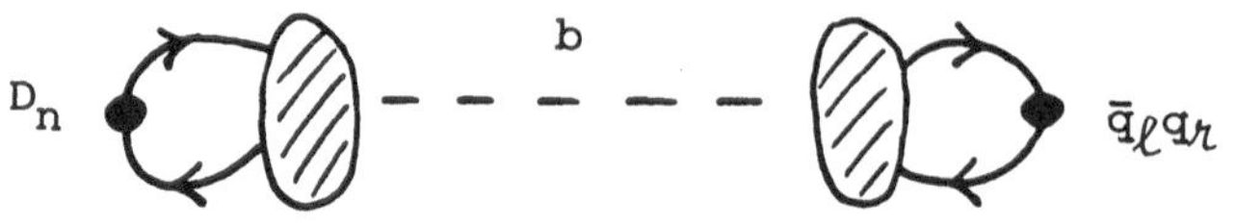

Fig.2 Pole diagrams considered in Eq.(4.28)

In the presence of the perturbation $\varepsilon H'$, the (n^2-1) bosons $\vec{b}$ associated with chiral SU(n)×SU(n) symmetry acquire masses which obey (4.27). Their contribution to (4.28) can be determined from current algebra. Let $F_\pi \simeq 93$ MeV be the decay constant for $\vec{b}$ to couple to its SU(n)×SU(n) current

$$\vec{I}_\mu = W(\phi)^{-1} \bar{q} \gamma_\mu \gamma_5 (\vec{\lambda}/2) q \, W(\phi) \tag{4.29}$$

where $W(\phi)$ is the chiral rotation (4.7) which relates $SU(n)_{ann}$ and $SU(n)_{L+R}$. Also, let

$$\vec{F} = \int d^3x \, \vec{I}_0(x) \tag{4.30}$$

be the corresponding chiral charge. Then soft-$\vec{b}$ formulas such as

$$\langle \vec{b} \,|\, (\bar{q}_L)_j (q_R)_i \,|\, vac \rangle = - i F_\pi^{-1} \langle vac | \, [\vec{F}, (\bar{q}_L)_j (q_R)_i] \, | vac \rangle + \\ + O(\varepsilon) \tag{4.31}$$

show that $\vec{b}$ poles almost always contribute to (4.28) in the limit $\varepsilon \to 0$.

However, there are some important exceptions:

(a) The operator

$$\Delta_R = \sum_{i=1}^{n} e^{i\phi_i} (\bar{q}_L q_R)_i \tag{4.32}$$

and its conjugate

$$\Delta_L = \Delta_R^+ \tag{4.33}$$

are $SU(n)_{ann}$ invariant, so to lowest order in ε, they do not couple to the (n^2-1) bosons $\vec{b}$:

$$\langle \vec{b} \,|\, \Delta_R | vac \rangle = O(\varepsilon) \tag{4.34}$$

(b) The mass operator $\varepsilon H'$ also has this property. The leading term in the soft-$\vec{b}$ limit is proportional to (4.12) and hence vanishes:

$$\langle \vec{b} \,|\, \sum_{i=1}^{n} m_i \bar{q}_i q_i \,|\text{vac}\rangle = -i F_\pi^{-1} \langle \text{vac}|\, [\vec{F}, \varepsilon H'] \,|\text{vac}\rangle + O(\varepsilon^2)$$
$$= O(\varepsilon^2) \qquad (4.35)$$

Obviously, the $\varepsilon \to 0$ limit of (4.26) becomes much simpler if contributions (4.28) to the last term in (4.26) can be avoided. Two steps are necessary if we are to ensure suppression of this last term:

(i) The indices (i,j) must be contracted to produce a suitable operator Δ_R, Δ_L, or $\varepsilon H'$, so that Eqs.(4.34) or (4.35) become applicable. This procedure ensures that the (n^2-1) SU(n)×SU(n) bosons $\vec{b}$ do not contribute.

(ii) We must assume the absence of a U(1) boson b_0 with mass $O(\sqrt{\varepsilon})$; i.e., we suppose that all contributions to the sum $\sum_b$ in (4.28) arise from SU(n)×SU(n) Nambu-Goldstone bosons.

The simplest procedure is to contract Eq.(4.26) with

$$e^{i\phi_i}\delta_{ij},$$

so that Eq.(4.34) becomes relevant. Afterwards, the result will be checked by contracting (4.26) with $m_i\,\delta_{ij}$.

The anomalous Ward identity

$$0 = 2 \sum_{i=1}^{n} e^{i\phi_i}(in\,\partial/\partial\theta - 1) \langle \text{vac}|\, (\bar{q}_L q_R)_i \,|\text{vac}\rangle$$
$$+ T \langle \text{vac}| \int d^4x D_n(x)\, \Delta_R \,|\text{vac}\rangle_{\text{CONN}} \qquad (4.36)$$

is a consequence of combining Eqs.(4.26) and (4.32). Because of Eq.(4.34) and the assumption (ii), the last term is negligible,

$$T \langle \text{vac}| \int d^4x\, D_n(x)\, \Delta_R \,|\text{vac}\rangle_{\text{CONN}} = O(\varepsilon) \qquad (4.37)$$

so substitution of Eq.(4.5) yields the constraint

$$O(\varepsilon) = \sum_{i=1}^{n} e^{i\phi_i}(in\,\partial/\partial\theta - 1)\, C e^{-i\phi_i} \qquad (4.38)$$

Since C does not depend on θ [Eq.(4.2)], Eq.(4.38) reduces to a differential equation for the phases ϕ_i:

$$\sum_{i=1}^{n} \{ n\, \partial\phi_i/\partial\theta - 1 \} = O(\varepsilon) \tag{4.39}$$

Before integrating (4.39), let us check its consistency with Eq.(4.35). When Eq.(4.26) is contracted with the mass matrix and symmetrized with respect to L $\leftrightarrow$ R, the result is

$$0 = 2(in\, \partial/\partial\theta - 1) \langle vac |\sum_{i=1}^{n} m_i (\bar{q}_L q_R)_i | vac \rangle$$
$$+ 2(in\, \partial/\partial\theta + 1) \langle vac |\sum_{i=1}^{n} m_i (\bar{q}_R q_L)_i | vac \rangle$$
$$+ \quad T \langle vac | \int d^4x\, D_n(x) \quad \varepsilon H' \quad | vac \rangle_{CONN} \tag{4.40}$$

According to the assumption (ii) and Eq.(4.35), the last term of (4.40) is $O(\varepsilon^2)$. The remainder of (4.40) combined with Eqs.(4.2) and (4.5) yields a condition

$$\sum_{i=1}^{n} (n\, \partial\phi_i/\partial\theta - 1)\, m_i \sin\phi_i = O(\varepsilon^2) \tag{4.41}$$

which is obviously consistent with Eqs.(4.18) and (4.39).

Eq.(4.18) and the solution of Eq.(4.39) constitute n equations for the n phases ϕ_i:

$$m_i \sin\phi_i = \lambda + O(\varepsilon^2) \tag{4.42}$$

$$\sum_{i=1}^{n} \phi_i = \theta + O(\varepsilon) \qquad (\text{mod } 2\pi) \tag{4.43}$$

In Eq. (4.42), the $O(\varepsilon)$ constant λ does not depend on the flavour index i. The integration constant in (4.43) has been fixed such that the P and T conserving solution

$$\phi_i = 0 \qquad (\text{mod } 2\pi) \tag{4.44}$$

corresponds to $\theta = 0$.

Remarks:

(a) The correction term $O(\varepsilon)$ in (4.43) depends crucially on assumptions about U(1) bosons. The first assumption (4.25) was that U(1) bosons are not massless. The second assumption was that U(1) bosons do not contribute pole terms (4.28) to the last term in (4.26) in the leading order in ε . In other words, if M_o denotes the mass of a boson with quantum numbers consistent with the identification "U(1)" (e.g. $\eta'(958)$), the term $O(\varepsilon)$ in (4.43) includes terms of the form

$$O(\varepsilon) \quad \rightarrow \quad \{\text{constant}\} \quad m_i/M_o^2 \tag{4.45}$$

Obviously, it is important that M_o^2 should <u>not</u> be proportional to ε.

(b) Eq.(4.43) has an obvious interpretation: the chiral U(1) angle $\sum_i \phi_i$ equals the θ parameter in the chiral SU(n)×SU(n) limit. However, as I have emphasized in remark (a) above, in Sect. 2, and in Refs. 2 and 8, this equality is <u>not</u> general. For example, it is incorrect to say that Eq.(4.43) can be derived directly from the anomalous divergence equation (2.10) without making assumptions about soft-meson amplitudes.

(c) A related and potentially confusing point is that Eqs. (4.42) and (4.43) turn out to be valid both for QCD and for the determinantal interaction

$$L_{det} = \sum_{i=1}^{n} \bar{q}_i(i\not{\partial} - m_i)q_i + K\{ e^{i\theta} \det \bar{q}_L q_R + \text{h.c.}\} \tag{4.46}$$

(for real K independent of θ and m_i) even though the two theories have completely different U(1) Ward identities and vacuum structure [3] . Many of the arguments which I have been criticizing are valid for L_{det} but not for QCD. Further discussion of (4.46) appears in Sect. 5.

(d) Eqs.(4.42) and (4.43) (and hence the assumptions on which they are based) are needed to justify analyses of θ-induced CP violating effects given by Baluni [16] and others [17,36] . See Sect. 6.

In general, the solutions of Eqs. (4.42) and (4.43) cannot be written in terms of elementary functions of n, m_i and θ , because high-degree polynomial equations with variable coefficients have to be solved. These polynomial equations are rationalized versions of the formulas

$$\prod_{i=1}^{n} \{\pm (1 - \lambda^2/m_i^2)^{1/2} + i\lambda/m_i\} = e^{i\theta} \tag{4.47}$$

Each choice of a set of signs $\pm$ in (4.47) yields a subset of solutions for λ and hence for $e^{i\bar{\phi}_i}$.

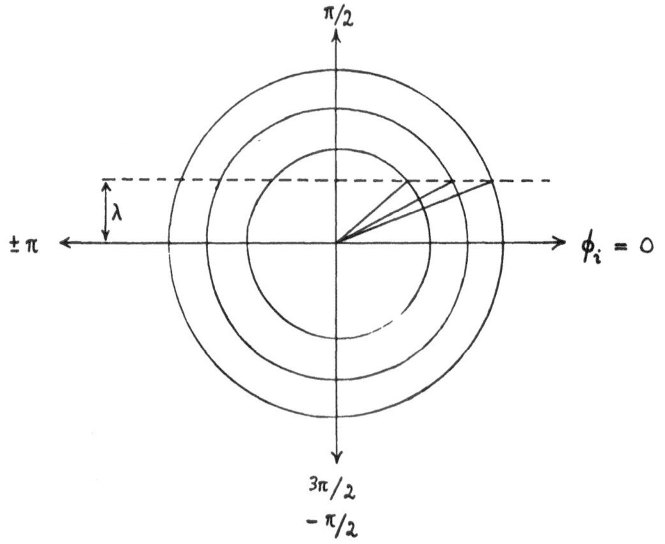

Fig.3 Geometrical method for the solution of Eq.(4.42). The circles have radii m_i. See Eqs.(4.48)-(4.50)

A qualitative picture of the behaviour of ϕ_i for arbitrary values of the mass ratios m_j/m_k can be obtained by inspecting the geometrical construction in Fig.3. Concentric circles are drawn with radii equal to m_i (in suitable units). There is a circle for each flavour i involved in the relevant SU(n)×SU(n) limit. Two or more flavours with equal mass parameters give rise to a corresponding set of circles with degenerate radii. The mass parameters are ordered as in Eq.(1.7), so the circle for flavour i lies inside (or is superimposed on) that for j if i is less than j. A horizontal line (the dashed line in Fig.3) is drawn a distance $|\lambda|$ above or below the $\phi_i = 0$ axis according to whether the quantity λ in (4.42) is positive or negative. This line should intersect all circles:

$$|\lambda| \leqq m_1 = m_u \tag{4.48}$$

Choose one intersection for each circle. Then the phase ϕ_i is the angle between the $\phi_i = 0$ axis and the radial line drawn to the ith intersection. The set of phases $\{\phi_i\}$ defined by this construction automatically satisfies Eq.(4.42). The choice of a value of λ and of a set of intersections depends on the value of θ being considered.

Each circle is intersected twice (unless ϕ_1 is $\pi/2$ or $-\pi/2$), so there are two choices of angle ϕ_i for each value of λ and each flavour i :

$$|\phi_i| \leq \pi/2 \quad \text{or} \quad \pi/2 \leq |\phi_i| \leq \pi \tag{4.49}$$

This corresponds to the ambiguity in sign $\pm$ for each factor in the product in Eq.(4.47). The two possibilities (4.49) are continuously connected only if i refers to the innermost circle - the corresponding phase(s) can be rotated from 0 to 2π by allowing θ to vary continuously. However, the remaining angles oscillate between bounds

$$|\phi_i| \leq \arcsin(m_1/m_i) < \pi/2$$

$$\text{or} \quad \pi/2 < \pi - \arcsin(m_1/m_i) < |\phi_i| \leq \pi \,, \qquad (m_1 < m_i) \tag{4.50}$$

depending on the choice (4.49). For example, there are 2^{n-1} continuously connected possibilities if the innermost circle is not degenerate ($m_1 < m_2$).

The number of solutions for ϕ_i as functions of θ and the ratios m_j/m_k is usually larger than the number of continuously connected possibilities. This is because λ or ϕ_1 need not be a single-valued function of θ for all values of θ. Even in the non-degenerate case $m_1 < m_2$, we shall see that θ is not necessarily a monotonically increasing function of ϕ_1. The situation becomes even more complicated if the lightest quark is degenerate. For example, if pairs of quarks are degenerate (even n), Eqs. (4.42) and (4.43) permit solutions

$$m_1 = m_2 \leq m_3 = m_4 \leq \dots \leq m_{n-1} = m_n \,,$$

$$\phi_1 = \pi - \phi_2 \,,\ \phi_3 = \pi - \phi_4 \,,\ \dots \,,\ \phi_{n-1} = \pi - \phi_n$$

$$\theta = n\pi/2 \pmod{2\pi} \tag{4.51}$$

which are continuously degenerate (ϕ_1=arbitrary) at an isolated value (0 or π) of θ.

However, many of these solutions are unphysical because, although they correspond to stationary points of (4.11) with respect to small SU(n)×SU(n) rotations (4.10), these points are not local minima. The correct ground state has positive σ-terms (4.13). A candidate state which produces a negative σ-term is unstable to the emission of SU(n)×SU(n) Nambu-Goldstone bosons; it will decay to lower-energy states until a stable state is reached.

We choose the basis given by Eqs.(4.29) and (4.30) to define σ-terms

$$\sigma^{ab} = - \langle vac | [F^a, [F^b, \varepsilon H']] | vac \rangle \quad , \qquad (a,b = 1, \ldots , n^2-1) \tag{4.52}$$

and a corresponding matrix $\Sigma = (\sigma^{ab})$. Substitutuion of Eqs.(3.19) and (4.5) in (4.52) yields anti-commutators of the traceless SU(n) matrices λ^a:

$$\sigma^{ab} = - C \sum_{i=1}^{n} m_i \cos \phi_i \{\tfrac{1}{2} [\lambda^a, \lambda^b]\}_{ii} + O(\varepsilon^2) \tag{4.53}$$

The physical relevance of Σ is that its eigenvalues Σ^a determine the masses M_a acquired by SU(n)×SU(n) Nambu-Goldstone bosons:

$$M_a^2 F_\pi^2 = \Sigma^a + O(\varepsilon^2) \tag{4.54}$$

According to Eq.(4.13), the quantity

$$\omega^a \omega^b \sigma^{ab} = (-C) \sum_{i=1}^{n} m_i \cos \phi_i \{(\vec{\omega}.\vec{\lambda})^2\}_{ii} + O(\varepsilon^2) \tag{4.55}$$

should be positive for all real $\vec{\omega}$. The factors (-C), m_i and

$$\{(\vec{\omega}.\vec{\lambda})^2\}_{ii} = \sum_j |(\vec{\omega}.\vec{\lambda})_{ij}|^2 \tag{4.56}$$

are all positive, so the sign of $\omega^a \omega^b \sigma^{ab}$ is mainly determined by the signs of $\{\cos \phi_i\}$.

Obviously, $\omega^a \omega^b \sigma^{ab}$ has to be positive if all of the cosines are positive.

It is equally obvious that two or more cosines cannot be negative. Let $\cos \phi_j$ and $\cos \phi_k$ be negative. Then we can choose $\vec{\omega}.\vec{\lambda}$ to be the traceless matrix $\vec{\omega}_0.\vec{\lambda}$ with all elements zero except for the jth and k^{th} diagonal elements which equal +1 or -1 respectively. The result

$$\omega_0^a \omega_0^b \sigma^{ab} = (-C)\{ m_j \cos \phi_j + m_k \cos \phi_k \} + O(\varepsilon^2) \tag{4.57}$$

is negative (for sufficiently small ε) and hence not acceptable.

The remaining case, in which only one cosine is negative, is more interesting. Let $\cos\phi_j$ be the negative cosine. The condition that the σ-term (4.57) be non-negative becomes

$$(m_j^2 - \lambda^2)^{1/2} \leq (m_k^2 - \lambda^2)^{1/2}$$

or equivalently

$$m_j \leq m_k \tag{4.58}$$

Since (4.58) has to be true for all k except j, m_j must be the smallest mass parameter (or one of the smallest parameters, if the lightest quark is degenerate). In other words, all phases ϕ_i should lie between $-\pi/2$ and $\pi/2$ (as in Fig.3) with the possible exception of a phase ϕ_j associated with the innermost circle. Let us call this last phase ϕ_1. When $\cos\phi_1$ is negative, its magnitude must satisfy the bound

$$m_1 |\cos\phi_1| \leq \sum_{i=2}^{n} m_i \cos\phi_i \sum_{k=1}^{n} |(\vec{\omega}.\vec{\lambda})_{ik}|^2 / \sum_{\ell=1}^{n} |(\vec{\omega}.\vec{\lambda})_{1\ell}|^2 \tag{4.59}$$

for all real $\vec{\omega}$ in order that negative values of $\omega^a \omega^b \sigma^{ab}$ are not produced.

Instead of (4.59), we want an equivalent condition which avoids the introduction of arbitrary vectors $\vec{\omega}$. This is obtained by minimizing the right-hand side of (4.59). We fix m_i and $\cos\phi_i$ and vary elements of the matrix $\vec{\omega}.\vec{\lambda}$ subject to the SU(n) constraint

$$\sum_{i=1}^{n} (\vec{\omega}.\vec{\lambda})_{ii} = 0 \tag{4.60}$$

The minimum value corresponds to choosing $\vec{\omega}.\vec{\lambda}$ to be diagonal, with elements $(\vec{\omega}.\vec{\lambda})_{kk}$ proportional to $(m_k \cos\phi_k)^{-1}$ for $k \neq 1$. Substitution of this minimizing matrix in (4.59) yields the desired result:

$$|m_1 \cos\phi_1|^{-1} \geq \sum_{i=2}^{n} (m_i \cos\phi_i)^{-1} \tag{4.61}$$

In summary, a stable ground state is obtained only if

(a) none of the cosines $\cos\phi_i$ is negative, or

(b) one cosine $\cos\phi_1$ is negative and obeys the inequality (4.61).

Note that in general, the condition (4.61) does not permit $\cos\phi_1$ to be negative if the lightest quark is degenerate. The only exception to this rule is the special case (1.8) (i.e., Eq. (4.51) for n=2) where either $\cos\phi_u$ or $\cos\phi_d$ may be negative:

$$\cos\phi_u = -\cos\phi_d \tag{4.62}$$

In this case, all $O(\varepsilon)$ contributions to σ^{ab} vanish. This means that at $\theta=\pi$, the squares of the pion masses M_a^2 have no $O(\varepsilon)$ contribution when ε is given the following meaning:

$$\varepsilon H' = \hat{m}\,(\bar{u}u + \bar{d}d)\,,$$

$$\lim_{\varepsilon\to 0} m_i \neq 0 \quad , \; (i = 3,\ldots, N) \tag{4.63}$$

Thus it is not surprising that to leading order in ε, the solution is continuously degenerate. [Actually, three continuous parameters are needed to fully describe the degeneracy because all three pions fail to acquire mass in this order. In this respect, the one-parameter solution (4.51) is misleading. The point is that the constraint (4.4), on which (4.51) is based, was derived by supposing that such degeneracy would be absent in the presence of $\varepsilon H'$.] Presumably *), positivity requirements for $\omega^a\omega^b\sigma^{ab}$ and M_a^2 are satisfied by $O(\varepsilon^2)$ contributions, so I see no contradiction between the continuous degeneracy of the solution and the non-conservation of the currents in (4.13).

Eq. (4.61) has a simple interpretation when the lightest quark is not degenerate. Eqs. (4.42) and (4.43) and the positivity of $\cos\phi_i$ for $i \neq 1$ imply the following relation for θ as a function (modulo 2π) of ϕ_1:

$$\theta = \phi_1 + \sum_{i=2}^{n} \arcsin\{(m_1/m_i)\sin\phi_1\} + O(\varepsilon) \tag{4.64}$$

Differentiating,

$$\partial\theta/\partial\phi_1 = 1 + m_1\cos\phi_1 \sum_{i=2}^{n} (m_i\cos\phi_i)^{-1} + O(\varepsilon) \tag{4.65}$$

---

*) In fact, the positivity of $\omega^a\omega^b\sigma^{ab}$ can be rigorously proven only for $O(\varepsilon)$ contributions. As noted in Sect. 5 of Ref. 2, the T-product in (4.13) needs an $O(\varepsilon^2)$ subtraction because of singular short-distance behaviour. However, there is no need to introduce this subtlety in order to justify the present discussion.

we see that the inequality (4.61) is the condition that θ should be a monotonically increasing function of ϕ_1.

Positivity of σ terms is certainly necessary for stability of the ground state but not sufficient to specify it uniquely. For example, let us suppose that the three circles in Fig.3 are not degenerate (n = 3) and have radii which are almost equal:

$$m_u \lesssim m_d \lesssim m_s \tag{4.66}$$

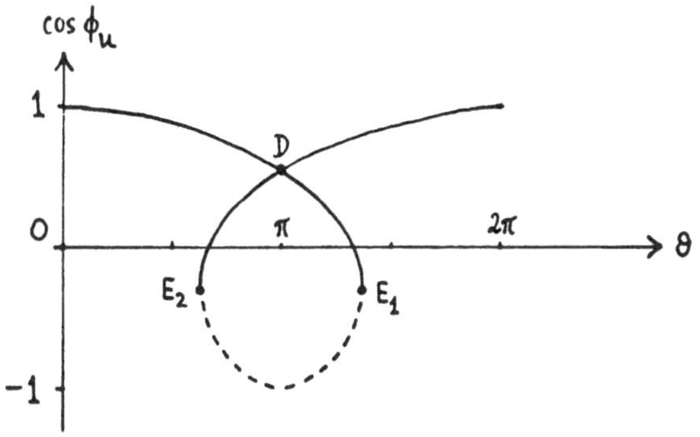

Fig. 4 A solution for the case (4.66). The solid lines indicate positive σ-terms. The intersection at D ($\theta=\pi$) results in two stable degenerate vacua.

As ϕ_u increases from zero, θ (given by the sum $\phi_u+\phi_d+\phi_s$) passes through π (the point D in Fig.4) when ϕ_u is slightly larger than $\pi/3$ but before it reaches $\pi/2$. The point of equality in (4.61) is achieved at a value of ϕ_u between $\pi/2$ and π (the point E_1 in Fig.4). Beyond that, σ terms are negative, so the solution (indicated by the dashed line in Fig.4) is certainly unphysical until it reaches a second point E_2 at which equality in (4.61) is again achieved at a value of ϕ_u between π and $3\pi/2$. Larger values of ϕ_u yield positive σ terms again, and the point D is passed a second time just before ϕ_u reaches $5\pi/3$.

So we see that there can be a range of θ values on either side of π in which there are two solutions with positive σ terms. Either two independent ground states exist, or one of them is unstable for reasons not evident from studying σ terms alone. I shall argue that the solutions given by the arcs DE_1 and E_2D in Fig. 4 are unphysical, but that <u>both</u> solutions at D are physical.

The simplest argument is that we expect the ground state to give the smallest value of the expectation value of $\varepsilon H'$ (generalizing to arbitrary n):

$$\langle vac|\ \varepsilon H'\ |vac\rangle = 2C \sum_{i=1}^{n} m_i \cos\phi_i + O(\varepsilon^2), \quad (C<0) \qquad (4.67)$$

Obviously this eliminates solutions of the type DE_1 or E_2D. At D, there is a two-fold degeneracy of $\langle\varepsilon H'\rangle$ corresponding to reversal of sign of the phases:

$$\phi_i \longrightarrow -\phi_i, \quad \theta = \pi \pmod{2\pi} \qquad (4.68)$$

Thus there are two stable ground states at $\theta=\pi$. They are related (up to an unspecified phase factor $e^{i\beta}$) by a chiral SU(n) operator $\hat{U}$ which I choose to be right-handed:

$$e^{i\beta}|vac\rangle_2 = \hat{U}|vac\rangle_1, \quad \theta = \pi \pmod{2\pi} \qquad (4.69)$$

$$\hat{U} = \exp i \sum_{i=1}^{n} \phi_i \int d^3x\, q_i^+(1+\gamma_5)q_i, \quad (\sum_j \phi_j = \pi) \qquad (4.70)$$

The existence of a two-fold vacuum degeneracy and hence of spontaneous CP violation was observed long ago by Dashen [10] in a study of chiral SU(3)×SU(3) symmetry breaking.

In general, the phases ϕ_i in (4.70) are complicated functions of the mass ratios m_j/m_k. In fact, there exist values of these ratios for which there is no non-trivial solution for $\hat{U}$ and hence no vacuum degeneracy at $\theta=\pi$. For example, consider Fig.3 for the case

$$m_u \ll m_d,\ m_s,\ \ldots \qquad (4.71)$$

As ϕ_u increases from 0 to 2π, the other phases perform a small oscillation about zero. Since ϕ_u is the dominant term in the sum

$$\sum_i \phi_i,$$

it increases monotonically with θ; i.e., unlike the case (4.66) illustrated in Fig.4, there is no loop in the graph of $\cos\phi_u$ against θ. Therefore, θ passes through π at only one physical value of the phases:

$$\phi_u = \pi; \quad \phi_i = 0, \quad (i \geqq 2) \qquad (4.72)$$

The result of substituting (4.72) into the formula (4.70) for $\hat{U}$ is proportional to the identity operator because it commutes with all other operators.

Eq.(1.6) is the general condition for the existence of a non-trivial operator $\hat{U}$. It can be derived from the inequality

$$\sin\left(\sum_{i=2}^{n} \phi_i\right) \leq \sum_{i=2}^{n} \sin\phi_i \quad , \quad (0 \leq \phi_i \leq \pi/2) \tag{4.73}$$

which is easily proven by induction in n. We require the formula

$$\pi - \phi_1 = \sum_{i=2}^{n} \phi_i \tag{4.74}$$

to be satisfied for non-trivial ϕ_i (i.e. different from (4.72)). Eqs.(4.73) and (4.74) imply a result

$$\sin\phi_1 \leq \sum_{i=2}^{n} \sin\phi_i \tag{4.75}$$

which, when combined with (4.42), yields the condition (1.6).

The converse theorem is also valid: if the up-quark mass is sufficiently small compared with other SU(n)×SU(n) masses,

$$m_1^{-1} > \sum_{i=2}^{n} m_i^{-1} \tag{4.76}$$

there is no non-trivial solution for $\hat{U}$. The proof is that (4.76) requires the inequality in (4.75) to be reversed for $0 \leq \phi_1 < \pi$, and that this contradicts Eqs.(4.73) and (4.74).

As noted in Eq.(1.3), a small variation in θ should correspond to insertion of the topological charge density at zero momentum. This operation is uniquely defined if the ground state is not degenerate, and finite if there are no zero-mass particles coupled to F.F*. We therefore expect the phases ϕ_i to be smooth functions of θ at all points except those at which the two-fold vacuum degeneracy appears or σ-terms vanish. This is indeed the case. For example, in Fig. 4, the physical solution from $\phi_u=0=\theta$ to D and back to $\phi_u=2\pi=\theta$ is smooth except for the reversal of phases at the point D. The two-fold vacuum degeneracy at D results in a two-fold ambiguity in $\partial/\partial\theta$: the left-handed derivative is minus the right-handed derivative. The perturbation (1.9) can be used to pick out one of the vacua at D by an appropriate choice of sign for $(\pi-\theta)$ as $\theta \to \pi$. Because of this mechanism, the arcs DE_1 and E_2D are avoided.

We are now in a position to discuss vacuum structure in the presence of a mass perturbation. As long as the ground state is not degenerate, the analysis in Sect.7 of Ref. 2 can be taken over without change - the θ dependence of non-degenerate vacua is given by a Fourier series or integral. It is only at the degenerate point $\theta=\pi$ (mod 2π) for the case (1.6) that any modification is required. Then there are two ground states, so we must find representations for both of them.

Briefly, the idea of the analysis [2] is to derive the representation

$$|\mathrm{vac}>_\theta = \int\!\!\!\!\!\!\sum_m e^{im\theta}\,|m> , \text{ (non-deg.cases)} \tag{4.77}$$

directly from Eqs. (1.3), (2.8) and (2.18) <u>without</u> introducing pure-gauge configurations or otherwise restricting infra-red behaviour. Here $|m>$ are θ-independent eigenstates*) of the winding-number operators K_+ with eigenvalues m. The result (4.77) follows from the fact that $i\partial/\partial\theta$ can be replaced by the θ-independent operator $(K_+ - K_-)$. Because of the absence of assumptions about infra-red behaviour, the analysis does not involve restricting the values of m in any way: hence the symbol $\int\!\!\!\!\!\!\sum$ is used in (4.77). This is essential because we have to incorporate the conclusion of Ref.3 (to be reinforced shortly) that the spectrum of topological charge

$$\nu = m_{out} - m_{in} \tag{4.78}$$

necessarily involves fractional values which depend on the mass ratios m_j/m_k. Here, "spectrum of ν " means the set of values of ν over which summation must be performed in order to obtain physical amplitudes. The phase convention in (4.77) is such that $\theta = 0$ implies CP invariance.

If (4.76) holds, there is no operator $\hat{U}$ and hence no vacuum degeneracy at $\theta = \pi$ (mod 2π). Therefore the representation (4.77) is valid for <u>all</u> θ . The solution is periodic in θ with period 2π , so we conclude that the topological charge spectrum

$$\nu = \text{integer} , \quad [\text{case } (4.76)] \tag{4.79}$$

suffices to generate the amplitudes (4.5).

However, if Eq. (1.6) is valid, the argument used to obtain (4.79) fails because the representation (4.77) does not display the required two-fold degeneracy at $\theta = \pi$ (mod 2π).

---

*) In general, eigenstates $|m>_+$ and $|m>_-$ of K_+ and K_- should be distinguished [2,8] . However, $_+<m'|m>_-$ becomes diagonal in the leading order of ε , so the distinction is irrelevant for the present discussion.

A representation of the form (4.77) is applicable only to non-degenerate vacua characterized by a continuous range of θ values. Hence, for each value of the integer k characterizing the range

$$(2k-1)\pi < \theta < (2k+1)\pi \tag{4.80}$$

there is a formula similar to (4.77), but different phase conventions (e.g. at CP invariant points $\theta = 2k\pi$) are required in different ranges. The simplest procedure is to adopt the representation (4.77) for the range $-\pi<\theta<\pi$, and then use the fact that the solution has to be periodic in θ with period 2π at non-degenerate points:

$$|\text{vac}\rangle_\theta = \sum\!\!\!\!\!\!\int_m e^{im\{\theta(\text{mod } 2\pi)\}}|m\rangle, \qquad -\pi < \theta(\text{mod } 2\pi) < \pi \tag{4.81}$$

Representations for the two vacua at $\theta = \pi\,(\text{mod } 2\pi)$ can be deduced by considering the effect of the perturbation (1.9). If θ approaches π from below in (4.81), $|\text{vac}\rangle_1$ is picked out, while $|\text{vac}\rangle_2$ is obtained by letting θ tend to π from above:

$$|\text{vac}\rangle_1 = \sum\!\!\!\!\!\!\int_m e^{im\pi}|m\rangle,$$

$$|\text{vac}\rangle_2 = \sum\!\!\!\!\!\!\int_m e^{-im\pi}|m\rangle, \qquad \theta = \pi(\text{mod } 2\pi) \tag{4.82}$$

When Eq.(1.6) is obeyed, the complete representation of vacua is given by Eqs.(4.81) and (4.82) instead of (4.77).

We now see that if Eq.(1.6) is valid, the topological charge ν <u>cannot</u> be restricted to integer values. It is essential for self-consistency of the solution that the two vacua (4.82) be <u>different</u> states: we must find

$$|\text{vac}\rangle_1 \neq e^{i\xi}|\text{vac}\rangle_2 \tag{4.83}$$

whatever the value of the real c-number ξ. Therefore, the inequality

$$\sum\!\!\!\!\!\!\int_m \{e^{im\pi} - e^{i\{\xi-m\pi\}}\}|m\rangle \neq 0 \tag{4.84}$$

must be valid for all m-independent constants ξ. This condition is satisfied only if there exist two values (m,m') of the winding number which contribute topological charge

$$\nu = m' - m \neq \text{integer}, \qquad [\text{case } (1.6)] \tag{4.85}$$

An equivalent representation of vacua for the case (1.6) can be obtained by making use of the chiral SU(n) operator $\hat{U}$. The fact that $\hat{U}$ relates the two degenerate vacua (4.82) has already been noted in (4.69). Since $\hat{U}$ commutes with K_+, $\hat{U}|m\rangle$ must also be an eigenstate of K_+ with eigenvalue m. In principle, there could be many K_+ eigenstates with the same eigenvalue; however, Eqs. (4.69) and (4.82) require $\hat{U}|m\rangle$ to be proportional to the eigenstate $|m\rangle$ which appears in (4.81):

$$\hat{U}|m\rangle = e^{i\beta} e^{-2\pi i m} |m\rangle \tag{4.86}$$

Therefore, the representation (4.81) for vacua at non-degenerate points can be written

$$|vac\rangle_\theta = e^{-ik\beta} \int\!\!\!\!\!\!\sum_m e^{im\theta}\, \hat{U}^k\, |m\rangle \tag{4.87}$$

where k is the integer defined by Eq. (4.80).

If a power of $\hat{U}$ is the identity operator, it is not difficult to determine which fractional values of ν are needed to generate the amplitude (4.5).

Let p be the smallest integer for which the equation

$$\hat{U}^p = I \tag{4.88}$$

is valid. According to Eqs. (4.70) and (4.88), the phases ϕ_i which define $\hat{U}$ must obey the constraints

$$e^{2i\phi_i p} = 1, \qquad \sum_i \phi_i = \pi \tag{4.89}$$

The general solution is found by considering all partitions of p into n positive integers

$$p_1 \geq p_2 \geq \dots \geq p_n > 0\ ; \quad p = \sum_{i=1}^{n} p_i \geq n \tag{4.90}$$

such that the set $\{p_i\}$ has no common factor. Then, for each partition, the set of phases

$$\phi_i = \pi p_i/p \quad (\mathrm{mod}\ 2\pi) \tag{4.91}$$

defines an operator $\hat{U}$ with the property (4.88). The condition (4.91) corresponds to a special choice of the mass parameters:

$$m_i = \lambda \operatorname{cosec}(\pi p_i/p)\ , \quad (i = 1, \dots, n) \tag{4.92}$$

[Once again, the case n=2 is special. The states $|vac\rangle_1$ and $|vac\rangle_2$ picked out by the perturbation (1.9) with $m_u=m_d$ are related by the operator $\hat{U}$ given by the partition

$$p = 2,\ p_1 = 1 = p_2 \tag{4.93}$$

Other partitions exist for $p > 2 = n$, but these are associated with the continuously degenerate set of states which interpolate between $|vac>_1$ and $|vac>_2$ at $\theta = \pi$ (mod 2π).]

Consider powers of $\hat{U}$ applied to the state $|vac>_1$ (for example):

$$|\Psi_r> \equiv \hat{U}^r |vac>_1 = e^{i\beta r} \sum_m e^{-i\pi m(2r-1)} |m> , \quad (r = 0,1,\ldots,p-1) \tag{4.94}$$

The identity

$$\{\hat{U}\}^{-r} (\bar{q}_L)_j (q_R)_i \hat{U}^r = e^{2i\phi_i} (\bar{q}_L)_j (q_R)_i \tag{4.95}$$

and Eq.(4.91) imply that, for any pair of integers $r \neq r'$ between 0 and p-1, flavour indices i=j can be found with the property

$$<\Psi_r| (\bar{q}_L)_j (q_R)_i |\Psi_r> \neq <\Psi_{r'}| (\bar{q}_L)_j (\bar{q}_R)_i |\Psi_{r'}> \tag{4.96}$$

Hence no two states in the list (4.94) are the same:

$$|\Psi_r> \neq e^{i\eta} |\Psi_{r'}> , \quad (r \neq r') \tag{4.97}$$

[Note that only two members of the list are vacua: r = 0,1.]

It is now easy to see that the property (4.88) requires topological charge ν to be quantized in units p^{-1}. Eqs. (4.86) and (4.88) require

$$m = \beta/2\pi + p^{-1} \{\text{integer}\}_m \tag{4.98}$$

where β does not depend on m. Eq.(4.97) shows that the integers in (4.98) must differ by unity. Therefore, there have to be contributions to amplitudes with topological charge

$$\nu = p^{-1} \quad [\text{case (4.88)}] \tag{4.99}$$

and integer multiples thereof.

The conclusions (4.79),(4.85) and (4.99) reinforce and generalize the conclusion of Ref. 3 that spontaneously broken SU(n)×SU(n) symmetry without a small-mass U(1) boson requires a spectrum of topological charge which depends sensitively on the mass ratios m_i/m_j and is quantized in units n^{-1} in the degenerate-mass case

$$m_1 = m_2 = \ldots\ldots = m_n = O(\varepsilon) \tag{4.100}$$

The improvement on Ref. 3 is that there is now no difficulty in understanding how different fractional ν values for different values of n can be reconciled with each other. Because of the action of the operator $\hat{U}$ in (for example) Eq. (4.87), the θ period is 2π whatever the spectrum of topological charge is. It must be emphasized that the connection between the θ period and the ν spectrum can be broken only if there is a value of θ for which at least two stable ground states exist.

[Strictly speaking, the statement that a function is "periodic" implies that it is single-valued, so I should qualify my remarks in the following way. The solution is rigorouslv periodic for the case (4.76):

$$e^{i\phi_i(\theta)} = e^{i\phi_i(\theta+2\pi)} \tag{4.101}$$

For the case (1.6), (4.101) is valid for all θ except $\theta = \pi$ (mod 2π) where the solution is double-valued. The same two values are obtained, whatever the value of the integer k in $\theta = \pi\,(2k+1)$.]

Special cases will now be used to illustrate some features of the general analysis. We begin with the case (4.100) in which the quarks involved in the SU(n)×SU(n) limit are degenerate in mass, while the other (N-n) quarks remain massive in the limit $\varepsilon \to 0$.

Eqs. (4.100) and (4.42) require $\sin\phi_i$ to be independent of i:

$$\phi_i = \phi_j \text{ or } (\pi-\phi_j), \qquad (i,j = 1, \ldots, n) \tag{4.102}$$

Let ℓ angles ϕ_i have a common value ϕ, while the remaining $(n-\ell)$ angles equal $(\pi-\phi)$, with $\ell \geq n/2$. The possibility $\ell = \frac{1}{2}n$ (even n) yields the continuous set of solutions (4.51) at $\theta = n\pi/2$ (mod 2π). Elsewhere, ℓ lies in the range $\frac{1}{2}n < \ell \leq n$, and Eq. (4.43) yields $(2\ell-n)$ solutions for ϕ:

$$\phi = \{\theta + (2j+\ell-n)\pi\}/(2\ell-n) \quad (\text{mod } 2\pi), \qquad (j=0,1,\ldots,2\ell-n-1) \tag{4.103}$$

Apart from the continuously degenerate point, the number of solutions of (4.42) and (4.43) is

$$\sum_{\ell>\frac{1}{2}n}^{n} \binom{n}{\ell}(2\ell-n) = \begin{cases} n!/(\frac{1}{2}n)!\,(\frac{1}{2}n-1)!\,, & (n \text{ even}) \\ n!/\{(\frac{n-1}{2})!\}^2\,, & (n \text{ odd}) \end{cases} \tag{4.104}$$

However, positivity of σ-terms eliminates all cases except $\ell = n$ and the special case (1.8).

The $\ell = n$ solutions are

$$\phi_i = \phi = (\theta+2\pi j)/n , \quad (j=\text{integer}) \tag{4.105}$$

Again, σ-term positivity imposes a constraint

$$-^{\pi}/2 \leqq \phi_i (\text{mod } 2\pi) \leqq \pi/2 \tag{4.106}$$

so positive ϕ_i (in the range $0 < \phi_i < ^{\pi}/2$) have to jump to negative values at some point. Of course, this is the point $\theta = \pi$ (mod 2π) at which ϕ_i jumps from π/n to $-\pi/n$. In the notation of Eq. (4.81), the vacuum expectation value of the mass perturbation is

$$\langle vac \,|\, \varepsilon H' \,|\, vac \rangle = 2nC\hat{m} \cos [\{\theta \text{ (mod } 2\pi)\} /n] + O(\varepsilon) \tag{4.107}$$

where $\hat{m}$ is the common value of the quark masses (4.100). The operator $\hat{U}$ which reverses the phases at $\theta = \pi$ (mod 2π) is an element of the right-handed chiral centre group Z_n. It relates different ℓ=n solutions.

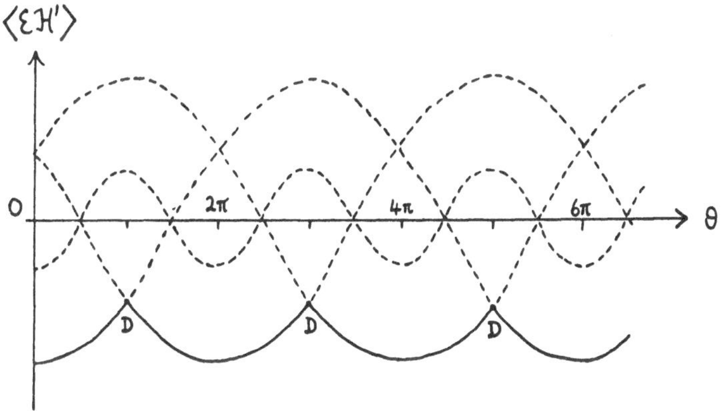

Fig. 5 Expectation value of the mass operator $\varepsilon H'$ for the SU(3)×SU(3) case with $m_u = m_d = m_s$. All solutions (4.103) are displayed, with unstable states indicated by dashed lines. The physical ground state, with a two-fold degeneracy at D, is given by the solid line [Eq. (4.107)] . Topological charge is quantized in units 1/3.

The situation for n=3 is shown in Fig.5. The solid line represents Eq.(4.107). The dashed lines which analytically continue the solid line have period 6π ; they are the $\ell = n = 3$ solutions (4.105). The two-fold degeneracy occurs at the points D (as in Fig.4), where two of the solutions intersect. The dotted line with period 2π represents three unphysical solutions with $\ell = 2$ in (4.103).

Another interesting special case is n = 2. Only m_u and m_d are considered to be $O(\varepsilon)$, so the solution of Eqs.(4.42) and (4.43) involves only one mass ratio:

$$0 < m_u/m_d \leqq 1 \tag{4.108}$$

As a result, the solution can be expressed in terms of elementary functions.*)

The leading terms in the $\varepsilon \to 0$ limit obey equations

$$m_u \sin \phi_u = m_d \sin \phi_d \tag{4.109}$$

$$\phi_u + \phi_d = \theta \quad (\text{mod } 2\pi) \tag{4.110}$$

with solutions

$$\begin{aligned} \tan \phi_u &= m_d \sin \theta / (m_u + m_d \cos \theta) , \\ \tan \phi_d &= m_u \sin \theta / (m_d + m_u \cos \theta) \end{aligned} \tag{4.111}$$

In the corresponding formulas for cosines,

$$\cos \phi_u = (m_u + m_d \cos \theta) / \sqrt{m_u^2 + m_d^2 + 2m_u m_d \cos \theta} \tag{4.112}$$

(with $\cos \phi_d$ given by $m_u \leftrightarrow m_d$), only the positive sign of the square root is taken. [The negative sign gives negative σ terms.]

Results for the case in which m_u/m_d is slightly less than 1 are shown in Fig. 6. The phases ϕ_u and ϕ_d are smooth functions of θ with period 2π . The near degeneracy of m_u and m_d produces rapid variations in the phases in the vicinity of $\theta = \pi$ (mod 2π). However, the condition (4.76) is satisfied, so there is <u>no</u> vacuum degeneracy at $\theta = \pi$ (mod 2π) and topological charge is quantized in <u>integer</u> units.

---

*) I thank P. di Vecchia for this observation.

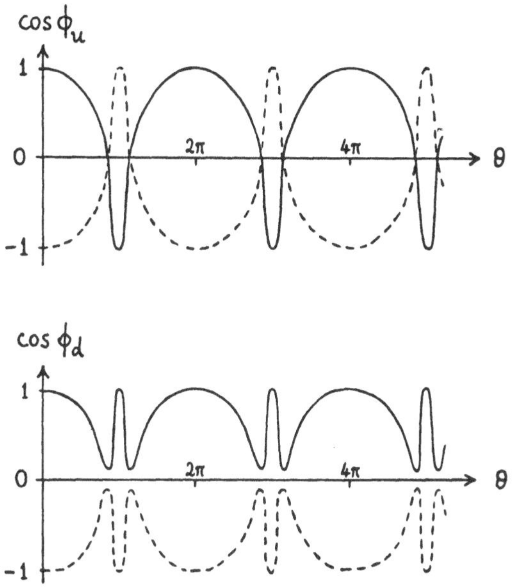

Fig. 6 Solutions for the SU(2)×SU(2) limit (n=2) with $m_u/m_d \lesssim 1$. The physical solution (4.112) is represented by the solid line.

When n is 2, it is only for the degenerate case $m_u = m_d$ that Eq.(1.6) is satisfied. The solutions (shown in Fig.7) can be obtained as a smooth limit

$$m_u/m_d \longrightarrow 1 \quad , \text{(fixed } \theta) \tag{4.113}$$

of Eq.(4.112):

$$\begin{aligned} \cos\phi_u = \cos\phi_d &= |\cos {}^{\theta}/2| \quad , \ \theta \neq \pi \ (\text{mod } 2\pi) \\ \cos\phi_u &= -\cos\phi_d \quad , \quad \theta = \pi \ (\text{mod } 2\pi) \end{aligned} \tag{4.114}$$

If $\hat{m}$ denotes the common value of m_u and m_d, Eq.(4.107) becomes

$$<\text{vac}|\varepsilon H'|\ \text{vac}> \ = \ 4C\,\hat{m}\,|\cos {}^{\theta}/2| \ + \ O(\varepsilon^2) \tag{4.115}$$

The phases ϕ_u and ϕ_d jump from $^{\pi}/2$ to $-\,^{\pi}/2$ as θ passes through π(mod 2π). The corresponding operator $\hat{U}$ is the non-trivial member of the chiral centre group Z_2, so topological charge is quantized in half-integer units.[The vertical lines at $\theta = \pi$ (mod 2π) in Fig.7 represent the continuously degenerate solution associated with the case (1.8); see Eqs. (4.51),(4.62) and (4.63).]

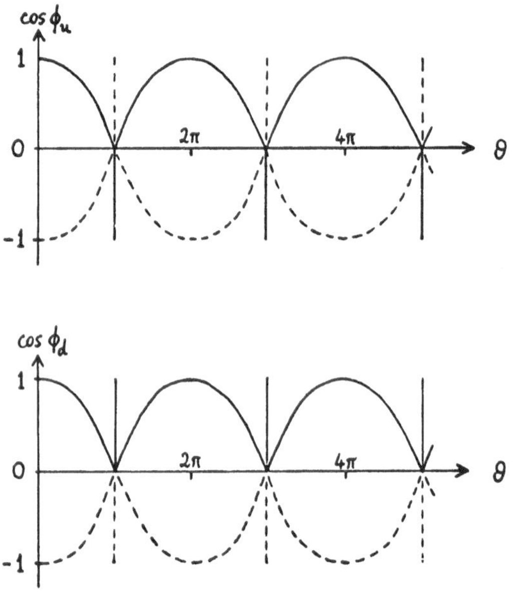

Fig.7 Solutions for the case n=2 with $m_u=m_d$. The solid and dashed lines are the result of applying the limit (4.113) to the corresponding curves in Fig. 6.

We see that the spectrum of topological charge changes abruptly as m_u/m_d reaches 1. Presumably, this is also a feature of the solution for arbitrary n. As discussed previously [2], discontinuities in mass dependence of the ν spectrum can be attributed to the coupling of SU(n)×SU(n) Nambu-Goldstone bosons to the operator F.F*. At the same time, the appearance of the two-fold vacuum degeneracy (4.82) allows amplitudes at fixed θ to display the smooth mass dependence required by chiral perturbation theory.

Briefly, the conclusions of this Section concerning the spectrum $\{\nu\}$ of the topological charge operator

$$(g^2/32\pi^2) \int d^4x\, F.F^*$$

are:

(a) If m_1^{-1} is greater than

$$\sum_{i=2}^{n} m_i^{-1} ,$$

ν takes <u>integer</u> values. For each value of θ, the ground state is uniquely fixed by the mass perturbation $\epsilon H'$.

(b) If m_1^{-1} is less than or equal to

$$\sum_{i=2}^{n} m_i^{-1},$$

ν cannot be quantized in integer units. There are two stable vacua at $\theta = \pi \pmod{2\pi}$ related by a chiral SU(n) transformation $\hat{U}$.

(c) If we choose m_i to be proportional to cosec $(\pi p_i/p)$, the quantization unit for ν is p^{-1}. Here $\{p_i;\ i = 1,\ldots, n\}$ is a set of positive integers with no common factor and with sum

$$p = \sum_{i=1}^{n} p_i.$$

The integer p is the smallest for which $\hat{U}^p$ is the identity operator.

(d) For the equal-mass case $m_i = \hat{m}$ (with i = 1, ..., n), topological charge is quantized in units n^{-1}. This is the SU(n)×SU(n) limit considered in Ref.3. The operator $\hat{U}$ becomes an element of the chiral centre group Z_n.

(e) For quark masses constrained as in (b) but not proportional to cosec $(\pi p_i/p)$, ν cannot be quantized in rational units. Sums over ν are written $\sum\!\!\!\!\!\!\int_\nu$ to indicate the possibility (depending on the precise values of m_i/m_j) of discrete and continuous contributions to the spectrum. None of the powers of $\hat{U}$ is equal to the identity operator.

5. SELF-CONSISTENT METHODS

The most striking feature of the results of Ref.3 and of Section 4 is the conclusion that the spectrum of the topological charge operator depends in detail on the ratios of the quark mass parameters. This conclusion is general: it is a consequence of anomalous Ward identities and the assumption that SU(n)×SU(n) symmetry is spontaneously broken without unwanted U(1) bosons being produced. It means that the gluonic functional integral is dominated by configurations whose infra-red behaviour is strongly dependent on m_k.

This poses a new problem for self-consistent calculations. In addition to quark self-energies, one must solve self-consistently for the infra-red boundary conditions of the gluons. In other words, the boundary conditions in gluonic space have to be determined dynamically, not on the basis of quasi-perturbative requirements like finite action.

These remarks run counter to the conventional practice of regarding the spontaneous breaking of SU(n)×SU(n) symmetry as a fermionic problem to be solved after problems of the gluonic sector (such as confinement) have been solved. However, I see little rationale for this convention. It is only in the short-distance region that it makes sense to treat quarks and gluons as essentially independent systems which perturb each other. Elsewhere, quarks and gluons interact strongly, especially in the low-momentum region where the Nambu-Goldstone mechanism is supposed to become important. The point of the present analysis is that anomalies provide a definitive, model-independent demonstration of this strong coupling.

I should emphasize that the terms "topological charge" and "winding number" are used in a general quantum-mechanical sense, and not in the specialized sense of mathematicians. The mathematical definitions refer to topological spaces with special properties. In our language, these definitions introduce hypotheses about infra-red behaviour which are physically unacceptable. The present analysis makes no assumptions about gluonic or fermionic infra-red behaviour beyond the requirement that the Nambu-Goldstone mechanism is permitted. Its conclusions concern the spectrum $\{\nu\}$ of the operator $(g^2/32\pi^2)\int d^4x\, F.F^*(x)$. For want of better names, I call this operator the "topological charge operator" and its values ν "topological charges". In this general sense, it is not a contradiction to talk about fractional topological charge. The spectrum $\{\nu\}$ refers only to those values ν which must be included in sums $\sum\!\!\!\!\int_\nu$ when constructing physical amplitudes.

Mention of "non-integer values of topological charge" can be found in several places in the literature, but it is apparent that different authors give this phrase a variety of interpretations - often, the meaning is not more profound than the statement that the dilute-gas approximation is not applicable. To place the conclusions of the present analysis in perspective, I should emphasize the point that fractional values or continuous distributions in the spectrum $\{\nu\}$ imply functional boundary conditions which are radically different from those encountered so far in explicit calculations. "Large quantum fluctuations", "infinitely dense instanton gases", or the absence of any instanton interpretation are perfectly consistent with topological charge being quantized in integer units.

Massless $(QED)_2$ (two-dimensional quantum electrodynamics) is a good example. The topological charge operator is $(e/4\pi)\times \int d^4x\, \varepsilon^{\mu\nu} F_{\mu\nu}$, but there are no instantons in quark-gluon space. Thus amplitudes and hence values ν of the topological charge operator are entirely generated by quantum fluctuations. Nevertheless, the spectrum $\{\nu\}$ is quantized in integer units [30].

If one attempts an analysis using ideas of instanton physics, it appears that continuous values of ν should be generated: the finite-action condition

$$F_{\mu\nu} \rightarrow 0 , (x_\mu \rightarrow 0) \qquad (5.1)$$

in two dimensions permits a rescaling $A_\mu \rightarrow cA_\mu$ and hence $\nu \rightarrow c\,\nu$, where c is any real number [37] . The finite-action condition is legitimate because it is a dynamical fact that ratios (3.18) for this model can be computed in compactified space [38] ; (i.e., Euclidean space can be smoothly mapped onto the sphere S_2 without boundary). However, the point [38,39] is that it is the presence of compactifiable fermion fields which forces the quantization of ν in integer units.

Fermions play a smilar role in massive $(QED)_2$. The θ parameter (given by a term $- \theta(e/4\pi)\varepsilon^{\mu\nu} F_{\mu\nu}$ in the Lagrangian) is linearly related to a background electric field

$$F = \frac{1}{2} \langle vac| \varepsilon^{\mu\nu} F_{\mu\nu} |vac\rangle \qquad (5.2)$$

The magnitude of F cannot exceed $e/2$ because for larger values, there would be an instability due to pair creation of fermions[40]:

$$F = (e/2\pi)\{\theta \pmod{2\pi}\} ,$$
$$- \pi < \theta \pmod{2\pi} < \pi \qquad (5.3)$$

At $\theta = \pi \pmod{2\pi}$, it appears [40] that there is a unique ground state (with F = 0) in the strong coupling limit $m \ll e$, but for $e \gg m$, there is a two-fold vacuum degeneracy ($F = \pm e/2$) and "half-asymptotic" particles exist.

For strong coupling, topological charge is quantized in integer units, because there is only one ground state for each θ and the θ period is 2π (because of the presence of fermions). Discontinuous θ dependence at $\pi \pmod{2\pi}$ is due to the fact that the sum in the Fourier representation for the vacuum functional

$$Z(\theta) = Z(0) \exp - \frac{i}{2} V F^2(\theta), \quad (V= \text{volume of space time}) \qquad (5.4)$$

is infinite:

$$Z(\theta) = \sum_{\nu=-\infty}^{\infty} Z_\nu e^{-i\nu\theta} , \quad (\text{integer } \nu,\ m \ll e) \qquad (5.5)$$

For weak coupling $e \ll m$, the spectrum $\{\nu\}$ involves non-integer values of ν if semi-classical arguments [40] for the existence of two stable vacua at $\theta = \pi \pmod{2\pi}$ are valid.

Remarks:

(a) Some recent accounts confuse θ with ν. The fact that F can be continuously varied (by changing θ) has nothing to do with continuous topological charge. For example, consider one-instanton QCD calculations of the 't Hooft type [15] in the presence of quark masses. The vacuum expectation value of $(g^2/32\pi^2)F.F^*$ is proportional to sin θ, so it varies continuously with θ ; but ν is ± 1.

(b) In the large-N limit of CP^{N-1} models in two dimensions, the vacuum expectation value of the topological charge density is proportional to [37,41] θ/N. The situation is similar to that in massive $(QED)_2$, except that we have no information about what happens at θ = π (mod 2π). Presumably, the pair-creation mechanism observed in $(QED)_2$ is also operative here, but in some non-leading order of the large-N expansion. It is not possible to say anything about {ν} without analyzing this point. Is the θ-period 2π? If so, is there a unique ground state at θ = π(mod 2π)?

(c) A remark of Coleman [33] that functional measure is dominated by infinite-action configurations has led to unjustified comments that fluctuations produce unquantized ν. Coleman's remark refers to the absolute normalization of the measure. This cancels out when the ratio (3.18) is considered. For example, compactification (and hence integer ν) is assumed in many calculations, but always for this ratio; there is no conflict with Coleman's remark.

(d) In a very interesting criticism of the dilute-gas approximation for models with instantons, it was shown (for CP^1) that the thermodynamic limit and dimensional transmutation are fully realized only if infinitely dense gases are considered [32] . The calculation was performed on the sphere S_2, with the large-size limit (infinite radius) considered at the end. Thus theorems for compact surfaces apply and ν is quantized in integer units.

This discussion digresses from the subject of self-consistent calculations, but its conclusion is relevant. Conventional ideas about boundary conditions in functional space are not adequate for the spontaneous breaking of SU(n)×SU(n) symmetry in QCD. They cannot reproduce the quark-mass dependence of ν spectra found in Section 4.

Self-consistent calculations will have to involve dynamically determined infra-red conditions for the gluon field. I can think of two general lines of investigation:

(i) Try to deduce explicit "chiral" boundary conditions for gluonic space by considering self-consistent calculations for fermions in an arbitrary gluon field and requiring the absence of unwanted U(1) bosons. If successful, find solutions of the Yang-Mills equations of motion which obey these unusual boundary conditions. Approximate gluonic space in the generating functional of QCD by computing fluctuations about these solutions. Treat the remaining fermionic integral self-consistently, e.g. by using a mean-field expansion.

(ii) Avoid any attempt to characterize gluonic boundary conditions explicitly. Instead, treat gluons and fermions in the same way. Solve the coupled Yang-Mills equations of motion for the quark self-energy and gluon vertex functions in a self-consistent approximation.

Migdal's approach [42] to the problem is of type (ii). Following Halpern [43], he considers the replacement

$$- \frac{1}{4} F^2 \longrightarrow - \frac{1}{4} F^2 + \frac{1}{2} G^{\mu\nu} F_{\mu\nu}[A] \tag{5.6}$$

in the QCD Lagrangian, where functional integration over both $G_{\mu\nu}$ and A_μ is understood. The integral over A_μ is then quadratic, so it can be performed explicitly. Migdal suggests mean-field methods for the resulting integral over $G_{\mu\nu}$, q and $\bar{q}$. The feature of this suggestion which I especially favour is the absence of constraints on the values of $(g^2/32\pi^2) \int d^4x\, F.F^*$.

As is evident from the above discussion, I maintain my claim [3] that instantons are not the answer to the U(1) problem. It is important to carry out a check of this conclusion by performing a self-consistent calculation with gluon space approximated by fluctuations about mixtures of instantons and anti-instantons. The prediction is that the unwanted U(1) boson will appear if $(\bar{q}_L)_j (q_R)_i$ develops a vacuum expectation value.

The calculation is formulated in the following way. Consider the result of performing the gluon integral in the Euclidean generating functional $Z[\eta,\eta^+,j_\mu]$ of QCD, where η,η^+, j_μ are sources for q^+,q,A_μ . Direct integration over q and q^+ with conventional boundary conditions must be avoided because a self-consistent solution is sought.

Fluctuations about configurations A^{cl} of instantons and anti-instantons are included, but cubic and quartic quantum vertices are ignored. These fluctuations contribute an effective quark interaction of the form

$$\int d^4x\, L_{fluct} = \iint d^4x\, d^4y\, \{g\, q^+ \gamma_\mu \tau q - (D_\lambda F_{\lambda\mu})^{cl} + j_\mu\}^a_x\, \Delta^{ab}_{\mu\nu}(x,y)' .$$
$$.\{g\, q^+ \gamma_\nu \tau q - (D_\eta F_{\eta\nu})^{cl} + j_\nu\}^b_y$$
$$+ \{\text{ zero-mode terms}\} \tag{5.7}$$

where $\Delta^{ab}_{\mu\nu}$ is the gluon propagator in the presence of the background field A^{cl}. The prime for the symbol $\Delta'_{\mu\nu}$ indicates that zero modes of the gluon fluctuations must be subtracted in a manner consistent with that in which collective coordinates are introduced. The "zero-mode terms" in (5.7) refer to the logarithm of the corresponding Jacobian. It is necessary to include mixtures of instantons and anti-instantons, so $(D_\lambda F_{\lambda\nu})^{cl}$ should not be set equal to zero.

In the corresponding Abelian case [44] , the four-quark interaction can be simplified in the usual way by introducing an auxiliary field

$$\sigma(x,y) = q(x)\bar{q}(y) \tag{5.8}$$

and expanding about its mean value. However, the presence of the zero-mode terms in (5.7) has so far prevented a similar analysis being performed for QCD.

Clearly, (5.7) is the correct effective interaction for self-consistent calculations, not the determinantal interaction (4.46). Eq.(4.46) is valid only for one-(anti-)instanton contributions to connected amplitudes in the cluster expansion (provided that for each connected piece, all fermionic chiralities have the same handedness). This is a leading contribution to the dilute-gas approximation - overlap between instantons and anti-instantons is not considered. But for the quark self-energy problem, we consider multi-instanton contributions to connected amplitudes. These are non-leading, overlap-dependent terms so even if the validity of the dilute-gas approximation is accepted, Eq.(4.46) is not valid. Non-leading orders involve instanton interactions associated with non-zero modes; there are (2N+4),(2N+8),... point fermionic interactions [3] as well as the 2N-point interaction (4.46) due to zero modes. Exponentiation of zero-mode one-instanton terms does not give the correct multi-instanton contribution except for the cluster limit mentioned above.

Lee and Bardeen [45] have observed that fermion propagation through mixtures of instantons and anti-instantons involves interference between zero modes and the continuum of non-zero modes. This is the same effect. An extension of their analysis to $\Delta'_{\mu\nu}$ is necessary if (5.7) is to be given a self-consistent analysis in the dilute-gas approximation.

As noted in the Introduction, Ref. 3 contains a second argument showing that self-consistent calculations with the determinantal interaction (4.46) are not self-consistent for QCD. In anomalous Ward identities, $F.F^*$ has zero U(1) chirality. Therefore, an effective Lagrangian will correctly reproduce these identities only if, under infinitesimal "effective" $U(1)_{ax}$ transformations (to be distinguished from γ_5 transformations (2.20) in QCD), the second variation vanishes [3]:

$$\delta_{ax} L_{eff} \neq 0 , \qquad \delta^2_{ax} L_{eff} = 0 \tag{5.9}$$

In other words, there must be no $U(1)_{ax}$ σ term.

Obviously, the determinantal interaction (4.46) fails this test. Its second variation is proportional to the interaction itself. This illegitimate σ-term is responsible for the mass of the U(1) boson in self-consistent calculations with (4.46). If the U(1) boson in QCD acquires a mass, it is certainly not via a $U(1)_{ax}$ σ term: such terms do not exist in correct calculations.

Effective Lagrangians with a $U(1)_{ax}$ breaking term of the form [42,46]

$$\Delta L_{trial} = Q(x)\, s(x) \tag{5.10}$$

were proposed, where Q(x) represents the topological charge density and s(x) is the pseudoscalar quark bilinear in the divergence of the U(1) axial vector current. Eq.(5.10) also fails the test (5.9).

However, an improvement of (5.10) has now been proposed [9,47] which is acceptable from the point of view of Ref. 3 *). The proposed breaking term

$$\Delta L = (\text{constant})\; Q(x)\; \{\ell n \det \bar{q}_L q_R - \ell n \det \bar{q}_R q_L\} \tag{5.11}$$

obviously obeys (5.9). Note that the derivation of this effective Lagrangian is based on anomalous Ward identities in which it is already assumed that massless U(1) bosons are not present. Therefore a calculation of a U(1) boson mass should certainly give a non-zero result (otherwise the effective Lagrangian is inconsistent) but this does not imply that the same result will be obtained in a self-consistent QCD calculation.

6. CURRENT ALGEBRA FOR CP VIOLATION DUE TO θ

The analysis of Section 4 can be applied to the problem of estimating CP violating effects in strong interactions [16,17,36], if the parameter θ is supposed to be small but not equal to the P and T conserving value zero.

The definition of CP conjugation depends on what we choose to call a quark. The definition of q_i in Section 4 is fixed by

---

*) But I do not agree with the claim of Ref.47 that this has something to do with the interaction in (4.46).

requiring the mass matrix in the Lagrangian (1.2) to be real, diagonal, and γ_5 free. The non-zero phases ϕ_i of Eq.(4.5) for $\theta \neq 0$ indicate that, with this definition of CP, the vacuum state is not CP invariant. This is inconvenient for current-algebraic purposes; we would like to have a definition of CP in terms of γ_5 rotated quarks for which hadronic states such as the nucleon have definite CP properties. The small first-order effect of $\theta \neq 0$ could then be represented as an operator perturbation. In other words, we look for a definition of CP such that to lowest order in θ, CP is <u>not</u> spontaneously broken.

The solution of this problem is obviously unique (apart from rearrangements of flavour indices in $SU(n)_{ann}$ space). The phases in (4.5) must be absorbed into the new definition of quarks:

$$(\Psi_R)_i = e^{i\phi_i/2}(q_R)_i \qquad (6.1)$$

The CP operation defined in terms of Ψ_i leaves the vacuum invariant:

$$\langle vac|\,(\bar{\Psi}_L)_j\,(\Psi_R)_i\,|vac\rangle = C\,\delta_{ij} + O(\varepsilon) \qquad (6.2)$$

Therefore, to first order in θ and to lowest order in the chiral symmetry breaking parameter ε, we are permitted to give hadronic states the definite CP properties characteristic of the CP conserving theory if we work with the new quarks Ψ_i.

The first-order CP-violating perturbation

$$\delta L_{CP} = \delta L_{CP}(\Psi, \bar{\Psi}) \qquad (6.3)$$

is found by substituting (6.1) into the Lagrangian (1.2). The only term which matters is the mass term

$$\sum_{i=1}^{n} m_i \bar{q}_i q_i = \sum_{i=1}^{n} m_i \cos\phi_i\, \bar{\Psi}_i \Psi_i - i\lambda \sum_{i=1}^{n} \bar{\Psi}_i \gamma_5 \Psi_i + O(\varepsilon^2) \qquad (6.4)$$

where Eq.(4.42) has been used. For small θ, λ is fixed by Eq.(4.43) to be

$$\lambda = \theta\Big(\sum_{i=1}^{n} m_i^{-1}\Big)^{-1} + O(\varepsilon^2, \theta^2) \qquad (6.5)$$

so the answer is

$$\delta L_{CP} \simeq i\,\theta\Big(\sum_{i=1}^{n} m_i^{-1}\Big)^{-1} \bar{\Psi}\gamma_5\Psi \qquad (6.6)$$

The idea of describing first-order CP violation in terms of an operator perturbation is due to Baluni [16] . The present discussion differs on two points:

(a) Baluni's expression (derived for n=3) contained an extra factor 3. Eq. (6.6) agrees with the normalization used in Refs. 17 and 36.

(b) Baluni's discussion of vacuum structure corresponds to the model (4.46) rather than QCD. When the mass perturbation $\varepsilon H'$ is absent, he considers the vacuum degeneracy to be $SU(n)\times SU(n)/SU(n)$ instead of (3.12). In fact, the model (4.46) gives the same answer (6.6), even though the derivations are different; e.g., θ is not a vacuum angle and there is no analogue of (1.3) in this model. Note that the QCD derivation of (6.6) involves an additional assumption: Eq.(4.43) and hence Eq.(6.5) is valid only if we assume the absence of a U(1) boson with mass $O(\sqrt{\varepsilon})$.

According to Eq.(4.29), the (n^2-1) Nambu-Goldstone bosons of $SU(n)\times SU(n)$ couple to $\bar{\Psi}\gamma_\mu\gamma_5(\vec{\lambda}/2)\Psi$, so the usual soft-meson formalism of current algebra is applicable. Contraction of a soft meson yields a commutator with the corresponding generator

$$F_5^a = \int d^3x \; \Psi^+\gamma_5(\lambda^a/2)\Psi \tag{6.7}$$

multiplied by $(iF_\pi)^{-1}$. Thus the $\eta \to \pi\pi$ decay rate [16,17] and the CP violating πNN coupling [17] can be related to θ in a model-independent way.

However, the bounds on θ implied by experiment for these quantities are very poor compared with that available from the experimental limit [48]

$$|D_n| \lesssim 10^{-24} \text{ cm} \tag{6.8}$$

for the electric dipole moment of the neutron D_n. In Ref. 17, it was shown that D_n can be related to θ in a model-independent way if just the leading logarithm

$$D_n = O(m_\pi^2 \ln m_\pi) \qquad (\text{small } m_\pi) \tag{6.9}$$

is retained. Then the only process which contributes is that shown in Fig. 8. It yields a bound

$$|\theta| \lesssim 2\times 10^{-9} \tag{6.10}$$

similar to Baluni's bag-model estimate [16] . Details of the phenomenology are given in Ref. 17.

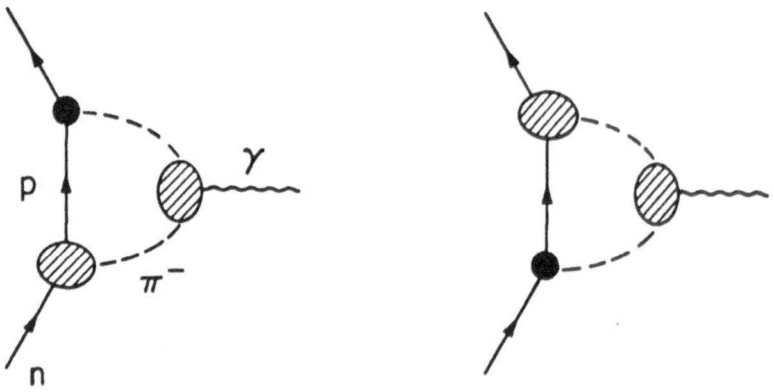

Fig. 8 Mechanism responsible for the $O(m_\pi^2 \ln m_\pi)$ contribution to the electric dipole moment of the neutron. The dark blob indicates a CP violating πNN interaction induced by the parameter θ.

ACKNOWLEDGEMENTS

I thank N. Christ, P. di Vecchia, G. Veneziano, E. Witten, and especially S. Coleman for stimulating discussions. This work was completed at the University of Bern with support from the Schweizerische Nationalfond.

REFERENCES

1. S. L. Glashow, in "Hadrons and their Interactions", Academic Press Inc. (New York 1968), p. 83.
2. R. J. Crewther, "Status of the U(1) Problem", in "Instantons in Field Theory", Riv.Nuovo Cimento 2, No 8 (1979), p. 63.
3. R. J. Crewther, Phys.Letters 70B, 349 (1977).
4. G. 't Hooft, Phys.Rev.Letters 37, 8 (1976).
5. C. G. Callan, Jr., R. F. Dashen and D. J. Gross, Phys.Letters 63B, 334 (1976).
6. R. J. Crewther, Proceedings of the 1979 International Workshop on High-Energy Physics, Serpukhov (to be published).
7. R. J. Crewther, Univ.Bern preprint (to appear).
8. R. J. Crewther, "Effects of Topological Charge in Gauge Theories", in "Facts and Prospects of Gauge Theories", Schladming 1978, ed.P. Urban, Acta Phys.Austriaca Suppl. 19, 47 (1978).
9. P. Di Vecchia, G. Veneziano, and E. Witten, (to appear).
10. R. F. Dashen, Phys.Rev. D3, 1879 (1969).
11. R. Jackiw and K. Johnson, Phys.Rev. 182, 1459 (1969).
12. S. L. Adler and D. G. Boulware, Phys.Rev. 184, 1740 (1969).
13. S. L. Adler, Phys.Rev. 177, 2426 (1969).
14. W. A. Bardeen, Nucl.Phys. 75, 246 (1974).
15. G. 't Hooft, Phys.Rev. D14, 3432 (1976); (E) D18, 2199 (1978).
16. V. Baluni, Phys.Rev. D19, 2227 (1979).

17. R. J. Crewther, P. Di Vecchia, G. Veneziano, and E. Witten, CERN preprint TH.2735(1979).
18. K. G. Wilson, Phys.Rev. 179, 1499 (1969).
19. Y. Nambu and G. Jona-Lasinio, Phys.Rev. 122, 345 (1961).
20. R. D. Peccei and H. R. Quinn, Phys.Rev.Letters 38, 1440 (1977); Phys.Rev. D16, 1791 (1977).
21. Y. Hosotani, Prog.Theor.Phys. 61, 1452 (1979).
22. W. F. Palmer and S. S. Pinsky, Ohio State preprint COO-1545-254 (1979).
23. J. Kogut and L. Susskind, Phys.Rev. D11, 3594 (1975).
24. S. P. de Alwis, Phys.Letters, 86B, 67 (1979).
25. M. Ida, Prog.Theor.Phys. 61, 618, 1784 (1979); Kyoto University preprint RIFP-375 (1979).
26. S. L. Glashow and S. Weinberg, Phys.Rev.Letters 20, 224 (1968).
27. M. Gell-Mann, R. J. Oakes and B. Renner, Phys.Rev. 175, 2195 (1968).
28. S. L. Glashow, R. Jackiw and S.-S. Shei, Phys.Rev. 187, 1916 (1969).
29. M. Gell-Mann, Proc.Third Topical Conference on Particle Physics, Honolulu 1969, eds.W. A. Simmonds and S. F. Tuan, Western Periodicals (Los Angeles, 1970), p.1.
30. J. Lowenstein and J. A. Swieca, Ann.Phys. (N.Y.) 68, 172 (1971).
31. D. G. Caldi, Phys.Rev.Lett. 39, 121 (1977).
32. V. A. Fateyev, I. V. Frolov and A. S. Schwarz, preprint of Landau Institute for Theoretical Physics (1979); B. Berg and M. Lüscher, Comm.Math.Phys. 69, 57 (1979).
33. S. Coleman, 1977 Erice Lectures, Harvard preprint HUTP-78/A004 (1978).
34. K. Uhlenbeck, quoted by M. Atiyah.
35. J. Nuyts, Phys.Rev.Lett. 26, 1604 (1971); 27, 361 (E) (1971).
36. M. A. Shifman, A. I. Vainshtein, and V. I. Zacharov, preprint ITEP-64 (1979).
37. E. Witten, Nucl.Phys. B149, 285 (1979).
38. N. K. Nielsen and B. Schroer, Nucl.Phys. B120, 62 (1977); Phys.Lett. 66B, 373 (1977).
39. J. Kiskis, Phys.Rev. D16, 2535 (1977).
40. S. Coleman, Ann.Phys. (N.Y.) 101, 239 (1976).
41. M. Lüscher, Phys.Lett. 78B, 465 (1978); A. D'Adda, M. Lüscher and P. di Vecchia, Nucl.Phys. B146, 63 (1978); B152, 125 (1979).
42. A. A. Migdal, Phys.Lett. 80B, 275 (1979); 81B, 37 (1979).
43. M. B. Halpern, Phys.Rev. D16, 1798, 3515 (1977).
44. H. Kleinert, Phys.Lett. 62B, 429 (1976); Erice 1976, "Understanding the Fundamental Constituents of Matter", ed. A. Zichichi (Plenum 1978), p. 289; Fortschr.Phys. 26, 565 (1978).
45. C. Lee and W. A. Bardeen, University of Michigan preprint UM HE 78-55 (1978).

46. E. Witten, Nucl.Phys. B156, 269 (1979); G. Veneziano, Nucl. Phys. B159, 213 (1979); P. di Vecchia, CERN preprint TH. 2680 (1979).
47. C. Rosenzweig, J. Schechter and G. Trahern, Syracuse University preprint COO-3533-148 (1979).
48. W. B. Dress et al., Phys.Rev. D15, 9 (1977).

INDEX